Springer Series in Solid-State Sciences

Volume 188

The Springer Series in Solid-State Sciences consists of fundamental scientific books prepared by leading researchers in the field. They strive to communicate, in a systematic and comprehensive way, the basic principles as well as new developments in theoretical and experimental solid-sate physics.

More information about this series at http://www.springer.com/series/682

Pierre Deymier • Keith Runge

Sound Topology, Duality, Coherence and Wave-Mixing

An Introduction to the Emerging New Science of Sound

Springer

Pierre Deymier
Dept. of Materials Science
and Engineering
University of Arizona
Tucson, Arizona
USA

Keith Runge
Dept. of Materials Science and Engineering
University of Arizona
Tucson, Arizona
USA

ISSN 0171-1873 ISSN 2197-4179 (electronic)
Springer Series in Solid-State Sciences
ISBN 978-3-319-62379-5 ISBN 978-3-319-62380-1 (eBook)
DOI 10.1007/978-3-319-62380-1

Library of Congress Control Number: 2017946710

Printed on acid-free paper

This Springer imprint is published by Springer Nature
The registered company is Springer International Publishing AG
The registered company address is: Gewerbestrasse 11, 6330 Cham, Switzerland

To Léonard Dobrzynski

Preface

A new science of sound is emerging based on the scientific principles of symmetry breaking and interactions. This new science of sound focuses on aspects of wave phenomena that have not been emphasized in traditional instruction and in particular the four paradigm-changing scientific notions of sound topology, duality, coherence, and wave mixing.

Topology When sound waves propagate in media under symmetry breaking conditions, they may exhibit amplitudes $A(k) = A_0 e^{i\eta(k)}$ that acquire a geometric phase η leading to non-conventional topology. Broken symmetry phenomena lead to the concept of symmetry-protected topological order. Topological acoustic waves promise designs and new device functionalities for acoustic systems that are unique, robust, and avoid the loss of coherence.

Duality The self-interaction of a wave through its supporting medium creates acoustic wave states determined by self-interference phenomena. In the phonon representation of sound waves, these self-interference phenomena uncover the notion of duality in the quantum statistics (*i.e.*, boson versus fermion characterized by the symmetry of multiple particle states). It also enables the development of analogies with quantum mechanics and Quantum Field Theory.

Coherence Interactions in nonlinear elastic media cause multiple scattering and resonances of sound waves. Individually, these resonances may lead to sound waves with non-conventional topologies. However, in the case of multiple channels for phonon scattering, nonlinearity leads to the loss of phase coherence while retaining a broken time-reversal symmetry demonstrated by acoustic wave amplitude degradation. Mechanical energy can be directed in a one-way, irreversible, and targeted fashion from a linear system to a nonlinear system in coupled nonlinear and linear vibrational systems. Targeted energy transfer offers strategies for vibration and sound management.

Wave Mixing Media supporting different types of waves and their sources can coherently convert energy between sound and other physical and biological waves.

Topological effects have been shown for biological waves (calcium signals) resulting from acoustic wave-driven spatiotemporal modulation of cell membrane conductance in biological tissue. Coherent phonons can bestow non-conventional topological characteristics to electronic wave functions and topological spin waves can arise from the spatiotemporal modulation of the spin coupling constant in ferromagnetic media due to phonons.

These notions are taken further and related to the development of *acoustic analogues* of other physical phenomena ranging from quantum mechanics to general relativity. These analogues offer perspectives for applications and technological developments of the new science of sound.

This book gives an introduction to these scientific notions and analogues by attempting to present a number of "simple" models. The authors have developed these models in the spirit of Jacques Friedel's approach to theoretical research. Friedel's research utilized extensively approximate but simple models that can be understood by non-specialists: "J'ai essayé de développer des modèles approximatifs mais simples, compréhensibles et même utilisables par des non-spécialistes" [1].

This book is not an exhaustive review of models relative to the four notions introduced above. These models serve as demonstrations of these notions and provide only a partial foundation for their unification into a new science of sound.

Most of the simple models presented here represent contributions of the authors and many of their collaborators, students and postdoctoral fellows, whom we would like to acknowledge at this time. In particular, Prof. Jerome Vasseur, who proofread portions of several chapters. This book paints a picture of the emerging new science of sound that reflects the biased perception and understanding of the authors alone. We would also like to acknowledge partial support from the US National Science Foundation (NSF) as well as the French Centre National de la Recherche Scientifique (CNRS) through the Laboratoire International Associé "Materials and Optics."

The book is divided into six chapters with their own set of references, each of which is intended to be self-contained and able to be read independently. In case there are overlapping concepts, we refer to the appropriate subsection of other chapters.

We hope that this book will stimulate future interest in the emerging field of sound and will initiate new developments in the four scientific notions of topology, duality, coherence, and wave mixing.

Tucson, AZ

Pierre A. Deymier
Keith Runge

Reference

[1] Entrevue avec Jacques Friedel, Paris, 17 octobre 2001, Hervé Arribart et Bernadette Bensaude-Vincent. http://authors.library.caltech.edu/5456/1/hrst.mit.edu/hrs/materials/public/index.html

Contents

1 **Introduction to Spring Systems** 1
1.1 Introduction 1
1.2 One-Dimensional Monatomic Harmonic Crystal 1
1.3 Phase and Group Velocity 3
1.4 One-Dimensional Diatomic Harmonic Crystal 4
1.5 One-Dimensional Monatomic Crystal with Spatially Varying Stiffness 6
1.6 Green's Function Approach 8
1.7 Monatomic Crystal with a Mass Defect 10
1.7.1 Monatomic Harmonic Crystal with a General Perturbation 12
1.7.2 Locally Resonant Structure 13
1.8 Interface Response Theory 16
1.8.1 Fundamental Equations of the Interface Response Theory 16
1.8.2 Green's Function of the Cleaved 1-D Monatomic Crystal 18
1.8.3 Finite Monatomic Crystal 20
1.8.4 1-D Monatomic Crystal with One Side Branch 22
1.8.5 1-D Monatomic Crystal with Multiple Side Branches 24
1.9 Conclusion 28
Appendix 1: Code Based on Green's Function Approach 28
References 36

2 **Phase and Topology** 37
2.1 Introduction 37
2.2 Overview 38
2.3 Harmonic Oscillator Model Systems 40
2.3.1 Geometric Phase and Dynamical Phase of the Damped Harmonic Oscillator 40
2.3.2 Geometric Phase of the Driven Harmonic Oscillator 42

2.3.3 Topological Interpretation of the Geometric Phase 44
2.4 Elastic Superlattice Model System 48
2.4.1 Geometric Phase of a One-Dimensional Elastic Superlattice: Zak Phase 48
2.4.2 Topological Interpretation of the Zak Phase 52
2.5 Green's Function Approach 53
2.5.1 Green's Functions and Berry Connection 53
2.5.2 The One-Dimensional Harmonic Crystal 55
2.5.3 The One-Dimensional Harmonic Crystal with Side Branches 58
2.6 Topological Modes at Interfaces Between Media with Different Zak Phases 63
2.7 Conclusion 65
Appendix 1: Eigen Values and Eigen Vectors in One-Dimensional Elastic Superlattice 66
Appendix 2: Discrete One-Dimensional Monatomic Crystal with Spatially Varying Stiffness 71
Appendix 3: Introduction to Green's Function Formalism 74
References 78

3 Topology and Duality of Sound and Elastic Waves 81
3.1 Introduction 81
3.2 Overview 82
3.3 Intrinsic Topological Phononic Structures 83
3.3.1 Two Coupled Mass-Spring Harmonic Crystals in the Long-Wavelength Limit 83
3.3.2 Single Harmonic Crystal Grounded to a Rigid Substrate in the Long-Wavelength Limit 88
3.3.3 Single Discrete Harmonic Crystal Grounded to a Rigid Substrate Beyond the Long-Wavelength Limit 92
3.4 Extrinsic Topological Phononic Structure 109
3.4.1 Time-Dependent Elastic Superlattice 109
3.4.2 Multiple time Scale Perturbation Theory of the Time-Dependent Superlattice 112
3.4.3 Topology of Elastic Wave Functions 115
3.5 Mixed Intrinsic and Extrinsic Topological Phononic Structure . . . 120
3.5.1 One-Dimensional Mass-Spring Harmonic Crystal Grounded to a Substrate with Spatio-Temporal Modulation 120
3.6 Separable and Non-separable States in Elastic Structures 127
3.6.1 Separability of Systems Composed of Harmonic Oscillators 128
3.6.2 Partitioning of a Harmonic Crystal into Normal Modes . . . 129
3.6.3 Uncoupled 1-D Harmonic Crystals 130
3.6.4 Separability and Non-separability of the States of Coupled Harmonic Crystals 133

Appendix 1: Berry Phase of the Spinor Amplitude for the Discrete Harmonic Crystal Grounded to a Rigid Substrate 139
Appendix 2: Rotational Modes in a Two-Dimensional Phononic Crystal 141
Appendix 3: Molecular Dynamics (MD) and Spectral Energy Density (SED) Methods 147
Appendix 4: Mutiple Time-Scale Perturbation Theory 156
References 158

4 Coherence 163
4.1 Introduction 163
4.2 The Anharmonic One-Dimensional Monatomic Crystal 164
4.2.1 Multiple-Time Scale Perturbation Theory of the Monatomic Anharmonic Crystal 164
4.2.2 Self-Interaction 166
4.2.3 Three-Wave Interactions 169
4.2.4 Molecular Dynamics (MD) Simulation and Spectral Energy Density (SED) Approaches 181
4.3 Nonlinear Phonon Modes in Second-Order Anharmonic Coupled Monatomic Chains 188
4.3.1 Model Systems 189
4.3.2 Multiple Time Scale Perturbation Theory 190
4.3.3 Self-Interaction 192
4.3.4 Three Wave Interactions 201
4.3.5 Numerical Results 203
4.4 Multi-Phonon Scattering Processes in One-Dimensional Anharmonic Biological Superlattices 207
4.4.1 Model of Nonlinear Hydrated Collagen 208
4.4.2 Model and Simulation Methods 214
4.5 Composite Media with Controllable Nonlinearity 221
4.6 Calcium Waves in Chains of Endothelial Cells with Nonlinear Reaction Dynamics: A Metaphor for Wave Propagation in Bi-stable Media 225
4.6.1 Background for Calcium Waves in Endothelial Cells 226
4.6.2 Linear and Nonlinear Reaction Dynamics 228
4.6.3 Reaction-Diffusion Problem with Linear or Nonlinear Reaction Dynamics 232
4.6.4 Propagation and Propagation Failure in a One-Dimensional Chain of Cells 235
4.7 Nonlinear Interactions Between Elastic Waves and Multiple Spatio-Temporal Modulations of Stiffness in a One-Dimensional Waveguide 240
References 255

5 Wave Mixing 261
5.1 Introduction 261
5.2 Effect of Sound on Gap-Junction Based Intercellular Signaling: Topological Calcium Waves Under Acoustic Irradiation 262
5.2.1 Background 262
5.2.2 Nonlinear Reaction Diffusion Model of Gap Junction Based Intercellular Calcium Waves 264
5.2.3 Linear Reaction Diffusion Model of Calcium Waves 271
5.3 Topological Interpretation of Phonon Drag 276
5.4 Phonon-Magnon Resonant Processes with Relevance to Acoustic Spin Pumping 281
5.4.1 Magnon Background 281
5.4.2 Model 283
5.4.3 Multiple Time Scale Perturbation Method Applied to Case I 287
5.4.4 Solutions of the Equations of Motion in Case I 289
5.4.5 Discussion of Results for Case I 296
5.4.6 Analysis of Case II 301
5.4.7 Phonon-Magnon Interaction Summary 302
Appendix 1: Fortran 90 Program Used to Model the Nonlinear Reaction-Diffusion Dynamics of a Chain of Cells Subjected to a Spatio-Temporal Modulation of Intercellular Diffusion Coefficient Induced by a Sound Wave 303
Appendix 2: Multiple Time Scale Perturbation Theory Analysis of Calcium Wave in an Acoustic Field 310
References 315

6 Acoustic Analogues 319
6.1 Introduction 319
6.2 Analogue of Quantum Bit (qubit): Phase Bit (φ-bit) 320
6.3 Gauge Fields Analogue for Phonons 325
6.4 Particle-Wave Duality with Acoustic Bubbles 328
6.4.1 Dynamical Model of a Single Bubble in an Acoustic Standing Wave Field 328
6.4.2 Bubble in a Chain of Interacting Bubbles 331
6.4.3 Acousto-hydrodynamics of a Bubble 334
6.5 General Relativity Analogues 336
6.5.1 Acoustic Waves in a Moving Fluid 337
6.5.2 Lagrangian Approach 342
6.5.3 Second-Order Partial Differential Equation 345
6.5.4 Time-Dependent Superlattice 346

Appendix 1: Laplacian and d'Alembertian in a General Coordinate System 354
Appendix 2: Geometrical Representation of the Dynamics of One-Dimensional Harmonic Systems 354
References 360

Subject Index 361

Chapter 1
Introduction to Spring Systems

1.1 Introduction

The new science of sound focuses on aspects of wave phenomena that have not been emphasized in traditional instruction. To elucidate these new aspects of wave phenomena, particularly the phase, in a clear exposition, we will rely on a number of simple models. We first define phase and group velocities using the one-dimensional monatomic harmonic crystal. Then, we advance to the diatomic one-dimension harmonic crystal and the one-dimensional harmonic crystal with alternating stiffness. We introduce the Green's function approach to solving the wave equation, which will prove to be an indispensable tool in exploring phase related wave behavior. Three simple systems provide the basics of the Green's function formalism, monatomic harmonic crystals with (1) a single mass defect, (2) a general perturbing potential, and (3) locally resonant structures. The introduction concludes with the introduction of Interface Response Theory (IRT) where we present the fundamental equations, introduce the cleavage operator, and demonstrate its use in a few examples. The Appendix 1 includes a Fortran77 code that will allow the reader to further explore the concepts presented here and in other chapters in greater detail.

1.2 One-Dimensional Monatomic Harmonic Crystal

The one-dimensional (1-D) monatomic harmonic crystal consists of an infinite chain of masses, m, with nearest neighbor interaction modeled by harmonic springs with spring constant, β. The separation distance between the masses at rest is defined as a. This model system is illustrated in Fig. 1.1.

In absence of external forces, the equation describing the motion of atom "n" is given by Newton's second law:

P. Deymier, K. Runge, *Sound Topology, Duality, Coherence and Wave-Mixing*,
Springer Series in Solid-State Sciences 188, DOI 10.1007/978-3-319-62380-1_1

Fig. 1.1 Schematic illustration of one 1-D monatomic harmonic crystal

$$m\frac{d^2u_n}{dt^2} = \beta(u_{n+1} - u_n) - \beta(u_n - u_{n-1}) \tag{1.1}$$

In this equation, u_n represents the displacement of the mass "n" with respect to its position at rest. The first term on the right-hand side of the equal sign is the harmonic force on mass "n" resulting from the spring on its right. The second term is the force due to the spring on the left of "n". The dynamics of the 1-D monatomic harmonic crystal can, therefore, be studied by solving (1.2):

$$m\frac{d^2u_n}{dt^2} = \beta(u_{n+1} - 2u_n + u_{n-1}) \tag{1.2}$$

The next subsections aim at seeking solutions of (1.2).

We seek solutions to (1.2) in the form of propagating waves:

$$u_n = Ae^{ikna}e^{i\omega t} \tag{1.3}$$

where k is a wave number and ω is an angular frequency. Inserting solutions of the form given by (1.3) into (1.2) and dividing both sides by $Ae^{ikna}e^{i\omega t}$, one obtains the relation between angular frequency and wave number:

$$\omega^2 = -\frac{\beta}{m}\left(e^{\frac{ika}{2}} - e^{-\frac{ika}{2}}\right)^2 \tag{1.4}$$

We use the relation $2i\sin\theta = e^{i\theta} - e^{-i\theta}$ and the fact that ω is a positive quantity to obtain the so-called dispersion relation for propagating waves in the 1-D harmonic crystal:

$$\omega(k) = \omega_0\left|\sin\left(k\frac{a}{2}\right)\right| \tag{1.5}$$

with $\omega_0 = 2\sqrt{\frac{\beta}{m}}$ representing the upper limit for angular frequency. Since the monatomic crystal is discrete and waves with wavelength $\lambda = \frac{2\pi}{k}$ larger than $2a$ are physically equivalent to those with wavelength smaller than $2a$, the dispersion relation of (1.5) needs only be represented in the symmetrical interval $k\epsilon\left[-\frac{\pi}{a}, \frac{\pi}{a}\right]$ (see Fig. 1.2). This interval is called the first Brillouin zone of the 1-D monatomic periodic crystal.

Fig. 1.2 Illustration of the dispersion relation for propagating waves in 1-D monatomic harmonic crystal

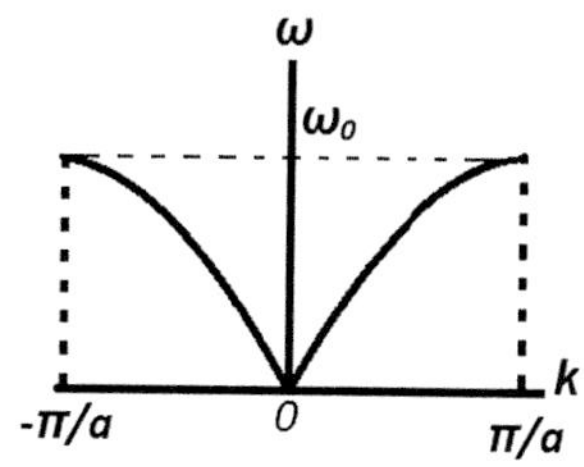

1.3 Phase and Group Velocity

The velocity at which the phase of a wave with wave vector, k, and angular frequency, ω, propagates is defined as:

$$v_\varphi = \frac{\omega}{k} \tag{1.6}$$

The group velocity is defined as the velocity at which a wave packet (a superposition of propagating waves with different values of wave number ranging over some interval) propagates. It is easier to understand this concept by considering the superposition of only two waves with angular velocities, ω_1 and ω_2, and wave vectors, k_1 and k_2. Choosing, $\omega_1 = \omega - \frac{\Delta\omega}{2}$ and $\omega_2 = \omega + \frac{\Delta\omega}{2}$, and, $k_1 = k - \frac{\Delta k}{2}$ and $k_2 = k + \frac{\Delta k}{2}$. The superposition of the two waves, assuming that they have the same amplitude, A, leads to the displacement field at mass "n":

$$u_n^s = 2Ae^{ikna}e^{i\omega t}cos\left(\frac{\Delta k}{2}na + \frac{\Delta\omega}{2}t\right) \tag{1.7}$$

The first part of the right-hand side of (1.7) is a traveling wave that is modulated by the cosine term. This latter term represents a beat pulse. The velocity at which this modulation travels is the group velocity and is given by:

$$v_g = \frac{\Delta\omega}{\Delta k} \tag{1.8}$$

In the limit of infinitesimally small differences in wave number and frequency, the group velocity is expressed as a derivative of the dispersion relation:

$$v_g = \frac{d\omega(k)}{dk} \tag{1.9}$$

In the case of the 1-D harmonic crystal, the group velocity is given by: $v_g = \omega_0 \frac{a}{2} cos k\frac{a}{2}$.

We now open a parenthesis concerning the group velocity and show that it is also equal to the velocity of the energy transported by a propagating wave. To that effect, we calculate the average energy density as the sum of the potential energy and the kinetic energy averaged over a time period. The average energy is given by:

$$\langle E\rangle = \frac{1}{2}\beta(u_n - u_{n-1})(u_n - u_{n-1})^* + \frac{1}{2}m\dot{u}_n\dot{u}_n^* \tag{1.10}$$

In (1.10), the * denotes the complex conjugate and $\dot{u}$ the time derivative of the displacement (i.e., the velocity of the mass "n"). Inserting into (1.10) the displacements given by (1.3) and the dispersion relation given by (1.5) yields the average energy density:

$$\langle e\rangle = \frac{\langle E\rangle}{a} = 4A^2\frac{\beta}{a}\sin^2 k\frac{a}{2} \tag{1.11}$$

We now calculate the energy flow through one unit cell of the 1-D crystal in the form of the real part of the power, Φ, defined as the product of the force on mass "n" due to one spring and the velocity of the mass:

$$\Phi = Re\{\beta(u_{n+1} - u_n)\dot{u}_n^*\} = \beta A^2\omega_0\left|\sin k\frac{a}{2}\right|\sin ka \tag{1.12}$$

The velocity of the energy, v_e, is therefore the ratio of the energy flow to the average energy density which after using trigonometric relations yields: $v_e = \omega_0\frac{a}{2}\cos k\frac{a}{2}$. This expression is the same as that of the group velocity. In summary, the group velocity also represents the velocity of the energy transported by the propagating waves in the crystal.

1.4 One-Dimensional Diatomic Harmonic Crystal

The 1-D diatomic harmonic crystal is illustrated in Fig. 1.3.

The equations of motion of two adjacent odd and even atoms are:

$$\begin{cases} m_1\ddot{u}_{2n} = \beta(u_{2n+1} - 2u_{2n} + u_{2n-1}) \\ m_2\ddot{u}_{2n+1} = \beta(u_{2n+2} - 2u_{2n+1} + u_{2n}) \end{cases} \tag{1.13}$$

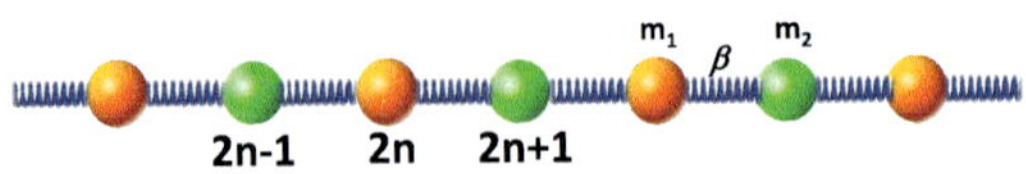

Fig. 1.3 Schematic illustration of the 1-D diatomic harmonic crystal. The atoms with even labels have a mass, m_1, and the odd atoms have a mass, m_2. The force constant of the springs is β. The periodicity of the crystal is $2a$

We seek solutions in the form of propagating waves with different amplitudes for odd or even atoms as their masses are different:

$$\begin{cases} u_{2n} = Ae^{i\omega t}e^{ik2na} \\ u_{2n+1} = Be^{i\omega t}e^{ik(2n+1)a} \end{cases} \tag{1.14}$$

Inserting these solutions into the (1.13) leads, after some algebraic manipulations and using the definition of the cosine in terms of complex exponentials, to the set of two linear equations in A and B:

$$\begin{cases} (2\beta - m_1\omega^2)A - 2\beta\cos ka\, B = 0 \\ -2\beta\cos ka\, A + (2\beta - m_1\omega^2)B = 0 \end{cases} \tag{1.15}$$

This is an eigen value problem in ω^2. This set of equations admits non-trivial solutions (i.e., $A \neq 0, B \neq 0$) when the determinant of the matrix composed of the linear coefficient in (1.15) is equal to zero, that is:

$$\begin{vmatrix} 2\beta - m_1\omega^2 & -2\beta\cos ka \\ -2\beta\cos ka & 2\beta - m_2\omega^2 \end{vmatrix} = 0 \tag{1.16}$$

Setting, $\alpha = \omega^2$, (1.16) takes the form of the quadratic equation:

$$\alpha^2 - 2\beta\left(\frac{1}{m_1} + \frac{1}{m_2}\right)\alpha + \frac{4\beta^2}{m_1 m_2}\sin^2 ka = 0 \tag{1.17}$$

which admits two solutions:

$$\omega^2 = \alpha = \beta\left(\frac{1}{m_1} + \frac{1}{m_2}\right) \pm \sqrt{\beta^2\left(\frac{1}{m_1} + \frac{1}{m_2}\right)^2 - \frac{4\beta^2}{m_1 m_2}\sin^2 ka} \tag{1.18}$$

These two solutions are periodic in wave number, k, with a period of $\frac{\pi}{a}$ and are represented graphically in the band structure of Fig. 1.4 over the interval, $k\epsilon\left[0, \frac{\pi}{2a}\right]$. This interval is the smallest interval, so-called irreducible Brillouin zone, for representing the band structure. The complete band structure is reconstructed by mirror symmetry with respect to a vertical line passing though the origin.

The frequencies ω_1, ω_2 and ω_3 are given by:

$$\omega_1 = \sqrt{\frac{2\beta}{m_1}}, \omega_2 = \sqrt{\frac{2\beta}{m_2}}, \text{and } \omega_3 = \sqrt{2\beta\left(\frac{1}{m_1} + \frac{1}{m_2}\right)} \text{ if one chooses } m_1 > m_2 \tag{1.19}$$

The band structure of Fig. 1.4 exhibits two branches since the unit cell of the 1-D diatomic crystal contains two atoms. These branches are separated by a gap in the interval of frequency $[\omega_1, \omega_2]$. The low frequency branch is called the acoustic

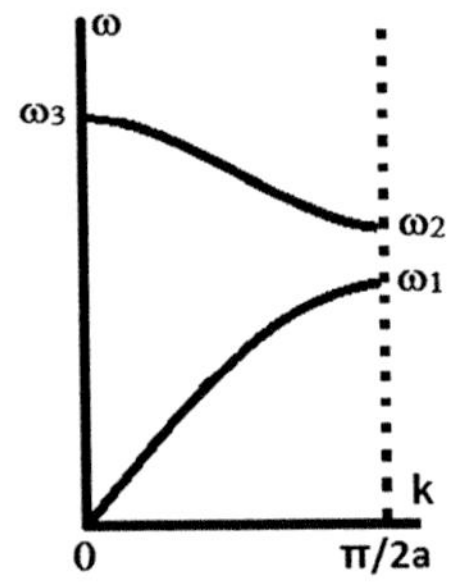

Fig. 1.4 Schematic representation of the band structure of the 1-D diatomic harmonic crystal in the irreducible Brillouin zone

branch and the high frequency branch is called the optical branch. In the limit, $m_1 = m_2 = m$, the diatomic crystal reduces to a monatomic crystal. The band structure of Fig. 1.4 becomes that of the 1-D monatomic harmonic crystal. It is interesting to note that in contrast to the acoustic branch, the optical branch has a negative slope, i.e., a negative group velocity. The group velocity and energy velocity point in a direction opposite to the direction of the wave vector and of the phase velocity. This observation is particularly important when dealing with the concept of negative refraction. However, since the diatomic crystal is one-dimensional, we cannot address the phenomenon of refraction yet. We rewrite the real part of the displacement of a superposition of waves given by (1.7) in the form:

$$u_n^s = 2A\cos\left(k\left(na + v_\varphi t\right)\right)\cos\left(\frac{\Delta k}{2}\left(na + v_g t\right)\right) \tag{1.20}$$

Equation (1.20) shows that the envelope of the wave packet appears to propagate in the opposite direction of the superposition of waves when the phase velocity and the group velocity have opposite signs.

1.5 One-Dimensional Monatomic Crystal with Spatially Varying Stiffness

The one-dimensional monatomic crystal with two different springs is illustrated in Fig. 1.5.

The equations of motion of two adjacent odd and even atoms are:

$$\begin{cases} m\ddot{u}_{2n} = \beta_2\left(u_{2n+1} - u_{2n}\right) - \beta_1\left(u_{2n} - u_{2n-1}\right) \\ m\ddot{u}_{2n+1} = \beta_1\left(u_{2n+2} - u_{2n+1}\right) - \beta_2\left(u_{2n+1} - u_{2n}\right) \end{cases} \tag{1.21}$$

We want to contrast (1.21) with the equation of motion of the diatomic harmonic crystal (1.13), which we recall here for convenience:

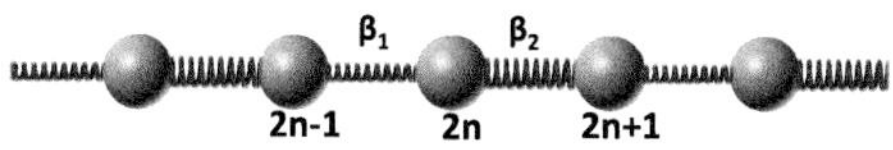

Fig. 1.5 Schematic illustration of 1-D monatomic harmonic crystal with varying stiffness. The interatomic spacing is a and the period $2a$

$$\begin{cases} m_1\ddot{u}_{2n} = \beta(u_{2n+1} - 2u_{2n} + u_{2n-1}) \\ m_2\ddot{u}_{2n+1} = \beta(u_{2n+2} - 2u_{2n+1} + u_{2n}) \end{cases}$$

Both set of (1.13) and (1.21) satisfy time-reversal symmetry owing to the second derivatives with respect to time on the right-hand side. Indeed, these equations are invariant upon application of a time-reversal operation, i.e., $t \to -t$. Equation (1.13) satisfy inversion symmetry, that is, the equations of motion are invariant upon the transformation of position index $p \to -p$. For example the first equation $m_1\ddot{u}_{2n} = \beta(u_{2n+1} - 2u_{2n} + u_{2n-1})$ transforms to $m_1\ddot{u}_{-2n} = \beta(u_{-2n-1} - 2u_{-2n} + u_{-2n+1}) = \beta(u_{-2n+1} - 2u_{-2n} + u_{-2n-1})$, which corresponds to a simple shift of origin. In contrast, the first equation for the monatomic crystal with alternating stiffness, $m\ddot{u}_{2n} = \beta_2(u_{2n+1} - u_{2n}) - \beta_1(u_{2n} - u_{2n-1})$ transforms to $m\ddot{u}_{-2n} = \beta_2(u_{-2n-1} - u_{-2n}) - \beta_1(u_{-2n} - u_{-2n+1})$. This transformation changes the unit cell of the system from one having the springs with β_1 and β_2 ordered from left to right to a unit cell with the reverse ordering. In Chap. 2, we will discuss in detail the effect of the breaking of inversion symmetry on the properties of the displacement. We now seek solutions in the form of propagating waves with different amplitudes for odd or even atoms:

$$\begin{cases} u_{2n} = Ae^{i\omega t}e^{ik2na} \\ u_{2n+1} = Be^{i\omega t}e^{ik(2n+1)a} \end{cases} \tag{1.22}$$

Inserting these solutions into the (1.21) leads to the set of two linear equations in A and B:

$$\begin{cases} (\beta_1 + \beta_2 - m\omega^2)A - (\beta_1 e^{-ika} + \beta_2 e^{+ika})B = 0 \\ -(\beta_1 e^{+ika} + \beta_2 e^{-ika})A + (\beta_1 + \beta_2 - m\omega^2)B = 0 \end{cases} \tag{1.23}$$

Non-trivial solutions (i.e., $A \neq 0, B \neq 0$) exist when the determinant of the matrix composed of the linear coefficient in (1.23) is equal to zero, that is:

$$(\beta_1 + \beta_2 - m\omega^2)^2 - (\beta_1 e^{-ika} + \beta_2 e^{+ika})(\beta_1 e^{+ika} + \beta_2 e^{-ika}) = 0 \tag{1.24}$$

Setting, $\alpha = \omega^2$, (1.24) takes the form of the quadratic equation:

$$\alpha^2 - 2\frac{\beta_1 + \beta_2}{m}\alpha + \frac{4\beta_1\beta_2}{m^2}\sin^2 ka = 0 \tag{1.25}$$

which admits two solutions:

$$\omega^2 = \alpha = \frac{\beta_1 + \beta_2}{m} \pm \frac{1}{m}\sqrt{(\beta_1 + \beta_2)^2 - 4\beta_1\beta_2 \sin^2 ka} \tag{1.26}$$

The band structure is schematically similar to that of Fig. 1.4 but with

$$\omega_1^2 = \frac{\beta_1 + \beta_2}{m} - \frac{1}{m}\sqrt{(\beta_1 + \beta_2)^2 - 4\beta_1\beta_2}$$
$$\omega_2^2 = \frac{\beta_1 + \beta_2}{m} + \frac{1}{m}\sqrt{(\beta_1 + \beta_2)^2 - 4\beta_1\beta_2}$$

and

$$\omega_3^2 = 2\frac{\beta_1 + \beta_2}{m}$$

1.6 Green's Function Approach

In anticipation of subsequent sections where Green's function approaches will be used to shed light on the vibrational and phase behavior of more complex harmonic structures, we present here the Green's function formalism applied to the 1-D monatomic crystal. Considering harmonic solution with angular frequency, ω, the equation of motion (1.2) can be recast in the form:

$$\frac{1}{m}\left[\beta u_{n+1} + \left(m\omega^2 - 2\beta\right)u_n + \beta u_{n-1}\right] = 0 \tag{1.27}$$

We now rewrite (1.27) in matrix form when applying it to all masses in the 1-D monatomic crystal:

$$\overleftrightarrow{H}_0\,\vec{u} = \frac{1}{m}\begin{bmatrix} \ddots & \vdots & \vdots & \vdots & \vdots & \vdots & \vdots & \vdots & \vdots \\ \cdots & 0 & \beta & -\gamma & \beta & 0 & 0 & 0 & \cdots \\ \cdots & 0 & 0 & \beta & -\gamma & \beta & 0 & 0 & \cdots \\ \cdots & 0 & 0 & 0 & \beta & -\gamma & \beta & 0 & \cdots \\ \vdots & \vdots & \vdots & \vdots & \vdots & \vdots & \vdots & \vdots & \ddots \end{bmatrix}\begin{bmatrix} \vdots \\ u_{n-1} \\ u_n \\ u_{n+1} \\ \vdots \end{bmatrix} = \begin{bmatrix} \vdots \\ 0 \\ 0 \\ 0 \\ \vdots \end{bmatrix} \tag{1.28}$$

where $\gamma = 2\beta - m\omega^2$. The operator, $\overleftrightarrow{H}_0$, is a more compact representation of the dynamic matrix in (1.28), and $\vec{u}$ is the vector which components are the displacements of the masses in the crystal. With this notation, the Green's function, $\overleftrightarrow{G}_0$, associated with $\overleftrightarrow{H}_0$ is defined by the relation:

$$\overleftrightarrow{H}_0 \overleftrightarrow{G}_0 = \overleftrightarrow{I} \tag{1.29}$$

In this equation, $\overleftrightarrow{I}$, is the identity matrix. Equation (1.29) is written in component form as:

$$\Sigma_{n''} H_0(n, n'') G_0(n'', n') = \delta_{nn'} \tag{1.30}$$

Here, we have used the Kroenecker symbol, $\delta_{nn'}$ to represent the components of the identity matrix, that is 1 when $n = n'$ and 0 when $n \neq n'$. Since $\overleftrightarrow{H}_0$ is tridiagonal (harmonic interactions are limited to first nearest neighbors), (1.29) becomes:

$$\frac{1}{m}[\beta G_0(n+1, n') - \gamma G_0(n, n') + \beta G_0(n-1, n')] = \delta_{nn'} \tag{1.31}$$

From a physical point of view, the Green's function, $G_0(n, n')$ is the displacement of mass "n" when a unit external force is applied at the site of mass "n'". The solution of (1.31) is known [1] and has the general form:

$$G_0(n, n') = \frac{m}{\beta} \frac{t^{|n-n'|+1}}{t^2 - 1} \tag{1.32}$$

The quantity, t, is determined by inserting this general solution into (1.31) and choosing $n = n'$. In that case, we obtain the simple quadratic equation:

$$t^2 - 2\xi t + 1 = 0 \tag{1.33}$$

with $\xi = \frac{\gamma}{2\beta} = 1 - \frac{m\omega^2}{2\beta} = 1 - \frac{2\omega^2}{\omega_0^2}$. The resolution of the quadratic equation yields:

$$t = \begin{cases} \xi - \left(\xi^2 - 1\right)^{1/2} \; if \; \xi > 1 \\ \xi + \left(\xi^2 - 1\right)^{1/2} \; if \; \xi < -1 \\ \xi + i\left(1 - \xi^2\right)^{1/2} \; if \; -1 \leq \xi \leq 1 \end{cases} \tag{1.34}$$

We note that for, $\omega \epsilon [0, \omega_0]$ and $\xi \epsilon [-1, 1]$ t is a complex quantity. We introduce some wave number, k, and write this complex quantity, $t = e^{ika}$. We equate the real part and the imaginary part of this quantity with those of the third form of the solution in (1.34) and using standard trigonometric relations, we obtain the dispersion relation given by (1.5). We therefore recover the propagating wave solution in the crystal. For, $\omega > \omega_0$ and $\xi < -1$ $t \epsilon [-1, 0]$, if we introduce a wave number, k, we can rewrite $\xi = -\cosh ka$, then $t = -e^{-ka}$ represents an evanescent wave.

As a final note, we recast the operator, $\overset{\leftrightarrow}{H}_0$, as the difference, $\overset{\leftrightarrow}{H}'_0 - \omega^2 \overset{\leftrightarrow}{I}$, where the operator, $\overset{\leftrightarrow}{H}'_0$, depends on the spring constant β only. Equation (1.29) then states that:

$$\overset{\leftrightarrow}{G}_0 = \overset{\leftrightarrow}{I}\left(\overset{\leftrightarrow}{H'}_0 - \omega^2\right)^{-1} \tag{1.35}$$

meaning that the poles (zeros of the denominator) of the Green's function are the eigenvalues of the operator, $\overset{\leftrightarrow}{H}'_0$. According to (1.32), the poles of the Green's function of the 1-D monatomic harmonic crystal, are, therefore, given by the equation:

$$t^2 - 1 = 0 \tag{1.36}$$

This condition is met when $t = e^{ika} = \cos ka + i \sin ka$. If we consider the case where, $\omega \epsilon [0, \omega_0]$, $t = \xi + i(1 - \xi^2)^{1/2}$ $if - 1 \leq \xi \leq 1$ and we can, subsequently write $\cos ka = \xi = 1 - \frac{2\omega^2}{\omega_0^2}$ which, using trigonometric relations, reduces to the dispersion relation of propagating waves in the crystal (1.5).

1.7 Monatomic Crystal with a Mass Defect

We investigate the propagation of waves in a 1-D monatomic harmonic crystal with a single mass defect. This system will have relevance to our future discussion of non-reciprocal propagation of elastic waves and topological immunity to backscattering. The system is constructed by substituting one atom with mass, m by another atom with mass, m'. The diatomic crystal may subsequently be created as a periodic substitution of atoms with different masses. We address the following question: does the gap originate from the scattering of propagating waves by mass defects independently of their periodicity or does the gap originate from the periodic arrangement of the mass defects?

The defected monatomic crystal is illustrated in Fig. 1.6.

The equations of motion of the atoms in the defected crystal are:

$$\begin{cases} -m\omega^2 u_n = \beta(u_{n+1} - 2u_n + u_{n-1}) \ for \ n \neq 0 \\ -m'\omega^2 u_o = \beta(u_1 - 2u_0 + u_{-1}) \end{cases} \tag{1.37}$$

Let us consider an incident wave (i) propagating from the left of the crystal:

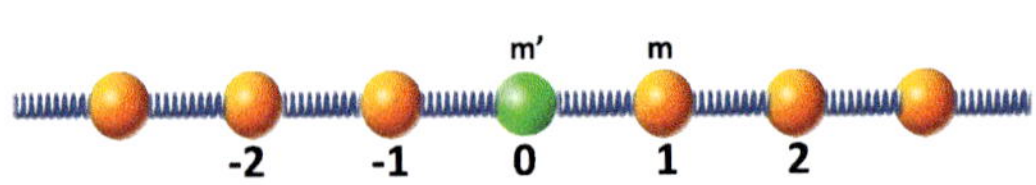

Fig. 1.6 Schematic illustration of the 1-D monatomic harmonic crystal with a single mass defect at site 0. The springs are all identical with the same spring constant

$$u_n^{(i)} = A_i e^{ikna} \; for \;\; n \leq -1 \tag{1.38}$$

Part of this wave will be reflected by the mass defect. Another part of the incident wave will be transmitted through the defect. We write the displacements associated with these reflected and transmitted waves in the form:

$$\begin{aligned} u_n^{(r)} &= A_r e^{-ikna} \quad for \; n \leq -1 \\ u_n^{(t)} &= A_t e^{ikna} \quad for \; n \geq 1 \end{aligned} \tag{1.39}$$

In (1.39), the upper-scripts (r) and (t) stand for reflected and transmitted waves, respectively. The total displacement on the left of the defect is the sum of the incident and reflected displacement. The displacement on the right of the defect consists only of the transmitted wave. The total displacement is therefore given by:

$$\begin{aligned} u_n &= u_n^{(i)} + u_n^{(r)} \quad for \; n \leq -1 \\ u_n &= u_n^{(t)} \quad for \; n \geq 1 \end{aligned} \tag{1.40}$$

The continuity of the displacement at the defected site "0" imposes the condition:

$$u_0 = u_0^{(i)} + u_0^{(r)} = u_0^{(t)} \tag{1.41}$$

Substituting (1.38) and (1.39) into the condition (1.41) yields a relation between the amplitudes of the incident, the reflected and the transmitted waves:

$$A_i + A_r = A_t \tag{1.42}$$

We now substitute (1.38), (1.39) and (1.41) into the (1.37) for the motion of the mass m'. After some algebra, this equation becomes:

$$\left(-m'\omega^2 + 2\beta - \beta e^{ika}\right)A_T = A_i \beta e^{-ika} + A_R \beta e^{ika} \tag{1.43}$$

Equations (1.42) and (1.43) constitute a set of linear equations in the amplitudes of the incident, reflected and transmitted waves. We can express the amplitude of the reflected and transmitted waves in terms of the amplitude of the incident wave to define a transmission coefficient and a reflection coefficient:

$$\begin{aligned} T &= \frac{A_t}{A_i} = \frac{\beta 2i \sin ka}{(m' - m)\omega^2 + \beta 2i \sin ka} \\ R &= \frac{A_r}{A_i} = \frac{-(m' - m)\omega^2}{(m' - m)\omega^2 + \beta 2i \sin ka} \end{aligned} \tag{1.44}$$

To obtain (1.44), we have used the fact that for the 1-D monatomic harmonic crystal, the dispersion relation of (1.5) can be recast in the form: $m\omega^2 = 2\beta(1 - \cos ka)$.

To analyze the behavior of the defected crystal further, we calculate the square of the modulus of the transmission coefficient:

$$T^2 = TT^* = \frac{4\beta^2 \sin^2 ka}{(m' - m)^2 \omega^4 + 4\beta^2 \sin^2 ka} \tag{1.45}$$

We note that when $m' = m$, the incident wave propagates without reflection i.e., the transmission coefficient (1.45) is equal to 1. We also note that for $k = \frac{\pi}{2a}$, i.e., the edge of the Brillouin zone for the diatomic crystal), the transmission coefficient simplifies to: $T^2 = \frac{4\beta^2}{(m'-m)^2 \omega^4 + 4\beta^2}$. The transmission coefficient decreases monotonically as a function of frequency showing no sign of resonance or any other localized vibration phenomenon. In absence of such a resonant phenomenon, the band structure of the diatomic harmonic crystal can, therefore, be ascribed to the periodicity of the structure, alone. The presence of an acoustic branch and of an optical branch separated by a gap results from scattering of waves by the periodic crystal, namely, Bragg's scattering.

1.7.1 *Monatomic Harmonic Crystal with a General Perturbation*

The approach of Sect. 1.7 is generalized by introducing a frequency dependent perturbation, $V(\omega)$, of the 1-D monatomic crystal at site 0. The equations of motion of the atoms in this defected crystal are:

$$\begin{cases} -m\omega^2 u_n = \beta(u_{n+1} - 2u_n + u_{n-1}) \ for \ n \neq 0 \\ -m\omega^2 u_o = \beta(u_1 - 2u_0 + u_{-1}) + V(\omega) u_0 \end{cases} \tag{1.46}$$

Following, the derivation of the transmission and reflection coefficients in the previous section, we obtain:

$$\begin{aligned} T &= \frac{\beta 2i \sin ka}{V(\omega) + \beta 2i \sin ka} \\ R &= \frac{V(\omega)}{V(\omega) + \beta 2i \sin ka} \end{aligned} \tag{1.47}$$

We note that if $V(\omega) = \infty$, then an incident wave is totally reflected. Such a condition may arise from a local resonance. This case will be discussed in the next subsection.

1.7.2 Locally Resonant Structure

Here, we are interested in the behavior of a monatomic crystal with a structural perturbation taking the form of a side branch. The side branch is composed of L' atoms of mass m' interacting via harmonic springs with force constant, β'. The side branch is attached to the monatomic crystal at site "0" via a spring with stiffness, β_I. We assume that the lattice parameter is the same in the side branch and the infinite monatomic crystal. This structure is illustrated in Fig. 1.7.

The derivation of an expression for the perturbation potential, V, begins with the equations of motion of atoms in the side branch:

$$\begin{cases} -m'\omega^2 u_{n'} = \beta'(u_{n'+1} - 2u_{n'} + u_{n'-1})\, for\ n' \neq 1', L' & (a) \\ -m'\omega^2 u_{L'} = -\beta'\,(u_{L'} - u_{L'-1}) & (b) \\ -m'\omega^2 u_{1'} = -\beta_I(u_{1'} - u_0) + \beta'(u_{2'} - u_{1'}) & (c) \end{cases} \quad (1.48)$$

This set of equations is complemented by the equation of motion of site "0":

$$-m\omega^2 u_o = \beta(u_1 - 2u_0 + u_{-1}) + \beta_I(u_{1'} - u_0) \quad (1.49)$$

To find the perturbation potential, we are interested in coupling (1.48) and (1.49) to obtain an effective equation taking the form of (1.40) for site "0". Rewriting (1.49) as: $\left(-m\omega^2 + \beta_I\left(1 - \frac{u_{1'}}{u_0}\right)\right)u_0 = \beta(u_1 - 2u_0 + u_{-1})$ yields:

$$V = -\beta_I\left(1 - \frac{u_{1'}}{u_0}\right) \quad (1.50)$$

The ratio of displacements in (1.50) is found by considering the general solution to (1.48)(a):

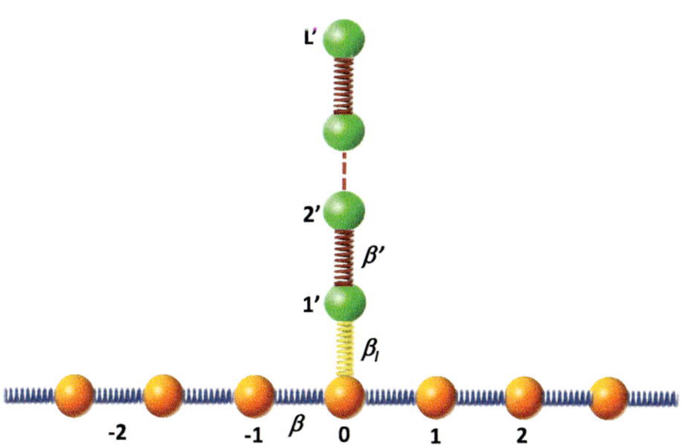

Fig. 1.7 Illustration of the 1-D monatomic crystal perturbed by a side branch

$$u_{n'} = A' e^{ik'n'a} + B' e^{-ik'n'a} \tag{1.51}$$

Inserting this solution in (1.48)(a) gives:

$$m'\omega^2 = 2\beta'(1 - \cos k'a) \tag{1.52}$$

Should site N' have been in an infinite monatomic crystal, its equation of motion would have been:

$$-m'\omega^2 u_{L'} = \beta'(u_{L'+1} - 2u_{L'} + u_{L'-1}) \tag{1.53}$$

Subtracting (1.53) and (1.48)(b) gives:

$$\beta'(u_{L'+1} - u_{L'}) = 0 \tag{1.54}$$

This equation serves as a boundary condition on site N' in the side branch. We define the displacement, $u_{L'+1}$, at a fictive site "L' + 1" as support for the boundary condition (1.54). Similarly subtracting the equation of motion (1.48)(c) and that of site "1′" if it were embedded in an infinite monatomic crystal leads to the boundary condition:

$$-\beta_I(u_{1'} - u_0) + \beta'(u_{1'} - u_{0'}) = 0 \tag{1.55}$$

Fictive site "0′" is only used to impose the boundary condition. The two boundary conditions at sites "*1′*" and "*L′*" form the set of equations:

$$\begin{cases} u_{L'} - u_{L'+1} = 0 \\ (\beta_I - \beta')u_{1'} + \beta' u_{0'} = 0 \end{cases} \tag{1.56}$$

We insert the general solution (1.51) into (1.56) and obtain the set of linear equations:

$$\begin{cases} A' e^{ik'L'a}\left(1 - e^{ik'a}\right) + B' e^{-ik'L'a}\left(1 - e^{-ik'a}\right) = 0 \\ A'\left[(\beta_I - \beta')e^{ik'a} + \beta'\right] + B'\left[(\beta_I - \beta')e^{-ik'a} + \beta'\right] = \beta_I u_0 \end{cases} \tag{1.57}$$

Solving (1.57) gives:

$$\begin{aligned} A' &= -\beta_I u_0 e^{-ik'L'a}\left(1 - e^{-ik'a}\right)/\Delta \\ B' &= \beta_I u_0 e^{ik'L'a}\left(1 - e^{ik'a}\right)/\Delta \end{aligned} \tag{1.58}$$

where

$$\Delta = -4i \sin \frac{k'a}{2}\left[(\beta_I - \beta')\cos k'\left(L' - \frac{1}{2}\right)a + \beta' \cos k'\left(L' + \frac{1}{2}\right)a\right] \tag{1.59}$$

To obtain (1.59) we have used a variety of trigonometric relations.

It is worth noting that in the limit of $\beta_I = 0$, the set of (1.57) can be used to find the displacement of an isolated finite segment of monatomic crystal. The existence of non-trivial solutions for the amplitudes, $A'$and B', is ensured by the condition $\Delta = 0$. This condition is rewritten as: $\cos k'\left(L' - \frac{1}{2}\right)a + \cos k'\left(L' + \frac{1}{2}\right)a = -2 \sin k' L'a \sin \frac{k'a}{2} = 0$ or $\sin k'L'a = 0$. These solutions correspond to vibrational modes of the finite crystal of length L' i.e., standing waves with wave vectors: $k' = \frac{p\pi}{L'a}$ where p is an integer.

Finally, to find the perturbation, V, we use (1.57) and (1.51) to obtain the displacement of atom "1′" which we subsequently insert into (1.50). After several algebraic and trigonometric manipulations, the perturbation becomes:

$$V(\omega) = \frac{2\beta'\beta_I \sin \frac{k'a}{2} \sin L'k'a}{(\beta_I - \beta')\cos k'\left(L' - \frac{1}{2}\right)a + \beta' \cos k'\left(L' + \frac{1}{2}\right)a} \tag{1.60}$$

The effect of the side branch on the propagation of waves along the infinite crystal is most easily understood by considering the limiting case: $\beta = \beta_I = \beta'$ and $m = m'$ such that $k = k'$. In this case the side branch is constituted of the same material as the infinite crystal and (1.60) becomes:

$$V(\omega) = \frac{2\beta \sin \frac{ka}{2} \sin L'ka}{\cos k\left(L' + \frac{1}{2}\right)a} \tag{1.61}$$

with the dispersion relation $\omega(k) = \omega_0 \left| \sin k\frac{a}{2} \right|$ [i.e., (1.5)]. At the frequency (wave number) corresponding to the standing wave modes of the side branch, the perturbation, $V = 0$. The transmission and reflection coefficients given by (1.57) are equal to 1 and 0, respectively. Zeros of transmission and complete reflection occur when $V = \infty$, that is, when $\cos k\left(L' + \frac{1}{2}\right)a = 0$ or $k = (2p+1)\frac{\pi}{(2L'+1)a}$. These conditions correspond to resonances with the side branch. For instance, for a single atom side branch, i.e., $L' = 1$, there is one zero of transmission in the irreducible Brillouin zone of the monatomic crystal at $k = \frac{\pi}{3a}$. For a two-atom side branch, $L' = 2$, there are two zeros of transmission in the irreducible Brillouin zone of the monatomic crystal at $k = \frac{\pi}{5a}$ and $k = \frac{3\pi}{5a}$. The number of zeros of transmission scales with the number of atoms in the side branch. Therefore, in contrast to the scattering by a mass defect which did not introduce any zeros of transmission, the side branch leads to perturbations of the band structure of the supporting infinite 1-D monatomic crystal. These perturbations arise from resonances ($V = \infty$) of the side branch. The alterations to the band structure of the monatomic crystal due to the side branch may be visualized as infinitesimally narrow band gaps. The crystal with a single side branch is not periodic and the perturbed band structure result only from local resonances. In the next sections, we develop the formalism necessary to shed light on the interplay between Bragg's scattering and local resonances on the band structure of a 1-D monatomic crystal with periodic arrangements of side branches. This formalism is based on a Green's function approach called the Interface Response Theory [2].

1.8 Interface Response Theory

1.8.1 Fundamental Equations of the Interface Response Theory

In this section, we review the fundamental equations of the Interface Response Theory (IRT) for discrete systems [2]. This formalism allows the calculation of the Green's function of a perturbed system in terms of the Green's functions of unperturbed systems. We recall (1.29) and (1.28) defining the Green's function, $\overleftrightarrow{G_0}$, by:

$$\overleftrightarrow{H_0}\,\overleftrightarrow{G_0} = \overleftrightarrow{I}$$

The operator, $\overleftrightarrow{H_0}$, is the infinite tridiagonal dynamic matrix:

$$\overleftrightarrow{H_0} = \frac{1}{m}\begin{bmatrix} \ddots & \vdots & \vdots & \vdots & \vdots & \vdots & \vdots & \vdots & \vdots \\ \cdots & 0 & \beta & -\gamma & \beta & 0 & 0 & \cdots & \cdots \\ \cdots & \cdots & 0 & \beta & -\gamma & \beta & 0 & \cdots & \cdots \\ \cdots & \cdots & \cdots & 0 & \beta & -\gamma & \beta & 0 & \cdots \\ \vdots & \vdots & \vdots & \vdots & \vdots & \vdots & \vdots & \vdots & \ddots \end{bmatrix} \tag{1.62}$$

where $\gamma = 2\beta - m\omega^2$. We initially consider a type of perturbation that cleaves the 1-D monatomic harmonic crystal by severing a bond between two neighboring atoms (Fig. 1.8).

The equations of motion of the atoms 0 and 1 are:

$$\begin{cases} \dfrac{1}{m}(-\alpha u_0 + \beta u_{-1}) = 0 \\ \dfrac{1}{m}(-\alpha u_1 + \beta u_2) = 0 \end{cases} \tag{1.63}$$

with $\alpha = m\omega^2 - \beta$.

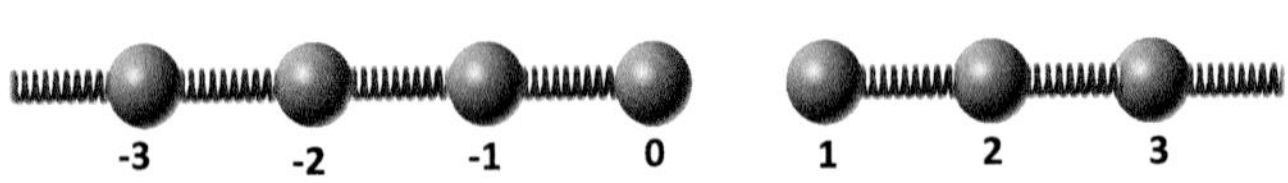

Fig. 1.8 Schematic illustration of the 1-D monatomic harmonic crystal cleaved between atoms 0 and 1

The dynamical operator for the cleaved crystal is written as:

$$\overleftrightarrow{h}_0 = \begin{bmatrix} \overleftrightarrow{h}_{S1} & \overleftrightarrow{0} \\ \overleftrightarrow{0} & \overleftrightarrow{h}_{S2} \end{bmatrix} = \frac{1}{m} \begin{array}{c} \begin{array}{ccccccccc} \cdots & -3 & -2 & -1 & 0 & 1 & 2 & 3 & \cdots \end{array} \\ \left[\begin{array}{ccccccccc} \beta & -\gamma & \beta & 0 & 0 & 0 & 0 & 0 & 0 \\ 0 & \beta & -\gamma & \beta & 0 & 0 & 0 & 0 & 0 \\ 0 & 0 & \beta & -\gamma & \beta & 0 & 0 & 0 & 0 \\ 0 & 0 & 0 & \beta & -\alpha & 0 & 0 & 0 & 0 \\ 0 & 0 & 0 & 0 & 0 & -\alpha & \beta & 0 & 0 \\ 0 & 0 & 0 & 0 & 0 & \beta & -\gamma & \beta & 0 \\ 0 & 0 & 0 & 0 & 0 & 0 & \beta & -\gamma & \beta \end{array} \right] \end{array} \begin{array}{c} \vdots \\ -3 \\ -2 \\ -1 \\ 0 \\ 1 \\ 2 \\ \vdots \end{array} \tag{1.64}$$

In (1.64), $\overleftrightarrow{h}_0$, is a block diagonal matrix composed of two independent matrices, $\overleftrightarrow{h}_{S1}$ and $\overleftrightarrow{h}_{S2}$, corresponding to the two semi-infinite crystals on the left and right of the cleaved bond, respectively. The Green's function of the perturbed system, $\overleftrightarrow{g}_0$, is therefore defined through the relation:

$$\overleftrightarrow{h}_0 \overleftrightarrow{g}_0 = \overleftrightarrow{I} \tag{1.65}$$

Since the dynamical matrix of the cleaved system is a block diagonal matrix, its associated Green's function is also a block diagonal matrix:

$$\overleftrightarrow{g}_0 = \begin{bmatrix} \overleftrightarrow{g}_{S1} & \overleftrightarrow{0} \\ \overleftrightarrow{0} & \overleftrightarrow{g}_{S2} \end{bmatrix} \tag{1.66}$$

We define the perturbation operator or cleavage operator, as the difference between the dynamical matrices of the cleaved and unperturbed crystals:

$$\overleftrightarrow{V}_0 = \overleftrightarrow{h}_0 - \overleftrightarrow{H}_0 \tag{1.67}$$

Using the matrix representation, the cleavage operator is a 2×2 matrix limited to the sites 0 and 1 of the crystal:

$$\overleftrightarrow{V}_0 = \begin{pmatrix} V_0(0,0) & V_0(1,0) \\ V_0(0,1) & V_0(1,1) \end{pmatrix} = \frac{1}{m} \begin{pmatrix} \beta & -\beta \\ -\beta & \beta \end{pmatrix} \tag{1.68}$$

We rewrite (1.65) in the form: $\overleftrightarrow{g}_0 \overleftrightarrow{h}_0 = \overleftrightarrow{I}$ by using the commutative property of the product of a matrix with its inverse. Introducing (1.7) into this latter relation, multiplying both sides of the equal sign by $\overleftrightarrow{G}_0$, applying the distributive property of the product of matrices and finally using (1.29) yields:

$$\overleftrightarrow{g}_0\left(\overleftrightarrow{I}+\overleftrightarrow{V}_0\overleftrightarrow{G}_0\right)=\overleftrightarrow{g}_0\left(\overleftrightarrow{I}+\overleftrightarrow{A}_0\right)=\overleftrightarrow{G}_0 \tag{1.69}$$

Equation (1.69) is called Dyson's equation. It enables the determination the Green's function of a perturbed system in terms of the perturbation operator and the Green's function of the unperturbed system. In (1.69), we have defined the surface operator:

$$\overleftrightarrow{A}_0=\overleftrightarrow{V}_0\overleftrightarrow{G}_0 \tag{1.70}$$

The Green's function of the perturbed system is then given by:

$$\overleftrightarrow{g}_0=\overleftrightarrow{G}_0\left(\overleftrightarrow{I}+\overleftrightarrow{A}_0\right)^{-1} \tag{1.71}$$

The poles of $\overleftrightarrow{g}_0$ (i.e., the eigenvalues of the operator $\overleftrightarrow{h}_0$) are the zeros of $\overleftrightarrow{I}+\overleftrightarrow{A}_0$.

1.8.2 Green's Function of the Cleaved 1-D Monatomic Crystal

We apply (1.71) to the calculation of the Green's function of the semi-infinite crystal on the right of the cleaved bond in Fig. 1.8 (i.e., $n \geq 1$). The components of the surface operator defined by (1.70) are written as:

$$A_{S2}(n,n')=\sum_{n''}V_0(n,n'')G_0(n'',n') \quad with \quad n,n'\geq 1 \tag{1.72}$$

The only non-zero components of the cleavage operator are for $n, n' \epsilon [0,1]$ so (1.72) reduces to:

$$A_{S2}(1,n')=V_0(1,0)G_0(0,n')+V_0(1,1)G_0(1,n'), \quad n'\geq 1 \tag{1.73}$$

Inserting the terms in (1.68) and (1.32) into (1.73) results in:

$$A_{S2}(1,n')=\frac{t^{n'}-t^{n'+1}}{t^2-1} \tag{1.74}$$

We now write (1.69) in component form:

$$g_{S2}(n,n') + g_{S2}(n,1)A_{S2}(1,n') = G_0(n,n'), \quad n,n' \geq 1 \tag{1.75}$$

Expressing (1.75) at site $n' = 1$ and using the relation (1.74) gives:

$$g_{S2}(n,1) = \frac{m}{\beta}\frac{t^n}{t-1}$$

We can now combine that relation with (1.74), (1.32) and (1.75) to obtain the function sought:

$$g_{S2}(n,n') = \frac{m}{\beta}\frac{t^{|n-n'|+1} + t^{n+n'}}{t^2-1}, \quad n,n' \geq 1 \tag{1.76}$$

The procedure used in this section to find the Green's function of the perturbed system can be generalized to obtain the universal equation of the IRT. All matrices in (1.69) are defined for, $n' \epsilon[-\infty, \infty]$. We now consider the space D for $n\,, n' \geq 1$ and rewrite (1.69) as:

$$\overleftrightarrow{g}_{S2}(D,D) + \overleftrightarrow{g}_{S2}(D,M)\overleftrightarrow{A}_{S2}(M,D) = \overleftrightarrow{G}_{S2}(D,D) \tag{1.77}$$

The index S specifies that all functions are limited to the space of a semi-infinite truncated chain. Equation (1.75) is a particular case of the general equation (1.77) where we have specified the space corresponding to the location of the perturbation by M. In the case of the cleavage of the monatomic crystal, $M = 1$. A particular form of (1.77) is:

$$\overleftrightarrow{g}_{S2}(D,M) + \overleftrightarrow{g}_{S2}(D,M)\overleftrightarrow{A}_{S2}(M,M) = \overleftrightarrow{G}_{S2}(D,M) \tag{1.78}$$

We combine (1.78) and (1.77) to obtain the universal equation of the IRT:

$$\overleftrightarrow{g}_{S2}(D,D) = \overleftrightarrow{G}_{S2}(D,D) + \overleftrightarrow{G}_{S2}(D,M)\overleftrightarrow{\Delta}^{-1}(M,M)\overleftrightarrow{A}_{S2}(M,D) \tag{1.79}$$

where

$$\overleftrightarrow{\Delta}\,(M,M) = \overleftrightarrow{I}\,(M,M) + \overleftrightarrow{A_{S2}}\,(M,M) \tag{1.80}$$

Equation (1.80) introduces the diffusion matrix, $\overleftrightarrow{\Delta}$.

The displacement vector, $\vec{u}(D)$, is related to the Green's function, $\overleftrightarrow{g}_S$, via the relation:

$$\vec{u}(D) = \vec{f}(D)\overleftrightarrow{g}_{S2}(D,D) \tag{1.81}$$

where, $\vec{f}$, is some force distribution applied in the space D. Inserting (1.79) into (1.81), we obtain the displacement vector of the perturbed system in terms of the displacement vector of the unperturbed system, U, as:

$$\vec{u}(D) = \vec{U}(D) - \vec{U}(M)\overleftrightarrow{\Delta}^{-1}(M, M)\overleftrightarrow{A}_{S2}(M, D) \tag{1.82}$$

Applying (1.82) to the right side of the cleaved monatomic crystal yields:

$$u(n') = U(n') - U(1)\Delta^{-1}(1,1)A_{S2}(1,n') \ \mathit{for} \ \ n' \geq 1$$

with $\Delta^{-1}\,(1,1) = \frac{1}{1+A_{S2}(1,1)} = \frac{t^2-1}{t-1}$ and $A_{S2}(1,n') = \frac{t^{n'}-t^{n'+1}}{t^2-1}$. The displacement is therefore:

$$u(n') = U(n') + U(1)t^{n'} \quad n' \geq 1$$

If we choose, $U(n') = t^{-n'}$, corresponding to an incident waves coming from $n' = +\infty$, the displacement field in the semi-infinite chain takes the form:

$$u(n') = t^{-n'} + t^{n'-1} = e^{-ikn'a} + e^{ik(n'-1)a}$$

This is a wave resulting from the superposition of an incident wave and a reflected wave. We can also obtain this result by writing the equation of motion at site 1 of the cleaved crystal:

$$-m\omega^2 u_1 = \beta(u_2 - u_1)$$

This equation implies that $u_1 - u_0 = 0$ where u_0 is the displacement of the site 0 taken as a fictive site imposing a zero force boundary condition on site 1. We assume that the displacement in the semi-infinite crystal is the sum of a reflected wave and a transmitted wave:

$$u_n = A_i e^{-ikna} + A_r e^{ikna}$$

Inserting this general solution into the boundary condition leads to the relation between the incident and reflected amplitudes: $A_r = A_i e^{-ika}$ leading to the displacement, $u(n) = A_i(e^{-ikna} + e^{ik(n-1)a})$.

1.8.3 Finite Monatomic Crystal

The finite 1-D monatomic crystal is formed by cleaving an infinite crystal at two separate locations. This doubly cleaved system is illustrated in Fig. 1.9.

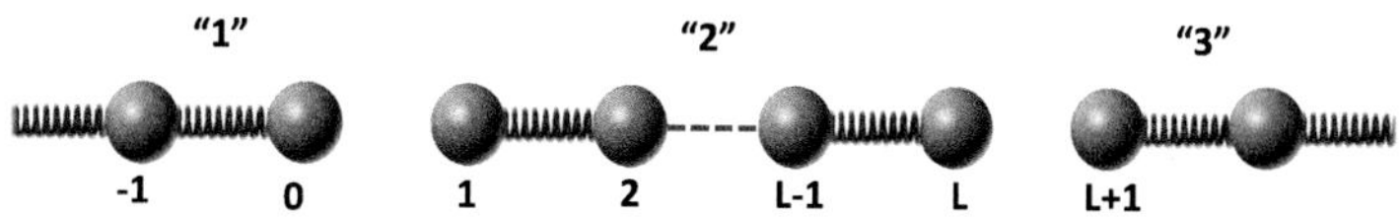

Fig. 1.9 Monatomic crystal cleaved between sites (0, 1) and (L, L + 1) to obtain a finite crystal composed of atoms [1, L]

The cleavage operator is a 4×4 matrix expressed in the space of the perturbed sites (0, 1) and (L, L + 1):

$$\overleftrightarrow{V_0} = \frac{1}{m} \begin{array}{c} \begin{matrix} 0 & 1 & L & L+1 \end{matrix} \\ \begin{bmatrix} \beta & -\beta & 0 & 0 \\ -\beta & \beta & 0 & 0 \\ 0 & 0 & \beta & -\beta \\ 0 & 0 & -\beta & \beta \end{bmatrix} \end{array} \begin{matrix} 0 \\ 1 \\ L \\ L+1 \end{matrix} \tag{1.83}$$

The dynamical matrix is composed of three separate blocks corresponding to the three uncoupled regions of the cleaved system of Fig. 1.9, namely regions "1", "2" and "3". Similarly, the Green's function and the surface operators are also block diagonal matrices. Using (1.70), the non-zero components of the surface operator matrix corresponding to the block of the finite segment of crystal, "2" are:

$$\begin{aligned} A_{S2}(1,n') &= V_0(1,0)G_0(0,n') + V_0(1,1)G_0(1,n') \\ A_{S2}(L,n') &= V_0(L,L)G_0(L,n') + V_0(L,L+1)G_0(L+1,n') \end{aligned} \quad n' \epsilon [1,L] \tag{1.84}$$

The Green's function of the infinite crystal given by (1.32) is inserted into (1.84) to obtain:

$$\begin{aligned} A_{S2}(1,n') &= \frac{-t^{n'}}{t+1} \\ A_{S2}(L,n') &= \frac{-t^{L-n'+1}}{t+1} \end{aligned} \quad n' \epsilon [1,L] \tag{1.85}$$

To apply the universal equation of the IRT, we need the block "2" of the surface operator matrix in the space of the corresponding perturbed sites, $M \epsilon [1,L]$, that is:

$$\overleftrightarrow{A_{S2}}(M,M) = \begin{bmatrix} A_{S2}(1,1) & A_{S2}(1,L) \\ A_{S2}(L,1) & A_{S2}(L,L) \end{bmatrix} = \frac{-1}{t+1} \begin{bmatrix} t & t^L \\ t^L & t \end{bmatrix} \tag{1.86}$$

The Green's function of the finite segment of crystal takes the form:

$$\begin{aligned} g_{S2}(n,n') = {} & G_0(n,n') - G_0(n,1)\Delta^{-1}(1,1)A_{S2}(1,n') - G_0(n,1)\Delta^{-1}(1,L)A_{S2}(L,n') \\ & - G_0(n,L)\Delta^{-1}(L,1)A_{S2}(1,n') - G_0(n,L)\Delta^{-1}(L,L)A_{S2}(L,n'), \quad n,n' \epsilon [1,L] \end{aligned} \tag{1.87}$$

In (1.87),

$$\overleftrightarrow{\Delta}^{-1}(M,M) = \frac{1}{W}\frac{1}{t+1}\begin{bmatrix} 1 & t^L \\ t^L & 1 \end{bmatrix} \tag{1.88}$$

with $W = det\,\overleftrightarrow{\Delta} = \frac{1-t^{2L}}{(t+1)^2}$ and according to (1.80) $\overleftrightarrow{\Delta}(M,M) = \overleftrightarrow{I}(M,M) + \overleftrightarrow{A_{S2}}(M,M)$.

Inserting the expressions given by (1.32), (1.85), and (1.88) into (1.87) yields the Green's function of the finite crystal (for $n\,,n'\epsilon[1,L]$):

$$g_{S2}(n,n') = \frac{m}{\beta}\left[\frac{t^{|n-n'|+1} + t^{n+n'}}{t^2-1} + \frac{t^{2L+1}}{(t^2-1)(1-t^{2L})}\left(t^{n'-n} + t^{n-n'} + t^{1-n-n'} + t^{n+n'-1}\right)\right] \tag{1.89}$$

According to (1.71), the poles of the Green's function are also those of $\overleftrightarrow{\Delta}^{-1}$. Here, these poles are the zeros of W. The Eigen values of the finite crystals are, therefore, given by the condition: $1 - t^{2L} = 0$. This condition may be rewritten as: $t^L - t^{-L} = 0$. For angular frequencies, $\omega\epsilon[0,\omega_0]$, $t = e^{ika}$ and the modes of the finite crystal are given by: $e^{ikLa} - e^{-ikLa} = \sin kLa = 0$. These modes correspond to standing waves with wave number conditioned by: $k = \frac{p\pi}{La}$ with p being an integer. The displacement field of these standing waves is obtained from (1.82). In components form, (1.82) becomes:

$$\begin{aligned} u(n') &= U(n') - U(1)\Delta^{-1}(1,1)A_{S2}(1,n') - U(1)\Delta^{-1}(1,L)A_{S2}(L,n') \\ &-U(L)\Delta^{-1}(L,1)A_{S2}(1,n') - U(L)\Delta^{-1}(L,L)A_{S2}(L,n'), \qquad n,n'\epsilon[1,L] \end{aligned} \tag{1.90}$$

Employing a reference displacement, $U(n') = t^{n'}$, (1.90) gives:

$$u(n') = t^{n'} + t^{n'}\frac{t}{1-t^{2L}} + t^{-n'}\frac{t^{2L+2}}{1-t^{2L}} + t^{n'}\frac{t^{2L}}{1-t^{2L}} + t^{-n'}\frac{t^{2L+1}}{1-t^{2L}}$$

This expression diverges when $1 - t^{2L} = 0$. It is therefore necessary to obtain a finite displacement by renormalizing the previous expression by W. The renormalized displacement reduces then to: $u(n') = t^{n'} + t^{-n'+1}$. This expression is that of the displacement of standing waves in the finite crystal.

1.8.4 1-D Monatomic Crystal with One Side Branch

The calculation of the displacement in a system composed of a 1-D monatomic crystal with a finite crystal branch coupled to its side via a spring with constant, β_I, as illustrated in Fig. 1.7, begins with the block matrix describing Green's function of the uncoupled system ($\beta_I = 0$):

$$\overleftrightarrow{G}_S = \begin{pmatrix} \overleftrightarrow{G}_0 & \overleftrightarrow{0} \\ \overleftrightarrow{0} & \overleftrightarrow{g}'_{S2} \end{pmatrix} \tag{1.91}$$

where $\overleftrightarrow{G}_0$ is the Green's function of the infinite crystal [its components are given by (1.32)] and where $\overleftrightarrow{g}'_{S2}$ is the Green's function of the finite side branch given by (1.89). This latter Green's function is labeled with a "prime" sign to indicate that the spring constants and masses, m' and β' of the finite crystal may be different from those of the infinite crystal, m and β. The difference between the dynamic matrix of the coupled systems and of the dynamic matrix of the uncoupled system defines a coupling operator:

$$\overleftrightarrow{V}_I = \begin{pmatrix} V_I(0,0) & V_I(0,1') \\ V_I(1',0) & V_I(1',1') \end{pmatrix} = \begin{pmatrix} \frac{-\beta_I}{m} & \frac{\beta_I}{m} \\ \frac{\beta_I}{m'} & \frac{-\beta_I}{m'} \end{pmatrix} \tag{1.92}$$

We note that if the masses in the finite and infinite crystals were the same, the coupling operator would simply be the opposite of the cleavage operator of (1.68). We now use the fundamental equation of the IRT to derive an expression for the displacement field in the coupled system in terms of the Green's function of the constituent crystals making up the uncoupled system and the perturbation operator of (1.92).

To that effect, we first write expressions for the surface operator:

$$\overleftrightarrow{A}_0(MD) = \begin{pmatrix} A(0,n) \\ A(0,n') \\ A(1',n) \\ A(1',n') \end{pmatrix} = \begin{pmatrix} V_I(0,0)G_0(0,n) \\ V_I(0,1')g'_{S2}(1',n') \\ V_I(1',0)G_0(0,n) \\ V_I(1',1')g'_{S2}(1',n') \end{pmatrix} \tag{1.93}$$

In (1.93), n and n' refer to sites in the infinite crystals and the finite side branch, respectively. The diffusion matrix takes then the form of a 2×2 matrix in the space of the interface sites, M:

$$\begin{aligned} \overleftrightarrow{\Delta}(MM) &= \begin{pmatrix} 1+A(0,0) & A(0,1') \\ A(1',0) & 1+A(1,1') \end{pmatrix} \\ &= \begin{pmatrix} 1+V_I(0,0)G_0(0,0) & V_I(0,1')g'_{S2}(1',1') \\ V_I(1',0)G_0(0,0) & 1+V_I(1',1')g'_{S2}(1',1') \end{pmatrix} \end{aligned} \tag{1.94}$$

The inverse of the diffusion matrix is then:

$$\overleftrightarrow{\Delta}^{-1}(MM) = \frac{1}{det\,\overleftrightarrow{\Delta}} \begin{pmatrix} 1+V_I(1',1')g'_{S2}(1',1') & -V_I(0,1')g'_{S2}(1',1') \\ -V_I(1',0)G_0(0,0) & 1+V_I(0,0)G_0(0,0) \end{pmatrix} \tag{1.95}$$

We use (1.82) to obtain the displacement field. For this we also need to assume a form for the reference displacement, $U(D) = t^n$. This displacement corresponds to a wave propagating in the in the infinite crystal and launched from $n = -\infty$. The displacement inside the side crystal is also assumed to be equal to zero. The displacement in the space of the perturbed sites, [0, 1′], take the form:

$$U(M) = (U(0), U(1')) = (1, 0) \tag{1.96}$$

The displacement field at a site $n \geq 1$ along the infinite crystal (i.e., on the right side of the grafted branch) is therefore determined from the equation:

$$u_n = t^n - (1,0)\begin{pmatrix} \Delta^{-1}(0,0) & \Delta^{-1}(0,1') \\ \Delta^{-1}(1',0) & \Delta^{-1}(1',1') \end{pmatrix}\begin{pmatrix} A(0,n) \\ A(1',n) \end{pmatrix} \tag{1.97}$$

Inserting (1.95) and (1.93) into (1.97) yields:

$$u_n = t^n - \frac{1}{det\overleftrightarrow{\Delta}}\overleftrightarrow{V}_I(0,0)G_0(0,n) \tag{1.98}$$

One then combines (1.32), (1.89), (1.92) and (1.98) to obtain:

$$u_n = t^n\left(1 + \frac{\beta_I}{\beta}\frac{1}{det\overleftrightarrow{\Delta}}\frac{t}{t^2-1}\right) = t^n T \tag{1.99}$$

with

$$det\,\overleftrightarrow{\Delta} = 1 - \frac{\beta_I}{\beta'}\frac{t' + t'^{2L'}}{(t'-1)\left(1 - t'^{2L'}\right)} - \frac{\beta_I}{\beta}\frac{t}{t^2-1} \tag{1.100}$$

In (1.99), T is the transmission coefficient. We can rewrite (1.100) in the form:

$$det\,\overleftrightarrow{\Delta} = -\frac{1}{V} - \frac{\beta_I}{\beta}\frac{1}{2i\sin ka} \tag{1.101}$$

where $-\frac{1}{V} = 1 - \frac{\beta_I}{\beta'}\frac{t'+t'^{2L'}}{(t'-1)\left(1-t'^{2L'}\right)}$. To obtain (1.101), we also defined $t = e^{ika}$, with $t' = e^{ik'a}$, one can show that the quantity V is that given by (1.60).

1.8.5 1-D Monatomic Crystal with Multiple Side Branches

We now consider N_c side branches of various lengths grafted along an infinite 1-D monatomic crystal. The spaces, D and M for this system are defined as:

$$D = \{-\infty,\ldots,-1,0,1,\ldots\infty\} \times \cup\{\{1',2',\ldots L'\},\{1'',2'',\ldots,L''\},\{1^{(3)},2^{(3)},\ldots,L^{(3)}\}\ldots,\{1^{(N_c)},2^{(N_c)},\ldots,L^{(N_c)}\}\}$$

and

$$M = \left\{p_1 = 0, 1', p_2, 1'', p_3, 1^{(3)}, \ldots, p_{N_c}, 1^{(N_c)}\right\}$$

We have located the first finite side branch at site $p_1 = 0$ of the infinite crystal. The second finite side branch is located at site $p_2 > p_1$ of the infinite crystal. The third finite side branch is located at $p_3 > p_2$, etc. In this case, the coupling operator is a $2N_c \times 2N_c$ matrix of the form:

$$\overleftrightarrow{V}_I = \frac{\beta_I}{m}\begin{pmatrix} -1 & 1 & 0 & 0 & \ldots & 0 & 0 \\ 1 & -1 & 0 & 0 & \ldots & 0 & 0 \\ 0 & 0 & -1 & 1 & \ldots & 0 & 0 \\ 0 & 0 & 1 & -1 & \ldots & 0 & 0 \\ \vdots & \vdots & \vdots & \vdots & \ldots & \vdots & \vdots \\ 0 & 0 & 0 & 0 & 0 & -1 & 1 \\ 0 & 0 & 0 & 0 & 0 & 1 & -1 \end{pmatrix} \quad (1.102)$$

To calculate $\overleftrightarrow{\Delta}(MM) = \overleftrightarrow{I}(MM) + \overleftrightarrow{V}_I(MM)\overleftrightarrow{G}_S(MM)$, one needs the Green's function of the uncoupled system, $\overleftrightarrow{G}_S(MM)$, which takes the form:

$$\overleftrightarrow{G}_s(MM) = \begin{pmatrix} G_0(p_1p_1) & 0 & G_0(p_1p_2) & 0 & G_0(p_1p_3) & 0 & \ldots & G_0(p_1p_{N_c}) & 0 \\ 0 & g_s(1'1') & 0 & 0 & 0 & 0 & \ldots & 0 & 0 \\ G_0(p_2p_1) & 0 & G_0(p_2p_2) & 0 & G_0(p_2p_3) & 0 & \ldots & G_0(p_2p_{N_c}) & 0 \\ 0 & 0 & 0 & g_s(1''1'') & 0 & 0 & \ldots & 0 & 0 \\ G_0(p_3p_1) & 0 & G_0(p_3p_2) & 0 & G_0(p_3p_3) & 0 & \ldots & G_0(p_3p_{N_c}) & 0 \\ 0 & 0 & 0 & 0 & 0 & g_s(1^{(3)}1^{(3)}) & \ldots & 0 & 0 \\ \vdots & \vdots & \vdots & \vdots & \vdots & \vdots & \vdots & \vdots & \vdots \\ G_0(p_{N_c}p_1) & 0 & G_0(p_{N_c}p_2) & 0 & G_0(p_{N_c}p_3) & 0 & \ldots & G_0(p_{N_c}p_{N_c}) & 0 \\ 0 & 0 & 0 & 0 & 0 & 0 & \ldots & 0 & g_s(1^{(N_c)}1^{(N_c)}) \end{pmatrix}.$$

In this matrix, the odd entries (rows or columns) correspond to locations along the infinite crystal in M and the even entries correspond to the position of the first atom of the finite side branches (also in the space M). From (1.32) and (1.89), the elements of this matrix are therefore:

$$G_0(p_ip_j) = \frac{\beta}{m}\frac{t^{|p_i-p_j|+1}}{t^2-1} \quad (1.104)$$

and

$$g_s\left(1^{(i)}1^{(i)}\right) = \frac{\beta'}{m'} \frac{t' + t'^{L'_i}}{(t' - 1)\left(1 - t'^{2L'_i}\right)} \tag{1.105}$$

We use (1.82) to obtain the displacement field. For this, we again need to assume a form for the reference displacement, $U(D) = t^n$. This displacement corresponds to a wave propagating in the in the infinite crystal and launched from $n = -\infty$. The displacement inside the side branch is also assumed to be equal to zero. The displacement in the space M takes the form:

$$\begin{aligned} U(M) &= \left(U(0), U(1'), U(p_2), U(1''), \ldots, U\left(p_{N_c}\right), U\left(1^{N_c}\right)\right) \\ &= \left(1, 0, t^{p_2}, 0, \ldots, t^{p_{N_c}}, 0\right) \end{aligned} \tag{1.106}$$

The displacement field at a site $n \geq p_{N_c}$ along the infinite crystal (i.e., on the right side of the last grafted finite side branch) is therefore determined from the (1.82) where we use:

$$\overleftrightarrow{A}(M,n) = \begin{pmatrix} V_I(0,0)G_0(0,n) \\ V_I(1',0)G_0(0,n) \\ V_I(p_2,p_2)G_0(p_2,n) \\ V_I(1'',p_2)G_0(p_2,n) \\ \vdots \\ V_I\left(p_{N_c},p_{N_c}\right)G_0\left(p_{N_c},n\right) \\ V_I\left(1^{p_{N_c}},p_{N_c}\right)G_0\left(p_{N_c},n\right) \end{pmatrix} = \frac{\beta_I}{m}\frac{\beta}{m}\frac{t^n}{t^2 - 1} \begin{pmatrix} -t \\ t \\ -t^{1-p_2} \\ t^{1-p_2} \\ \vdots \\ -t^{1-pN_c} \\ t^{1-pN_c} \end{pmatrix} \tag{1.107}$$

A transmission coefficient is subsequently defined as the ratio, $T = u_n/t^n$. For a large number of grafted finite side branches, one has to resort to numerical calculation of the transmission coefficient inserting (1.102)–(1.107) into (1.82). For the sake of illustration, we have performed such calculations using the limiting case: $\beta = \beta_I = \beta'$ and $m = m'$ (i.e., $t = t'$). The numerical calculation involves the following steps for a series of values of the angular frequency $\omega \leq \omega_0$:

(a) calculating $\xi = 1 - \frac{m\omega^2}{2\beta}$,
(b) calculating $t = \xi + i(1 - \xi^2)^{1/2}$ *since* $-1 \leq \xi \leq 1$
(c) inserting t into (1.102)–(1.107)
(d) calculating the transmission coefficient, $T(\omega)$

This algorithm is implemented in Appendix 1.

Figure 1.10 illustrates the formation of band gap by (1) local resonances [upper panels in (a) and (b)] and (2) band folding effects from Bragg scattering [lower panels in (b)] in the transmission coefficient as a function of frequency for $L' = 1$, $\beta = \beta_I = 1$, $m = m' = 1$, with these conditions, $\omega_0 = 2$. A single one-atom side branch produces one resonant zero of transmission at $\omega = 1$. As one increases the number of side branches, spaced regularly by one inter-atomic spacing the periodicity of the infinite chain is conserved and the resonant zero of transmission

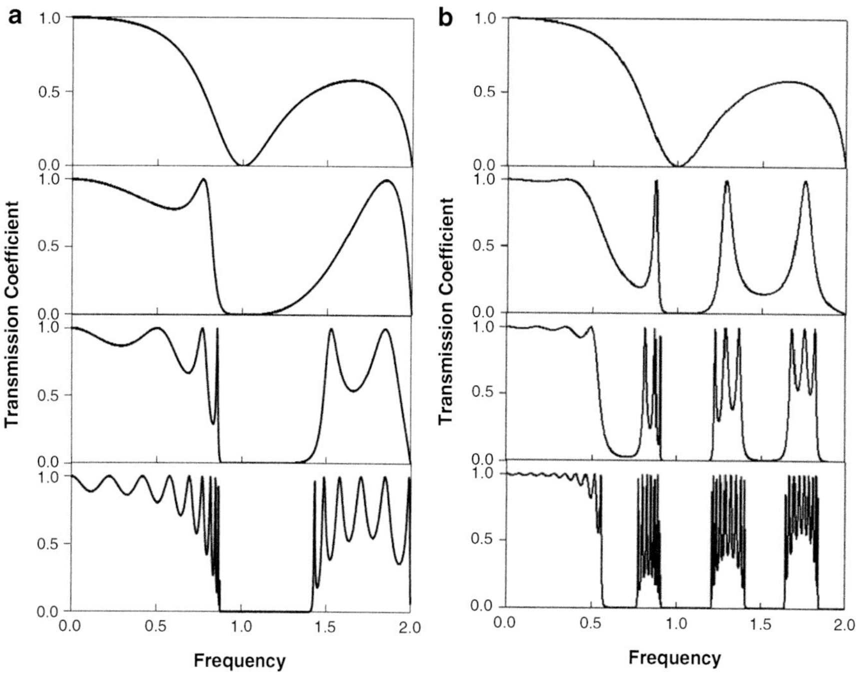

Fig. 1.10 Infinite monatomic crystal with (**a**) from *top* to *bottom* a one-atom ($L' = 1$) side branch located at $p_1 = 0$, two single-atom side branches located at $p_1 = 0$ and at $p_2 = 1$, four single-atom side branches at $p_1 = 0$, $p_2 = 1$, $p_3 = 2$, $p_4 = 3$, and ten one-atom side branches at $p_1 = 0$,. . ., $p_{10} = 9$ and (**b**) from *top* to *bottom* a one-atom ($L' = 1$) side branch located at $p_1 = 0$, two single-atom side branches located at $p_1 = 0$ and at $p_2 = 4$, four single-atom side branches at $p_1 = 0$, $p_2 = 4$, $p_3 = 8$, $p_4 = 12$, and ten one-atom side branches at $p_1 = 0$,. . ., $p_{10} = 36$. $\beta = \beta_I = 1$, $m = m' = 1$

broadens into a stop band. Two additional dips in transmission on both sides of the resonant stop band form if the side branches are spaced by four atomic spacings. For a large number of side branches spaced by four lattice parameters, these dips broaden and deepen approaching the band gaps that would result from the multiple scattering of waves by a periodic array of side branches.

This example clearly illustrates the contribution of local resonance to wave propagation as well as the contribution of scattering by a periodic array of scatterers. The former mechanism is the foundation of locally resonant structures that determine the properties of acoustic metamaterials. The latter is associated with Bragg scattering which is the fundamental mechanism underlying the properties of phononic crystals.

1.9 Conclusion

In this chapter, we have introduced a number of simple model systems that will serve as exemplars throughout this book. The wave equation has been introduced in the context of the monatomic and diatomic harmonic crystal, allowing us to distinguish between the phase and group velocities. Green's functions have been introduced and applied to a variety of modifications to the monatomic harmonic crystal. Finally, we have introduced the Interface Response Theory and used the cleavage operator to demonstrate both local resonant and periodic scattering and their effect on band structure. The Appendix 1 that follows provides a Fortran77 code that the reader can use to explore these topics more deeply and develop an understanding of some aspects of the new science of sound.

Appendix 1: Code Based on Green's Function Approach

Fortran 77 code used to calculate the transmission, density of state and transmission phase of an infinite chain with grafted finite side chains (For relevant equations see Sects. 1.8.4, 1.8.5 and 2.5). This code was used to generate the data of Figs. 1.9, 2.6, and 2.7.

```
C *************************
C  definition of variables
C *************************
      IMPLICIT REAL*8(A-H,O-X)
      IMPLICIT COMPLEX*16(Y-Z)
      PARAMETER (NM=4)
C NM needs to be fixed to NM=2*NC where NC is the number of chains
 C W is a arrays containing the various values of frequency
 C ETA is an array containing the reflection phase as a function of
frequency
 C DEN is an array containing the difference of density of state
between reference system (uncoupled) and coupled system
      DIMENSION W(15000),ETA(15000),DEN(15000)
C ZGREEN is an array containing the complex Green's function
```

C ZDEL is an array containing the matrix $\overleftrightarrow{\Delta}$ (MM)

C VI is the coupling operator: $\overleftrightarrow{V_I}$ (MM)

```
      DIMENSION ZGREEN(NM,NM),ZDEL(NM,NM),VI(NM,NM)
      DIMENSION DIS(NM),YDEL(NM,NM),INDX(NM),XL(NM)
      DIMENSION YDENT(NM,NM)
```

C YAMD is $\overleftrightarrow{A}\ (M,n)$

```
      DIMENSION YAMD(NM),YDM1AMD(NM)
C YU is the vector  U(M)
       DIMENSION YU(NM)
C Output: file "TRANS.DAT" contains the transmission coefficient,
transmission amplitude and transmission phase as functions of fre-
quency
     OPEN(UNIT=2,FILE='TRANS.DAT',STATUS=OLD
C OUTPUT: file "DENS.DAT" contains the reflection phase and the
density of states as functions of frequency
      OPEN(UNIT=3,FILE='DENS.DAT',STATUS=OLD)

C ************
C  Parameters
C ************
c NC : number of finite side chains
      NC=2
c Number of frequency intervals (NTOT)
c Size of frequency interval (DELO)
       NTOT=1500
       DELO=0.0015D00
C number Pi

      PI=ACOS(-1.D00)
c XL : length of finite side chains (in terms of number of masses)
      XL(1)=5.D00
      XL(2)=5.D00
C Spring constants of the infinite chain (BE1), side chains(BE2)
C and coupling spring (BEI)
      BE1=1.D00
      BE2=1.D00
      BEI=1.D00
C Mass in infinite chain (AM1) and mass in side chains (AM2)
      AM1=1.D00
      AM2=1.D00
c DIS(I) : position of side chains along the infinite chain in
units of inter-mass spacing.
      DIS(1)=0.D00
      DIS(2)=4.D00
```

```
C Normalization of arrays
      DO I=1,2*NC
      INDX(I)=0.D00
      DO J=1,2*NC
      ZGREEN(I,J)=DCMPLX(0.D00,0.D00)
      VI(I,J)=0.D00
      END DO
      END DO
c VI : constructing the coupling operator
      DO I=1,NC
      IC=(I-1)*2
      VI(IC+1,IC+1)=-BEI/AM1
      VI(IC+1,IC+2)=BEI/AM1
      VI(IC+2,IC+1)=BEI/AM2
      VI(IC+2,IC+2)=-BEI/AM2
      END DO
C ********************************************
C  Main loop over frequencies
C ********************************************
      W(1)=0.00000001D00
      DO 1 IM=2,NTOT
      W(IM)=W(IM-1)+DELO
      WW=W(IM)**2
```

C Calculating $\xi = 1 - \frac{m\omega^2}{2\beta}$ for infinite chain and side chains

```
      CZ1=1.D00-AM1*WW/(2.D00*BE1)
      CZ2=1.D00-AM2*WW/(2.D00*BE2)
```

C Calculating $t = \xi + i(1-\xi^2)^{1/2}$ $since -1 \leq \xi \leq 1$ for infinite chain and side chains

```
      IF(ABS(CZ1).LT.1.D00) THEN
          AZ1=DSQRT(1.D00-CZ1*CZ1)
          ZT1=DCMPLX(CZ1,AZ1)
      ELSE
          SQR1=DSQRT(CZ1*CZ1-1.D00)
          IF(CZ1.GT.1.D00) ZT1=DCMPLX((CZ1-SQR1),0.D00)
          IF(CZ1.LT.-1.D00) ZT1=DCMPLX((CZ1+SQR1),0.D00)
      ENDIF
C
      IF(ABS(CZ2).LT.1.D00) THEN
          AZ2=DSQRT(1.D00-CZ2*CZ2)
          ZT2=DCMPLX(CZ2,AZ2)
```

```
      ELSE
          SQR2=DSQRT(CZ2*CZ2-1.D00)
          IF(CZ2.LT.-1.D00) ZT2=DCMPLX((CZ2+SQR2),0.D00)
          IF(CZ2.GT.1.D00)  ZT2=DCMPLX((CZ2-SQR2),0.D00)
      ENDIF
```

C Constructing the reference Green's function, $\overleftrightarrow{G_S}(MM)$
C Components of $\overleftrightarrow{G_S}(MM)$ from infinite chain (1.98)

```
      DO I=1,NC
      DO J=1,NC
      IC=(I-1)*2
      JC=(J-1)*2
      POS1=DIS(I)
      POS2=DIS(J)
      CALL GR(ZT1,POS1,POS2,AM1,BE1,ZG0)
      ZGREEN(IC+1,JC+1)=ZG0
      END DO
      END DO
```

C Components of $\overleftrightarrow{G_S}(MM)$ from side chains (1.99)

```
      DO I=1,NC
      ZT21=ZT2**(2.0D00*XL(I))
      ZT22=(AM2/BE2)*(ZT2+ZT21)/((ZT2-1.0D00)*(1.0D00-ZT21))
      IC=(I-1)*2
      ZGREEN(IC+2,IC+2)=ZT22
      END DO
```

C Calculation of the matrix $\overleftrightarrow{\Delta}(MM)$

```
      DO I=1,2*NC
      DO J=1,2*NC
      ZDEL(I,J)=DCMPLX(0.D00,0.D00)
      DO K=1,2*NC
      ZDEL(I,J)=ZDEL(I,J)+VI(I,K)*ZGREEN(K,J)
      END DO
      END DO
      END DO

      DO I=1,2*NC
      ZDEL(I,I)=DCMPLX(1.D00,0.D0)+ZDEL(I,I)
      END DO
```

C Calculation of $\overleftrightarrow{\Delta}^{-1}(MM)$

```
      NX=2*NC
      DO I=1,NX
      DO J=1,NX
      YDEL(I,J)=ZDEL(I,J)
      END DO
      END DO
C Identity matrix
      DO I=1,NX
      DO J=1,NX
        YDENT(I,J)=DCMPLX(0.D0,0.D0)
       END DO
        YDENT(I,I)=DCMPLX(1.0D0,0.D0)
      END DO
      CALL LUDCMP(YDEL,NX,NX,INDX,D)
C LUDCMP is a subroutine for the lower-upper decomposition of
matrices from "Numerical Recipes" see reference [3]
C YDEL contains the LU decomposition

C*******************************************
C Calculation of the transmission phase (see Chap. 2 for details)
```

C calculation of the determinant of the matrix $\overleftrightarrow{\Delta}(MM)$ from its LU decomposition

```
      ZD=DCMPLX(D,0.D00)
      DO I=1,NX
      ZD=ZD*YDEL(I,I)
      END DO

      ZN1=ZD
C Calculation of the transmission phase from the determinant for
each frequency (See Chap. 2)
      IF(ZN1.NE.DCMPLX(0.D00,0.D00))THEN
      ZN2=CDLOG(ZN1)
      PII=1.D00/PI
      ETA(IM)=-PII*DIMAG(ZN2)
      ELSE
      ETA(IM)=0.D00
       ENDIF
C*********************************************
```

C Calculation of the inverse matrix $\overleftrightarrow{\Delta}^{-1}(MM)$

```
C LUBKSB calculates the inverse from the LU decomposition (see
reference [3])
       DO J=1,NX
        CALL LUBKSB(YDEL,NX,NX,INDX,YDENT(1,J))
       END DO
```

C the matrix YDENT now contains the matrix $\overleftrightarrow{\Delta}^{-1}(MM)$

C calculating $\overleftrightarrow{A}\,(M,n)$

```
       DO I=1,NC
      IC=(I-1)*2
      POS1=DIS(I)
      POWER=1.D0-POS1
      ZTEMP=(AM1/BE1)*(ZT1**POWER)/(ZT1**2.D00-1.D00)
       YAMD(IC+1)=VI(IC+1,IC+1)*ZTEMP
       YAMD(IC+2)=VI(IC+2,IC+1)*ZTEMP
      ENDDO
C calculating the transmission coefficient
      DO I=1,NX
      YDM1AMD(I)=DCMPLX(0.D0,0.D0)
       DO J=1,NX
       YDM1AMD(I)=YDM1AMD(I)+YDENT(I,J)*YAMD(J)
      ENDDO
      ENDDO
       DO I=1,NC
      IC=(I-1)*2
      POS1=DIS(I)
      YU(IC+1)=ZT1**POS1
      YU(IC+2)=DCMPLX(0.D0,0.D0)
      ENDDO

      YTR=DCMPLX(0.D0,0.D0)
      DO I=1,NX
      YTR=YTR+YU(I)*YDM1AMD(I)
      ENDDO
      YTR=DCMPLX(1.0D0,0.D0)-YTR

      RZTR=DREAL(YTR)
      SZTR=DIMAG(YTR)
C Transmission coefficient
      TRANS=RZTR**2+SZTR**2
C Writing transmission coefficient, TRANS, transmission amplitude
(RZTR, SZTR), and transmission phase ATAN(SZTR/RZTR) as functions
```

```
of frequency to file
      WRITE(2,*)W(IM),TRANS, RZTR,SZTR,ATAN(SZTR/RZTR)

1     CONTINUE

C#########################
C  Calculation of the derivative of the reflection phase with
respect to the eigen value (square of frequency) to obtain the
variation in density of state
C#########################
      DO 2 I=1,NTOT-1
      DIFF=ETA(I+1)-ETA(I)
      DEN(I)=DIFF/(W(I+1)**2.0D00-W(I)**2.0D00)
      IF((DEN(I).GT.100.D00).OR.(DEN(I).LT.-100.D00)) GOTO 2
C Write reflection phase ETA and density of states DEN as functions
of frequency W.
      WRITE(3,*)W(I),ETA(I),DEN(I)
2     CONTINUE
C*************************************************************
      STOP
      END
C
*************************************************************
*******

        SUBROUTINE  GR(ZT1,POS1,POS2,AM1,BE1,ZG0)
        IMPLICIT REAL*8(A-H,O-Y)
        IMPLICIT COMPLEX*16(Z)
        POWER=DABS(POS1-POS2)+1.D00
        ZINT=(ZT1**POWER)/(ZT1**2.D00-1.D00)
        ZG0=(AM1/BE1)*ZINT
        RETURN
        END

        SUBROUTINE LUDCMP(Y,N,NP,INDX,D)
        IMPLICIT REAL*8(A-H,O-X,Z)
        IMPLICIT COMPLEX*16(Y)
        PARAMETER (NMAX=1000,YTINY=DCMPLX(1.0D-32,0.d00))
        DIMENSION Y(NP,NP),VV(NMAX)
        DIMENSION INDX(N)
        D=1.D00
        DO 12 I=1,N
        AAMAX=0.D00
        DO 11 J=1,N
        IF (CDABS(Y(I,J)).GT.AAMAX) AAMAX=CDABS(Y(I,J))
```

```
11      CONTINUE
        IF (AAMAX.EQ.0.) PAUSE 'Singular matrix.'
        VV(I)=1./AAMAX
12    CONTINUE
        DO 19 J=1,N
        IF (J.GT.1) THEN
        DO 14 I=1,J-1
        YSUM=Y(I,J)
        IF (I.GT.1) THEN
        DO 13 K=1,I-1
        YSUM=YSUM-Y(I,K)*Y(K,J)
13            CONTINUE
        Y(I,J)=YSUM
        ENDIF
14        CONTINUE
        ENDIF
        AAMAX=0.d00
        DO 16 I=J,N
        YSUM=Y(I,J)
        IF (J.GT.1) THEN
        DO 15 K=1,J-1
        YSUM=YSUM-Y(I,K)*Y(K,J)
15          CONTINUE
        Y(I,J)=YSUM
        ENDIF
        DUM=VV(I)*CDABS(YSUM)
        IF (DUM.GE.AAMAX) THEN
        IMAX=I
        AAMAX=DUM
        ENDIF
16       CONTINUE
        IF (J.NE.IMAX) THEN
        DO 17 K=1,N
        YUM=Y(IMAX,K)
        Y(IMAX,K)=Y(J,K)
        Y(J,K)=YUM
17        CONTINUE
        D=-D
        VV(IMAX)=VV(J)
        ENDIF
        INDX(J)=IMAX
        IF(CDABS(Y(J,J)).EQ.0.d00) Y(J,J)=YTINY
        IF(J.NE.N) THEN
        YUM=1.d00/Y(J,J)
        DO 18 I=J+1,N
```

```
          Y(I,J)=Y(I,J)*YUM
18          CONTINUE
          ENDIF
19         CONTINUE
          IF(CDABS(Y(N,N)).EQ.0.d00) Y(N,N)=YTINY
          RETURN
          END

          SUBROUTINE LUBKSB(Y,N,NP,INDX,YB)
          IMPLICIT REAL*8(A-H,O-X,Z)
          IMPLICIT COMPLEX*16(Y)
          DIMENSION Y(NP,NP),INDX(N),YB(N)
          II=0
          DO 12 I=1,N
             LL=INDX(I)
             YSUM=YB(LL)
             YB(LL)=YB(I)
             IF(II.NE.0)THEN
                DO 11 J=II,I-1
                  YSUM=YSUM-Y(I,J)*YB(J)
11              CONTINUE
              ELSE IF (YSUM.NE.DCMPLX(0.D0,0.D0)) THEN
                 II=I
              ENDIF
              YB(I)=YSUM
12        CONTINUE
          DO 14 I=N,1,-1
            YSUM=YB(I)
              DO 13 J=I+1,N
               YSUM=YSUM-Y(I,J)*YB(J)
13            CONTINUE
             YB(I)=YSUM/Y(I,I)
14          CONTINUE
         RETURN
         END
```

References

1. E.N. Economou, *Green's Functions in Quantum Physics* (Springer, New York, 1990)
2. L. Dobrzynski, Interface response theory of composite systems. Surf. Sci. **200**, 435 (1988)
3. W.H. Press, S.A. Teukolsky, W.T. Vetterling, B.P. Flannery, *Numerical Recipes in Fortran 77*, 2nd edn. (Cambridge University Press, New York, 1992)

Chapter 2
Phase and Topology

2.1 Introduction

The primary objective of this chapter is to present and review the introductory concepts related to phase and topology that will serve as foundation for the development of the more complex concepts to be presented in subsequent chapters. In particular, we illustrate the concept of geometric phase in the case of two prototypical elastic systems, namely the one-dimensional harmonic oscillator and a one-dimensional binary superlattice. We demonstrate formally the relationship between the variation of the geometric phase in the spectral and wave number domains and the parallel transport of a vector field along paths on curved manifolds possessing helicoidal twists which exhibit non-conventional topology. The formal mapping of the evolution of the geometric phase on the spectral or wave number domains onto the parallel transport of a vector field on curved manifold spanning the frequency and wave number spaces will be used to help interpret topological features of elastic waves in more complex media such as two-dimensional or three-dimensional phononic crystals and acoustic metamaterials.

We relate the notion of geometric phase and Green's functions. In particular, we show for Hermitian operators that the Berry connection is proportional to the imaginary part of its Green's function. We illustrate this notion in the case of simple elastic systems composed of mass and springs, namely the one-dimensional harmonic crystal and the one-dimensional harmonic crystal with a side branch. In the latter case, the notions of Friedel phase and transmission (reflection) phase are introduced and related to the notion of geometric phase.

Finally, we sketch how topological modes arise at interfaces between media with different topologies.

P. Deymier, K. Runge, *Sound Topology, Duality, Coherence and Wave-Mixing*, Springer Series in Solid-State Sciences 188, DOI 10.1007/978-3-319-62380-1_2

2.2 Overview

From a historical perspective, our scientific understanding of sound and vibrations dates back to Sir Isaac Newton's *Principia* [1], which examined its first mathematical theory. The mid-nineteenth century book *The Theory of Sound* by Lord Rayleigh [2] still constitutes the foundation of our modern theory of vibrations, whereas the quantum theory of phonons followed in the early part of the twentieth century [3]. During this nearly 300-year period, our understanding of sound and elastic waves has been nourished essentially by the paradigm of the plane wave and its periodic counterpart (the Bloch wave) in periodic media. This paradigm relies on the four canonical characteristics of waves: frequency (ω); wave vector ($\boldsymbol{k}$); amplitude ($\boldsymbol{A}$); and phase (ϕ). Over the past two decades, the fields of phononic crystals and acoustic metamaterials have developed in which researchers manipulate the spectral and refractive properties of phonons and sound waves through their host material by exploiting ω and $\boldsymbol{k}$ [4]. The spectral properties of elastic waves include phenomena such as the formation of stop bands in the transmission spectrum due to Bragg-like scattering or resonant processes, as well as the capacity to achieve narrow band spectral filtering by introducing defects in the material's structure. Negative refraction, zero-angle refraction and other unusual refractive properties utilize the complete characteristics of the dispersion relations of the elastic waves, $\omega(\boldsymbol{k})$, over both frequency and wave number domains. More recently, renewed attention has been paid to the amplitude and the phase characteristics of the elastic waves. Indeed, it is in the canonical characteristic realms of $\boldsymbol{A}$ and ϕ where non-conventional new forms of elastic waves reside. This new realm opens gateways to non-conventional forms of elastic wave- or phonon-supporting media. In the most general form of the complex amplitude, $A = A_0 e^{i\varphi}$, elastic oscillations, vibrations and waves can acquire a geometric phase φ with spectral or wave vector dependency can be described in the context of topology. Electronic waves [5] or electromagnetic waves [6–8] with non-conventional topology have been shown to exhibit astonishing properties such as the existence of unidirectional, backscattering-immune edge states. Phononic structures have also been shown recently to possess non-conventional topology as well as topologically constrained propagative properties. These properties have been achieved by breaking time-reversal symmetry through internal resonance or symmetry breaking structural features (*e.g.*, chirality) [9–19] and without addition of energy from the outside. Energy can also be added to extrinsic topological elastic systems to break time reversal symmetry [20–26]. Topological elastic oscillations, vibrations and waves promise designs and new device functionalities which require a deeper insight into the relationship between geometric phase and topology. It is the objective of this chapter to shed light on this relation. In particular, we first employ two prototypical elastic model systems, namely the one-dimensional harmonic oscillator and a one-dimensional elastic binary superlattice to demonstrate analytically and formally the relationship between the variation of the geometric phase in the spectral and wave number domains and its topological interpretation in terms of the parallel transport of a vector field along paths in frequency or wave vector on some curved manifold, namely strips containing a local helicoidal twist.

In Sect. 2.3, we introduce the formalism to describe the geometric phase of the amplitude of a one-dimensional harmonic oscillator in its spectral domain. A detailed topological interpretation of this phase in a curved space is also derived. More specifically, the well-known phase change of π of the displacement amplitude of a driven harmonic oscillator as the driving frequency crosses its characteristic frequency is interpreted topologically in terms of the parallel transport of a vector field along a curved manifold constituted of a strip containing a helicoidal twist. The twist occurs at the characteristic frequency.

In Sect. 2.4, we consider a one-dimensional binary superlattice and its dispersion characteristics. We analyze the amplitude of elastic wave supported by this superlattice in the wave number domains and pay particular attention to elastic bands that accumulate a non-zero geometric phase within the Brillouin zone. The topological interpretation of the evolution of the phase along a path in wave number space (*i.e.*, Brillouin zone) is formally established. In particular, an elastic superlattice is known to possess dispersion bands along which the displacement amplitude changes sign and therefore exhibits a change in phase of π. In this periodic system, the geometric phase (also known as the Zak phase) is now a periodic function of the wave number. The change in phase along a path in the Brillouin zone of the superlattice is interpreted topologically in terms of the parallel transport of a vector field along a manifold constituted of a closed twisted strip. The twist occurs at the wave number where the amplitude of elastic waves changes sign.

In Sect. 2.5, we relate the notions of geometric phase and the Green's function of Hermitian operators. The Berry connection is shown to be proportional to within a sign to the imaginary part of its Green's function. We illustrate this notion in the case of simple elastic systems composed of mass and springs, namely the one-dimensional harmonic chains and the one-dimensional harmonic chain with side branches. In the first case, we determine the change in phase of elastic waves as one moves in the space of frequency and wave number. The harmonic chain with side branches introduces the notion of transmission (and reflection) phase as well as the notion of Friedel phase. The transmission and reflection phases are functions of frequency and are equivalent to Berry phases averaged at fixed ω over all wave numbers within the Brillouin zone of the system.

In Sect. 2.6, we use Green's functions to illustrate the emergence of topological states at an interface between media with different topologies. In particular, we consider the case of media with broken inversion symmetry. We construct schematically interfaces between a medium with a zero Zak phase and a medium with a Zak phase of π and show the existence of interfacial modes with topologically protected one-way propagation.

2.3 Harmonic Oscillator Model Systems

In this section, we consider, two model systems, namely a simple one-dimensional harmonic oscillator and the driven harmonic oscillator. In both cases we illustrate the concept of geometric phase and develop the formalism necessary to interpret it in the context of topology.

2.3.1 *Geometric Phase and Dynamical Phase of the Damped Harmonic Oscillator*

The dynamics of the damped harmonic oscillator is given by:

$$\frac{\partial^2 \tilde{u}(t)}{\partial t^2} + \mu \frac{\partial \tilde{u}(t)}{\partial t} + \omega_0^2 \tilde{u}(t) = 0 \tag{2.1}$$

Here, μ is the damping coefficient and ω_0 is the characteristic frequency. $\tilde{u}(t)$ is the displacement of the oscillator. We rewrite this equation in the form:

$$\frac{\partial^2 u(\xi,t)}{\partial t^2} + \mu \frac{\partial u(\xi,t)}{\partial t} = -i \frac{\partial u(\xi,t)}{\partial \xi} \tag{2.2}$$

To obtain (2.2), we have defined: $u(\xi,t) = \tilde{u}(t)e^{-i\omega_0^2 \xi}$. We generalize (2.2) further by introducing the equation:

$$\frac{\partial^2 u(\xi,t)}{\partial t^2} - i\varepsilon\phi(\xi) \frac{\partial u(\xi,t)}{\partial t} = -i \frac{\partial u(\xi,t)}{\partial \xi} \tag{2.3}$$

The damped oscillator is recovered when $i\varepsilon\phi(\xi) = -\mu$. Here ε and ϕ are a parameter and a function. In the limit of small ε (*i.e.*, to first order), we can perform the following substitution:

$$\frac{\partial^2 u(\xi,t)}{\partial t^2} - i\varepsilon\phi(\xi) \frac{\partial u(\xi,t)}{\partial t} \sim \left(\frac{\partial}{\partial t} - i\varepsilon \frac{\phi(\xi)}{2}\right)^2 u(\xi,t) \tag{2.4}$$

With this substitution, (2.4) takes the form of the one-dimensional Schroedinger equation in the presence of a magnetic field:

$$-i \frac{\partial u(\xi,t)}{\partial \xi} = \left(\frac{\partial}{\partial t} - i\varepsilon \frac{\phi(\xi)}{2}\right)^2 u(\xi,t) \tag{2.5}$$

where ξ plays the role of time and t plays the role of position. ϕ acts as a single component vector potential associated with the magnetic field. The term in parenthesis plays the role of the canonical momentum of a charged particle in a magnetic field. If we choose a solution of the form:

$$u(\xi,t) = v(\omega(\xi),t)e^{-i\omega_0^2\xi} \text{ with } v(\omega(\xi),t) = \tilde{v}\,(\omega(\xi))e^{i\omega(\xi)t} \tag{2.6}$$

and insert it into (2.5), we obtain:

$$\omega_0^2 = \left[\omega(\xi) - \varepsilon\frac{\phi(\xi)}{2}\right]^2 \tag{2.7}$$

Equation (2.7) states that $\omega(\xi) = \omega_0 + \varepsilon\frac{\phi(\xi)}{2}$. The function $\phi(\xi)$ offers a mechanism for tuning/driving the frequency of the oscillator around its characteristic frequency.

We now assume that the solution to (2.5) may carry a phase $\eta(\omega(\xi))$ that depends on the frequency. This solution is therefore rewritten in the form:

$$u_\eta(\xi,t) = u(\xi,t)e^{i\eta(\omega(\xi))} = v(\omega(\xi),t)e^{-i\omega_0^2\xi}e^{i\eta(\omega(\xi))} \tag{2.8}$$

Inserting this solution into (2.5) yields:

$$i\left(\frac{\partial u}{\partial\xi}e^{i\eta} + u_\eta i\frac{\partial\eta}{\partial\omega}\frac{\partial\omega}{\partial\xi}\right) = \left[\omega(\xi) - \varepsilon\frac{\phi(\xi)}{2}\right]^2 u_\eta \tag{2.9}$$

We multiply both sides of this equation by the complex conjugate: $u_\eta^* = u^* e^{-i\eta}$. After some manipulations we get:

$$\frac{\partial\eta}{\partial\omega} = iu^*\frac{\partial u}{\partial\xi}\frac{\partial\xi}{\partial\omega} - \left[\omega(\xi) - \varepsilon\frac{\phi(\xi)}{2}\right]^2\frac{\partial\xi}{\partial\omega}$$

This equation reveals the change in phase of the oscillator:

$$d\eta = iu^*\frac{\partial u}{\partial\omega}d\omega - \left[\omega(\xi) - \varepsilon\frac{\phi(\xi)}{2}\right]^2 d\xi \tag{2.10}$$

The first term on the right hand side of (2.10) contains the Berry connection defined as $-iu^*\frac{\partial u}{\partial\omega}$ [27]. Indeed, since $u(\xi,t) = \tilde{u}\,(t)e^{-i\omega_0^2\xi}$, then $iu^*\frac{\partial u}{\partial\omega} = i\tilde{u}^*\frac{\partial\tilde{u}}{\partial\omega}$ where $\tilde{u}$ is the solution of (2.1). Equation (2.10) can be integrated along a path in eigen value space driven by the parameter ξ.

$$\int_{\xi_1}^{\xi_2} d\eta = \int_{\omega(\xi_1)}^{\omega(\xi_2)} i\tilde{u}^* \frac{\partial \tilde{u}}{\partial \omega} d\omega - \int_{\xi_1}^{\xi_2} \left[\omega(\xi) - \varepsilon \frac{\phi(\xi)}{2} \right]^2 d\xi \tag{2.11}$$

The second term on the right-hand side of (2.11) is the dynamical phase. The first term on the right-hand side of (2.11) is the geometric phase. Here we have used the parameter ξ to vary the frequency of the oscillator. In the Sect. 2.3.2, we will use a driving force to achieve the same result, *i.e.*, we will consider the case of the driven harmonic oscillator. Both approaches provide a similar description of the evolution of the phase of the propagating waves in the space of the eigen values of the system.

2.3.2 Geometric Phase of the Driven Harmonic Oscillator

The dynamics of the driven harmonic oscillator is given by:

$$\frac{\partial^2 u}{\partial t^2} + \omega_0^2 u = ae^{i\omega t} \tag{2.12}$$

where u is the displacement, ω_0 is again the characteristic frequency of the oscillator, ω is the angular frequency of the driving function and the parameter a has the dimension of an acceleration. To solve this equation, we seek solutions of the form:

$$u(t) = u_0(\omega)e^{i\omega t} \tag{2.13}$$

Inserting (2.13) into (2.12), leads to:

$$\left(-\omega^2 + \omega_o^2\right)u_0 = a \tag{2.14}$$

We note that (2.12) is the spectral decomposition of the following equation:

$$\left(\frac{\partial^2}{\partial t^2} + \omega_0^2\right)U = a\delta(t) \tag{2.15}$$

with $\delta(t) = \int_{-\infty}^{+\infty} e^{i\omega t} d\omega$ and $U(t) = \int_{-\infty}^{+\infty} u_0(\omega)e^{i\omega t} d\omega$. U in (2.15) is a Green's function. $u_0(\omega)$ is then its spectral representation.

From (2.14), we get:

$$u_0\left(\omega^2\right) = \frac{1}{\left(\omega_o^2 - \omega^2\right)} \sim \frac{1}{\left(\omega_o^2 - \omega^2 - i\varepsilon\right)} = \frac{1\left(\omega_o^2 - \omega^2 + i\varepsilon\right)}{\left(\omega_o^2 - \omega^2\right)^2 + \varepsilon^2} \tag{2.16}$$

In (2.16) we have analytically continued the solution into the complex plane by introducing an imaginary term $-i\varepsilon$ with $\varepsilon \to 0$. It is important to keep in mind that the eigen values are now denoted $E = \omega^2$. To calculate the Berry connection, $BC(E)$, we use the first term on the right hand side of (2.10) where ω^2 is replaced by E:

$$BC(E) = -i\widehat{u}_0^*(E)\frac{d\widehat{u}_0(E)}{dE} = \frac{-\varepsilon}{\left(\omega_o^2 - \omega^2\right)^2 + \varepsilon^2} \tag{2.17}$$

$\widehat{u}_0$ in (2.17) is the normalized Green's function. We will see in Sect. 2.5 that one can determine the Berry connection directly from the knowledge of the Green's function. We will show later that the Berry connection is proportional (to within a sign) to the imaginary part of the Green's function of Hermitian operators. It is easy to verify this relation by comparing the imaginary part of (2.16) and the expression for the Berry connection given by (2.17).

It is interesting to take the limit of (2.17) when $\varepsilon \to 0$. For this we can use the identity: $\lim\limits_{\varepsilon\to 0}\frac{\varepsilon}{x^2+\varepsilon^2} = \pi\delta(x)$. In that limit, the Berry connection becomes:

$$BC(E) = -\pi\delta\left(\omega_0^2 - E\right) \tag{2.18}$$

This expression can be reformulated in terms of frequencies by using the identity:

$\delta\left(x^2 - b^2\right) = \frac{1}{2b}(\delta(x-b) + \delta(x+b))$ for $b > 0$. In the positive frequency range, the Berry connection becomes:

$$BC(E) = -\pi\frac{1}{2\omega_0}\delta(\omega - \omega_0) \tag{2.19}$$

Now using (2.10), we can determine the phase change from the relation:

$$BC(E) = \frac{d\eta(E)}{dE} = \frac{d\eta(\omega)}{2\omega d\omega} = -\pi\frac{1}{2\omega_0}\delta(\omega - \omega_0) \tag{2.20}$$

so we obtain

$$\frac{d\eta(\omega)}{d\omega} = -\pi\frac{\omega}{\omega_0}\delta(\omega - \omega_0) \tag{2.21}$$

The variation in phase of the displacement amplitude, u_0, over some range of frequency: $[\omega_1, \omega_2]$ is now obtained by integration [see (2.11)]:

$$\Delta\eta_{1,2} = -\pi \int_{\omega_1}^{\omega_2} d\omega \frac{\omega}{\omega_0} \delta(\omega - \omega_0) \tag{2.22}$$

There is no phase change for intervals with both frequencies below the characteristic frequency and for intervals with both frequencies above the characteristic frequency, as well. However, by tuning the driving frequency from below the characteristic frequency to above, ω_0, the amplitude of the oscillation accumulates a -π phase difference. The oscillator changes from being in phase to being out of phase with the driving force. This means that the amplitude of the oscillation changes sign at the characteristic frequency (this is clear from (2.16) in the limit $\varepsilon \to 0$).

2.3.3 *Topological Interpretation of the Geometric Phase*

In Sect. 2.3.3, we construct a manifold whose topology leads to the same geometric phase characteristics as the driven harmonic oscillator as shown in (2.21). We consider first a three-dimensional helicoid manifold (see Fig. 2.1) with a parametric equation is given by:

$$\vec{r}(r,\phi) = X(r,\phi)\vec{i} + Y(r,\phi)\vec{j} + Z(r,\phi)\vec{k} = r\cos\phi\vec{i} + r\sin\phi\vec{j} + c\phi\vec{k} \tag{2.23}$$

The parameter c is the pitch of the helicoid.

An element of length on the manifold is:

$$d\vec{s} = dX\vec{i} + dY\vec{j} + dZ\vec{k} = dr\left(\cos\phi\vec{i} + \sin\phi\vec{j}\right) + d\phi\left(-r\sin\phi\vec{i} + r\cos\phi\vec{j} + c\vec{k}\right)$$
$$= dr\,\vec{e}_r + d\phi\,\vec{e}_\phi \tag{2.24}$$

Where the vectors $\vec{e}_r$ and $\vec{e}_\phi$ are the tangent vectors of the helicoid. We normalize these tangent vectors, and we introduce the vector $\vec{e}_n = \vec{e}_r \times \vec{e}_\phi$ to form the helicoidal coordinate system:

$$\vec{e}_r = \cos\phi\vec{i} + \sin\phi\vec{j} \tag{2.25a}$$

$$\vec{e}_\phi = \frac{1}{\sqrt{r^2 + c^2}}\left(-r\sin\phi\vec{i} + r\cos\phi\vec{j} + c\vec{k}\right) \tag{2.25b}$$

$$\vec{e}_n = \frac{1}{\sqrt{r^2 + c^2}}\left(c\sin\phi\vec{i} - c\cos\phi\vec{j} + r\vec{k}\right) \tag{2.25c}$$

The affine connection is defined through the derivative in the manifold of the coordinate basis vector projected onto the tangent vectors, namely [28]:

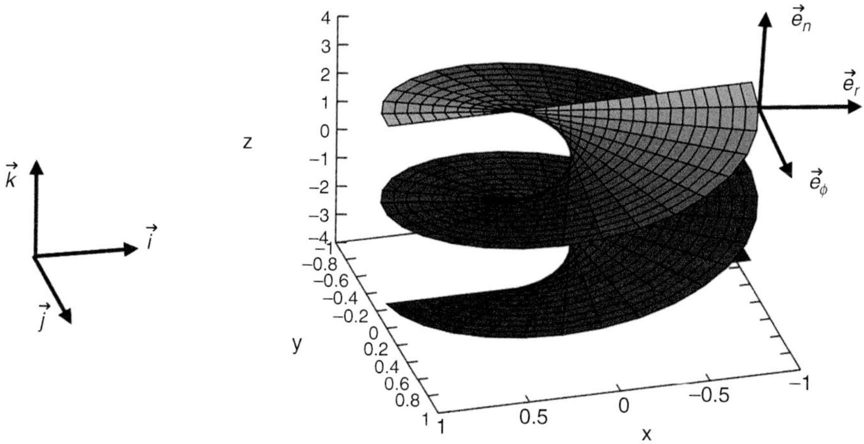

Fig. 2.1 Schematic representation of a helicoid. $\left(\vec{i}, \vec{j}, \vec{k}\right)$ is a fixed Cartesian coordinate system and $(\vec{e}_r,\ \vec{e}_\phi,\ \vec{e}_n)$ is the local coordinate system

$$\frac{\partial \vec{e}_\alpha}{\partial \beta} = \Gamma^{\gamma}_{\alpha\beta}\vec{e}_\gamma \tag{2.26}$$

where $\alpha, \beta, \gamma = r, \phi$. In (2.26), we have used the Einstein notation where summation on the repeating indices (here γ) is implicit.

We now calculate the connection component, $\Gamma^{\phi}_{\phi r}$:

$$\Gamma^{\phi}_{\phi r} = \vec{e}_\phi(r,\phi).\vec{e}_r(r,\phi + d\phi) \cong \vec{e}_\phi(r,\phi).\vec{e}_r(r,\phi) + \vec{e}_\phi(r,\phi)\frac{\partial \vec{e}_r(r,\phi)}{\partial \phi}d\phi + \dots \tag{2.27}$$

The first term on the right hand side of (2.27) is zero by virtue of the orthogonality of the coordinate system. The derivative in the second term can be determined in the fixed Cartesian coordinate system $\left(\vec{i}, \vec{j}, \vec{k}\right)$ and converted in the $(\vec{e}_r,\ \vec{e}_\phi,\ \vec{e}_n)$ coordinate system:

$$\frac{\partial \vec{e}_r(r,\phi)}{\partial \phi} = \frac{r}{\sqrt{r^2 + c^2}}\vec{e}_\phi(r,\phi) - \frac{c}{\sqrt{r^2 + c^2}}\vec{e}_n \tag{2.28}$$

This leads to the connection:

$$\Gamma^{\phi}_{\phi r} \cong \vec{e}_\phi(r,\phi)\frac{\partial \vec{e}_r(r,\phi)}{\partial \phi}d\phi = \frac{r}{\sqrt{r^2 + c^2}}d\phi \tag{2.29}$$

As illustrated in Fig. 2.2, we note that $\vec{e}_\phi(r,\phi)\cdot\vec{e}_r(r,\phi + d\phi) = \sin(d\eta) \cong d\eta$. Here $d\eta$ is the change in angle of the vector $\vec{e}_r$ as one varies the parameter ϕ.

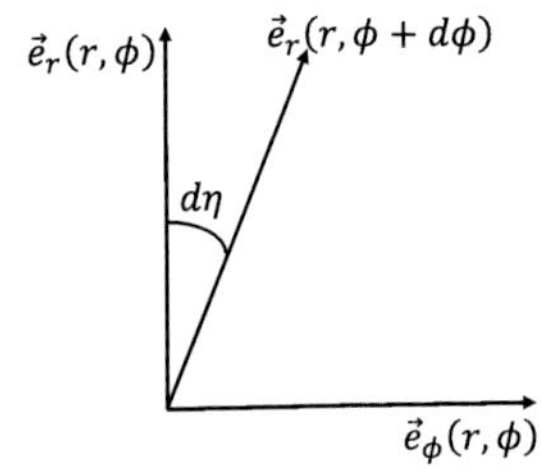

Fig. 2.2 Schematic illustration of the connection $d\eta \cong \Gamma_{\phi r}^{\phi}$ in the system of coordinate ($\vec{e}_r$, $\vec{e}_\phi$, $\vec{e}_n$)

Therefore, we can write:

$$d\eta \cong \Gamma_{\phi r}^{\phi} \cong \frac{r}{\sqrt{r^2+c^2}} d\phi$$

or

$$\frac{d\eta}{d\phi} \cong \frac{r}{\sqrt{r^2+c^2}} \tag{2.30}$$

We now construct the manifold of interest out of a helicoid with pitch $c = 2\Delta\omega$ by introducing a parametrization in terms of the frequency, ω: $\phi(\omega) = \frac{\pi}{\Delta\omega}\left(\omega - \left(\omega_0 - \frac{\Delta\omega}{2}\right)\right)$ for $\omega_0 - \frac{\Delta\omega}{2} \leq \omega \leq \omega_0 + \frac{\Delta\omega}{2}$ and $\phi(\omega)$ is a constant otherwise. The limit of this function when $c = \Delta\omega \to 0$ is the Heaviside function whose derivative is the Dirac delta function. This construction leads to the manifold illustrated in Fig. 2.3.

This manifold may be visualized as a strip with one single half-turn twist. The segment of helicoid represents the twisted region. In the limit $c = \Delta\omega \to 0$ the twisted region becomes infinitesimally narrow.

With this parametrization, the angle η changes according to: $\frac{d\eta}{d\omega} \cong \frac{r}{\sqrt{r^2+c^2}} \frac{d\phi}{d\omega} = \frac{r}{\sqrt{r^2+c^2}} \frac{\pi}{\Delta\omega}$ for $\omega_0 - \frac{\Delta\omega}{2} \leq \omega \leq \omega_0 + \frac{\Delta\omega}{2}$ and $\frac{d\eta}{d\omega} = 0$ otherwise. In the limit $c = \Delta\omega \to 0$, the angle variation becomes:

$$\frac{d\eta}{d\omega} \cong \pi\delta(\omega - \omega_0) \tag{2.31}$$

To within an unimportant sign, this equation is isomorphic to (2.21) that described the change in phase of a harmonic oscillator through resonance along the space of its eigen values. The topology of the eigen vectors of the harmonic oscillator is therefore isomorphic to that of a manifold constituted of a twisted strip with an infinitesimally narrow twist.

The topology of a system with multiple resonances may be visualized by a manifold with a sequence of twists along the frequency axis. The properties of the phase of the displacement of the harmonic oscillator can be visualized by the parallel transport of a vector field parallel to the twisted strip manifold. This point is illustrated below. Let consider some parametric curve, **C**, on the helicoid

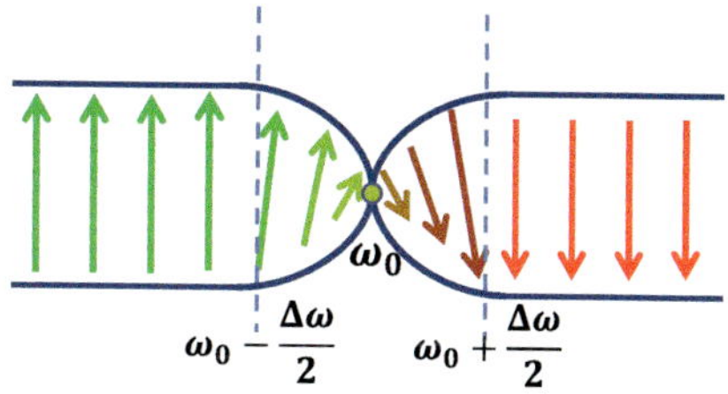

Fig. 2.3 Schematic illustration of a manifold with a single half-turn twist, its topology is isomorphic to that of the eigen vectors of the harmonic oscillator near resonance

manifold, $x^{\alpha}(\omega) = (r(\omega), \phi(\omega))$ with $\alpha = r, \phi$. The parameter ω enables us to move along the curve. Let us also consider some vector field $\vec{v}(\omega) = v^{\alpha}(\omega)\vec{e}_{\alpha}(\omega)$ at any point along the curve **C**. Here $\vec{e}_{\alpha}(\omega)$ correspond to the coordinate basis vectors at a point on the curve. The derivative of the vector, $\vec{v}$, along the curve if given by:

$$\frac{d\vec{v}}{d\omega} = \frac{dv^{\alpha}}{d\omega}\vec{e}_{\alpha} + v^{\alpha}\frac{d\vec{e}_{\alpha}}{d\omega} = \frac{dv^{\alpha}}{d\omega}\vec{e}_{\alpha} + v^{\alpha}\frac{\partial \vec{e}_{\alpha}}{\partial x^{\beta}}\frac{dx^{\beta}}{d\omega} \tag{2.32}$$

Substituting for $\frac{\partial \vec{e}_{\alpha}}{\partial x^{\beta}}$ using (2.26), we can write (2.32) in terms of the connection:

$$\frac{d\vec{v}}{d\omega} = \frac{dv^{\alpha}}{d\omega}\vec{e}_{\alpha} + v^{\alpha}\Gamma^{\gamma}_{\alpha\beta}\vec{e}_{\gamma}\frac{dx^{\beta}}{d\omega}$$

The dummy indices α and γ can be interchanged such that we can factor out the basis vectors:

$$\frac{d\vec{v}}{d\omega} = \left(\frac{dv^{\alpha}}{d\omega} + v^{\gamma}\Gamma^{\alpha}_{\gamma\beta}\frac{dx^{\beta}}{d\omega}\right)\vec{e}_{\alpha} \tag{2.33}$$

The term in parentheses is defined as the absolute derivative

$$\frac{Dv^{\alpha}}{D\omega} = \frac{dv^{\alpha}}{d\omega} + v^{\gamma}\Gamma^{\alpha}_{\gamma\beta}\frac{dx^{\beta}}{d\omega} \tag{2.34}$$

Let us suppose that the condition: $\frac{d\vec{v}}{d\omega} = 0$ is always satisfied along the curve **C**. This condition defines the notion of parallelism of the vector field $\vec{v}$ as the vector is transported along the curve. In the case of the manifold of Fig. 2.3 with a segment of helicoid connected to two flat strips, if we choose $\vec{v} = v^{r}\vec{e}_{r}$ (*i.e.*, $v^{r} = 1$ and $v^{\phi} = 0$) then we can show that $\frac{Dv^{r}}{D\omega} = 0 + \Gamma^{r}_{r\phi}\frac{dx^{\phi}}{d\omega} + \Gamma^{r}_{rr}\frac{dx^{r}}{d\omega}$. The last term in this expression $\frac{dx^{r}}{d\omega} = \frac{dr}{d\omega}$ is zero because r is independent of ω. We also have $\frac{dx^{\phi}}{d\omega} = \frac{d\phi}{d\omega} \neq 0$ for $\omega_0 - \frac{\Delta\omega}{2} \leq \omega \leq \omega_0 + \frac{\Delta\omega}{2}$ and from (2.26): $\Gamma^{r}_{r\phi} = 0$. By consequence, $\frac{d\vec{e}_{r}}{d\omega} = 0$ for $\omega_0 - \frac{\Delta\omega}{2} \leq \omega \leq \omega_0 + \frac{\Delta\omega}{2}$, that is $\vec{e}_{r}$ satisfies the condition for parallel transport along the segment of helicoid in Fig. 2.3. Outside the interval:

$\omega_0 - \frac{\Delta\omega}{2} \le \omega \le \omega_0 + \frac{\Delta\omega}{2}, \frac{d\phi}{d\omega} = 0, \frac{d\vec{e}_r}{d\omega} = 0$ because the manifold is a planar strip. The parallel transported vector is illustrated in Fig. 2.3 as colored arrows.

2.4 Elastic Superlattice Model System

2.4.1 *Geometric Phase of a One-Dimensional Elastic Superlattice: Zak Phase*

The geometric phase that characterizes the property of bulk bands in one-dimensional (1D) periodic systems is also known as the Zak phase [29]. In this section, we illustrate the concept of Zak phase in the case of a 1D elastic superlattice [30, 31]. In the Appendix 1, we consider a 1D elastic superlattice constituted of layers composed of alternating segments of material 1 and material 2 with density and speed of sound ρ_1, ρ_2 and c_1, c_2. The lengths of the alternating segments are d_1 and d_2, respectively. The period of the superlattice is $L = d_1 + d_2$. We find solutions for the displacement inside segment 1 in layer n in the form:

$$u_1(x,t) = e^{iqnL}\left(A_+ e^{ik_1(x-nL)} + A_- e^{-ik_1(x-nL)}\right)e^{i\omega t} \tag{2.35}$$

with the amplitudes

$$A_+ = \frac{1}{2}\left(F - \frac{1}{F}\right)\sin k_1 d_1 \sin k_2 d_2 + \frac{i}{2}\left(F - \frac{1}{F}\right)\cos k_1 d_1 \sin k_2 d_2 \tag{2.36a}$$

$$A_- = i\left[\sin k_1 d_1 \cos k_2 d_2 + \frac{1}{2}\left(F + \frac{1}{F}\right)\cos k_1 d_1 \sin k_2 d_2 - \sin qL\right] \tag{2.36b}$$

and the dispersion relation, $\omega(q)$, given by the relation:

$$\cos qL = \cos k_1 d_1 \cos k_2 d_2 - \frac{1}{2}\left(F + \frac{1}{F}\right)\sin k_1 d_1 \sin k_2 d_2 \tag{2.37}$$

In these equations, $F = \frac{k_1 \rho_1 c_1^2}{k_2 \rho_2 c_2^2}$ with $k_1 = \frac{\omega}{c_1}$ and $k_2 = \frac{\omega}{c_2}$. The wave number $q \in \left[\frac{-\pi}{L}, \frac{\pi}{L}\right]$.

From (2.36a), one observes that when $\sin k_2 d_2 = 0$, the amplitude $A_+ = 0$. Let us consider an isolated band in the band structure of the superlattice for which this condition is satisfied. Under this condition the dispersion relation simplifies to:

$$\cos qL = \cos(k_1 d_1 + k_2 d_2) \tag{2.38}$$

To obtain (2.38), we used the trigonometric relation:

$$\cos k_1 d_1 \cos k_2 d_2 - \sin k_1 d_1 \sin k_2 d_2 = \cos(k_1 d_1 + k_2 d_2)$$

Under this same condition the amplitude A_- reduces to:

$$A_- = i[\sin k_1 d_1 \cos k_2 d_2 - \sin qL]$$

or using standard trigonometric relations

$$A_- = i[\sin(k_1 d_1 + k_2 d_2) - \sin qL] \tag{2.39}$$

When the wave number is in the positive half of the Brillouin zone, *i.e.*, $qL \in [0, \pi]$, (2.38) is satisfied when $k_1 d_1 + k_2 d_2 = qL + m2\pi$ with m being an integer. In this case, $\sin(k_1 d_1 + k_2 d_2) - \sin qL = 0$, that is $A_- = 0$. Therefore, we conclude that when $\sin k_2 d_2 = 0$ and $q>0$ both amplitudes A_+ and A_- becomes zero (so does the displacement field).

When the wave number is negative, *i.e.*, $qL \in [-\pi, 0]$, (2.38) is satisfied when $k_1 d_1 + k_2 d_2 = |q|L + 2m\pi$ (note that $k_1 d_1 + k_2 d_2 > 0$). In this case, $\sin(k_1 d_1 + k_2 d_2) = \sin(|q|L + 2m\pi) = \sin|q|L$ and $\sin(k_1 d_1 + k_2 d_2) - \sin qL \neq 0$, the amplitude $A_- \neq 0$ and the displacement field does not vanish.

Let us define the point along the dispersion curve where the displacement amplitudes vanish by (q_0, ω_0). We have at this point $k_2 d_2 = \frac{\omega_0(q_0)}{c_2} d_2 = m\pi$ where m is an integer. We now calculate the slope of A_+ and A_- as functions of q. Using (2.36a, 2.36b) and the dispersion relation (2.37) as well as its derivative, we obtain after numerous steps:

$$\frac{dA_+}{dq} = \frac{1}{2}\left(F + \frac{1}{F}\right) \times \left\{ d_1 \frac{dk_1}{dq} (\cos k_1 d_1 - i \sin k_1 d_1) \sin k_2 d_2 + d_2 \frac{dk_2}{dq} (\sin k_1 d_1 + i \cos k_1 d_1) \cos k_2 d_2 \right\}$$

and

$$\frac{dA_-}{dq} = L \cos qL \left\{ \frac{\sin qL - \frac{1}{L} \cos k_1 d_1 \sin k_2 d_2 \left(d_2 \frac{dk_2}{dq} + \frac{1}{2}\left(F + \frac{1}{F}\right) d_1 \frac{dk_1}{dq} \right)}{\sin k_1 d_1 \cos k_2 d_2} - 1 \right\}$$

At the point (q_0, ω_0), $\sin k_2 d_2 = 0$ and $\sin k_1 d_1 \cos k_2 d_2 = \sin(k_1 d_1 + k_2 d_2)$ and

$$\left.\frac{dA_+}{dq}\right|_{q_0} = \frac{1}{2}\left(F + \frac{1}{F}\right) \left\{ \frac{dk_2}{dq} (\sin k_1 d_1 + i \cos k_1 d_1)(-1)^m \right\}$$

and

$$\left.\frac{dA_-}{dq}\right|_{q_0} = L\cos qL\left\{\frac{\sin qL}{\sin\left(k_1d_1 + k_2d_2\right)} - 1\right\}$$

We have $\left.\frac{dA_+}{dq}\right|_{q_0} \neq 0$ and $\left.\frac{dA_-}{dq}\right|_{q_0} = 0$ on one side of the Brillouin zone (at q_0) where $\sin(k_1d_1 + k_2d_2) = \sin qL$. Therefore, when following a path along the dispersion curve, the amplitude A_+ changes sign when crossing (q_0, ω_0) and therefore its phase changes by π. Along the same path, the amplitude A_- does not change sign.

In Fig. 2.4, we illustrate the concept of Zak phase for a particular case. We have chosen the following parameters: $\frac{d_2}{c_2} = 1.2\frac{d_1}{c_1}$ and $F = 2$. The band structure of the superlattice is shown in Fig. 2.4a with its usual band folding and formation of band gaps at the origin and the edges of the Brillouin zone. The band structure is obtained by solving for qL at various values of reduced frequency $\omega\frac{d_1}{c_1}$ using (2.37). In Fig. 2.4b and c, we have plotted the real part and imaginary part of A_+ and the imaginary part of A_- for two isolated dispersion branches, namely the second and third branches. One notices that the amplitude A_+ as functions of $qL \in [-\pi, \pi]$ cross and change sign in the case of the second branch, at $q_0L = 0.524$. The amplitude A_- reaches zero there but does not change sign (its slope is zero). The amplitudes do not cross at a value of 0 in Fig. 2.4c. This behavior repeats for the 4th, 5th etc. bands.

The amplitudes A_+ and A_- are now expanded in a series around the point q_0:

$$A_+(q) = A_+(q_0) + \left.\frac{dA_+}{dq}\right|_{q_0}(q - q_0) + \ldots \approx \left.\frac{dA_+}{dq}\right|_{q_0}\delta q \tag{2.40}$$

and

$$A_-(q) = A_-(q_0) + \left.\frac{dA_-}{dq}\right|_{q_0}(q - q_0) + \left.\frac{d^2A_-}{dq^2}\right|_{q_0}(q - q_0)^2 + \ldots \approx \left.\frac{d^2A_-}{dq^2}\right|_{q_0}\delta q^2 \tag{2.41}$$

The first amplitude is a linear function of the deviation from the wave number q_0 while the second amplitude is a quadratic function of the wave number deviation. The periodic part of the displacement field is given in the Appendix 1 for a layer n so for the layer n = 0, we have:

$$u_1(q, x) = e^{-iqx}\left(A_+e^{ik_1x} + A_-e^{-ik_1x}\right) \tag{2.42}$$

And expansion of this expression around q_0 gives:

$$u_1(q, x) = u_1(q_0, x) + \left.\frac{du_1}{dq}\right|_{q_0}(q - q_0) + \ldots \tag{2.43}$$

Inserting (2.42), (2.40) and (2.41) into (2.43) yields:

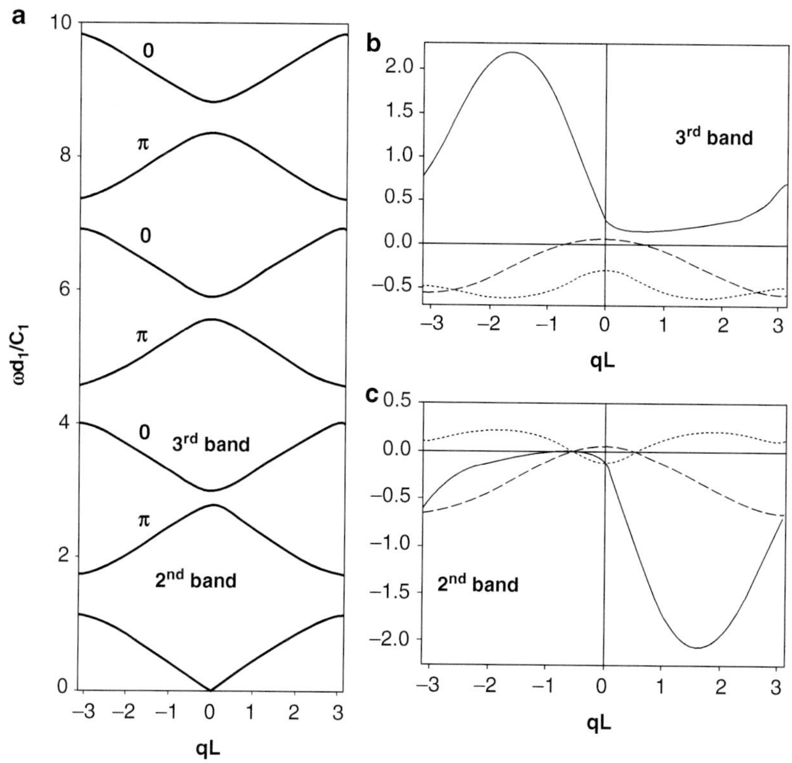

Fig. 2.4 (**a**) Band structure of one-dimensional superlattice (see text for details), real part of A_+ (*dotted line*), imaginary part of A_+ (*dashed line*) and the imaginary part of A_- (*solid line*) for (**b**) the second dispersion branch and (**c**) the third branch. The geometric phase accumulated by the elastic wave as one follows a closed path in qL space is indicated on the band structure

$$u_1(q,x) \approx e^{-iq_0 x} e^{ik_1^{(0)}x} \frac{dA_+}{dq}\bigg|_{q_0} (q-q_0) \tag{2.44}$$

The Berry connection, $BC(q)$, in wave number space, q, is calculated from the relation:

$$\frac{1}{u_1^* u_1} u_1^* \frac{du_1}{dq} = \frac{1}{(q-q_0)+i\varepsilon} \tag{2.45}$$

In that expression, $u_1^* u_1$ is a normalizing factor. Equation (2.45) is analytically continued into the complex plane by introducing the quantity $\varepsilon \to 0$. The imaginary part of the Berry connection is the phase change, namely

$$\frac{\delta\eta}{\delta q} = \lim_{\varepsilon\to 0} \frac{-\varepsilon}{(q-q_0)^2+\varepsilon^2} = -\pi\delta(q-q_0) \tag{2.46}$$

This expression is valid only in the vicinity of the point (q_0, ω_0). The amplitude does not change sign elsewhere, so we anticipate that the phase change $\delta\eta = 0$ everywhere else but at q_0. Thus we extend the use of expression (2.46) to the entire Brillouin zone. Since there is only one point (q_0, ω_0) along the second branch in the band structure of Fig. 2.4, the integral of (2.46) over the Brillouin zone, $\eta = \int_{-\frac{\pi}{L}}^{\frac{\pi}{L}} dq \frac{\delta\eta}{\delta q}$ gives a Zak phase of $-\pi$. The third band does not possess a point (q_0, ω_0) and therefore, the Zak phase is zero. Similarly, the fourth, sixth, etc. bands exhibit a π geometric phase while the fifth, seventh, etc. bands have a geometric phase equal to zero.

The origin of a non-zero Zak phase can be found in the breaking of inversion symmetry of the elastic waves in the binary superlattice. We observed a similar breaking of inversion symmetry in Chap. 1 when we considered a one-dimensional monatomic crystal with spatially varying stiffness. We complement the analysis of the Zak phase in superlattices by investigating the phase of elastic waves in the discrete superlattice with broken inversion symmetry in Appendix 2. In Chap. 3, we will examine the breaking of four other types of symmetries. Appendix 2 also illustrate the dependency of the Zak phase on the choice of the origin of the system.

2.4.2 *Topological Interpretation of the Zak Phase*

This topological interpretation of the geometric phase derived for the harmonic oscillator (Sect. 2.3.3) can also be applied to the Zak phase of bands in the band structure of superlattices. For instance, (2.46) is isomorphic to (2.31) where the frequency, ω, is replaced by the wave number, q. The major difference though lies in the fact that in a periodic system, such as a superlattice, the dispersion relations are periodic functions of the wave number, $q \in \left[\frac{-\pi}{L}, \frac{\pi}{L}\right]$. In this case, the topological interpretation of the Zak phase in terms of a manifold is given in Fig. 2.5.

The primary difference between Figs. 2.5 and 2.3 is that the manifold is formed of a closed strip in the latter case because of the periodicity in wave number, q space. The twist may be visualized as a segment of helicoid with infinitesimally small width in q space. The arrows in Fig. 2.5 illustrate parallel transport of a vector field on a closed loop on the manifold. Upon spanning the Brillouin zone once (closed path of length $\frac{2\pi}{L}$), the transported vector accumulates a phase of π. One needs to complete two turns in q space *i.e.*, follow a closed path which length $\frac{4\pi}{L}$ to recover the original orientation of the parallel transported vector, *i.e.*, accumulate a phase of 2π.

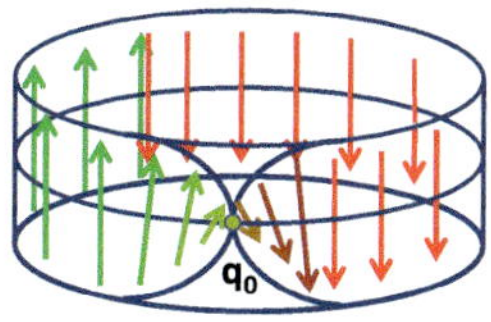

Fig. 2.5 Schematic illustration of a closed manifold with a single half-turn twist this topology is isomorphic to that of the eigen vectors of a superlattice along a band with a Zak phase of π

2.5 Green's Function Approach

2.5.1 *Green's Functions and Berry Connection*

This section attempts to clarify the nature of the phase η of the Green's function of Hermitian operators. A brief review of definition and properties of Green's functions can be found in Appendix 3. For this, we consider a normalized Green's function $\widehat{G} = \frac{G}{\sqrt{G^* G}}$ where G^* is the complex conjugate of G. Let us calculate the derivative of this normalized function with respect to the eigen value E:

$$\frac{d\widehat{G}(E)}{dE} = \frac{1}{\sqrt{G^* G}} \frac{dG}{dE} - \frac{G}{2(G^* G)^{3/2}} \frac{d(G^* G)}{dE} \tag{2.47}$$

We now define the Berry connection as the following quantity:

$$BC(E) = -i\widehat{G}^*(E) \frac{d\widehat{G}(E)}{dE} \tag{2.48}$$

The quantity $BC(E)$ can be understood as describing the local variation of the phase of the Green's function along a path in the space of eigen values. Indeed, consider the two eigen values E and $E + dE$. The normalized Green's function at $E + dE$ can be expanded to first order:

$$\widehat{G}(E + dE) \approx \widehat{G}(E) + \frac{d\widehat{G}(E)}{dE} dE \tag{2.49}$$

Multiplying both sides of (2.49) by $\widehat{G}^*(E)$ gives:

$$\widehat{G}^*(E)\widehat{G}(E + dE) \approx \widehat{G}^*(E)\widehat{G}(E) + \widehat{G}^*(E) \frac{d\widehat{G}(E)}{dE} dE = 1 + iA(E)dE$$

Taking the trace of both sides of the previous relation results in:

$$Tr\left(\widehat{G}^*(E)\widehat{G}(E+dE)\right) = 1 + Tr\left(\widehat{G}^*(E)\frac{d\widehat{G}(E)}{dE}\right)dE + \ldots \tag{2.50}$$

The normalized Green's function can be now written as $\widehat{G}(E+dE) = e^{i\eta(E+dE)} \sim e^{i\left(\eta(E)+\frac{d\eta(E)}{dE}dE\right)} = e^{i\eta(E)}e^{i\frac{d\eta(E)}{dE}dE}$. Since $\widehat{G}^*(E) \propto e^{-i\eta(E)}$, we can rewrite (2.50) in terms of the phase η:

$$e^{i\frac{d\eta(E)}{dE}dE} \sim 1 + i\frac{d\eta(E)}{dE}dE = 1 + iBC(E)dE \tag{2.51}$$

Using the definition of (2.48), comparison of the two imaginary terms in (2.51) leads to

$$\frac{d\eta(E)}{dE} = BC(E) \tag{2.52}$$

From the definition of $\eta(E)$, one notes that $BC(E)$ must be real. It is informative to obtain the relation given by (2.52) by directly calculating $BC(E)$ using (2.47):

$$BC(E) = -i\widehat{G}^*(E)\frac{d\widehat{G}(E)}{dE} = -i\left\{\frac{G^*\,dG}{G^*GdE} - \frac{1}{2G^*G}\frac{d(G^*G)}{dE}\right\} \tag{2.53}$$

By ensuring that $BC(E)$ is real, it is easy to show that:

$$BC(E) = Im\left(\frac{1}{G}\frac{dG}{dE}\right) = Im\left(\frac{dlnG}{dE}\right) \tag{2.54}$$

Taking the trace of (2.54), we obtain:

$$\begin{aligned} TrBC(E) &= Tr\left(Im\left(\frac{dlnG}{dE}\right)\right) = Im\left(Tr\left(\frac{dlnG}{dE}\right)\right) = Im\left(\frac{d}{dE}(TrlnG)\right) \\ &= -ImTrG \end{aligned} \tag{2.55}$$

To derive (2.55), we have used Green's functions identities presented in Appendix 3 and in particular the properties of the Green's function in the space of its Eigen values. From (2.144), it is now straight forward to show that

$$TrBC(E) = Tr\left(\frac{d\eta(E)}{dE}\right) = -ImTrG = \pi n(E) \tag{2.56}$$

2.5.2 The One-Dimensional Harmonic Crystal

The one-dimensional (1-D) monatomic harmonic crystal is used as an illustrative example. This system consists of an infinite chain of masses, m, with nearest neighbor interaction modeled by harmonic springs with spring constant, β. (see Sect. 1.2). The separation distance between the masses at rest is defined as a. We recall the equations defining the Green's function, $\overleftrightarrow{G_0}$, of the one-dimensional harmonic crystal by:

$$\overleftrightarrow{H_0}\overleftrightarrow{G_0} = \overleftrightarrow{I} \tag{2.57}$$

The operator, $\overleftrightarrow{H_0}$, is the infinite tridiagonal dynamic matrix:

$$\overleftrightarrow{H_0} = \frac{1}{m}\begin{bmatrix} \ddots & \vdots & \vdots & \vdots & \vdots & \vdots & \vdots & \vdots & \vdots \\ \cdots & 0 & \beta & -\gamma & \beta & 0 & 0 & \cdots & \cdots \\ \cdots & \cdots & 0 & \beta & -\gamma & \beta & 0 & \cdots & \cdots \\ \cdots & \cdots & \cdots & 0 & \beta & -\gamma & \beta & 0 & \cdots \\ \vdots & \vdots & \vdots & \vdots & \vdots & \vdots & \vdots & \vdots & \ddots \end{bmatrix} \tag{2.58}$$

where $\gamma = 2\beta - m\omega^2$. In component form, (2.53) reads:

$$\frac{1}{m}[\beta G_0(n+1,n') - \gamma G_0(n,n') + \beta G_0(n-1,n')] = \delta_{nn'} \tag{2.59}$$

Since we are interested in calculating the Berry connection for this system, we need to use the trace of the Green's function. For this, we consider only diagonal terms and take $n' = n$. In this case (2.59) simplifies to

$$\frac{1}{m}[\beta G_0(n+1,n) - \gamma G_0(n,n) + \beta G_0(n-1,n)] = 1 \tag{2.60}$$

To solve that equation, we choose $G_0(n+1,n) = G_0(n,n)e^{ika}$ and $G_0(n-1,n) = G_0(n,n)e^{-ika}$. Inserting these relations in (2.60), gives:

$$G_0(n,n) = \frac{-m}{2\beta}\frac{1}{\zeta - \cos ka} \tag{2.61}$$

with $\zeta = 1 - 2\frac{\omega^2}{\omega_0^2}$ and $\omega_0^2 = 4\frac{\beta}{m}$.

Note that this relation differs from the 1-D harmonic crystal Green's function reported in Chap. 1. In Chap. 1, the Green's function was defined as the integral over wave vectors, k.

For the sake of simplicity, we take $\beta = 1\ N/m$, $m = 1kg$, and $\zeta = 1 - \frac{\omega^2}{2}$. Using the trigonometric relation $(\sin x)^2 = \frac{(1-\cos 2x)}{2}$, the Green's function becomes: $G_0(n,n) = \frac{+1}{\omega^2 - 4sin^2\left(\frac{ka}{2}\right)}$.

We expand the Green's function into the complex plane by analytical continuation ($\omega^2 \to \omega^2 + i\varepsilon$):

$$G_0^+(n,n) = \frac{+1}{\omega^2 - 4sin^2\left(\frac{ka}{2}\right) + i\varepsilon} \tag{2.62}$$

In the limit of small ε, the Berry connection is then obtained in the form:

$$\begin{aligned} \frac{d\eta}{d(\omega^2)} = -ImG_0^+(n,n) &= +\lim_{\varepsilon \to 0} \frac{\varepsilon}{\left(\omega^2 - 4sin^2\left(\frac{ka}{2}\right)\right)^2 + \varepsilon^2} \\ &= +\pi\delta\left(\omega^2 - 4sin^2\left(\frac{ka}{2}\right)\right) \end{aligned} \tag{2.63}$$

We can then express the evolution of the phase along a path in frequency at a fixed wave number k:

$$\left(\frac{\partial \eta}{\partial \omega}\right)_k = +\pi \frac{\omega}{\left|2\sin\left(\frac{ka}{2}\right)\right|} \left\{\delta\left(\omega - \left|2\sin\left(\frac{ka}{2}\right)\right|\right) + \delta\left(\omega + \left|2\sin\left(\frac{ka}{2}\right)\right|\right)\right\} \tag{2.64}$$

The elastic waves acquire a phase of $+\pi$ every time one crosses the dispersion curve of the 1D harmonic crystal along a path of increasing positive ω at a fixed positive or negative value of k. This behavior is reminiscent of the behavior of a single harmonic oscillator. The topology of this system is also reminiscent of the harmonic oscillator's topology.

We now note that (2.53) defining the Berry connection along a path in the space of eigen values, namely: $BC(E) = -i\widehat{G}^*(E)\frac{d\widehat{G}(E)}{dE} = Im\left(\frac{1}{G}\frac{dG}{dE}\right)$, can be generalized to characterizing the evolution of the phase along paths determined by some other parameter, such as for instance the wave number, k. In this case, we write:

$$A(k) = -i\widehat{G}^*(k)\frac{d\widehat{G}(k)}{dk} = Im\left(\frac{1}{G}\frac{dG}{dk}\right) \tag{2.65}$$

The derivatives in (2.65) are taken at a fixed frequency ω.

Using the Green's function given by (2.62), we determine in the limit $\varepsilon \to 0$:

$$A(k) = +\pi\delta\left(\omega^2 - 4sin^2\left(\frac{ka}{2}\right)\right)\frac{\partial}{\partial k}\left(-4sin^2\left(\frac{ka}{2}\right)\right) \tag{2.66}$$

We now use the following property of the delta function: $\delta(f(x)) = \sum_i \frac{\delta(x-x_i)}{|f'(x_i)|}$ where x_i are the roots of f. Here, we have $f(k) = \omega^2 - 4sin^2\left(\frac{ka}{2}\right)$ with the roots: $\omega = \pm 2\sin\left(\frac{ka}{2}\right)$ and $f'(k) = \frac{\partial}{\partial k}\left(-4sin^2\left(\frac{ka}{2}\right)\right)$. With these, (2.66) becomes:

$$\begin{aligned}\left(\frac{\partial\eta}{\partial k}\right)_\omega = A(k) = +\pi\Bigg\{&\frac{1}{\left|\frac{\partial}{\partial k}\left(-4sin^2\left(\frac{ka}{2}\right)\right)\right|_{\omega=+2\sin\left(\frac{ka}{2}\right)}}\delta\left(\omega-\left|2\sin\left(\frac{ka}{2}\right)\right|\right)\\ &+\frac{1}{\left|\frac{\partial}{\partial k}\left(-4sin^2\left(\frac{ka}{2}\right)\right)\right|_{\omega=-2\sin\left(\frac{ka}{2}\right)}}\delta\left(\omega+\left|2\sin\left(\frac{ka}{2}\right)\right|\right)\Bigg\}\\ &\times\frac{\partial}{\partial k}\left(-4sin^2\left(\frac{ka}{2}\right)\right)\end{aligned} \tag{2.67}$$

At fixed positive ω, the elastic waves accumulate a phase of $\pm\pi$ when crossing a dispersion branch of the 1D harmonic crystal's band structure by moving along the positive line of k. From a topological point of view, this can be visualized again in the form of the parallel transport of a vector on a manifold with a ½ twist occurring on the dispersion branch.

Let us calculate the complete phase change by integrating (2.66) over the Brillouin zone:

$$\Delta\eta = \int_{\frac{-\pi}{a}}^{\frac{\pi}{a}} dk\left(\frac{\partial\eta}{\partial k}\right)_\omega$$

We note that $\frac{\frac{\partial}{\partial k}\left(-4sin^2\left(\frac{ka}{2}\right)\right)}{\left|\frac{\partial}{\partial k}\left(-4sin^2\left(\frac{ka}{2}\right)\right)\right|_{\omega=+2\sin\left(\frac{ka}{2}\right)}} = -1$ for a value of $k>0$ that satisfies $\delta\left(\omega - \left|2\sin\left(\frac{ka}{2}\right)\right|\right)$ and $\frac{\frac{\partial}{\partial k}\left(-4sin^2\left(\frac{ka}{2}\right)\right)}{\left|\frac{\partial}{\partial k}\left(-4sin^2\left(\frac{ka}{2}\right)\right)\right|_{\omega=-2\sin\left(\frac{ka}{2}\right)}} = +1$ for the value of $k<0$ that also satisfies $\delta\left(\omega - \left|2\sin\left(\frac{ka}{2}\right)\right|\right)$. The preceding integral is therefore zero. There is no accumulation of phase as one twice crosses the dispersion band by varying the wave number across the Brillouin zone.

We can now consider the phase change as one follows a path that takes us twice across the dispersion curve (once by moving along the positive frequency line at fixed wave number and then keeping the frequency constant and moving along the positive line of wave numbers) by considering:

$$d\eta = \left(\frac{\partial \eta}{\partial k}\right)_{\omega} dk + \left(\frac{\partial \eta}{\partial \omega}\right)_{k} d\omega \tag{2.68}$$

For a path that takes us along the dispersion curve by a series of increments of dk and the corresponding $d\omega$ (with appropriate sign determined by the group velocity $\frac{d\omega}{dk}$) the elastic wave of the one-dimensional harmonic crystal accumulates a net phase of zero.

2.5.3 The One-Dimensional Harmonic Crystal with Side Branches

In Chap. 1, we presented the Green's function of the 1D harmonic crystal in the form:

$$G_0(n,n') = G_0(n,n',\omega) = \frac{m}{\beta}\frac{t^{|n-n'|+1}}{t^2-1}. \tag{2.69}$$

with

$$t = \begin{cases} \xi - \left(\xi^2-1\right)^{1/2} \; if \; \xi > 1 \\ \xi + \left(\xi^2-1\right)^{1/2} \; if \; \xi < -1 \\ \xi + i\left(1-\xi^2\right)^{1/2} \; if \; -1 \le \xi \le 1 \end{cases}$$

and $\xi = \frac{\gamma}{2\beta} = 1 - \frac{m\omega^2}{2\beta} = 1 - \frac{2\omega^2}{\omega_0^2}$. We note that when $\omega\epsilon[0,\omega_0]$ and $\xi\epsilon[-1,1]$, this corresponds to the region of propagative waves. We used this function to calculate the Green's function of a 1D harmonic crystal with finite harmonic crystal grafted as side branches. The Green's function given by (2.69) can be written as:

$$G_0(n,n',\omega) = \frac{m}{\beta}\frac{1}{2\pi}a\int_{-\frac{\pi}{a}}^{\frac{\pi}{a}} dk \frac{e^{ika(n-n')}}{\xi - \cos ka} \tag{2.70}$$

We make the change of variable $\alpha = ka$ and (2.70) becomes:

$$G_0(n,n',\omega) = \frac{m}{\beta}\frac{1}{2\pi}\int_{-\pi}^{\pi} d\alpha \frac{e^{i\alpha(n-n')}}{\xi - \cos\alpha} \tag{2.71}$$

By making the additional substitution: $w = e^{i\alpha}$, the integral in (2.71) becomes an integral over the unit circle:

$$G_0(n,n',\omega)=\frac{m}{\beta}\frac{1}{2\pi i}\oint dw\frac{w^{|n-n'|}}{w^2+2\xi w+1} \tag{2.72}$$

In (2.72), the integral only depends on the absolute value: $|n-n'|$.

The roots of the denominator: $w^2+2\xi w+1=0$ are $w_+=-\xi+\sqrt{\xi^2-1}$ and $w_-=-\xi-\sqrt{\xi^2-1}$. We rewrite the integral in (2.72) as

$$\oint dw\frac{w^{|n-n'|}}{w^2+2\xi w+1}=\oint dw\frac{w^{|n-n'|}}{(w-w_+)(w-w_-)} \tag{2.73}$$

If $\xi>1$, the root w_+ lies within the unit circle ($|w_+|<1$) and the root w_- lies outside ($|w_-|>1$). The method of residues is used to calculate the integral.

$$\oint dw\frac{w^{|n-n'|}}{(w-w_+)(w-w_-)}=2\pi i\frac{w_+^{|n-n'|}}{w_+-w_-}$$

By noting that $w_-=\frac{1}{w_+}$ and setting $w_+=t$, then one shows that (2.72) reduces to (2.69).

Similarly, if $\xi<-1$, the root w_- lies within the unit circle ($|w_-|<1$) and the root w_+ lies outside ($|w_+|>1$). The method of residues gives the integral:

$$\oint dw\frac{w^{|n-n'|}}{(w-w_+)(w-w_-)}=2\pi i\frac{w_-^{|n-n'|}}{w_--w_+}$$

By noting that $w_-=\frac{1}{w_+}$ and setting $w_-=t$, then one shows that (2.72) reduces to (2.69).

If $-1\le\xi\le 1$, the roots lie on the unit circle. Economou [32] shows that (2.72) also reduces to (2.69). For this we have to follow a limiting procedure. We replace w_+ by $w_++i\varepsilon$ and w_- by $w_-+i\varepsilon$ with $\varepsilon\to 0$. With this, $w_++i\varepsilon$ is now outside the unit circle contour and $w_-+i\varepsilon$ is inside the contour. Applying the method of residues gives:

$$\oint dw\frac{w^{|n-n'|}}{(w-w_+-i\varepsilon)(w-w_--i\varepsilon)}=2\pi i\frac{(w_-+i\varepsilon)^{|n-n'|}}{w_--w_+}\to 2\pi i\frac{w_-^{|n-n'|}}{w_--w_+}$$

This leads to (2.69). We can address the other limit by replacing w_+ by $w_+-i\varepsilon$ and w_- by $w_--i\varepsilon$ with $\varepsilon\to 0$, $w_+-i\varepsilon$ is now inside the unit circle. The contour integral gives $2\pi i\frac{w_+^{|n-n'|}}{w_+-w_-}$ and subsequently (2.69).

We now consider a system composed of a 1D finite length side branch (5 masses) grafted onto an infinite 1D harmonic crystal. Similar systems have been discussed in Chap. 1. We calculate the transmission amplitude, T, of this system using the code of Appendix, Chap. 1. For this calculation, we have taken all masses and spring constants equal to 1. The transmission amplitude is given by (1.93):

$$u_n = t^n T$$

Here t^n corresponds to a plane wave launched into the side of the infinite chain to the left of the grafted branch. u_n is the wave transmitted to the right of the side branch. In Fig. 2.6a and b, we report the transmission coefficient, T^2, the transmission phase, θ^T, as functions of normalized frequency. The transmission amplitude can be expressed in terms of its modulus $|T|$ and its phase θ^T, $T = |T|e^{i\theta^T}$. We also illustrate in Fig. 2.6c, the path of the transmission amplitude (imaginary part versus real part of the transmission amplitude) as a function of frequency. The transmission coefficient as expected exhibits five zeros of transmission corresponding to resonances with the modes of the finite side chain.

Focusing on the transmission phase, we note that the phase goes through a series of continuous variations between each pair of adjacent resonances. For instance, between two zeros of transmission, as frequency increases the wave accumulates a phase of π continuously. However, at each resonance *i.e.*, zero of transmission, there is an abrupt phase slip of π. The transmission phase, $\theta^T(\omega)$ is actually the sum of two phases:

$$\theta^T(\omega) = \theta^f(\omega) + \theta(\omega) \tag{2.74}$$

where $\frac{d\theta^f(\omega)}{d(\omega^2)} = -\pi\Delta n(\omega)$ is the Friedel phase [33]. The derivative of the Friedel phase is directly related to the variation in density of states. The Friedel phase accumulates π each time the frequency crosses a bound state in the side branch. This is a continuous function of ω since the density of states is a continuous function of frequency. The discontinuous phase change $\theta(\omega)$ is the phase change associated with the zeros of the transmission amplitude. As frequency increases, each time the transmission amplitude passes through a zero, the real part of the amplitude, Re(T), becomes zero and the imaginary part, Im(T), changes sign. The ratio $\frac{Imag(T)}{Real(T)} = \tan\theta^T(\omega)$ diverges with a sign change across the zero of transmission. The transmission phase then exhibits a phase slip of π. This behavior has been investigated extensively, for instance, in the case of coherent electron transport and scattering by quantum dots [34–36].

We now use the Green's function of the infinite crystal (2.69) and that of a finite side branch derived in Chap. 1 (1.83) to calculate the difference in phase $\Delta\theta(\omega) = \frac{1}{\pi} Im \ln det\left(I - VG_0^+(\omega)\right) = arg det\left(I - VG_0^+(\omega)\right)$ [see (2.148)], and the density of states (2.126). Again, we use the code of Appendix, Chap. 1. Since the calculation is performed with the Green's function at the origin (location of the coupling between the infinite chain and the grafted chain), the quantity, $\Delta\theta$, actually represents the

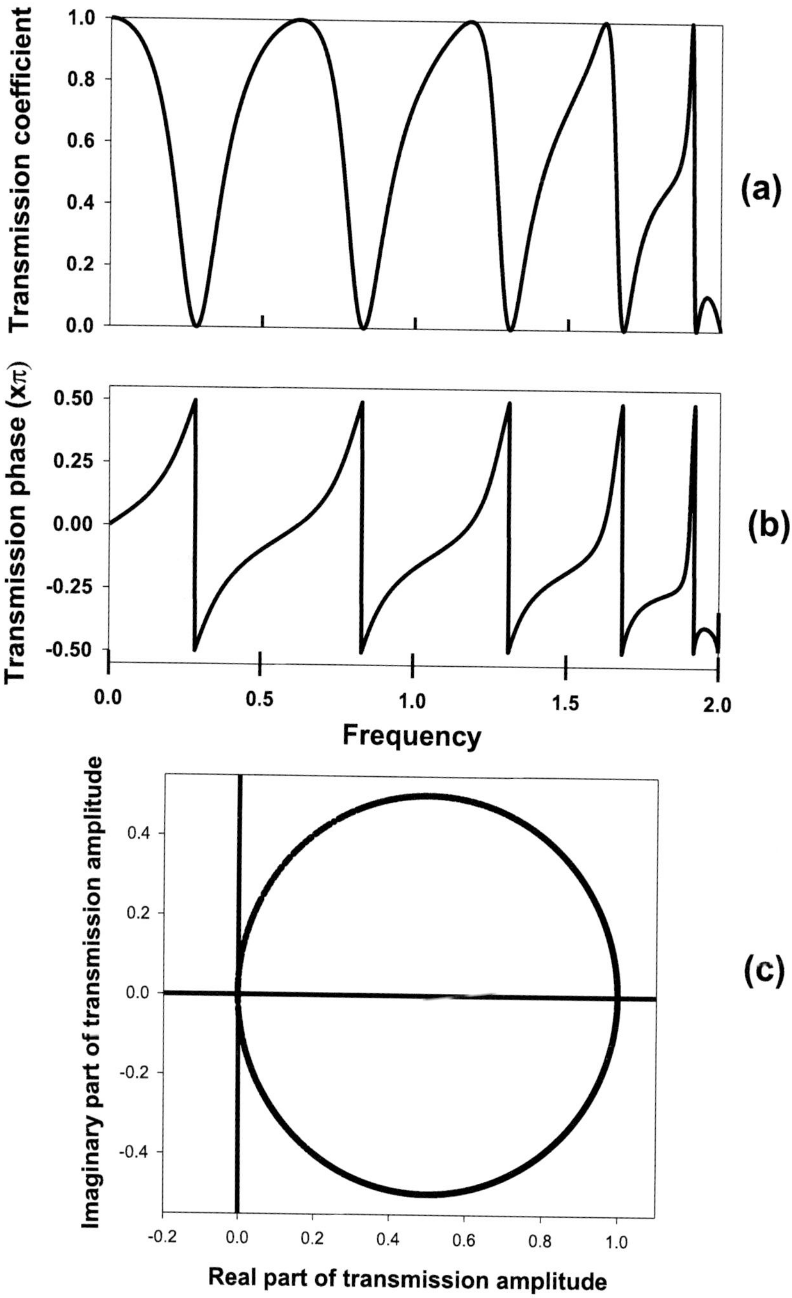

Fig. 2.6 Transmission spectrum (**a**), transmission phase (**b**) and transmission amplitude as functions of frequency of a 1D infinite harmonic crystal with one finite side branch containing five masses. All masses and spring constants have the value 1

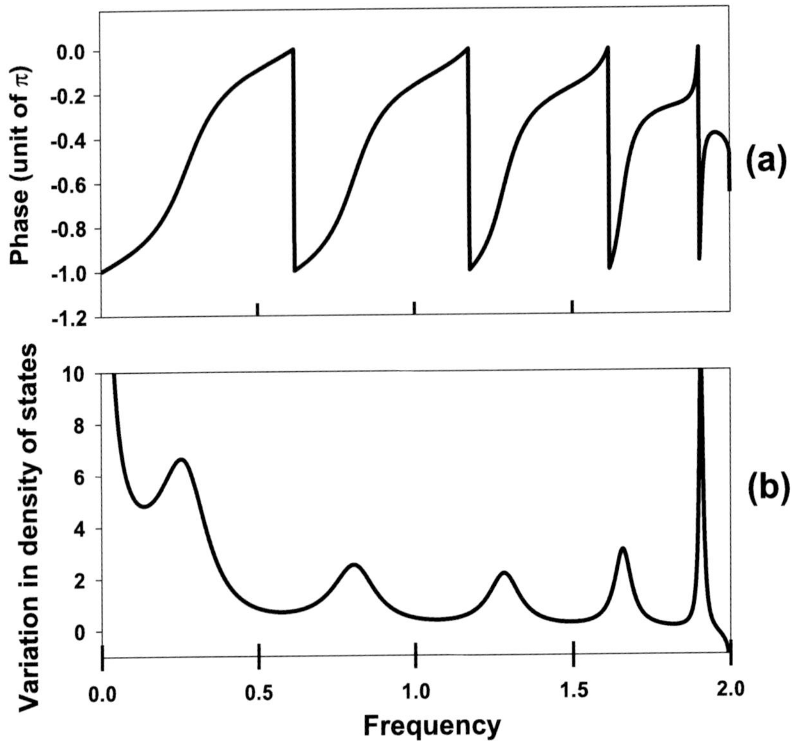

Fig. 2.7 Reflection phase (**a**) and variation in density of states phase (**b**) as functions of frequency of a 1D infinite harmonic crystal with one finite side branch containing five masses. All masses and spring constants have the value 1

reflection phase, $\Delta\theta = \theta^R = \tan^{-1}\frac{Im(T)}{1-Re(T)}$. Figure 2.7a shows the reflection phase as a function of frequency. $\theta^R(\omega)$ is now continuous across the zeros of transmission and exhibits phase slip of π between adjacent resonances where the transmission coefficient is 1 (the reflection coefficient is zero). We calculate the variation in density of states using the relation $\frac{d\theta^R(\omega)}{d(\omega^2)} = -\pi\Delta n(\omega)$ excluding the discontinuous phase slips. The density of states is reported in Fig. 2.7b. As expected it shows peaks at the resonant frequencies of the side branch.

We recall that all calculations performed to obtain the transmission amplitude and transmission (or reflection) phase using the Interface Response Theory [37], were all based on the 1D harmonic crystal Green's function given by (2.69) and (2.70):

$$G_0(n, n', \omega) = \frac{m}{\beta} \frac{t^{|n-n'|+1}}{t^2 - 1} = \frac{m}{\beta} \frac{1}{2\pi} a \int_{-\frac{\pi}{a}}^{\frac{\pi}{a}} dk \frac{e^{ika(n-n')}}{\xi - \cos ka}$$

This Green's function is the sum over wave numbers (k) at a fixed frequency of the function: $\frac{e^{ika(n-n')}}{\xi - \cos ka}$. We have used the trace of such a function (*i.e.*, $n = n'$), $G_0(n, n) = \frac{-m}{2\beta} \frac{1}{\zeta - \cos ka}$ [see (2.61)] with $\zeta = 1 - 2\frac{\omega^2}{\omega_0^2}$ to calculate the evolution of the Berry phase, η, along a path in frequency at a fixed wave number k, namely [see (2.64)]: $\left(\frac{\partial \eta}{\partial \omega}\right)_k = +\pi \frac{\omega}{2 \sin\left(\frac{ka}{2}\right)} \left\{\delta\left(\omega - 2 \sin\left(\frac{ka}{2}\right)\right) + \delta\left(\omega + 2 \sin\left(\frac{ka}{2}\right)\right)\right\}$. This equation shows that the elastic wave acquires a phase of $+\pi$ every time one crosses the dispersion curve of the 1D harmonic crystal along a path of increasing positive ω at a fixed value of k. Therefore, it is expected that the transmission and reflection phases reported in this section as functions of frequency are equivalent to Berry phases averaged at fixed ω over all wave numbers within the Brillouin zone of the system.

2.6 Topological Modes at Interfaces Between Media with Different Zak Phases

Here, we consider a three-dimensional continuous medium with broken inversion symmetry. The Green's function of this medium is given by:

$$(E - H_i(\vec{r}))G_i(E, \vec{r}, \vec{r}\,') = \delta(\vec{r} - \vec{r}\,') \tag{2.75}$$

E refers to the eigen value and the vector position is $\vec{r} = (x_1, x_2, x_3)$. The Green's function and the operator H are labelled with a subscript "i" to mark the dependency of the Zak phase on the choice of the origin of the system (see Appendix 2). When $i = 1$, the Zak phase is 0 and the Zak phase is equal to π when $i = 2$. We assume that the operator $H_i(\vec{r})$ can be separated into the following sum:

$$H_i(\vec{r}) = \frac{\partial^2}{\partial x_1^2} + \frac{\partial^2}{\partial x_2^2} + h_i(x_3) \tag{2.76}$$

Equation (2.76) allows for broken inversion symmetry in the x_3 direction through the operator $h_i(x_3)$. If $h_i(x_3) = \frac{\partial^2}{\partial x_3^2}$, then inversion symmetry would not be broken. Because of translational symmetry in the plane $\vec{r}_{\parallel} = (x_1, x_2)$, we write the Green's function as a two-dimensional Fourier transform:

$$G_i(E,\vec{r},\vec{r}')=\int\frac{d^2\vec{k}_\parallel}{(2\pi)^2}G_i\left(E,\vec{k}_\parallel,x_3,x_3'\right)e^{i\vec{k}_\parallel\left(\vec{x}_\parallel-\vec{x}_\parallel'\right)} \tag{2.77}$$

with $\vec{k}_\parallel\vec{x}_\parallel=k_1x_1+k_2x_2$.
Inserting (2.77) into (2.75), we obtain (2.75) in the two-dimensional space:

$$\left(E-k_\parallel^2-h_i(x_3)\right)G_i\left(E,\vec{k}_\parallel,x_3,x_3'\right)=\delta\left(x_3-x_3'\right) \tag{2.78}$$

To obtain (2.78), we have used $\delta(\vec{r}-\vec{r}')=\delta\left(\vec{x}_\parallel-\vec{x}_\parallel'\right)\delta\left(x_3-x_3'\right)=\int\frac{d^2\vec{k}_\parallel}{(2\pi)^2}e^{i\vec{k}_\parallel\left(\vec{x}_\parallel-\vec{x}_\parallel'\right)}\delta\left(x_3-x_3'\right)$. We posit that solutions (2.78) take the form:

$$G_i\left(E,\vec{k}_\parallel,x_3,x_3'\right)=A_i\left(E,\vec{k}_\parallel\right)f\left(E,\vec{k}_\parallel,x_3,x_3'\right) \tag{2.79}$$

The phase information is contained in A_i and the function $f\left(E,\vec{k}_\parallel,0,0\right)=f_0$ is a constant.

To construct the Green's function of a semi-infinite continuous medium, G_i^S with a free surface parallel to the plane (x_1,x_2) at $x_3=0$, we impose the Neumann boundary condition:

$$\left.\frac{\partial G_i^S}{\partial x_3}\right|_{x_3=0}=0 \tag{2.80}$$

This boundary condition affects only the function, f, we therefore write the Green's function of the semi-infinite medium in the form:

$$G_i^S\left(E,\vec{k}_\parallel,x_3,x_3'\right)=A_i\left(E,\vec{k}_\parallel\right)s\left(E,\vec{k}_\parallel,x_3,x_3'\right) \tag{2.81}$$

with $s\left(E,\vec{k}_\parallel,0,0\right)=s_0$ a constant. The Green's function of the semi-infinite medium G_i^S retains the information about the Zak phase. This Green's function is used to construct the Green's function of a composite medium constituted of the two semi-infinite media 1 and 2 separated by at interface at $x_3=0$. Medium 1 is at the left of the interface and medium 2 at its right (we denote this system 1–2). The Green's function of the composite system, g, can be evaluated at the interface itself. The Interface Response Theory of continuous media [38] states that at the interface:

$$\frac{1}{g\left(E,\vec{k}_{\parallel},0,0\right)}=\frac{1}{G_1^S\left(E,\vec{k}_{\parallel},0,0\right)}+\frac{1}{G_2^S\left(E,\vec{k}_{\parallel},0,0\right)} \tag{2.82}$$

or

$$g\left(E,\vec{k}_{\parallel},0,0\right)=\frac{G_1^S\left(E,\vec{k}_{\parallel},0,0\right)G_2^S\left(E,\vec{k}_{\parallel},0,0\right)}{G_1^S\left(E,\vec{k}_{\parallel},0,0\right)+G_2^S\left(E,\vec{k}_{\parallel},0,0\right)} \tag{2.83}$$

Interface states are poles of g, that is, they are obtained as solutions of $G_1^S\left(E,\vec{k}_{\parallel},0,0\right)+G_2^S\left(E,\vec{k}_{\parallel},0,0\right)=0$. Using (2.81), this condition becomes

$$A_1\left(E,\vec{k}_{\parallel}\right)+A_2\left(E,\vec{k}_{\parallel}\right)=0 \tag{2.84}$$

From this equation, we obtain the dispersion relation for interfacial modes in the form $E\left(\vec{k}_{\parallel}\right)$.

We note that A_1 and A_2 only differ by a sign which depends on the value of $\vec{k}_{\parallel}$. They do not differ in their magnitudes. Let us define $\vec{k}_{\parallel}^0$ the wave vector where both $A_i=0$. Because medium 1 has a zero Zak phase, A_1 does not change sign as one crosses $\vec{k}_{\parallel}^0$. Let us say for the sake of illustration that it is always a positive function. In contrast A_2 will change sign. For instance, let us assume that $A_2\left(E,\vec{k}_{\parallel}\right)=A_1\left(E,\vec{k}_{\parallel}\right)$ for $\vec{k}_{\parallel}<\vec{k}_{\parallel}^0$ and $A_2\left(E,\vec{k}_{\parallel}\right)=-A_1\left(E,\vec{k}_{\parallel}\right)$ for $\vec{k}_{\parallel}>\vec{k}_{\parallel}^0$. The condition $A_1\left(E,\vec{k}_{\parallel}\right)+A_2\left(E,\vec{k}_{\parallel}\right)=0$ has only solutions for $\vec{k}_{\parallel}>\vec{k}_{\parallel}^0$. There are no interfacial modes for $\vec{k}_{\parallel}<\vec{k}_{\parallel}^0$. These interfacial modes will propagate in only one direction.

If we now construct a composite system with medium 2 at the left of the interface and medium 1 and the right of the interface (composite system 2–1), then interfacial modes would exist (*i.e.*, (2.84) would have solutions) only for $\vec{k}_{\parallel}<\vec{k}_{\parallel}^0$. This interfacial mode would propagate only in the direction opposite to that of the composite 1–2. These interfacial modes exhibit non-reciprocity in their propagation along the interface.

2.7 Conclusion

In this chapter, we have used some of the simple models introduced in Chap. 1 to demonstrate the connection between various phases defined in the literature and the topology of the manifold of solutions for these models. The harmonic crystals that have been studied, here and in the appendices, show that phases can accumulate abruptly as a function of one of the variables of the system. For the one-dimensional driven harmonic oscillator, a π phase accumulates as the driving frequency is tuned through the characteristic frequency of the oscillator, *i.e.*, through resonance. In the case of the Zak phase, we have shown that for the one-dimensional elastic superlattice, elaborated in Appendix 1, the π phase accumulates for selected bands as a function of wave vector. The Berry connection has been associated with the imaginary part of the trace of the Green's function. The transmission phase has been analyzed in terms of the Friedel phase which arises from changes in the density of states and discontinuous phase jumps that are induced by side chain resonances. Finally, we have introduced a simple model that illustrates the propagation characteristics of interfacial modes that may form at an interface between two media differing only by their topology (*i.e.*, Zak phase). Related to the subject of topological surface and interface modes are surface avoiding modes [39, 40]. These are extended waves near the center and edge of the Brillouin zone of superlattices that have a tendency to avoid the boundaries, irrespective of the boundary conditions. A second set of related phenomena are topologically protected edge floppy modes in Maxwell lattices [41–43]. Maxwell lattices are constituted of point particles connected by central force bonds which are on the verge of instability due to coordination. Floppy modes arise from missing constraints on boundaries of the lattice and they can reside either on the boundary or the bulk.

The topology of the manifolds that support these accumulations of phase have been another subject of this chapter. In particular, it has been demonstrated that the abrupt phase changes that occur at resonances can be describes as helicoidal twists in the manifold of solutions to the displacement equation for the one-dimensional driven harmonic oscillator. The visualizations provided in Figs. 2.1, 2.3, and 2.5 provide a topological interpretation of the phases that have been the subject of this chapter. The variables of band structures, frequency and wave vector, have been used to elucidate the role of phase in understanding the topology of simple harmonic crystals including side branches. Focusing on the phase and amplitude, rather than frequency and wave vector, will be essential in our development of the new science of sound.

Appendix 1: Eigen Values and Eigen Vectors in One-Dimensional Elastic Superlattice

We consider a one-dimensional (1D) superlattice composed of alternating segments of material 1 and material 2 (Fig. 2.8). The density and speed of sound in the two types of materials are ρ_1, ρ_2 and c_1, c_2. The lengths of the alternating segments are d_1 and d_2, respectively. The 1D equation of propagation of longitudinal waves in a homogeneous medium with speed of sound c is:

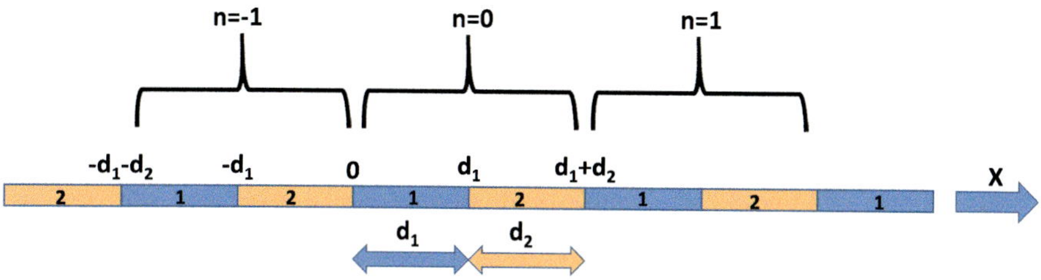

Fig. 2.8 Schematic representation of the one-dimensional superlattice. A layer, n, is composed of two adjacent segments. The period of the super lattice is $L = d_1 + d_2$

$$\frac{\partial^2 u(x,t)}{\partial t^2} = c^2 \frac{\partial^2 u(x,t)}{\partial x^2} \tag{2.85}$$

We seek solutions in the form: $u(x,t) = u(x)e^{i\omega t}$. Inserting into (2.85) gives:

$$-\omega^2 u(x) = c^2 \frac{\partial^2 u(x)}{\partial x^2} \tag{2.86}$$

The solution to (2.86) will take the general form of quasi-standing waves:

$$u(x) = A_+ e^{ikx} + A_- e^{-ikx} \tag{2.87}$$

with $k^2 = \frac{\omega^2}{c^2}$.

We expect the solution given by (2.87) to be a periodic function of position, x, with a period L. We therefore write the solution in the form of a Bloch wave, namely:

$$u(x) = e^{iqx} u(q,x) \tag{2.88}$$

where the quantity $q \in \left[\frac{-\pi}{L}, \frac{\pi}{L}\right]$. The periodic function $u(q,x)$ must meet the condition $u(q,x) = u(q, x+L)$. The periodic functions in the segments 1 and 2 in the nth layer are given by:

$$u_1(q,x) = e^{-iq(x-nL)}\left(A_+ e^{ik_1(x-nL)} + A_- e^{-ik_1(x-nL)}\right) \tag{2.89a}$$

$$u_2(q,x) = e^{iq(x-nL)}\left(B_+ e^{ik_2(x-nL-d_1)} + B_- e^{-ik_2(x-nL-d_2)}\right) \tag{2.89b}$$

with $k_1 = \frac{\omega}{c_1}$ and $k_2 = \frac{\omega}{c_2}$. $A_\pm$ and $B_\pm$ are the amplitude of the forward and backward propagating waves in media 1 and 2, respectively.

The solutions in the segment 1 and 2 in the nth layer of the superlattice are therefore given by:

$$u_1(x) = e^{iqnL}\left(A_+e^{ik_1(x-nL)} + A_-e^{-ik_1(x-nL)}\right) \tag{2.90a}$$

$$u_2(x) = e^{iqnL}\left(B_+e^{ik_2(x-nL-d_1)} + B_-e^{-ik_2(x-nL-d_1)}\right) \tag{2.90b}$$

To find the amplitudes, we use the conditions of continuity of displacement and of stress at the interfaces.

The condition of continuity of displacement at the interface between layer n and layer n − 1 (*i.e.*, location $x = nL$ between segment 1 in layer n and segment 2 in layer n − 1) states:

$$u_1(nL) = e^{iqnL}(A_+ + A_-) = u_2(nL) = e^{iq(n-1)L}\left(B_+e^{ik_2(L-d_1)} + B_-e^{-ik_2(L-d_1)}\right)$$

which reduces to

$$A_+ + A_- = e^{-iqL}\left(B_+e^{ik_2d_2} + B_-e^{-ik_2d_2}\right) \tag{2.91}$$

The stress in a medium "i" is given by $\rho_i c_i^2 \frac{\partial u_i}{\partial x}$ where $\rho_i c_i^2$ is the stiffness of the medium. The continuity of stress at the interface $x = nL$ is expressed:

$$k_1\rho_1c_1^2(A_+ - A_-) = e^{-iqL}k_2\rho_2c_2^2\left(B_+e^{ik_2d_2} - B_-e^{-ik_2d_2}\right) \tag{2.92}$$

Considering now the conditions of continuity of displacement and stress at the interface between media 1 and 2 in the same layer n, *i.e.*, at location $x = nL + d_1$ leads to:

$$A_+e^{ik_1d_1} + A_-e^{-ik_1d_1} = B_+ + B_- \tag{2.93}$$

$$k_1\rho_1c_1^2\left(A_+e^{ik_1d_1} - A_-e^{-ik_1d_1}\right) = k_2\rho_2c_2^2(B_+ - B_-) \tag{2.94}$$

Equations (2.91), (2.92), (2.93) and (2.94) form a system of four linear equations in the amplitudes:

$$\left\{\begin{array}{c} A_+e^{ik_1d_1} + A_-e^{-ik_1d_1} - B_+ - B_- = 0 \\ A_+Fe^{ik_1d_1} - A_-Fe^{-ik_1d_1} - B_+ + B_- = 0 \\ A_+ + A_- - B_+e^{-iqL}e^{ik_2d_2} - B_-e^{-iqL}e^{-ik_2d_2} = 0 \\ A_+F - A_-F - B_+e^{-iqL}e^{ik_2d_2} + B_-e^{-iqL}e^{-ik_2d_2} = 0 \end{array}\right. \tag{2.95}$$

where F is defined as $F = \frac{k_1\rho_1c_1^2}{k_2\rho_2c_2^2}$. This system has nontrivial solutions if the determinant of matrix: $\begin{pmatrix} \alpha_1 & \beta_1 & -1 & -1 \\ F\alpha_1 & -F\beta_1 & -1 & +1 \\ +1 & +1 & -e^{-iqL}\alpha_2 & -e^{-iqL}\beta_2 \\ F & -F & -e^{-iqL}\alpha_2 & +e^{-iqL}\beta_2 \end{pmatrix}$ is equal to zero.

This condition gives the eigen values of the system. In the matrix, we have introduced: $\alpha_i = e^{ik_i d_i} = \frac{1}{\beta_i}$. After a number of algebraic manipulations, the eigen values are obtained from the dispersion relation:

$$\cos qL = \cos k_1 d_1 \cos k_2 d_2 - \frac{1}{2}\left(F + \frac{1}{F}\right) \sin k_1 d_1 \sin k_2 d_2 \tag{2.96}$$

To solve for the Eigen values, we use the approach of transfer matrices. For this, (2.91) and (2.92) can be recast in the form:

$$\begin{pmatrix} 1 & 1 \\ F & -F \end{pmatrix} \begin{pmatrix} A_+ \\ A_- \end{pmatrix}_{n+1} = e^{-iqL} \begin{pmatrix} \alpha_2 & \beta_2 \\ \alpha_2 & -\beta_2 \end{pmatrix} \begin{pmatrix} B_+ \\ B_- \end{pmatrix}_n$$

where the indices n + 1 and n indicate that the amplitudes are in layers n + 1 and n. The preceding equation can be rewritten as:

$$\begin{pmatrix} A_+ \\ A_- \end{pmatrix}_{n+1} = \frac{1}{2F} e^{-iqL} \begin{pmatrix} (F+1)\alpha_2 & (F-1)\beta_2 \\ (F-1)\alpha_2 & (F+1)\beta_2 \end{pmatrix} \begin{pmatrix} B_+ \\ B_- \end{pmatrix}_n \tag{2.97}$$

Equations (2.93) and (2.94) can be reformulated as:

$$\begin{pmatrix} 1 & 1 \\ 1 & -1 \end{pmatrix} \begin{pmatrix} B_+ \\ B_- \end{pmatrix}_n = \begin{pmatrix} \alpha_1 & \beta_1 \\ F\alpha_1 & -F\beta_1 \end{pmatrix} \begin{pmatrix} A_+ \\ A_- \end{pmatrix}_n$$

And recast in the form:

$$\begin{pmatrix} B_+ \\ B_- \end{pmatrix}_n = \frac{1}{2} \begin{pmatrix} (1+F)\alpha_1 & (1-F)\beta_1 \\ (1-F)\alpha_1 & (1+F)\beta_1 \end{pmatrix} \begin{pmatrix} A_+ \\ A_- \end{pmatrix}_n \tag{2.98}$$

In (2.98), both sets of amplitudes are located within a layer n.
Finally, we can insert (2.98) into (2.97) to obtain:

$$\begin{pmatrix} A_+ \\ A_- \end{pmatrix}_{n+1} = \begin{pmatrix} t_{11} & t_{12} \\ t_{21} & t_{22} \end{pmatrix} \begin{pmatrix} A_+ \\ A_- \end{pmatrix}_n \tag{2.99}$$

The 2×2 matrix in (2.99) is the transfer matrix that relates the amplitudes between two adjacent layers. The components of the transfer matrix are given by:

$$t_{11} = \frac{1}{4F}\alpha_1 \left[(F+1)^2 \alpha_2 - (F-1)^2 \beta_2\right] \tag{2.100a}$$

$$t_{22} = -\frac{1}{4F}\beta_1 \left[(F-1)^2 \alpha_2 - (F+1)^2 \beta_2\right] \tag{2.100b}$$

$$t_{12} = -\frac{1}{4F}\beta_1(F+1)(F-1)(\alpha_2-\beta_2) \tag{2.100c}$$

$$t_{21} = \frac{1}{4F}\alpha_1(F+1)(F-1)(\alpha_2-\beta_2) \tag{2.100d}$$

Note that because the modes in the periodic superlattice are Bloch waves, we can write:

$$\begin{pmatrix} A_+ \\ A_- \end{pmatrix}_{n+1} = e^{iqL}\begin{pmatrix} A_+ \\ A_- \end{pmatrix}_n$$

With this condition, (2.99) can be recast in the form of an eigen value problem:

$$\left(\begin{pmatrix} t_{11} & t_{12} \\ t_{21} & t_{22} \end{pmatrix} - e^{iqL}\begin{pmatrix} 1 & 0 \\ 0 & 1 \end{pmatrix}\right)\begin{pmatrix} A_+ \\ A_- \end{pmatrix}_n = 0 \tag{2.101}$$

If e^{iqL} is an eigen value, the determinant of the matrix in the left hand side of (2.101) vanishes and the system of (2.101) reduces to a single equation:

$$\left(t_{11} - e^{iqL}\right)A_+ = -t_{12}A_-$$

which gives the unnormalized eigen vectors:

$$A_+ = -t_{12} \tag{2.102a}$$

$$A_- = t_{11} - e^{iqL} \tag{2.102b}$$

where we express the components of the transfer matrix in the form:

$$t_{11} = e^{ik_1d_1}\left[\cos k_2d_2 + \frac{i}{2}\left(F + \frac{1}{F}\right)\sin k_2d_2\right] \tag{2.103a}$$

and

$$t_{12} = -e^{-ik_1d_1}\left[\frac{i}{2}\left(F - \frac{1}{F}\right)\sin k_2d_2\right] \tag{2.103b}$$

Introducing $e^{\pm ik_1d_1} = \cos k_1d_1 \pm i\sin k_1d_1$ into (2.103a, 2.103b) and using the dispersion relation given by (2.96), one gets the complex amplitudes:

$$A_+ = \frac{1}{2}\left(F - \frac{1}{F}\right)\sin k_1d_1 \sin k_2d_2 + \frac{i}{2}\left(F - \frac{1}{F}\right)\cos k_1d_1 \sin k_2d_2 \tag{2.104a}$$

$$A_{-} = i\left[\sin k_1 d_1 \cos k_2 d_2 + \frac{1}{2}\left(F + \frac{1}{F}\right) \cos k_1 d_1 \sin k_2 d_2 - \sin qL \right] \quad (2.104b)$$

Appendix 2: Discrete One-Dimensional Monatomic Crystal with Spatially Varying Stiffness

We first recall some of the results of Sect. 1.5. The equations of motion of two adjacent odd and even atoms in a chain of identical masses connected by springs with alternately varying stiffness β_1 and β_2:

$$\begin{cases} m\ddot{u}_{2n} = \beta_2(u_{2n+1} - u_{2n}) - \beta_1(u_{2n} - u_{2n-1}) \\ m\ddot{u}_{2n+1} = \beta_1(u_{2n+2} - u_{2n+1}) - \beta_2(u_{2n+1} - u_{2n}) \end{cases}. \quad (2.105)$$

We noted before that the (2.105) break inversion symmetry. Inserting solutions in the form of propagating waves with different amplitudes for odd or even atoms into (2.105):

$$\begin{cases} u_{2n} = Ae^{i\omega t}e^{ik2na} \\ u_{2n+1} = Be^{i\omega t}e^{ik(2n+1)a} \end{cases}. \quad (2.106)$$

leads to the set of two linear equations in the amplitudes A and B:

$$\begin{cases} (\beta_1 + \beta_2 - m\omega^2)A - (\beta_1 e^{-ika} + \beta_2 e^{+ika})B = 0 \\ -(\beta_1 e^{+ika} + \beta_2 e^{-ika})A + (\beta_1 + \beta_2 - m\omega^2)B = 0 \end{cases}. \quad (2.107)$$

Non-trivial solutions (*i.e.*, $\neq 0, B \neq 0$) exist when the following condition is satisfied:

$$(\beta_1 + \beta_2 - m\omega^2)^2 - (\beta_1 e^{-ika} + \beta_2 e^{+ika})(\beta_1 e^{+ika} + \beta_2 e^{-ika}) = 0 \quad (2.108)$$

With the relation

$$(\beta_1 e^{-ika} + \beta_2 e^{+ika})(\beta_1 e^{+ika} + \beta_2 e^{-ika}) = (\beta_1 + \beta_2)^2 - 4\beta_1\beta_2 \sin^2 ka \quad (2.109)$$

and setting, $\alpha = \omega^2$, the condition (2.108) takes the form of the quadratic equation:

$$\alpha^2 - 2\frac{\beta_1 + \beta_2}{m}\alpha + \frac{4\beta_1 + \beta_2}{m^2}\sin^2 ka = 0, \quad (2.110)$$

The Eigen values of the problem are given by the two solutions of (2.110):

$$\omega^2 = \alpha = \frac{\beta_1 + \beta_2}{m} \pm \frac{1}{m}\sqrt{(\beta_1 + \beta_2)^2 - 4\beta_1\beta_2 \sin^2 ka}. \tag{2.111}$$

The solution with a "−" corresponds to the acoustic branch in the band structure. The Brillouin zone spans the values: $ka\,\epsilon\left[-\frac{\pi}{2}, \frac{\pi}{2}\right]$. The solution with a "+" represents the optical branch. We now solve for the Eigen vectors of the system.

We recast (2.111) in the form:

$$\beta_1 + \beta_2 - m\omega^2 = \mp\sqrt{(\beta_1 + \beta_2)^2 - 4\beta_1\beta_2 \sin^2 ka} \tag{2.112}$$

Using (2.112) and (2.109) into (2.107) leads to the complex amplitudes with a real part representing the displacement of the masses:

$$A = \sqrt{\beta_1 e^{-ika} + \beta_2 e^{+ika}} \tag{2.113}$$

$$B = \mp\sqrt{\beta_1 e^{+ika} + \beta_2 e^{-ika}} \tag{2.114}$$

The $\mp$ refers to the optical branch (−) and the acoustic branch (+) in the band structure. We define $\beta_1 = \beta - \Delta$ and $\beta_2 = \beta + \Delta$, when $\Delta > 0$ then $\beta_1 < \beta_2$. With $\Delta < 0$, we have $\beta_1 > \beta_2$ and the complex amplitudes become

$$A = \sqrt{2\beta \cos ka + i2\Delta \sin ka} \tag{2.115}$$

$$B = \mp\sqrt{2\beta \cos ka - i2\Delta \sin ka} \tag{2.116}$$

To get more insight into the phase, we evaluate the complex amplitude A at $ka = \frac{\pi}{2}$ and $ka = -\frac{\pi}{2}$:

$$A\left(\frac{\pi}{2}\right) = \sqrt{i2\Delta} = \sqrt{\Delta}(1 + i) \tag{2.117}$$

and

$$A\left(-\frac{\pi}{2}\right) = \sqrt{-i2\Delta} = -\sqrt{\Delta}(1 - i) \tag{2.118}$$

To obtain the preceding expressions, we have used $\sqrt{i} = \frac{1}{\sqrt{2}}(1 + i)$.

From (2.117) and (2.118), we note that

$$A\left(-\frac{\pi}{2}\right) = iA\left(\frac{\pi}{2}\right) \tag{2.119}$$

independently of the sign of Δ. The phase difference between the amplitude at the band edges, $A\left(-\frac{\pi}{2}\right)$ and $A\left(\frac{\pi}{2}\right)$, is therefore: $\varphi = \frac{\pi}{2}$. We also need to calculate the continuous phase change of A as the wave number spans the Brillouin zone. For this, we use the Berry connection:

$$BC(k) = -i\widehat{A}^*(k)\frac{\partial \widehat{A}(k)}{\partial k} \tag{2.120}$$

With the normalized amplitude defined by:

$$\widehat{A}(k) = \frac{A}{\sqrt{A^*A}} = \sqrt{\frac{\beta \cos ka + i\Delta \sin ka}{\sqrt{\beta^2 \cos^2 ka + \Delta^2 \sin^2 ka}}} \tag{2.121}$$

After numerous algebraic steps, one finds:

$$BC(k) = \frac{1}{2}\frac{\beta\Delta}{\beta^2 \cos^2 ka + \Delta^2 \sin^2 ka} a \tag{2.122}$$

We now integrate the Berry connection over the Brillouin zone, however, in light of the relation (2.119), we exclude one of the band edge and the interval of integration is taken as $\left[-\frac{\pi}{2a} + \varepsilon, \frac{\pi}{2a}\right]$. The Berry phase is then obtained as:

$$\begin{aligned} \eta &= \lim_{\varepsilon \to 0} \frac{1}{2} \int_{-\frac{\pi}{2a}+\varepsilon}^{\frac{\pi}{2a}} \frac{\beta\Delta}{\beta^2 \cos^2 ka + \Delta^2 \sin^2 ka} a dk \\ &= \lim_{\varepsilon \to 0} \frac{1}{2} \int_{-\frac{\pi}{2}+\varepsilon'}^{\frac{\pi}{2}} \frac{\beta\Delta}{\beta^2 \cos^2 \theta + \Delta^2 \sin^2 \theta} d\theta \end{aligned} \tag{2.123}$$

For positive Δ, we obtain $\eta = \frac{\pi}{2}$. For negative Δ, $\eta = -\frac{\pi}{2}$. This result can be visualized analytically in the particular case of small Δ as follows. The integrand of (2.123) is rewritten in the form:

$$\frac{\beta\Delta}{\beta^2 \cos^2 \theta + \Delta^2 \sin^2 \theta} = \frac{1}{|\sin \theta|} \times sgn(\Delta) \frac{\beta|\Delta||\sin \theta|}{\beta^2 \cos^2 \theta + \Delta^2 \sin^2 \theta} \tag{2.124}$$

In the limit $|\Delta||\sin\theta| \to 0^+$, the last term on the right-hand side of the preceding equation becomes:

$$\lim_{|\Delta||\sin \theta| \to 0^+} \frac{|\Delta||\sin \theta|}{\beta^2 \cos^2 \theta + \Delta^2 \sin^2 \theta} = \pi\delta(\beta \cos \theta) = \pi\frac{1}{\beta}\delta(\cos \theta) \tag{2.125}$$

In the interval $\theta \in \left[-\frac{\pi}{2}, \frac{\pi}{2}\right]$, the function $\cos\theta$ has only one zero at $\theta = \frac{\pi}{2}$, so $\delta(\cos \theta) = \frac{1}{|-\sin \theta|_{\theta=\frac{\pi}{2}}}\delta\left(\theta - \frac{\pi}{2}\right)$. The limit of (2.125) is now written as:

$$\lim_{|\Delta||\sin\theta|\to 0^+} \frac{\beta|\Delta||\sin\theta|}{\beta^2\cos^2\theta + \Delta^2\sin^2\theta} = \pi\delta\left(\theta - \frac{\pi}{2}\right) \tag{2.126}$$

Equation (2.124) simplifies to:

$$\frac{\beta\Delta}{\beta^2\cos^2\theta + \Delta^2\sin^2\theta} \to \frac{1}{|\sin\theta|} \times sgn(\Delta)\pi\delta\left(\theta - \frac{\pi}{2}\right) \tag{2.127}$$

Inserting (2.127) into (2.123) yields:

$$\eta = \lim_{\varepsilon'\to 0}\frac{1}{2}\int_{-\frac{\pi}{2}+\varepsilon'}^{\frac{\pi}{2}} \frac{1}{|\sin\theta|} \times sgn(\Delta)\pi\delta\left(\theta - \frac{\pi}{2}\right)d\theta = sgn(\Delta)\frac{\pi}{2}$$

The total phase change, *i.e.*, the Zak phase, $\eta_Z = \eta + \varphi$ which is equal to π if $\Delta > 0$ and 0 if $\Delta < 0$.

Note that this result is dependent on the choice of origin. For instance, consider Fig. 2.9 below.

In the system (a), the Zak phase is non-zero when $\beta_1 < \beta_2$. With a change of origin of a [system (b)], the Zak phase is non-zero when $\beta_1 > \beta_2$. The change of phase with the choice of a gauge origin is also familiar from classical electrodynamics where the gauge origin of the vector potential, whose curl is the magnetic field, is somewhat arbitrary. This is not to be confused with gauge transformations which address the fact that the curl of the vector potential is not affected by an arbitrary gradient of a scalar function.

Appendix 3: Introduction to Green's Function Formalism

Green's Function and the Dyson Equation

In this section, we briefly review the formalism of Green's function and derive a number of mathematical identities that will be useful in shedding light on the nature of elastic waves and in particular that will be useful in sections relating Green's functions and the phase of waves. Some of these identities may have been already

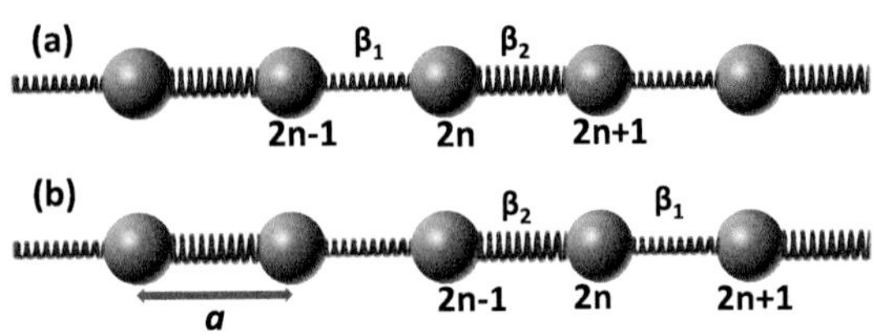

Fig. 2.9 Illustration of the effect of the choice of origin on the Zak phase. The lattice (**b**) is the same as (**a**) but for a translation of origin by one inter-mass spacing a

introduced in Chap. 1, but are repeated in this chapter for pedagogical reasons with the aim of making individual chapters nearly self-contained.

Let us consider an eigen value problem for a reference system "0" described by the following equation:

$$(EI - H_0)\varphi = 0 \tag{2.128}$$

where H_0 is some Hermitian operator (taking the form of a discrete matrix or a continuous operator), E represents the eigen values, I is the identity operator (identity or delta function), and φ represents the eigen functions (or vectors).

One defines the Green's function, G_0 by rewriting (2.128) in the form:

$$(EI - H_0)G_0 = I \tag{2.129}$$

This function is given as

$$G_0 = \frac{I}{(EI - H_0)} \tag{2.130}$$

Let us now consider a physical perturbation, V, of the system "0" which is now represented by the operator

$$H = H_0 + V \tag{2.131}$$

The eigen value problem *i.e.*, $(EI - H)\psi = 0$ is now restated in terms of the perturbed operator

$$(EI - H_0)\psi = V\psi \tag{2.132}$$

The solution of this equation is the sum of the solution of (2.128) in absence of perturbation and of a particular solution with the perturbed term on the right hand side of (2.132), namely:

$$\psi = \varphi + (EI - H_0)^{-1}V\psi \tag{2.133}$$

Using (2.130), this solution becomes:

$$\psi = \varphi + G_0V\psi \tag{2.134}$$

From (2.130), one sees that the eigen values of H_0 are given by the poles of G_0. If E is such that there are no solutions with $\varphi \neq 0$ to (2.133), (that is, if we impose $\varphi = 0$) then we have:

$$(I - G_0 V)\psi = 0 \tag{2.135}$$

The eigen values associated with ψ are given by:

$$\det(I - G_0(E)V) = 0 \tag{2.136}$$

where “det” refers to the determinant of the matrix operators.

Equation (2.136) enables us to obtain the new eigen values of the perturbed operator, H, which are not also eigen values of the unperturbed operator, H_0. The Green’s function associated with H can be defined as

$$G(E) = (EI - H)^{-1} = \frac{I}{EI - H_0 - V} \tag{2.137}$$

Starting from (2.137) and defining $A = -V$ and $B = EI - H_0$, we can use the matrix identity: $\frac{I}{A+B} = \frac{I}{A} - \frac{I}{A} B \frac{I}{A+B}$ to show that

$$G(E) = (I - G_0 V)^{-1} G_0 \tag{2.138}$$

Equation (2.138) is called the Dyson equation. It states that the poles of $G(E)$ which are not also poles of G_0 are given by (2.136).

Green’s Function and Density of States

By definition, the density of states of a system is the number of states with the same eigen value. We will label the states by some wavenumber, k. The density of states, associated with the eigen value E, of the unperturbed system is written as $n_0(E)$. We have:

$$n_0(E) = \sum_k \delta(E - E(k)) \tag{2.139}$$

In (2.139), we have used the delta function defined by: $\delta(x) = \begin{cases} 0 \; if \; x \neq 0 \\ \infty if \; x = 0 \end{cases}$. We also recall that $\int_{-\infty}^{+\infty} \delta(x)dx = 1$.

For matrix operators, the density of states given by (2.139) is rewritten in the form:

$$n_0(E) = Tr\delta(EI - H_0) \tag{2.140}$$

where Tr stands for the trace of a matrix.

We will now try to express $n_0(E)$ in terms of $G_0(E)$. Let us first consider the analytical continuation of the Green’s function in the complex plane:

$$G_0^+ = \lim_{\varepsilon \to 0^+} \frac{I}{EI - H_0 + i\varepsilon} \tag{2.141}$$

To progress further, we first consider the mathematical relation: $\frac{1}{x+i\varepsilon} = \frac{x}{x^2+\varepsilon^2} - i\frac{\varepsilon}{x^2+\varepsilon^2}$. In the limit of $\varepsilon \to 0$, the real part of the preceding relation is by definition the Cauchy's principle part, namely: $\lim\limits_{\varepsilon \to 0} \frac{x}{x^2+\varepsilon^2} = PP\left(\frac{1}{x}\right)$ and the imaginary term gives (Sokhotski's formula [44]): $\lim\limits_{\varepsilon \to 0} \frac{\varepsilon}{x^2+\varepsilon^2} = \pi\delta(x)$. It is instructive to verify the second identity by considering: $\delta(x) = \frac{1}{\pi} \lim\limits_{\varepsilon \to 0} \frac{\varepsilon}{x^2+\varepsilon^2}$ and integrating over all x. This gives: $\int_{-\infty}^{+\infty} \delta(x)dx = \frac{1}{\pi} \lim\limits_{\varepsilon \to 0^+} \int_{-\infty}^{+\infty} \frac{\varepsilon}{x^2+\varepsilon^2} = \frac{1}{\pi} \lim\limits_{\varepsilon \to 0^+} \tan^{-1}\frac{x}{\varepsilon}\Big|_{-\infty}^{+\infty} = 1$. With these identities, (2.141) can be rewritten as:

$$G_0^+ = PP\left(\frac{I}{EI - H_0}\right) - i\pi\delta(EI - H_0) \tag{2.142}$$

We have

$$ImG_0^+ = -\pi\delta(EI - H_0) \tag{2.143}$$

where Im stands for the imaginary part of G_0^+. Combining (2.140) and (2.143), one obtains the relation for the density of states:

$$n_0(E) = -\frac{1}{\pi} ImTrG_0^+(E) \tag{2.144}$$

The previous relation can be further modified by utilizing the properties of diagonalizable matrices. Let R be such a matrix, then: $ln\, detR = TrlnR$. Indeed, if we diagonalize R into R^0 then $detR = detR^0 = \prod_i R_{ii}^0$ and therefore: $lndetR = lndetR^0 = ln\prod_i R_{ii}^0 = \sum_i lnR_{ii}^0$. In light of the invariance of the trace and the determinant, one obtains: $ln\, detR = ln\, det\, R^0 = TrlnR$.

We also recall that any Green's function, G, satisfies the relation:

$$TrG = -\frac{d}{dE} lndetG$$

This can be demonstrated via the following identities: $-\frac{d}{dE} lndetG = -\frac{d}{dE} TrlnG = -Tr\frac{d}{dE} lnG = -Tr\frac{1}{G}\frac{dG}{dE} = -Tr\frac{1}{G}\left(-G^2\right) = TrG$. Here we have used the fact that $G = (EI - H)^{-1}$ (2.137).

We can finally show that the density of states given by (2.144) becomes

$$n_0(E) = +\frac{1}{\pi} Im \frac{d}{dE} ln det G_0^+(E) \tag{2.145}$$

A similar expression can be derived for the density of states of the perturbed system:

$$n(E) = +\frac{1}{\pi} Im \frac{d}{dE} ln det G^+(E) \tag{2.146}$$

When comparing the perturbed system with the unperturbed system, one can calculate the variation in density of states:

$$\Delta n(E) = n(E) - n_0(E) = \frac{1}{\pi} Im \frac{d}{dE} ln \frac{det G^+(E)}{det G_0^+(E)} \tag{2.147}$$

The Dyson equation (2.138) can now be used to show that

$$\Delta n(E) = \frac{1}{\pi} Im \frac{d}{dE} ln det\left(I - V G_0^+(E)\right) \tag{2.148}$$

If we define $det\left(I - V G_0^+(E)\right) = \rho e^{i\Delta\theta}$ where $\rho(E)$ and $\Delta\theta(E)$ are the modulus and the phase of $det\left(I - V G_0^+(E)\right)$, then $Im \frac{d}{dE} ln det\left(I - V G_0^+(E)\right) = \Delta\theta$ and the variation in density of states is

$$\Delta n(E) = -\frac{1}{\pi} \frac{d\Delta\theta(E)}{dE} \tag{2.149}$$

where $\Delta\theta(E) = arg det\left(I - G_0^+(E) V\right)$. $\Delta\theta$ is a difference in phase between the perturbed system and the unperturbed system.

References

1. I. Newton, *Principia—Book II, Imprimatur* (S. Pepys, Reg. Soc. Præses, London, 1686)
2. J.W.S. Rayleigh, *The Theory of Sound*, vol. 1 (Dover, New York, two vols., 1877–1878)
3. F. Schwabel, *Advanced Quantum Mechanics*, 4th edn. (Springer, Berlin, 2008)
4. P. A. Deymier (ed.), *Acoustic Metamaterials and Phononic Crystals*, Springer Series in Solid State Sciences 173 (Springer, Heidelberg, 2013)
5. M.Z. Hasan, C.L. Kane, Colloquium: topological insulators. Rev. Mod. Phys. **82**, 3045–3067 (2010)
6. A.B. Khanikaev, S.H. Mousavi, W.-K. Tse, M. Kargarian, A.H. MacDonald, G. Shvets, Photonic topological insulators. Nat. Mater. **12**, 233–239 (2013)
7. M.C. Rechtsman, J.M. Zeuner, Y. Plotnik, Y. Lumer, D. Podolsky, F. Dreisow, S. Nolte, M. Sergev, A. Szameit, Photonic Floquet topological insulators. Nature **496**, 196–200 (2013)
8. F.D.M. Haldane, S. Raghu, Possible realization of directional optical waveguides in photonic crystals with broken time-reversal symmetry. Phys. Rev. Lett. **100**, 013904 (2008)

9. P.A. Deymier, K. Runge, N. Swinteck, K. Muralidharan, Torsional topology and fermion-like behavior of elastic waves in phononic structures. C. R. Méc. **343**, 700–711 (2015)
10. P.A. Deymier, K. Runge, N. Swinteck, K. Muralidharan, Rotational modes in a phononic crystal with fermion-like behaviour. J. Appl. Phys. **115**, 163510 (2014)
11. E. Prodan, C. Prodan, Topological phonon modes and their role in dynamic instability of microtubules. Phys. Rev. Lett. **103**, 248101 (2009)
12. C.L. Kane, T.C. Lubensky, Topological boundary modes in isostatic lattices. Nat. Phys. **10**, 39–45 (2013)
13. S. Mousavi, A.B. Khanikaev, Z. Wang, Topologically protected elastic waves in phononic metamaterials. Nat. Commun. **6**, 8682 (2015)
14. B.G. Chen, N. Upadhyaya, V. Vitelli, Nonlinear conduction via solitons in a topological mechanical insulator. Proc. Natl. Acad. Sci. U.S.A. **111**, 13004–13009 (2014)
15. R. Süsstrunk, S.D. Huber, Observation of phononic helical edge states in a mechanical topological insulator. Science **349**, 47–50 (2015)
16. M. Xiao, G. Ma, Z. Yang, P. Sheng, Z.Q. Zhang, C.T. Chan, Geometric phase and band inversion in periodic acoustic systems. Nat. Phys. **11**, 240–244 (2015)
17. J. Paulose, B.G. Chen, V. Vitelli, Topological modes bound to dislocations in mechanical metamaterials. Nat. Phys. **11**, 153–156 (2015)
18. N. Berg, K. Joel, M. Koolyk, E. Prodan, Topological phonon modes in filamentary structures. Phys. Rev. E **83**, 021913 (2011)
19. R.K. Pal, M. Schaeffer, M. Ruzzene, Helical edge states and topological phase transitions in phononic systems using bi-layered lattices. J. Appl. Phys. **119**, 084305 (2016)
20. A.B. Khanikaev, R. Fleury, S.H. Mousavi, A. Alù, Topologically robust sound propagation in an angular-momentum-biased graphene-like resonator lattice. Nat. Commun. **6**, 8260 (2015)
21. G. Salerno, T. Ozawa, H.M. Price, I. Carusotto, Floquet topological system based on frequency-modulated classical coupled harmonic oscillators. Phys. Rev. B **93**, 085105 (2015)
22. J. Paulose, A.S. Meeussen, V. Vitelli, Selective buckling via states of self-stress in topological metamaterials. Proc. Natl. Acad. Sci. U.S.A. **112**, 7639–7644 (2015)
23. L.M. Nash, D. Kleckner, A. Read, V. Vitelli, A.M. Turner, W.T.M. Irvine, Topological mechanics of gyroscopic metamaterials. Proc. Natl. Acad. Sci. U.S.A. **112**, 14495–14500 (2015)
24. P. Wang, L. Lu, K. Bertoldi, Topological phononic crystals with one-way elastic edge waves. Phys. Rev. Lett. **115**, 104302 (2015)
25. Z. Yang, F. Gao, X. Shi, X. Lin, Z. Gao, Y. Chong, B. Zhang, Topological acoustics. Phys. Rev. Lett. **114**, 114301 (2015)
26. N. Swinteck, S. Matsuo, K. Runge, J.O. Vasseur, P. Lucas, P.A. Deymier, Bulk elastic waves with unidirectional backscattering-immune topological states in a time-dependent superlattice. J. Appl. Phys. **118**, 063103 (2015)
27. M.V. Berry, Quantal phase factors accompanying adiabatic changes. Proc. R. Soc. A **392**, 45–57 (1984)
28. M.P. Hobson, G. Estathiou, A.N. Lasenby, *General Relativity – An Introduction for Physicists* (Cambridge University Press, Cambridge, 2006)
29. J. Zak, Berry's phase for energy bands in solids. Phys. Rev. Lett. **62**, 2747 (1989)
30. R.E. RCamley, B. Djafari-Rouhani, L. Dobrzynski, A.A. Maradudin, Transverse elastic waves in periodically layered infinite and semi-infinite media. Phys. Rev. B **27**, 7318 (1983)
31. B. Djafari-Rouhani, L. Dobrzynski, O. Hardouin Duparc, R.E. Camley, A.A. Maradudin, Sagittal elastic waves in infinite and semi-infinite superlattices. Phys. Rev. B **28**, 1711 (1983)
32. E.N. Economou, *Green's Functions in Quantum Physics* (Springer, New York, 1990)
33. J. Friedel, The distribution of electrons round impurities in monovalent metals. Philos. Mag. **43**, 153 (1952)
34. H.-W. Lee, Generic transmission zeros and in-phase resonances in time-reversal symmetric single channel transport. Phys. Rev. Lett. **82**, 2358 (1999)

35. A. Yacoby, M. Heiblum, D. Mahalu, H. Shtrikman, Coherence and phase sensitive measurements in a quantum dot. Phys. Rev. Lett. **74**, 4047 (1995)
36. T. Taniguchi, T. Buttuker, Friedel phases and phases of transmission amplitudes in quantum scattering systems. Phys. Rev. B **60**, 13814 (1999)
37. L. Dobrzynski, Interface response theory of composite systems. Surf. Sci. **200**, 435 (1988)
38. L. Dobrzynski, J. Mendialdua, A. Rodriguez, S. Bolibo, M. More, Interface response theory of composite elastic media. J. Phys. **50**, 2563 (1989)
39. M. Trigo, T.A. Eckhause, M. Reason, R.S. Goldman, R. Merlin, Observation of surface-avoiding waves: a new class of extended states in periodic media. Phys. Rev. Lett. **97**, 124301 (2006)
40. R. Merlin, Raman scattering by surface-avoiding acoustic phonons in semi-infinite superlattices. Philos. Mag. Part B **70**, 761 (1994)
41. C. Kane, T. Lubensky, Topological boundary modes in isostatic lattices. Nat. Phys. **10**, 39 (2014)
42. T. Lubensky, C. Kane, X. Mao, A. Souslov, K. Sun, Phonons and elasticity in critically coordinated lattices. Rep. Prog. Phys. **78**, 073901 (2015)
43. S.D. Huber, Topological mechanics. Nat. Phys. **12**, 621 (2016)
44. V.S. Vladimirov, *Equations of Mathematical Physics* (Marcel Dekker, New York, 1971)

Chapter 3
Topology and Duality of Sound and Elastic Waves

3.1 Introduction

When sound waves propagate in media under symmetry breaking conditions, they may exhibit amplitudes $A(k) = A_0 e^{i\eta(k)}$ that acquire a geometric phase η leading to non-conventional topology. In the previous chapter, we considered the consequences of breaking inversion symmetry in discrete superlattices. Here, we present examples of phononic structures that break four types of symmetry, namely time-reversal symmetry, parity symmetry, chiral symmetry and particle-hole symmetry. The implications of symmetry breaking on the topology of the acoustic wave function in the space of its Eigen values are discussed. Particular attention is focused on the torsional topology of acoustic waves in periodic media in wave vector space. Two types of approach to achieve symmetry breaking are considered: (a) intrinsic topological phononic structures whereby symmetry breaking occurs from the internal structural characteristics, and (b) extrinsic topological phononic structures where external stimuli such as spatio-temporal modulations of the physical properties of the medium are used to break symmetry. Broken symmetry phenomena lead to the concept of symmetry protected topological order. Broken symmetry also underpins the concepts of non-reciprocal wave propagation. Topological acoustic waves promise designs and new device functionalities for acoustic systems that are unique, robust and avoid the loss of coherence. Furthermore, the self-interaction of an elastic wave in topological phononic structures creates states determined by self-interference phenomena, revealing a quantum mechanical analogy with the concept of particle–wave duality. In the phonon representation of sound and elastic waves, these self-interference phenomena uncover the notion of duality in the quantum statistics (*i.e.*, boson *vs.* fermion characterized by the symmetry of multiple particle states). Furthermore, we also consider the partitioning the phononic structures presented in this chapter into subsystems and the separability and non-separability of their wave functions into tensor products. Separability is shown to be relative to the choice of the subsystems in which one

P. Deymier, K. Runge, *Sound Topology, Duality, Coherence and Wave-Mixing*,
Springer Series in Solid-State Sciences 188, DOI 10.1007/978-3-319-62380-1_3

partitions the system of interest. The choice of the subsystems in turn may be dictated by the possible observables and measurements. The non-separability of these phononic structures is analogous to the quantum phenomenon of entanglement. Analogies with quantum phenomena interrogate mechanical waves in ways thought to be reserved for the microscopic realm.

3.2 Overview

A new frontier in wave propagation involves media that have broken time-reversal symmetry associated with non-conventional topology. Topological electronic [1], electromagnetic [2, 3], and phononic crystals [4–20] all have demonstrated unusual topologically constrained properties. In phononic crystals and acoustic metamaterials, symmetry breaking is linked to constraints on the topological form of acoustic wave functions. For instance, in the context of topology, for the well-known driven damped oscillator, we have seen in Chap. 2 that the amplitude of the wave function has properties isomorphic to the evolution of a field of parallel vectors tangent to a strip-like manifold with a twist and perpendicular to the length of the strip. The direction along the length of the strip represents frequency space and the strip has to exhibit a twist (vectors in the vector field change orientation) at the oscillator resonant frequency as the amplitude changes sign as one crosses the resonance (*i.e.*, the amplitude accumulates a π-phase shift). Dissipation aside, one of the most central elements to symmetry breaking and topology of elastic waves, is dispersion. The simple, linear one-dimensional (1D) harmonic monatomic crystal discussed in Chaps. 1 and 2 is a dispersive system, but one that obeys time-reversal symmetry and supports elastic waves with conventional topology along its dispersion curve. Perturbing the 1D harmonic crystal through linear or nonlinear coupling may create resonant phonon modes that are dispersive, but whose amplitude may depend on the frequency and wave vector. In this case, the interplay between the coupling and dispersion of the system may lead to symmetry breaking conditions and therefore non-conventional elastic wave topology. There exist two classes of phonon structures possessing non-conventional topology, namely intrinsic and extrinsic systems. Time-reversal symmetry in intrinsic systems [4–14] is broken through internal resonance or symmetry breaking structural features (*e.g.*, chirality) and without addition of energy from the outside. Energy is added to extrinsic topological systems to break time reversal symmetry [15–20]. A common example of an extrinsic approach is that of time-reversal symmetry breaking of acoustic waves by moving fluids [21–29]. Recently, extrinsic topological phononic crystals have demonstrated the astonishing property of non-reciprocity and backscattering-immune edge states and bulk states establishing classical equivalents of topological electronic insulators [1]. The non-conventional topology of elastic waves in an intrinsic topological phononic structure has been associated with the notion of duality in the quantum statistics of phonons (*i.e.*, boson vs. fermion) [4, 5]. In this chapter, we illustrate the topological properties of elastic waves in the two classes of topological phononic structures. We first uncover the spinor characteristics of elastic waves in crystals composed of connected masses and springs (Sect. 3). We also consider, in that section, the fermion-like characteristics of elastic waves with spinor characteristics. An externally applied spatio-temporal modulation of the spring stiffness in mass-spring systems can also be employed to break the symmetry of the system [30]. We illustrate this effect in the case of the 1D harmonic crystal with spatio-temporal modulation of the spring stiffness in Sect. 4. Finally, in Sect. 5, we combine the intrinsic and extrinsic approaches and show that a spatio-temporal modulation is able to tune the spinor part of the elastic wave function and therefore its topology.

3.3 Intrinsic Topological Phononic Structures

3.3.1 Two Coupled Mass-Spring Harmonic Crystals in the Long-Wavelength Limit

Let us consider a system composed of two coupled one-dimensional harmonic crystals as illustrated in Fig. 3.1.

In absence of external forces, the equations describing the motion of atoms at location n in the two coupled 1-D harmonic crystals are given by:

$$m\frac{\partial^2 u_n}{\partial t^2} = K_0(u_{n+1} - 2u_n + u_{n-1}) + K_I(v_n - u_n) = 0 \tag{3.1a}$$

$$M\frac{\partial^2 v_n}{\partial t^2} = K_0(v_{n+1} - 2v_n + v_{n-1}) - K_I(v_n - u_n) = 0 \tag{3.1b}$$

Here u and v represent the displacement in the upper and lower chains, respectively. These displacements can be visualized as being oriented along the chains. The side springs are illustrated for the sake of simplicity as vertical spring but

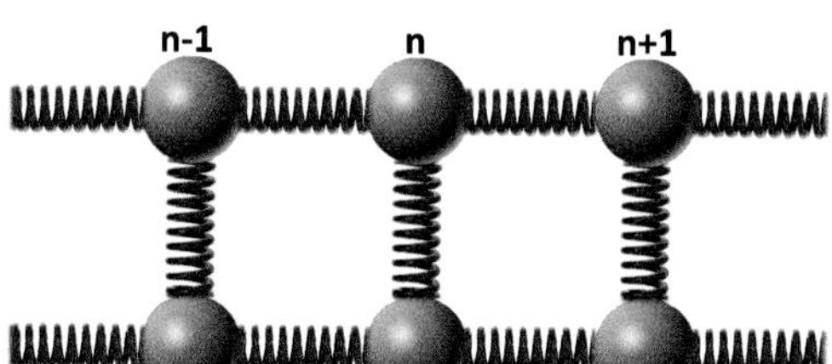

Fig. 3.1 Schematic illustration of the phononic structure composed of two coupled 1-D harmonic crystals. The atoms in the *lower* and *upper* 1-D harmonic crystals have mass m and M, respectively. The force constant of the springs of each 1-D harmonic crystal is taken to be the same, K_0. The force constant of the coupling springs is K_I. The periodicity of the crystal is h

physically they would couple masses between chains along the direction of the displacements.

In the long wavelength limit the discrete Lagrangian is expressed as a continuous second derivative of position. Taking $M = m$ for the sake of simplicity and mathematical tractability, the equations of motion (3.1a, 3.1b) can be rewritten as:

$$\left\{\left(\frac{\partial^2}{\partial t^2} - \beta^2 \frac{\partial^2}{\partial x^2}\right)\boldsymbol{I} + \alpha^2 \boldsymbol{D}\right\}\boldsymbol{U} = 0 \tag{3.2}$$

where $\boldsymbol{I}$ is the 2 × 2 identity matrix, $\boldsymbol{D} = \begin{pmatrix} 1 & -1 \\ -1 & 1 \end{pmatrix}$ and $U = \begin{pmatrix} u \\ v \end{pmatrix}$ is the displacement vector. We also have defined $\alpha^2 = \frac{K_I}{m}$ and $\beta^2 = \frac{K_0}{m}$. Equation (3.2) takes a form similar to the Klein–Gordon equation. Using an approach paralleling that of Dirac, one can factor (3.2) into the following form:

$$\left(\left[\boldsymbol{A}\frac{\partial}{\partial t} + \beta \boldsymbol{B}\frac{\partial}{\partial x}\right] - i\frac{\alpha}{\sqrt{2}}C\right)\left(\left[\boldsymbol{A}\frac{\partial}{\partial t} + \beta \boldsymbol{B}\frac{\partial}{\partial x}\right] + i\frac{\alpha}{\sqrt{2}}\boldsymbol{C}\right)\Psi = 0$$

In this equation, we have introduced the 4×4 matrices:

$$\boldsymbol{A} = \begin{pmatrix} 0 & 0 & 1 & 0 \\ 0 & 0 & 0 & 1 \\ 1 & 0 & 0 & 0 \\ 0 & 1 & 0 & 0 \end{pmatrix}, \boldsymbol{B} = \begin{pmatrix} 0 & 0 & 0 & 1 \\ 0 & 0 & 1 & 0 \\ 0 & -1 & 0 & 0 \\ -1 & 0 & 0 & 0 \end{pmatrix}, \boldsymbol{C} = \begin{pmatrix} 1 & -1 & 0 & 0 \\ -1 & 1 & 0 & 0 \\ 0 & 0 & 1 & -1 \\ 0 & 0 & -1 & 1 \end{pmatrix}.$$

The symbol i refers to $\sqrt{-1}$. The wave functions $\Psi = \begin{pmatrix} \psi_1 \\ \psi_2 \\ \psi_3 \\ \psi_4 \end{pmatrix}$ and $\overline{\Psi} = \begin{pmatrix} \overline{\psi}_1 \\ \overline{\psi}_2 \\ \overline{\psi}_3 \\ \overline{\psi}_4 \end{pmatrix}$ are 4-vector solutions of

$$\left(\left[\boldsymbol{A}\frac{\partial}{\partial t} + \beta \boldsymbol{B}\frac{\partial}{\partial x}\right] - i\frac{\alpha}{\sqrt{2}}\boldsymbol{C}\right)\Psi = 0 \tag{3.3a}$$

and

$$\left(\left[\boldsymbol{A}\frac{\partial}{\partial t} + \beta \boldsymbol{B}\frac{\partial}{\partial x}\right] + i\frac{\alpha}{\sqrt{2}}\boldsymbol{C}\right)\overline{\Psi} = 0 \tag{3.3b}$$

Ψ and $\overline{\Psi}$ are non-self-dual solutions. The equations (3.3a, 3.3b) do not satisfy time-reversal symmetry ($t \to -t$), T-symmetry, nor parity symmetry ($x \to -x$), P-symmetry, separately. That is, one does not recover the set of (3.3a) when the sign of time and position are changed individually. However, one notes that if the sign of t and x are changed simultaneously then, to within a phase, (3.3a) becomes (3.3b) and vice versa. Equations (3.3a, 3.3b) have simultaneous T-symmetry and P-symmetry or simply PT-symmetry.

For the sake of illustration, let us now seek solutions of (3.3a) $\left(\left[A\frac{\partial}{\partial t}+\beta B\frac{\partial}{\partial x}\right]-i\frac{\alpha}{\sqrt{2}}C\right)\Psi=0$ in the plane wave form: $\psi_j=a_je^{\pm ikx}e^{\pm i\omega t}$ with $j=1,2,3,4$. In the all positive case, $\psi_j=a_je^{+ikx}e^{+i\omega t}$ this form for the solution gives the eigen value problem:

$$\begin{cases} -\delta a_1+\delta a_2+\omega a_3+\beta k a_4=0 \\ \delta a_1-\delta a_2+\beta k a_3+\omega a_4=0 \\ \omega a_1-\beta k a_2-\delta a_3+\delta a_4=0 \\ -\beta k a_1+\omega a_2+\delta a_3-\delta a_4=0 \end{cases} \tag{3.4}$$

where $\delta=\frac{\alpha}{\sqrt{2}}$. We find two dispersion relations: $\omega=\pm\beta k$ and $\omega=\pm\sqrt{(\beta k)^2+2\alpha^2}$. The first set of dispersion relations corresponds to branches that start at the origin $k=0$ and relates to symmetric eigen modes. The second set of branches represents antisymmetric modes with a cut off frequency at $\mathrm{k}=0$ of $\alpha\sqrt{2}$. These dispersion curves are illustrated in Fig. 3.2 as part of the band structure of the discrete two-chain model.

Assuming that $a_1=a_2=a_F$ and that $a_3=a_4=a_B$, for the symmetric waves characterized by the dispersion relation, $\omega=\pm\beta k$, then the (3.4) reduce to two independent equations $(\omega+\beta k)a_B=0$ and $(\omega-\beta k)a_F=0$ which are satisfied by plane waves of arbitrary amplitudes, a_F and a_B, propagating in the forward (F) direction with $\omega=+\beta k$ or backward (B) direction with $\omega=-\beta k$, respectively. This is the conventional character of boson-like phonons. This bosonic behavior is even clearer when one considers the interchange the indices 1 and 3 and 2 and 4, which is equivalent to interchanging the top chain and the bottom chain in the coupled system. This leads to $\begin{pmatrix} a_3 \\ a_4 \\ a_1 \\ a_2 \end{pmatrix}=\begin{pmatrix} a_1 \\ a_2 \\ a_3 \\ a_4 \end{pmatrix}$ which is the original solution.

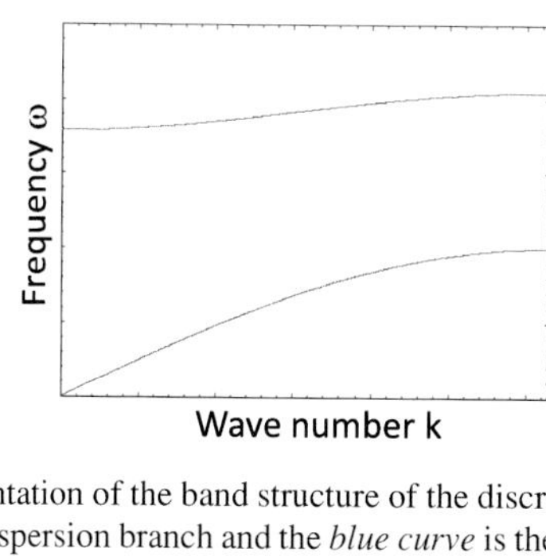

Fig. 3.2 Schematic representation of the band structure of the discrete coupled two-chain system. The *red curve* is the *upper* dispersion branch and the *blue curve* is the *lower* dispersion branch. The dispersion relations $\omega=+\beta k$ and $\omega=+\sqrt{(\beta k)^2+2\alpha^2}$ correspond to the long-wavelength of these dispersion relations

The solution of the coupled two-chain system is symmetric upon exchange of the chains.

We now look for the eigen vectors that correspond to the second dispersion relation. Let us use the positive eigen value as an illustrative example: $\omega = +\sqrt{(\beta k)^2 + 2\alpha^2}$. Another solution of (3.4) can be found when $a_1 = -a_3$ and $a_2 = -a_4$. Inserting $2\delta = +\sqrt{\omega^2 - \beta^2 k^2}$ (the + sign is chosen because δ represents a stiffness), into (3.4) yields:

$$\begin{cases} (\omega + \beta k)a_4 = \sqrt{\omega^2 - \beta^2 k^2} a_1 \\ (\omega - \beta k)a_3 = \sqrt{\omega^2 - \beta^2 k^2} a_2 \\ (\omega + \beta k)a_2 = \sqrt{\omega^2 - \beta^2 k^2} a_3 \\ (\omega - \beta k)a_1 = \sqrt{\omega^2 - \beta^2 k^2} a_4 \end{cases} \tag{3.5}$$

Substituting, $\omega + \beta k = \sqrt{\omega + \beta k}\sqrt{\omega + \beta k} = \sqrt{+}\sqrt{+}$, $\omega - \beta k = \sqrt{\omega - \beta k}\sqrt{\omega - \beta k} = \sqrt{-}\sqrt{-}$, and $\sqrt{\omega^2 - \beta^2 k^2} = \sqrt{\omega + \beta k}\sqrt{\omega - \beta k} = \sqrt{+}\sqrt{-} = \sqrt{-}\sqrt{+}$ into (3.5) leads to

$$\begin{cases} \sqrt{+}\sqrt{+}a_4 = \sqrt{+}\sqrt{-}a_1 \\ \sqrt{-}\sqrt{-}a_3 = \sqrt{+}\sqrt{-}a_2 \\ \sqrt{+}\sqrt{+}a_2 = \sqrt{+}\sqrt{-}a_3 \\ \sqrt{-}\sqrt{-}a_1 = \sqrt{+}\sqrt{-}a_4 \end{cases}$$

A possible solution of the set of (3.5) is:

$$\begin{pmatrix} a_1 \\ a_2 \\ a_3 \\ a_4 \end{pmatrix} = a_0 \begin{pmatrix} \sqrt{+}\sqrt{+} \\ -\sqrt{+}\sqrt{-} \\ -\sqrt{+}\sqrt{+} \\ \sqrt{+}\sqrt{-} \end{pmatrix} \tag{3.6}$$

Note that we could normalize (3.6) by $\widehat{a}_0 = a_0\sqrt{+}$ to obtain:

$$\begin{pmatrix} \widehat{a}_1 \\ \widehat{a}_2 \\ \widehat{a}_3 \\ \widehat{a}_4 \end{pmatrix} = \widehat{a}_0 \begin{pmatrix} \sqrt{+} \\ -\sqrt{-} \\ -\sqrt{+} \\ \sqrt{-} \end{pmatrix}$$

In that representation, we can interpret $\widehat{a}_1$ and $\widehat{a}_3$ as relating to the amplitude of the forward component of waves propagating in the top and bottom crystals in the coupled system. $\widehat{a}_2$ and $\widehat{a}_4$ relate to the amplitude of the backward component of waves propagating in the top and bottom crystals in the coupled system.

In (3.6), a_0 is an arbitrary constant. We note that interchanging the indices 1 and 3 and 2 and 4, which is equivalent to interchanging the top crystal and the bottom crystal in the coupled system leads to $\begin{pmatrix} a_3 \\ a_4 \\ a_1 \\ a_2 \end{pmatrix} = a_0 \begin{pmatrix} \sqrt{+}\sqrt{+} \\ -\sqrt{+}\sqrt{-} \\ -\sqrt{+}\sqrt{+} \\ \sqrt{+}\sqrt{-} \end{pmatrix}$ or $\begin{pmatrix} a_1 \\ a_2 \\ a_3 \\ a_4 \end{pmatrix} = a_0 \begin{pmatrix} -\sqrt{+}\sqrt{+} \\ \sqrt{+}\sqrt{-} \\ \sqrt{+}\sqrt{+} \\ -\sqrt{+}\sqrt{-} \end{pmatrix}$ which is (3.6) with a sign change (*i.e.*, within a global phase). This solution of the coupled two-crystal system is antisymmetric upon exchange of the crystals which is indicative of fermionic behavior.

One of the degenerate solutions of the system of four linear equation (3.4) is:

$$\begin{pmatrix} a_1 \\ a_2 \\ a_3 \\ a_4 \end{pmatrix} = a_0 \begin{pmatrix} \sqrt{\omega - \beta k} \\ -\sqrt{\omega - \beta k} \\ \sqrt{\omega + \beta k} \\ -\sqrt{\omega + \beta k} \end{pmatrix} \tag{3.7}$$

where a_0 is some arbitrary constant. Note that the negative signs reflect the antisymmetry of the displacement. Other solutions can be found by considering the complete set of plane wave solutions $\psi_j = a_j e^{\pm ikx} e^{\pm i\omega t}$ with $j = 1,2,3,4$ as well as the negative frequency eigen value.

The key result is that the upper dispersion curve in the band structure is associated with a wave function whose amplitude shows spinor character (3.7). In this case, the displacement of waves in the forward and backward directions along the two coupled harmonic crystals are constrained and related to each other. The directions of propagation of waves in the two-crystal system are not independent of each other and the two-crystal system breaks chiral symmetry. That is, one cannot apply a rotation onto a forward amplitude (or multiply a forward amplitude by a term taking the form $e^{i\theta}$ with θ being some arbitrary phase) without needing to apply the same rotation to the backward amplitude. For instance, at $k = 0$, the antisymmetric mode is represented by a standing wave which enforces a strict relation between the amplitude of a forward propagating wave and a backward propagating wave. This characteristic was shown [4, 5] to be representative of fermion-like behavior of phonons and will be address in a subsequent subsection. As $k \to \infty$, $\omega \to +\beta k$, the first two terms in (3.5) go to zero and only one direction of propagation (backward) is supported by the medium (third and fourth terms in 3.5). This example illustrates the difference in topology of elastic waves corresponding to the lower and upper bands in the band structure of the two-crystal system. The constraint on the amplitude of waves in the upper band imparts a nonconventional spinor topology to the eigen modes which does exists for modes in the lower band. This topology will be discuss below for a simpler system that supports elastic waves that still retains the spinor character of fermions.

Fig. 3.3 Schematic illustration of the harmonic crystal grounded to a substrate via side springs (system in Fig. 3.1 when $M \to \infty$)

3.3.2 Single Harmonic Crystal Grounded to a Rigid Substrate in the Long-Wavelength Limit

The topology of the upper band of the two-crystal system can be best visualized by taking the limit $M \to \infty$. In that case, the system of Fig. 3.1 becomes a single harmonic crystal grounded to a rigid substrate (see Fig. 3.3).

In the long-wavelength limit, the displacement v in (3.1a, 3.1b) is negligible. Equation (3.2) becomes the Klein–Gordon equation: $\frac{\partial^2 u}{\partial t^2} - \beta^2 \frac{\partial^2 u}{\partial x^2} + \alpha^2 u = 0$ with $\alpha^2 = K_l/m$ and $\beta^2 = K_0/m$. This equation describes only the displacement field, u. Equations (3.3a, 3.3b) can be written as the set of Dirac-like equations:

$$\left[\boldsymbol{\sigma}_x \frac{\partial}{\partial t} + i\beta\boldsymbol{\sigma}_y \frac{\partial}{\partial x} - i\alpha I\right]\Psi = 0 \tag{3.8a}$$

$$\left[\boldsymbol{\sigma}_x \frac{\partial}{\partial t} + i\beta\boldsymbol{\sigma}_y \frac{\partial}{\partial x} + i\alpha I\right]\overline{\Psi} = 0 \tag{3.8b}$$

where $\boldsymbol{\sigma}_x$ and $\boldsymbol{\sigma}_y$ are the 2×2 Pauli matrices: $\begin{pmatrix} 0 & 1 \\ 1 & 0 \end{pmatrix}$ and $\begin{pmatrix} 0 & -i \\ i & 0 \end{pmatrix}$ and I is the 2×2 identity matrix. Solutions of these equations are still non-self-dual. Taking the complex conjugate of (3.8a) results in (3.8b), indeed the first two terms are real and only the last term changes sign. In particular, this results from the negative sign of the second term in the original Klein–Gordon equation which requires the multiplicative imaginary number "i" on the second term of the Dirac-like equations. Then $\overline{\Psi} = \Psi^*$. So while Ψ is a solution of (3.8a), its complex conjugate is not solution of (3.8a). $\overline{\Psi}$ is solution of (3.8b). In the language of quantum field theory, Ψ and $\overline{\Psi}$ represent two different physical entities, namely "particles" and "antiparticles." These equation still break time-reversal (T) symmetry and parity (P) symmetry individually, though, they obey time and parity (PT) symmetry when applied simultaneously.

We now write our solutions in the form: $\Psi_k = \Psi(k, \omega_k) = c_0\xi_k(k, \omega_k)e^{(\pm)i\omega_k t} e^{(\pm)ikx}$ and $\overline{\Psi}_k = \overline{\Psi}(k, \omega_k) = c_0\overline{\xi}_k(k, \omega_k)e^{(\pm)i\omega_k t}e^{(\pm)ikx}$ where $\xi_k = \begin{pmatrix} a_1 \\ a_2 \end{pmatrix}$ and $\overline{\xi}_k = \begin{pmatrix} \bar{a}_1 \\ \bar{a}_2 \end{pmatrix}$ are two by one spinors. Inserting the various forms for these solutions in (3.8a, 3.8b) lead to the same eigen values that we obtained before for the upper band of the two-crystal system, namely by $\omega = \pm\sqrt{\alpha^2 + \beta^2 k^2}$. Again, let us note that the band structure has two branches corresponding to positive frequencies and negative frequencies. Negative frequencies can be visualized as representing waves that propagate in a direction opposite to that of waves with positive frequency. The spinor part of the solutions for the different plane waves is summarized in the Table 3.1 below. Negative and positive wave numbers k correspond to waves propagating in opposite direction.

The spinors in Table 3.1 are given to within any complex constant c_0.

We note that $\bar{\xi}_k = \boldsymbol{\sigma}_z \xi_k$ where $\boldsymbol{\sigma}_z = \begin{pmatrix} 1 & 0 \\ 0 & -1 \end{pmatrix}$ is the third Pauli matrix. This can be shown by considering (3.8a) and multiplying it on the left by $\boldsymbol{\sigma}_z$:

$$\left[\boldsymbol{\sigma}_z\boldsymbol{\sigma}_x \frac{\partial}{\partial t} + i\beta\boldsymbol{\sigma}_z\boldsymbol{\sigma}_y \frac{\partial}{\partial x} - i\alpha\boldsymbol{\sigma}_z I\right]\Psi = 0$$

Using the identities $\boldsymbol{\sigma}_z\boldsymbol{\sigma}_x = -\boldsymbol{\sigma}_x\boldsymbol{\sigma}_z$ and $\boldsymbol{\sigma}_z\boldsymbol{\sigma}_y = -\boldsymbol{\sigma}_y\boldsymbol{\sigma}_z$, the preceding equation becomes

$$\left[-\boldsymbol{\sigma}_x\boldsymbol{\sigma}_z \frac{\partial}{\partial t} - i\beta\boldsymbol{\sigma}_y\boldsymbol{\sigma}_z \frac{\partial}{\partial x} - i\alpha I\boldsymbol{\sigma}_z\right]\Psi = -\left[\boldsymbol{\sigma}_x \frac{\partial}{\partial t} + i\beta\boldsymbol{\sigma}_y \frac{\partial}{\partial x} + i\alpha I\right]\boldsymbol{\sigma}_z\Psi = 0$$

Which is effectively (3.8b) with $\overline{\Psi} = \boldsymbol{\sigma}_z\Psi$.

We note that if $\alpha = 0$, (3.8a) becomes: $\left[\boldsymbol{\sigma}_x \frac{\partial}{\partial t} + i\beta\boldsymbol{\sigma}_y \frac{\partial}{\partial x}\right]\psi = 0$. Using a plane wave solution with $\xi_k(k, \omega_k) = \begin{pmatrix} a_1 \\ a_2 \end{pmatrix}$, this equation reduces to the system: $\begin{cases} (\omega - \beta kh)a_1 = 0 \\ (\omega + \beta kh)a_2 = 0 \end{cases}$. We obtain two solutions for the angular velocity of the plane wave, $\omega = \pm\beta kh$. These correspond to plane waves propagating in the positive and negative directions. In this case, the components of the two-spinor, a_1 and a_2 are now independent of each other and independent of the wave number. The amplitude of the plane wave propagating in the positive direction is independent of that of the wave propagating in the opposite direction. When $\alpha \neq 0$, from Table 3.1, we see that the components of the two-spinor are not independent of each other. This indicates that the directions of propagation are not independent of each other anymore. It is the parameter α that couples those directions.

Table 3.1 Two by one spinor solutions of (3.8a) and (3.8b) for the different plane wave forms

	$e^{+ikx}e^{+i\omega_k t}$	$e^{-ikx}e^{+i\omega_k t}$	$e^{+ikx}e^{-i\omega_k t}$	$e^{-ikx}e^{-i\omega_k t}$
ξ_k	$\begin{pmatrix} \sqrt{\omega+\beta k} \\ \sqrt{\omega-\beta k} \end{pmatrix}$	$\begin{pmatrix} \sqrt{\omega-\beta k} \\ \sqrt{\omega+\beta k} \end{pmatrix}$	$\begin{pmatrix} -\sqrt{\omega-\beta k} \\ \sqrt{\omega+\beta k} \end{pmatrix}$	$\begin{pmatrix} -\sqrt{\omega+\beta k} \\ \sqrt{\omega-\beta k} \end{pmatrix}$
$\bar{\xi}_k$	$\begin{pmatrix} \sqrt{\omega+\beta k} \\ -\sqrt{\omega-\beta k} \end{pmatrix}$	$\begin{pmatrix} \sqrt{\omega-\beta k} \\ -\sqrt{\omega+\beta k} \end{pmatrix}$	$\begin{pmatrix} -\sqrt{\omega-\beta k} \\ -\sqrt{\omega+\beta k} \end{pmatrix}$	$\begin{pmatrix} -\sqrt{\omega+\beta k} \\ -\sqrt{\omega-\beta k} \end{pmatrix}$

Table 3.1 can be used to identify the symmetry properties of Ψ and $\overline{\Psi}$ in the allowed space: k, ω. We find the following transformation rules:

$$T_{\substack{\omega\to\omega\\ k\to -k}}(\Psi(\omega,k)) = \Psi(\omega,-k) \tag{3.9a}$$

$$T_{\substack{\omega\to -\omega\\ k\to k}}(\Psi(\omega,k)) = -i\overline{\Psi}(-\omega,k) \tag{3.9b}$$

which lead to the combined transformation

$$T_{\substack{\omega\to -\omega\\ k\to -k}}(\Psi(\omega,k)) = -i\overline{\Psi}(-\omega,-k) \tag{3.10}$$

We have defined $T_{\substack{\omega\to\omega\\ k\to -k}}$ and $T_{\substack{\omega\to -\omega\\ k\to k}}$ as transformations that change the sign of the frequency and wave number, respectively. As one crosses the gap at the origin $k = 0$, the multiplicative factor "i" indicates that the wave function accumulated a phase of $\frac{\pi}{2}$. We also note the orthogonality condition $\overline{\Psi}\boldsymbol{\sigma}_x\Psi = 0$. The topology of the spinor wave functions that reflects their symmetry properties is illustrated in Fig. 3.4.

We now shed more light on the properties of the spinor solutions by treating α in (3.8a, 3.8b) as a perturbation, ε. When $\alpha = 0$, (3.8a) reduces to the two independent equations:

$$\left(\frac{\partial}{\partial t} - \beta\frac{\partial}{\partial x}\right)\varphi_1^{(0)} = 0 \tag{3.11a}$$

$$\left(\frac{\partial}{\partial t} + \beta\frac{\partial}{\partial x}\right)\varphi_2^{(0)} = 0 \tag{3.11b}$$

whose solutions correspond to plane waves propagating in the forward direction, $\varphi_1^{(0)}$, with dispersion relation $\omega^+ = \beta k$ and the backward direction, $\varphi_2^{(0)}$ with dispersion relation $\omega^- = -\beta k$. We rewrite (3.11a, 3.11b) in the form

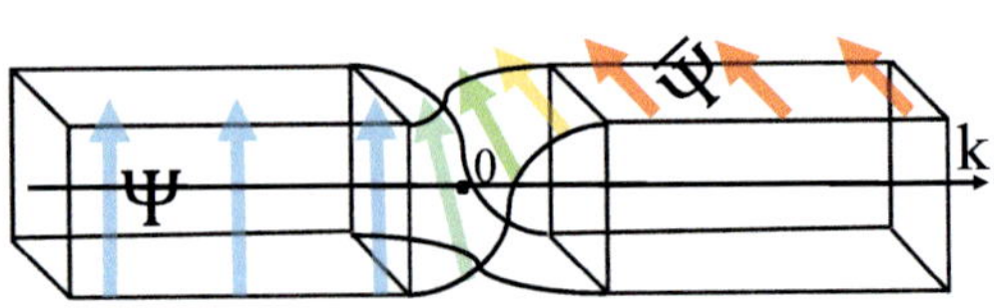

Fig. 3.4 Schematic representation of the manifold supporting Ψ and $\overline{\Psi}$. The manifold exhibits a local quarter-turn twist around $k = 0$ of the square cross section of the manifold reflects the orthogonality of Ψ and $\overline{\Psi}$. The *colored arrows* are parallel transported on the manifold along the direction of wave number. Their change in orientation is indicative of the phase change

$$\left(\frac{\partial}{\partial t} - \beta \frac{\partial}{\partial x}\right)\varphi_1 = i\varepsilon\varphi_2^{(0)} = i\varepsilon e^{ikx}e^{\omega^- t} \tag{3.12a}$$

$$\left(\frac{\partial}{\partial t} + \beta \frac{\partial}{\partial x}\right)\varphi_2 = i\varepsilon\varphi_1^{(0)} = i\varepsilon e^{ikx}e^{\omega^+ t} \tag{3.12b}$$

Seeking the particular solutions to first order, $\varphi_1^{(1)}$ and $\varphi_2^{(1)}$, which follow in frequency the driving terms on the right-side side of (3.12a, 3.12b), we find their respective amplitude to be:

$$a_1 = \frac{\varepsilon}{\omega^- - \beta k} = \frac{\varepsilon}{\omega^- - \omega^+} \tag{3.13a}$$

$$a_2 = \frac{\varepsilon}{\omega^+ + \beta k} = \frac{\varepsilon}{\omega^+ - \omega^-} = -a_1 \tag{3.13b}$$

Since $\omega^+ = \omega^-$ at k=0 only, the amplitude of the first order perturbed forward wave, a_1 changes sign as k varies from $-\infty$ to $+\infty$. A similar but opposite change of sign occurs for the backward perturbed amplitude. These changes of sign are therefore associated with changes in phase of π and $-\pi$ for the forward and backward waves as one crosses the origin k=0. These phase changes (sign changes) are characteristic of that occurring at a resonance. The gap that would occur at k=0 in the band structure of the harmonic crystal grounded to a substrate via side springs, should we push the perturbation theory to higher orders, may therefore be visualized as resulting from a resonance of the forward waves driven by the backward propagating waves and vice versa. However, since to first order the amplitudes given by (3.13a, 3.13b) diverge at the only point of intersection between the dispersion relations of the forward and backward waves, we have to use analytic continuation to expand them into the complex plane:

$$a_1^+ = \frac{\varepsilon}{\omega^- - \beta k - i\eta} = \frac{\varepsilon}{-2\beta k - i\eta} \tag{3.14a}$$

$$a_2^+ = \frac{\varepsilon}{\omega^+ + \beta k + i\eta} = \frac{\varepsilon}{+2\beta k + i\eta} \tag{3.14b}$$

where $\eta \to 0$ continues the eigen values ω^- and ω^+ into the complex plane.

By inspection, we can see that at the origin, $k = 0$, both amplitudes are pure imaginary quantities and therefore exhibit a phase of $\pm\frac{\pi}{2}$. This is expected as (3.14a, 3.14b) are representative of the amplitude of a driven damped harmonic oscillator which also shows a phase of $\frac{\pi}{2}$ with respect to the driving frequency at resonance. It is instructive to calculate the Berry connection [31] (or see Chap. 2) for these perturbed amplitudes. The Berry connection determines the phase change of a wave as some parameter takes the wave function along a continuous path on the manifold that supports it. Since the Berry phase applies to continuous paths, we cannot use that concept to determine the phase change across the gap of our system *i.e.*, between the positive and negative frequency branches of the band structure.

Therefore, we resort to calculating the Berry connection for the first order perturbed solution which still remains continuous but may capture the interaction between directions of propagation. Our intention is to characterize the topology of the spinor part of the wave function (3.14a, 3.14b), by calculating the change in phase of the waves as one crosses $k = 0$. It is important first to normalize the spinor: $\xi = \begin{pmatrix} a_1^+ \\ a_2^+ \end{pmatrix}$. This normalized spinor takes the form: $\tilde{\xi} = \begin{pmatrix} \tilde{a}_1^+ \\ \tilde{a}_2^+ \end{pmatrix} = \frac{\sqrt{4\beta^2k^2+\eta^2}}{\sqrt{2}} \begin{pmatrix} \frac{1}{-2\beta k - i\eta} \\ \frac{1}{+2\beta k + i\eta} \end{pmatrix}$. The Berry connection is given by $BC(k) = -i\tilde{\xi}^* \frac{\partial \tilde{\xi}}{\partial k} = -i\tilde{a}_1^{+*} \frac{\partial \tilde{a}_1^+}{\partial k} - i\tilde{a}_2^{+*} \frac{\partial \tilde{a}_2^+}{\partial k}$. After several analytical and algebraic manipulations, we obtain:

$$BC(k) = \beta \frac{\eta}{4\beta^2k^2 + \eta^2} + \beta \frac{\eta}{4\beta^2k^2 + \eta^2} \tag{3.15}$$

In (3.15), the contribution to the Berry connection of $\tilde{a}_1^+$ and $\tilde{a}_2^+$ are identical. Using the identities: $\lim_{\eta \to 0} \frac{\eta}{x^2+\eta^2} = \pi\delta(x)$ and $\delta(ax) = \frac{1}{a}\delta(x)$, leads to

$$BC(k) = \frac{\pi}{2}\delta(k) + \frac{\pi}{2}\delta(k) \tag{3.16}$$

The contribution of each direction of propagation to the spinor part of the wave function accumulates a $\frac{\pi}{2}$ phase shift as one crosses the origin k=0. This corresponds to parallel transport along a manifold that undergoes a ¼ turn.

3.3.3 *Single Discrete Harmonic Crystal Grounded to a Rigid Substrate Beyond the Long-Wavelength Limit*

3.3.3.1 Dirac-Like Equation for the Discrete Crystal and Spinor Solutions

For the discrete system of Fig. 3.3, the elastic equation takes the form of the discrete Klein–Gordon equation:

$$\frac{\partial^2 u_n}{\partial t^2} - \beta^2(u_{n+1} - 2u_n + u_{n-1}) + \alpha^2 u_n = 0 \tag{3.17}$$

Equation (3.17) involves the second derivatives with respect to continuous time and the discrete second derivative with respect to position of the angular degree of freedom. Here, following the approach of Dirac in linearizing the relativistic Klein–Gordon equation, we wish to derive a wave equation in terms of first order spatial and temporal derivatives. To do this, we need to rewrite the Laplacian, $\Delta u_n = u_{n+1} - 2u_n + u_{n-1}$, in a "square root" form: $\Delta u_n = D(Du_n)$. This can be done exactly by introducing the following first order differential operator:

$$D = \boldsymbol{e}_1\Delta^+ + \boldsymbol{e}_2\Delta^- = \begin{pmatrix} 0 & 1 \\ 0 & 0 \end{pmatrix}\Delta^+ + \begin{pmatrix} 0 & 0 \\ 1 & 0 \end{pmatrix}\Delta^- \tag{3.18}$$

In (3.18), $\Delta^+ u_n = u_{n+1} - u_n$ and $\Delta^- u_n = u_n - u_{n-1}$ are the forward and backward finite differences acting now on a two-vector. The 2×2 matrices $\boldsymbol{e}_1$ and $\boldsymbol{e}_2$ satisfy the conditions $\boldsymbol{e}_1\boldsymbol{e}_1 = \boldsymbol{e}_2\boldsymbol{e}_2 = 0$ and $\boldsymbol{e}_1\boldsymbol{e}_2 + \boldsymbol{e}_2\boldsymbol{e}_1 = \boldsymbol{I}$ with $\boldsymbol{I}$ representing the 2×2 identity matrix. This formalism permits the exact and formal definition of the "square root" of the discrete Laplacian.

The Dirac-like equation for the discrete system then takes the form:

$$\left[\boldsymbol{\sigma}_x \otimes \boldsymbol{I}\frac{\partial}{\partial t} + i\beta\boldsymbol{\sigma}_y \otimes \{\boldsymbol{e}_1\Delta^+ + \boldsymbol{e}_2\Delta^-\} \pm i\alpha\boldsymbol{I} \otimes \boldsymbol{I}\right]\psi_n = 0 \tag{3.19}$$

Where again $\boldsymbol{\sigma}_x$ and $\boldsymbol{\sigma}_y$ are the 2×2 Pauli matrices: $\begin{pmatrix} 0 & 1 \\ 1 & 0 \end{pmatrix}$ and $\begin{pmatrix} 0 & -i \\ i & 0 \end{pmatrix}$, respectively. Again, the parameter α plays the role of mass in the relativistic Dirac equation. Applying the outer product $\otimes$ leads to 4×4 matrices and ψ_n is a four-vector: $\psi_n = \begin{pmatrix} \psi_{1n} \\ \psi_{2n} \\ \psi_{3n} \\ \psi_{4n} \end{pmatrix}$. This four component representation is the consequence of the discrete nature of the Laplacian. In contrast, with a continuous Laplacian there is no distinction between forward and backward derivatives and one would only need to use a two component representation. This is the case in the long wavelength limit of the previous Sect. 3.2. In this limit having a two component spinor indicates that there is a coupling between waves propagating in opposite directions (positive or negative) along the chain of masses. When considering the short wavelength four component spinor solution, the first two components represent propagation of waves in the positive direction and the next two components propagation in the negative direction. The two components for the positive direction and the two components from the negative directions reflect a lifting of degeneracy due to asymmetry of the forward and backward finite different in the discrete Dirac equation. The solutions of (3.19) are automatically solutions of (3.8a, 3.8b), but the converse is not true. As will be seen later, the directions of propagation of the wave are expressed separately in the Dirac wave function. The $\pm$ in (3.19) corresponds to choices of the sign of the parameter α (*i.e.*, choice of positive or negative "mass" in Dirac's formalism). Let us first seek solutions of (3.19) with the negative value. Equation (3.19) becomes:

$$\left[\boldsymbol{C}\frac{\partial}{\partial t}+\beta\{\boldsymbol{A}\Delta^{+}+\boldsymbol{B}\Delta^{-}\}-i\alpha\boldsymbol{I}\right]\psi=0 \tag{3.20}$$

C, ***A***, ***B***, and ***I*** are the 4×4 matrices:

$$\begin{pmatrix}0&0&1&0\\0&0&0&1\\1&0&0&0\\0&1&0&0\end{pmatrix},\begin{pmatrix}0&0&0&1\\0&0&0&0\\0&-1&0&0\\0&0&0&0\end{pmatrix},\begin{pmatrix}0&0&0&0\\0&0&1&0\\0&0&0&0\\-1&0&0&0\end{pmatrix},\begin{pmatrix}1&0&0&0\\0&1&0&0\\0&0&1&0\\0&0&0&1\end{pmatrix}.$$

It is easily verifiable that $CC=I, AA=BB=0, AB+BA=-I$, and $C(A\Delta^{+}+B\Delta^{-})+(A\Delta^{+}+B\Delta^{-})C=0$, which are the conditions necessary to recover the Klein–Gordon wave equation (3.17) by applying the operator in (3.20) twice (with appropriate $\pm$ sign). Equation (3.20) is the basis for our discussion of elastic waves with non-trivial topologies in discrete structures.

The solutions take the form of plane waves $\psi_{jn}=a_j e^{\pm i\omega t}e^{\pm iknh}$. For illustrative purposes, let's seek solutions in a particular form of plane waves, $\psi_{jn}=a_j e^{-i\omega t}e^{iknh}$ with $j=1,2,3,4$. ω and k are the angular frequency and wave number, respectively. We remind the reader that "h" is the spacing distance between masses. Equation (3.20) yields the system of equations:

$$\begin{cases}-i\alpha a_1-i\omega a_3+\beta\left(e^{ikh}-1\right)a_4=0\\-i\alpha a_2+\beta\left(1-e^{-ikh}\right)a_3-i\omega a_4=0\\-i\omega a_1-\beta\left(e^{ikh}-1\right)a_2-i\alpha a_3=0\\-\beta\left(1-e^{-ikh}\right)a_1-i\omega a_2-i\alpha a_4=0\end{cases} \tag{3.21}$$

This system of equation admits two doubly degenerate Eigen values:

$$\omega=\pm\sqrt{\alpha^2-\beta^2\left(e^{ikh}-1\right)\left(1-e^{-ikh}\right)}=\pm\sqrt{\alpha^2+4\beta^2 sin^2\frac{kh}{2}} \tag{3.22}$$

The elastic wave at the origin, k=0, has a finite frequency. We note that (3.22) gives two branches with positive and negative frequencies that do not intersect at the origin unless $\alpha=0$. The dispersion relations are periodic and defined in the first

Brillouin zone: $k \in \left[-\frac{\pi}{h}, \frac{\pi}{h}\right]$. Choosing the positive or negative branches of the dispersion relations, we determine the four Eigen vectors:

$$\psi_1^{\pm} = \begin{pmatrix} a_1 \\ a_2 \\ a_3 \\ a_4 \end{pmatrix} e^{-i\omega t} e^{iknh} = a_0 \begin{pmatrix} ie^{+i\frac{kh}{4}}\left(\alpha - \beta\left(e^{i\frac{kh}{2}} - e^{-i\frac{kh}{2}}\right)\right) \\ \mp e^{-i\frac{kh}{4}}\sqrt{\alpha^2 - \beta^2\left(e^{i\frac{kh}{2}} - e^{-i\frac{kh}{2}}\right)^2} \\ \mp ie^{+i\frac{kh}{4}}\sqrt{\alpha^2 - \beta^2\left(e^{i\frac{kh}{2}} - e^{-i\frac{kh}{2}}\right)^2} \\ e^{-i\frac{kh}{4}}\left(\alpha - \beta\left(e^{i\frac{kh}{2}} - e^{-i\frac{kh}{2}}\right)\right) \end{pmatrix} e^{-i\omega t} e^{iknh} \tag{3.23a}$$

$$\psi_2^{\pm} = \begin{pmatrix} a_1 \\ a_2 \\ a_3 \\ a_4 \end{pmatrix} e^{-i\omega t} e^{iknh} = a_0 \begin{pmatrix} -ie^{+i\frac{kh}{4}}\left(\alpha + \beta\left(e^{i\frac{kh}{2}} - e^{-i\frac{kh}{2}}\right)\right) \\ \mp e^{-i\frac{kh}{4}}\sqrt{\alpha^2 - \beta^2\left(e^{i\frac{kh}{2}} - e^{-i\frac{kh}{2}}\right)^2} \\ \pm ie^{+i\frac{kh}{4}}\sqrt{\alpha^2 - \beta^2\left(e^{i\frac{kh}{2}} - e^{-i\frac{kh}{2}}\right)^2} \\ e^{-i\frac{kh}{4}}\left(\alpha + \beta\left(e^{i\frac{kh}{2}} - e^{-i\frac{kh}{2}}\right)\right) \end{pmatrix} e^{-i\omega t} e^{iknh} \tag{3.23b}$$

The upper signs in (3.23a) and (3.23b) correspond to the positive branch of the band structure ($\omega > 0$) and the lower signs to the negative branch ($\omega < 0$).

Similarly, we solve the equation:

$$\left[\boldsymbol{C}\frac{\partial}{\partial t} + \beta\{\boldsymbol{A}\Delta^+ + \boldsymbol{B}\Delta^-\} + i\alpha\boldsymbol{I}\right]\overline{\psi} = 0. \tag{3.24}$$

The operator in (3.24) is the complex conjugate of the operator in (3.20). So we seek solutions in the form of plane waves, $\overline{\psi}_{jn} = \bar{a}_j e^{i\omega t} e^{-iknh}$ with $j = 1,2,3,4$. We find the four Eigen vectors:

$$\overline{\psi}_1^{\pm} = \begin{pmatrix} \bar{a}_1 \\ \bar{a}_2 \\ \bar{a}_3 \\ \bar{a}_4 \end{pmatrix} e^{i\omega t} e^{-iknh} = a_0 \begin{pmatrix} ie^{-i\frac{kh}{4}}\left(\alpha - \beta\left(e^{i\frac{kh}{2}} - e^{-i\frac{kh}{2}}\right)\right) \\ \mp e^{+i\frac{kh}{4}}\sqrt{\alpha^2 - \beta^2\left(e^{i\frac{kh}{2}} - e^{-i\frac{kh}{2}}\right)^2} \\ \mp ie^{-i\frac{kh}{4}}\sqrt{\alpha^2 - \beta^2\left(e^{i\frac{kh}{2}} - e^{-i\frac{kh}{2}}\right)^2} \\ e^{+i\frac{kh}{4}}\left(\alpha - \beta\left(e^{i\frac{kh}{2}} - e^{-i\frac{kh}{2}}\right)\right) \end{pmatrix} e^{i\omega t} e^{-iknh} \tag{3.25a}$$

$$\overline{\psi}_2^{\pm} = \begin{pmatrix} \bar{a}_1 \\ \bar{a}_2 \\ \bar{a}_3 \\ \bar{a}_4 \end{pmatrix} e^{i\omega t} e^{-iknh} = a_0 \begin{pmatrix} -ie^{-i\frac{kh}{4}}\left(\alpha + \beta\left(e^{i\frac{kh}{2}} - e^{-i\frac{kh}{2}}\right)\right) \\ \mp e^{+i\frac{kh}{4}}\sqrt{\alpha^2 - \beta^2\left(e^{i\frac{kh}{2}} - e^{-i\frac{kh}{2}}\right)^2} \\ \pm ie^{-i\frac{kh}{4}}\sqrt{\alpha^2 - \beta^2\left(e^{i\frac{kh}{2}} - e^{-i\frac{kh}{2}}\right)^2} \\ e^{+i\frac{kh}{4}}\left(\alpha + \beta\left(e^{i\frac{kh}{2}} - e^{-i\frac{kh}{2}}\right)\right) \end{pmatrix} e^{i\omega t} e^{-iknh} \tag{3.25b}$$

3.3.3.2 Non-conventional Topology of Elastic Waves in the Discrete 1D Mass-Spring Single Crystal System

Now, we investigate the topology in k space of the solutions given by (3.23a, 3.23b) and (3.25a, 3.25b). It is easy to show the following relations:

$$\psi_{1,2}^{\pm}\left(k + \frac{2\pi}{h}\right) = e^{-i\frac{\pi}{2}}\overline{\psi}_{1,2}^{*\pm}(k) \tag{3.26a}$$

$$\psi_{1,2}^{\pm}\left(k + \frac{4\pi}{h}\right) = e^{-i\pi}\psi_{1,2}^{\pm}(k) \tag{3.26b}$$

$$\psi_{1,2}^{\pm}\left(k + \frac{6\pi}{h}\right) = e^{-i\frac{3\pi}{2}}\overline{\psi}_{1,2}^{*\pm}(k) \tag{3.26c}$$

$$\psi_{1,2}^{\pm}\left(k + \frac{8\pi}{h}\right) = e^{-i2\pi}\psi_{1,2}^{\pm}(k) \tag{3.26d}$$

In (3.26a–3.26d), the * represents the complex conjugate. We also note that ψ and $\overline{\psi}$ are orthogonal and satisfy the condition: $\overline{\psi}\psi = 0$. The relations given by

(3.26a–3.26d) indicate that the wave function has a periodicity of $\frac{8\pi}{h}$. The topology of the wave function is more easily illustrated by considering the parallel transport of two orthogonal vector fields along a closed loop corresponding to $k \in \left[0, \frac{8\pi}{h}\right]$. The first vector field represents the wave function ψ and the second vector field the complex conjugate of $\overline{\psi}$. This non-conventional torsional topology is illustrated in Fig. 3.5 in the form of a torus with square cross section that exhibits four 90° twists. Starting at the top of the figure, the red arrow and green arrow represent the two orthogonal wave functions ψ and $\overline{\psi}$, respectively. The red arrow points in a direction perpendicular to the plane of the figure and the green arrow point in a direction within the plane of the figure. As the vector fields are transported along the first twist, the green arrow is now pointing in the plane of the figure and the green arrow becomes perpendicular to that plane, that is ψ turns into $\overline{\psi}$. The red arrow transported through the first twist is making now a right angle with the red arrow at the starting point. This means that the wave function ψ has accumulated a phase of $\frac{-\pi}{2}$. The process is repeated through the second twist. The red arrow transported through the second twist is now parallel to the red arrow at the origin but is pointing in the opposite direction and satisfies the condition: $\psi\left(k + \frac{4\pi}{h}\right) = -\psi(k)$. This means that the wave function ψ has accumulated a phase of $-\pi$. The vector fields can be transported through the third and fourth twists (not illustrated in the figure) to close the loop in k space and reach the starting point. At that point: $\psi\left(k + \frac{8\pi}{h}\right) = \psi(k)$. The topological properties of the spinor solutions of the discrete Dirac equation for the harmonic crystal attached to a rigid substrate generalizes that of the same system in the long wavelength limit (Fig. 3.4) to the entire periodic Brillouin zone.

The topology of the spinor solution of the discrete system is further illuminated in Appendix 1 by calculating its Berry phase. The torsional topology of the wave function results from its spinor characteristics which endow it with fermion-like character. This character is analyzed in more detail in the next subsection. To simplify the mathematics, we will consider again only the long-wavelength limit.

3.3.3.3 Analysis of Fermion-Like Behavior of Single Harmonic Crystal Grounded to a Substrate in the Long-Wavelength Limit

3.3.3.3.1 Lagrangian Formalism

We recall that in the long wavelength limit, $k \to 0$, (3.17) reduces to the Klein–Gordon form

$$\frac{\partial^2 u}{\partial t^2} - \beta^2 \frac{\partial^2 u}{\partial x^2} + \alpha^2 u = 0 \tag{3.27}$$

The corresponding Dirac equations are given by (3.8a, 3.8b). We repeat these equations here

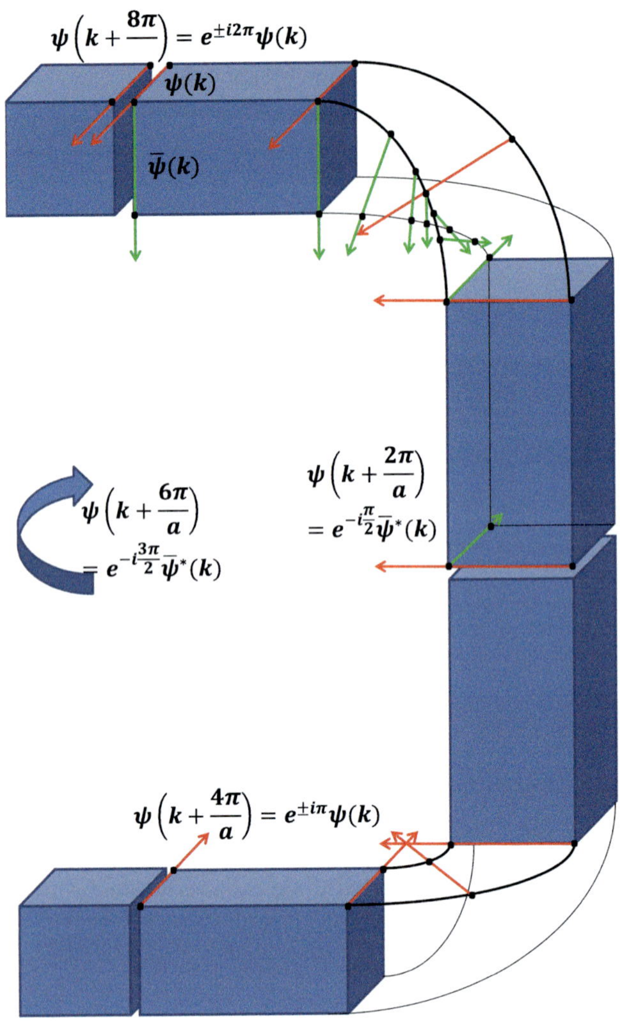

Fig. 3.5 Schematic illustration of the torsional topology of the wave function for modes in k space. The wave functions are represented in the form of two orthogonal vector fields (*red* and *green*) supported by a square cross section torus manifold possessing four 90° twists. Parallel transport of the vector field on the manifold shows the topological properties of the wave function

$$\left[\boldsymbol{\sigma}_x\frac{\partial}{\partial t}+i\beta\boldsymbol{\sigma}_y\frac{\partial}{\partial x}-i\alpha\boldsymbol{I}\right]\psi=0$$

$$\left[\boldsymbol{\sigma}_x\frac{\partial}{\partial t}+i\beta\boldsymbol{\sigma}_y\frac{\partial}{\partial x}+i\alpha\boldsymbol{I}\right]\overline{\psi}=0$$

where $\boldsymbol{\sigma}_x$ and $\boldsymbol{\sigma}_y$ are the 2×2 Pauli matrices and $\boldsymbol{I}$ is the 2×2 identity matrix.

The Dirac-like equation (3.8a) can be obtained from the Lagrangian:

$$L\left(\psi^{\dagger}, \frac{\partial \psi^{\dagger}}{\partial t}, \frac{\partial \psi^{\dagger}}{\partial x}\right) = \frac{\partial \psi^{\dagger}}{\partial t}\boldsymbol{\sigma}_{\mathbf{x}}\psi + i\beta\frac{\partial \psi^{\dagger}}{\partial x}\boldsymbol{\sigma}_{y}\psi + i\alpha\psi^{\dagger}\boldsymbol{I}\psi \tag{3.28a}$$

In (3.28a), $\psi^{\dagger}$ is the Hermitian conjugate of ψ. The Lagrangian (3.28a) is anti-Hermitian and its conjugate is given by

$$L^{\dagger}\left(\psi, \frac{\partial \psi}{\partial t}, \frac{\partial \psi}{\partial x}\right) = \psi^{\dagger}\boldsymbol{\sigma}_{\mathbf{x}}\frac{\partial \psi}{\partial t} + i\beta\psi^{\dagger}\boldsymbol{\sigma}_{y}\frac{\partial \psi}{\partial x} - i\alpha\psi\boldsymbol{I}\psi^{\dagger} \tag{3.28b}$$

The Euler–Lagrange's equation of motion for the field variables ψ is given by:

$$\frac{\partial}{\partial t}\left(\frac{\partial L}{\partial(\partial_t(\psi^{\dagger}))}\right) + \frac{\partial}{\partial x}\left(\frac{\partial L}{\partial(\partial_x\psi^{\dagger})}\right) - \frac{\partial L}{\partial(\psi^{\dagger})} = 0 \tag{3.29}$$

This recovers the Dirac equation (3.8a).

We also derive an equation of motion for the field $\psi^{\dagger}$ from (3.29) by replacing L by $L^{\dagger}$ and $\psi^{\dagger}$ by ψ.

$$\frac{\partial}{\partial t}\left(\frac{\partial L^{\dagger}}{\partial(\partial_t(\psi))}\right) + \frac{\partial}{\partial x}\left(\frac{\partial L^{\dagger}}{\partial(\partial_x\psi)}\right) - \frac{\partial L^{\dagger}}{\partial(\psi)} = 0$$

This equation becomes:

$$\left[\frac{\partial}{\partial t}\boldsymbol{\sigma}_x + i\beta\frac{\partial}{\partial x}\boldsymbol{\sigma}_y + i\alpha\boldsymbol{I}\right]\psi^{\dagger} = 0$$

This is the Hermitian conjugate of (3.8b). Its plane wave solutions are listed in the table below as well as those of (3.8a).

3.3.3.3.2 Energy and Anticommutation

Here we calculate the Hamiltonian density associated with the Lagrangian, $L^{\dagger}$. It is given by the Legendre transformation:

$$\mathrm{H}^{\dagger} = \Pi^{\dagger}\frac{\partial \psi}{\partial t} - L^{\dagger} \tag{3.30}$$

Where the momentum conjugate is defined by: $\Pi^{\dagger} = \frac{\partial L^{\dagger}}{\partial\left(\partial \psi/\partial t\right)} = \psi^{\dagger}\boldsymbol{\sigma}_x$. Inserting this expression in (3.30) and using (3.28b), one obtains the Hamiltonian density:

$$\mathrm{H}^{\dagger} = \psi^{\dagger}\boldsymbol{\sigma}_x \frac{\partial \psi}{\partial t} \tag{3.31}$$

We now promote the quantity ψ to be an operator (we will use the same symbol ψ to represent the operator) and use a representation whereby the field ψ is expanded into plane waves and their complex conjugate:

$$\psi(x) = \sum\nolimits_k \frac{1}{\sqrt{2\omega}} \left[a_k \xi_k e^{ikx} e^{-i\omega t} + a_k^* \xi_k^* e^{-ikx} e^{i\omega t}\right] \tag{3.32}$$

Similarly, we write the Hermitian conjugate of the ψ operator:

$$\psi^{\dagger}(x) = \sum\nolimits_k \frac{1}{\sqrt{2\omega}} \left[a_k^{\dagger} \xi_k^{\dagger} e^{-ikx} e^{i\omega t} + a_k^{\dagger *} \xi_k^{\dagger *} e^{ikx} e^{-i\omega t}\right] \tag{3.33}$$

The quantities: $a_k^{\dagger}$, $a_k^{\dagger *}$, a_k, and a_k^* are now creation and annihilation operators, respectively. Inserting (3.32) and (3.33) in (3.31) and integrating over all space results in

$$\begin{aligned}
E = \int dx\bar{H} &= \sum_k \sum_{k'} \frac{a_{k'}^{\dagger}}{\sqrt{2\omega'}} (-i\omega) \xi_{k'}^{\dagger} \boldsymbol{\sigma}_x \xi_k \frac{a_k}{\sqrt{2\omega}} e^{-i(\omega-\omega')t} \int dx e^{-i(k'-k)x} \\
&+ \sum_k \sum_{k'} \frac{a_{k'}^{\dagger}}{\sqrt{2\omega'}} (i\omega) \times \xi_{k'}^{\dagger} \boldsymbol{\sigma}_x \xi_k^* \frac{a_k^*}{\sqrt{2\omega}} e^{i(\omega+\omega')t} \int dx e^{-i(k'+k)x} \\
&+ \sum_k \sum_{k'} \frac{a_{k'}^{\dagger *}}{\sqrt{2\omega'}} (-i\omega) \xi_{k'}^{\dagger *} \boldsymbol{\sigma}_x \xi_k \frac{a_k}{\sqrt{2\omega}} e^{-i(\omega+\omega')t} \int dx e^{i(k'+k)x} \\
&+ \sum\nolimits_k \sum\nolimits_{k'} \frac{a_{k'}^{\dagger *}}{\sqrt{2\omega'}} (i\omega) \xi_{k'}^{\dagger *} \boldsymbol{\sigma}_x \xi_k^* \frac{a_k^*}{\sqrt{2\omega}} e^{-i(\omega'-\omega)t} \int dx e^{-i(k-k')x}
\end{aligned} \tag{3.34}$$

We now replace the spatial integrals in (3.34) by their respective delta functions, $\delta_{k',k}$, $\delta_{k',-k}$. Using Table 3.2, we can show that $\xi_k^{\dagger} \boldsymbol{\sigma}_x \xi_k = \xi_k^{\dagger *} \boldsymbol{\sigma}_x \xi_k^* = 2\omega c_0^2$ and $\xi_{-k}^{\dagger} \boldsymbol{\sigma}_x \xi_k^* = \xi_{-k}^{\dagger *} \boldsymbol{\sigma}_x \xi_k = 0$. Taking the constant $c_0^2 = -i$ to obtain an energy in the form of a real number, the energy of the system becomes:

$$E = \sum\nolimits_k \omega \left(a_k^{\dagger} a_k - a_k^{\dagger *} a_k^*\right) \tag{3.35}$$

To avoid the negative contribution to the energy in (3.35) and following the methods of quantum field theory, we need to impose anticommutation rules on the creation and annihilation operators, that is:

Table 3.2 Two by one spinor solutions of Dirac equation and its Hermitian conjugate for the different plane wave forms

	$e^{+ikx}e^{+i\omega_k t}$	$e^{-ikx}e^{+i\omega_k t}$	$e^{+ikx}e^{-i\omega_k t}$	$e^{-ikx}e^{-i\omega_k t}$
ξ_k	$\begin{pmatrix}\sqrt{\omega+\beta k}\\ \sqrt{\omega-\beta k}\end{pmatrix}$	$\begin{pmatrix}\sqrt{\omega-\beta k}\\ \sqrt{\omega+\beta k}\end{pmatrix}$	$\begin{pmatrix}-\sqrt{\omega-\beta k}\\ \sqrt{\omega+\beta k}\end{pmatrix}$	$\begin{pmatrix}-\sqrt{\omega+\beta k}\\ \sqrt{\omega-\beta k}\end{pmatrix}$
$\xi_k^\dagger$	$\begin{pmatrix}\sqrt{\omega-\beta k}\\ -\sqrt{\omega+\beta k}\end{pmatrix}$	$\begin{pmatrix}\sqrt{\omega+\beta k}\\ -\sqrt{\omega-\beta k}\end{pmatrix}$	$\begin{pmatrix}\sqrt{\omega+\beta k}\\ \sqrt{\omega-\beta k}\end{pmatrix}$	$\begin{pmatrix}\sqrt{\omega-\beta k}\\ \sqrt{\omega+\beta k}\end{pmatrix}$

Note that the Hermitian conjugate should be a row vector (1×2) but for compactness we render its transpose in the table

$$\left\{a_k^\dagger, a_{k'}\right\} = a_k^\dagger a_{k'} + a_{k'} a_k^\dagger = \delta_{k,k'} \tag{3.36}$$

With these anticommutation rules, the energy takes the form

$$E - E_0 = \sum\nolimits_k \omega\left(a_k^\dagger a_k + a_k^* a_k^{\dagger *}\right) \tag{3.37}$$

In (3.37), we have lumped the terms that arise from the delta function of (3.36) into a negative zero point energy, $E_0 = -\sum_k \omega(k)$.

The anticommutation rules given by (3.36) also imply that $a_k^\dagger a_k^\dagger = 0$. This is Pauli's exclusion principle stating that one cannot create more than one particle per state. The anticommutation rules indicate that the modes have fermion character. How can these anticommutation rules be interpreted physically? The fields given by (3.32) and (3.33) are composed of forward and backward traveling waves whose relative amplitudes are determined solely by the spinor part of the wave function $\xi_k e^{ikx} e^{-i\omega t}$. If one tries to add an additional traveling component to one of the quasi-standing solution via the operator $a_k^* a_k^{\dagger *}$, one would create a new state that could not be a solution of the Dirac equation (3.8a, 3.8b). Such a state is not allowed. This restriction provides a physical interpretation of Pauli's exclusion principle.

3.3.3.3.3 Number Operators

In the previous subsection, we have shown that the energy is an invariant. There exist other symmetries. In particular, the Dirac-like equations are invariant upon the transformation involving a phase shift. Here we show that as a consequence, the number operator:

$$N = \int dx \psi^\dagger \boldsymbol{\sigma}_x \psi \tag{3.38}$$

is also an invariant. First, we start from the field operator given by (3.32) and replace k by $-k$ in the second term of the summation. This leads to:

$$\begin{aligned}\psi(x) &= \sum_k \frac{1}{\sqrt{2\omega}}\left[a_k\xi_k e^{ikx}e^{-i\omega t} + a^*_{-k}\xi^*_{-k}e^{ikx}e^{i\omega t}\right] \\ &= \sum_k \frac{1}{\sqrt{2\omega}}\left[a_k\xi_k e^{ikx}e^{-i\omega t} + b_k\eta_k e^{ikx}e^{i\omega t}\right]\end{aligned} \tag{3.39}$$

Similarly, we rewrite the field

$$\begin{aligned}\psi^\dagger(x) &= \sum_k \frac{1}{\sqrt{2\omega}}\left[a_k^\dagger\xi_k^\dagger e^{-ikx}e^{i\omega t} + a^{\dagger *}_{-k}\xi^{\dagger *}_{-k}e^{-ikx}e^{-i\omega t}\right] \\ &= \sum_k \frac{1}{\sqrt{2\omega}}\left[a_k^\dagger\xi_k^\dagger e^{-ikx}e^{i\omega t} + b_k^\dagger\eta_k^\dagger e^{-ikx}e^{-i\omega t}\right]\end{aligned} \tag{3.40}$$

Inserting (3.39) and (3.40) into (3.38), and following the same approach as in Sect. 3.3.3.2., we obtain:

$$N = \sum_k \frac{1}{2\omega}\left(a_k^\dagger a_k \xi_k^\dagger \boldsymbol{\sigma}_x \xi_k + b_k^\dagger b_k \eta_k^\dagger \boldsymbol{\sigma}_x \eta_k\right) \tag{3.41}$$

Using Table 3.2 for the spinors and their Hermitian conjugate, we obtain

$$N = \sum_k \left(a_k^\dagger a_k + b_k^\dagger b_k\right) \tag{3.42}$$

This is clearly an invariant. The operator $\bar{a}_k a_k$ counts the number of waves propagating in a given directions and $b_k^\dagger b_k$ correspond to the number of waves propagating in the opposite direction. Considering a field representation that includes only one type of plane wave and therefore, looking at the first term of (3.42), we define the operator:

$$N_k = \frac{1}{2\omega} a_k^\dagger a_k \xi_k^\dagger \boldsymbol{\sigma}_x \xi_k \tag{3.43}$$

We recall that $\xi_k^\dagger \boldsymbol{\sigma}_x \xi_k = 2\omega = \omega + \beta kh + \omega - \beta kh$. Note that here, we have renormalized the spinors with respect to the arbitrary complex constant c_0. The last equality suggests, the introduction of direction switching operators $\boldsymbol{S}_+ = \frac{1}{2}(\boldsymbol{\sigma}_x + i\boldsymbol{\sigma}_y) = \begin{pmatrix} 0 & 1 \\ 0 & 0 \end{pmatrix}$ and $\boldsymbol{S}_- = \frac{1}{2}(\boldsymbol{\sigma}_x - i\boldsymbol{\sigma}_y) = \begin{pmatrix} 0 & 0 \\ 1 & 0 \end{pmatrix}$. These operators anticommute: $\boldsymbol{S}_+\boldsymbol{S}_- + \boldsymbol{S}_-\boldsymbol{S}_+ = \boldsymbol{I}$. Using this later relation, we reformulate (3.43) in the form:

$$N_k = \frac{1}{2\omega} a_k^\dagger a_k \xi_k^\dagger \boldsymbol{I}\boldsymbol{\sigma}_x \xi_k = a_k^\dagger a_k \xi_k^\dagger \left(\frac{1}{2\omega} S_+ S_- \sigma_x + \frac{1}{2\omega} S_- S_+ \sigma_x\right)\xi_k \tag{3.44}$$

The operator $\frac{1}{2\omega} S_+ S_- \sigma_x$ defines the occupancy of one of the directions along the chain of mass and springs. Its Eigen values are given by:

$n_k^+ = \xi_k^\dagger \frac{1}{2\omega} S_+ S_- \sigma_x \xi_k = \frac{\omega + \beta kh}{2\omega}$. The operator $\frac{1}{2\omega} S_- S_+ \sigma_x$ defines the occupancy of the other opposite direction. Its Eigen values are given by: $n_k^- = \xi_k^\dagger \frac{1}{2\omega} S_- S_+ \sigma_x \xi_k = \frac{\omega - \beta kh}{2\omega}$. Considering the dispersion relation for the positive branch of the band structure, $\omega = +\sqrt{\alpha^2 + \beta^2 (kh)^2}$, the Eigen values, n_k^+ and n_k^-, can be rewritten in terms of the wave vector k only:

$$n_k^\pm = \frac{1}{2} \pm \frac{\beta kh/\alpha}{2\sqrt{1 + \left(\frac{\beta kh}{\alpha}\right)^2}} \tag{3.45}$$

At the origin of the band structure, $k = 0$, and $n_k^+ = n_k^- = 1/2$. The direction occupancy is the same for the two opposite directions, hence the state of the system is described by a standing wave. For $k \to \infty$, $n_k^+ = 1$ and $n_k^- = 0$, the phononic structure supports a pure traveling wave with only one direction favored. For finite k, one has the superposition of traveling and standing waves. By spanning the dispersion relation, one spans the "direction" states. One can quantify the "amount" of traveling wave character of the wave function with the quantity: $(n_k^+ - n_k^-)$. When expressed as a function of frequency, this quantity takes the functional form: $(n_k^+ - n_k^-) = \frac{\sqrt{\omega^2 - \alpha^2}}{\omega}$. This clearly shows that the wave function character evolves from a pure standing wave at $\omega = \alpha$ to a pure traveling wave as the frequency increases to infinity.

The fermion-like behavior of the elastic wave opens up opportunities in the control of the direction of propagation of elastic waves. For instance, we have just seen that the two spinor solution represents coupling between wave propagating in opposite directions. The spinor part of the wave function can be projected on the orthonormal basis $\begin{pmatrix} 1 \\ 0 \end{pmatrix}$ and $\begin{pmatrix} 0 \\ 1 \end{pmatrix}$ representing the possible directions of propagation of the wave. This enables us to encode information in the relative weight (phase) of the directions of propagation by controlling the wavenumber, k. This control will enable applications in the field of information encoding and processing.

3.3.3.3.4 Measurement of the Spinor State and Information Encoding and Processing

The spinor part of the wave function when projected on the orthonormal basis $\begin{pmatrix} 1 \\ 0 \end{pmatrix}$ and $\begin{pmatrix} 0 \\ 1 \end{pmatrix}$ represents a superposition of states in the possible directions of propagation of the wave. In conventional quantum systems a measurement on a superposition of states would collapse the wave function into a pure state. A number of measurements is then required to obtain probabilistic information on the characteristics of the superposition of states. In this subsection, we show that one can use measurement of the transmission coefficient to determine the observables, n_k^+ and n_k^-, and more specifically the spinor part of the wave function.

Fig. 3.6 Schematic illustration of the device composite system constituted of a segment of mass-spring system with fermion-like behavior sandwiched between two semi-infinite harmonic crystals

For this, we consider a device which is composed of a segment of mass-spring system with side springs attached to a rigid substrate sandwiched between two semi-infinite harmonic crystals (Fig. 3.6).

The number of masses in the central region of the sandwich is denoted N_c. We assume that all the masses, m, have the same value. We also assume that the spring constants of the semi-infinite regions are the same as the mass-to-mass spring constant, κ_1. The spring constant of the side springs is κ_2.

We are particularly interested in the transmission of plane waves launched from the left semi-infinite chain through the fermion-like segment. For this, we calculate the transmission coefficient of the composite system of Fig. 3.6. We employ the methods of the Interface Response Theory (IRT) and the code given in Appendix, Chap. 1. This code calculates the transmission coefficient of an infinite harmonic crystal perturbed by attaching a number of harmonic finite side chains. Provided that the side chains contain only one atom with a very large mass, this code can reproduce the behavior of the system of Fig. 3.6 with side spring attached to a rigid substrate.

We report in Fig. 3.7, the transmission coefficient as a function of frequency for a segment composed of $N_c = 80$ masses with side springs. We have chosen $m = 1$ and $\kappa_1 = 1$, that is $\beta^2 = 1$. The side springs have stiffness $\kappa_2 = 0.1$ *i.e.*, $\alpha = 0.316$.

The transmission coefficient is zero for frequencies up to the value $\omega = \alpha = 0.316$. This zero transmission corresponds to the gap in the band structure. Waves that are launched and propagate in the right semi-infinite harmonic crystal cannot propagate through the central segment of the composite. Inside the passing band of the central segment, transmission occurs. The oscillatory nature of the transmission coefficient is due to the finite size and the discrete nature of the central segment. The frequency equal to 2 corresponds to the upper frequency of waves that can be supported by a discrete one-dimensional harmonic crystal. The decrease of the transmission coefficient as the frequency of 2 is approached results from the curvature (zero group velocity) of the band of the one-dimensional harmonic crystal. Therefore, one cannot establish a complete one-to-one correspondence between the long wavelength limit result (dotted line) and the transmission through a discrete system. However, the general trend of an increase in transmission as the frequency increases from the bottom of the passing band of the central segment structure is representative of both the discrete and long-wavelength limit fermion-like behavior. The most important observation is that one can establish a one-to-one correspondence between a measurable scalar quantity: transmission coefficient and

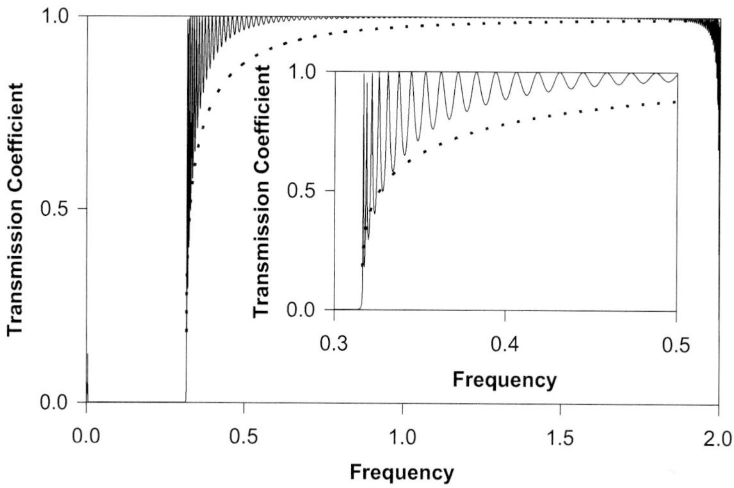

Fig. 3.7 Coefficient of transmission (*solid line*) of a plane wave launched in the right semi-infinite chain through the composite system illustrated in Fig. 3.6. The *inset* is a magnified region around the *bottom* of the positive branch of the band structure. The *dotted curve* represents the traveling wave character of the wave function in the long wavelength limit that is quantified by $\sqrt{n_k^+ - n_k^-}$

the spinor characteristics of the wave that can be supported by the central segment of the composite. For instance for an incident wave $e^{ikx}e^{i\omega t}$ at $\omega = \alpha$ ($k=0$) the transmission coefficient is zero and the spinor part of the wave function takes on the form: $\alpha\begin{pmatrix} 1 \\ 1 \end{pmatrix}$. The components of the spinor vary monotonically as the frequency is increased (*i.e.*, the wave number k is increased) until it takes a form approaching: $\sqrt{2\beta kh}\begin{pmatrix} 1 \\ 0 \end{pmatrix}$. A similar tuning of the spinor part of the wave function can be achieved by fixing the frequency of the incident waves and varying the stiffness κ_2 (*i.e.*, α) of the side springs relative to the stiffness of the 1D harmonic crystal. If we define a basis set $|0\rangle$ and $|1\rangle$ with a matrix representation of the form $\begin{pmatrix} 1 \\ 0 \end{pmatrix}$ and $\begin{pmatrix} 0 \\ 1 \end{pmatrix}$ we can create a superposition of states $u(\omega)|0\rangle + v(\omega)|1\rangle$ at fixed α or $u(\alpha)|0\rangle + v(\alpha)|1\rangle$ at fixed ω. Note that launching the incident wave from the right hand side of the composite harmonic system corresponds to an incident wave with $\omega < 0$. In that case, for instance for $\omega = \alpha$ ($k = 0$), the spinor takes the value $\alpha\begin{pmatrix} -1 \\ 1 \end{pmatrix}$. This spinor is equivalent to the value $\alpha\begin{pmatrix} 1 \\ -1 \end{pmatrix}$ to within a factor of -1. Therefore, for a given incident wave of frequency ω, one can visualize the operation of tuning α, between zero and $\omega = \alpha$ as that of a Hadamard transformation. As seen previously, the Hadamard transformation can then be physically achieved by modulating the strength of an external stimulus such as the magnitude of an external field. In its matrix representation, the Hadamard transformation takes the form $\begin{pmatrix} 1 & 1 \\ 1 & -1 \end{pmatrix}$ and transforms a wave launched from the left $\begin{pmatrix} 1 \\ 0 \end{pmatrix}$ into the standing wave whose spinor is proportional to $\begin{pmatrix} 1 \\ 1 \end{pmatrix}$. It also transforms a wave propagating from the right $\begin{pmatrix} 0 \\ 1 \end{pmatrix}$ into the standing wave whose spinor part is proportional to $\begin{pmatrix} 1 \\ -1 \end{pmatrix}$. The action of applying the Hadamard gate a second time is equivalent to detuning the parameter α between $\omega = \alpha$ and zero. This transforms the states $\begin{pmatrix} 1 \\ 1 \end{pmatrix}$ and $\begin{pmatrix} 1 \\ -1 \end{pmatrix}$ back into the traveling states $\begin{pmatrix} 1 \\ 0 \end{pmatrix}$ and $\begin{pmatrix} 0 \\ 1 \end{pmatrix}$. This transformation can be detected easily through the change in transmission from zero to 1. The control of the relative magnitude and sign of the spinor components of the wave function enables the encoding of information in the phase of the superposition of states.

3.3.3.4 Examples of Physical Realization of Elastic and Acoustic Systems Supporting Fermion-Like Waves

In addition to the mass-spring elastic system used to introduce the concept of elastic waves with fermion-like behavior, we present in the subsections below some examples of elastic and acoustic systems that exhibit similar behavior. These examples include elastic media supporting rotational waves as well as pressure waves in ducts of slowly varying cross-sectional area (such as horns).

3.3.3.4.1 Micromechanics Model Supporting Rotational Modes

Here, we describe a 1D mass-spring model that supports rotational waves. This model is based on a discrete linear one-dimensional micromechanics model that includes longitudinal, shear and rotational degrees of freedom [31, 32]. This 1D discrete lattice model consists of an infinite chain of square block elements connected with multiple harmonic springs. Each element in the model is considered to have two translational degrees of freedom (displacement in the x and y directions) and one rotational degree of freedom (rotation about an axis perpendicular to the xy-plane). Figure 3.7 shows the repeatable unit cells for a monoblock lattice models with periodicity (h).

Three different harmonic springs (spring constants k_0, k_1, and k_2) connect different parts of the block elements. The element in Fig. 3.8 has mass (m) and moment of inertia (I). The block constituting the nth unit cell has x-displacement

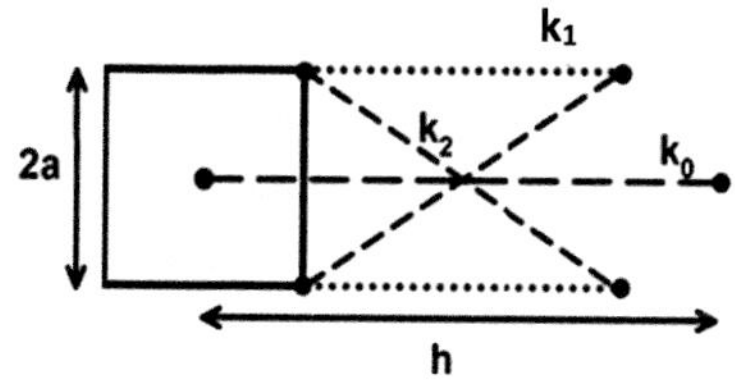

Fig. 3.8 Schematic illustration of a discrete micromechanics model that supports rotational waves. Unit cell of a monoblock model with elements (blocks) connected by three types of harmonic springs (spring constants k_0, k_1, and k_2). Each element possesses translational longitudinal (u), shear (v) and rotational (φ) degrees of freedom

(u_n), y-displacement (v_n) and rotation component (φ_n). u_n and v_n represent displacements associated with longitudinal and transverse vibrations, respectively. The potential energy associated with the elastic connections of elements (n) and ($n + 1$) in the monoblock chain is written as follows:

$$E_{n,n+1} = \frac{1}{2}K_0(u_{n+1} - u_n)^2 + \frac{1}{2}K_1\left[(v_{n+1} - v_n) + \frac{h}{2}(\varphi_{n+1} + \varphi_n)\right]^2 + \frac{1}{2}K_2(\varphi_{n+1} - \varphi_n)^2 \quad (3.46)$$

where $K_0 = \left(\frac{k_0}{h^2} + \frac{2k_1}{l^2} + \frac{2k_2 l^2}{l_d^4}\right)$, $K_1 = \left(\frac{2k_2(2a)^2}{l_d^4}\right)$, $K_2 = \left(\frac{2a^2 k_1}{l^2}\right)$, $l = h - (2a)$, $l_d = \sqrt{\left(l^2 + (2a)^2\right)}$. Accordingly, the equations of motion for the element in the nth unit cell of the monoblock lattice are written as:

$$m\frac{d^2u_n}{dt^2} = K_0(u_{n+1} - 2u_n + u_{n-1}) \quad (3.47a)$$

$$m\frac{d^2v_n}{dt^2} = K_1(v_{n+1} - 2v_n + v_{n-1}) + \frac{hK_1}{2}(\varphi_{n+1} - \varphi_{n-1}) \quad (3.47b)$$

$$I\frac{d^2\varphi_n}{dt^2} = K_2(\varphi_{n+1} - 2\varphi_n + \varphi_{n-1}) + \frac{hK_1}{2}(v_{n-1} - v_{n+1}) - \frac{h^2K_1}{4}(\varphi_{n+1} + 2\varphi_n + \varphi_{n-1}) \quad (3.47c)$$

The band structure of such a lattice would contain three bands corresponding to the three degrees of freedom (u, v, and φ). However, for the sake of simplicity, we now restrict this model to the propagation of rotational waves by allowing only rotation of the blocks about their center of mass and by constraining shear and longitudinal displacements in the lattice. The equation of motion associated with the rotational degrees of freedom (3.47c) then takes the simpler general form:

$$I\frac{\partial^2 \varphi_n}{\partial t^2} = K_1{}'(\varphi_{n+1} - 2\varphi_n + \varphi_{n-1}) - K_2{}'\varphi_n \tag{3.48}$$

$K_1{}' = K_2 - \frac{h^2 K_1}{4}$, and $K_2^{'} = h^2 K_1$. Dividing the equation by the moment of inertia, I, yields our rotational wave equation:

$$\frac{\partial^2 \varphi_n}{\partial t^2} - \beta^2(\varphi_{n+1} - 2\varphi_n + \varphi_{n-1}) + \alpha^2\varphi_n = 0 \tag{3.49}$$

with $\beta^2 = \frac{K_1{}'}{I}$ and $\alpha^2 = \frac{K_2{}'}{I}$.

Equation (3.49) takes the form of the discrete Klein–Gordon equation. An example of a two-dimensional phononic crystal supporting rotational waves is discussed in details in Appendix 2.

3.3.3.4.2 Acoustic Klein–Gordon Equation

The propagation of low frequency sound waves along a rigid duct of slowly varying cross section, $S(x)$, is described by Webster's horn equation [33]. The Webster equation can be written as:

$$\frac{\partial^2 P(x,t)}{\partial t^2} - c^2\left\{\frac{1}{S(x)}\frac{\partial}{\partial x}\left(S(x)\frac{\partial P(x,t)}{\partial x}\right)\right\} = 0 \tag{3.50}$$

$P(x, t)$ is the excess pressure and c is the constant speed of sound in the ambient medium. Introducing the variable:

$$\psi(x,t) = P(x,t)S(x)^{1/2} \tag{3.51}$$

into (3.50) leads to [34, 35]:

$$\frac{\partial^2 \psi}{\partial t^2} - \beta^2\left\{\frac{\partial^2 \psi}{\partial x^2} - V(x)\psi\right\} = 0 \tag{3.52}$$

with $V(x) = (d^2S(x)^{1/2}/dx^2)/S(x)^{1/2}$. For exponentially varying grading $S(x) = e^{ax}$ where a is some geometrical coefficient, the Webster equation becomes the Klein–Gordon:

$$\frac{\partial^2 \psi}{\partial t^2} - \beta^2\frac{\partial^2 \psi}{\partial x^2} + \alpha^2\psi = 0 \tag{3.53}$$

Acoustic waves in ducts of slowly varying cross section may therefore possess non-conventional topology that can be revealed by applying Dirac formalism to (3.53).

3.4 Extrinsic Topological Phononic Structure

Extrinsic topological phononic structures can be created by applying external stimuli that break symmetry. Here, we consider the periodic spatial modulation of the stiffness of a one-dimensional elastic medium and its directed temporal evolution that breaks time reversal symmetry. The bulk elastic states of this time-dependent superlattice do not possess the conventional mirror symmetry in momentum space (*i.e.*, particle-hole symmetry is broken) leading to non-reciprocity in the direction of propagation of the waves. The wave functions of bulk elastic waves are supported by a manifold in momentum space that has the non-conventional torsional topology of a strip with a single twist. Immunity to back-scattering arises from the non-conventional topology and the non-reciprocity in the propagation of these elastic waves.

3.4.1 Time-Dependent Elastic Superlattice

To realize the time-dependent elastic superlattice, we consider the propagation of longitudinal elastic waves along a one-dimensional material supporting a spatial and temporal sinusoidal modulation of its stiffness. Unique properties of some materials such as the giant photo-elastic effects in chalcogenide glasses [36] can be exploited to practically achieve the desired stiffness modulations by, for instance, illuminating the material with light of spatially and temporally varying intensity. It has been shown that illuminating Ge-Se chalcogenide glasses with near bandgap laser radiation of increasing power results in a reduction of the longitudinal elastic constant (C_{11}) by nearly 50%. This photo-softening is athermal and reversible making it ideal as a means to realize time-dependent modulations. The stiffness modulation may also be achieved by various other means such as the application of time-space dependent magnetic fields to a magneto-elastic medium, or the modulation of voltage applied to a medium composed of piezoelectric elements or the mechanical stimulation of a non-linear elastic medium. Here we consider the medium to be composed of a Ge-Se chalcogenide glass of composition $GeSe_4$ [36]. Depending on the power of the laser irradiating the glass, C_{11} values for $GeSe_4$ can vary between 9.2 GPa (full-power) to 18.4 GPa (zero-power) [36]. We assume constant density for $GeSe_4$ (4361 kg/m^3), therefore the minimum and maximum values of C_{11} coincide with sound velocities of 1452 m/s and 2054 m/s, respectively. By itself, a block of $GeSe_4$ is merely a homogeneous medium with constant elastic properties. If, however, the glass were placed under an array of lasers, the elastic properties of the material could be modulated in space and/or time by dynamically adjusting the power of each element in the laser array. This configuration is the basis for the time-dependent elastic superlattice described hereafter. The vibrational properties of this system are investigated numerically. We represent the time-dependent elastic superlattice by a discrete one-dimensional mass-spring system with a spatial sinusoidal modulation of the stiffness of the springs that propagates in time with the velocity $\pm V$ (Fig. 3.9a). Individual masses ($m = 4.361 \cdot 10^{-9}$ kg) are equally spaced by $a = 0.1$ mm. The masses are connected by springs which spring constant varies between 920,000 and 1,840,000 kgs^{-2}. The study of the dynamics of the discretized time-dependent model superlattice is amenable to the method of molecular dynamics (MD). For the calculation of the elastic band structure of the superlattice, we use a one-dimensional chain that contains $N = 3200$ masses with Born–Von Karman boundary conditions. The system takes the form of a ring. We have chosen the value of 100 inter-mass spacings for the period of the stiffness modulation, L. The dynamical trajectories generated by the MD simulation are analyzed within the framework of the Spectral Energy Density (SED) method [37] for generating the elastic band structure of the model superlattice. The SED method is detailed in Appendix 3. To ensure adequate sampling of the system's phase-space our reported SED calculations represent an average over 15 individual MD simulations each with time step 1.5 ns and total simulation time of 2^{22} time steps. We report in Fig. 3.9b and c, the calculated band structure of the superlattice for two velocities of the spatial modulation, namely 0 and 350 m/s.

The band structure of the time-independent superlattice (Fig. 3.9b) exhibits the usual band folding features with gaps forming at the edge of the Brillouin zone. The band structure has the mirror symmetry in momentum space about the origin characteristic of time reversal symmetry. In this case, since the system studied takes the form of a ring, it supports degenerate counter-propagating elastic eigen modes. One consequence of the time-dependence of the stiffness modulation is the loss of the mirror symmetry in k-space which is indicative of breaking both time reversal symmetry and particle-hole symmetry (Fig. 3.9c). Particle-hole symmetry is broken when the condition $\omega^{+}(k) = -\omega^{-}(k)$ is not satisfied. $\omega^{+}(k)$ and $\omega^{-}(k)$ are positive (particle) and negative (hole) eigen values of the system, respectively. Here, in addition to the presence of bands reminiscent of the time-independent bands, the band structure of the time-dependent superlattice contains a series of faint frequency shifted bands. The frequency shift amounts to multiples of $\Omega = \frac{2\pi V}{L}$. The intensity of these bands decreases as the shift in frequency increases. More remarkable is the formation of hybridization gaps between the frequency-shifted bands and the original time-independent bands. Two such gaps appear in the positive frequency-wavenumber quadrant of the first Brillouin zone at the same wavenumber $+k_g$. Such gaps do not appear in the positive frequency-negative wave number quadrant (*i.e.*, at $-k_g$) thus the loss of mirror symmetry. Changing the sign of the modulation velocity leads to a horizontal flip of the band structure. In the frequency range corresponding to the band gaps, the time-dependent ring-like

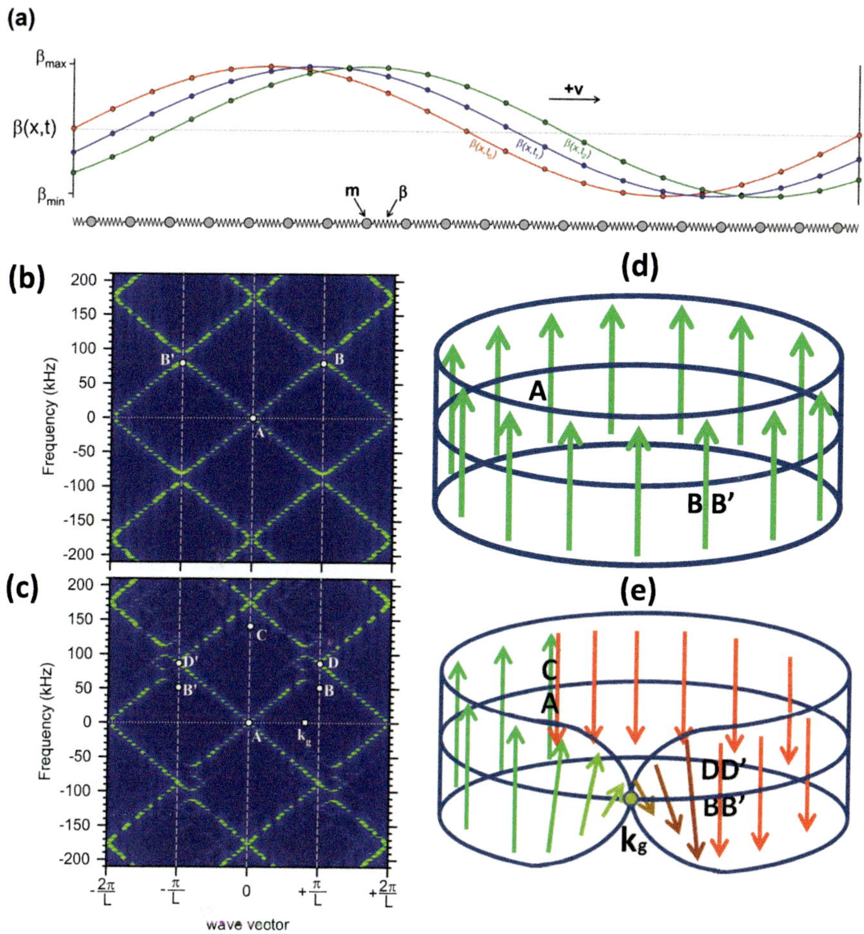

Fig. 3.9 Elastic waves in time-dependent elastic superlattice. (**a**) One-dimensional harmonic mass-spring system with spatial and temporal sinusoidal modulation of the spring stiffness: $\beta(x,t)$ as a realization of an elastic time-dependent superlattice. The spatial modulation propagates in time with the velocity $\pm V$. (**b**) and (**c**) Calculated elastic wave band structure in the cases of modulation velocities of 0 (time-independent superlattice) and 350 m/s, respectively. (**d**) Illustration of the conventional momentum space (k-space) manifold supporting Bloch waves in the time-independent superlattice. Parallel transport of a vector field along a $\frac{2\pi}{L}$ closed path in k space that starts and finishes at the origin of the band structure A and goes through points B and B′ shows no accumulation of phase as the tangent vector do not change orientation along the path. (**e**) Illustration of the k-space manifold supporting elastic waves in the time-dependent superlattice. The manifold takes the form of a strip with a single twist centered on the wavenumber k_g corresponding to the band gaps in the band structure. The accumulation of phase along a $\frac{4\pi}{L}$ long closed path in k space that starts and ends at the origin A and goes through points B, B′, C, D, and D′, is shown through parallel transport of a vector field. The amplitude of the wave function accumulates a π phase every time the path crosses k_g. The phase change of the amplitude is represented by the change in orientation of the vector tangent to the manifold as it is transported along the closed path

superlattice does not support degenerate counter-propagating elastic eigen modes anymore. At these frequencies, the degeneracy in the direction of propagation is lifted and the time-dependent mass-spring ring supports left-handed or right-handed modes depending upon the velocity of the modulation.

3.4.2 *Multiple time Scale Perturbation Theory of the Time-Dependent Superlattice*

To illustrate the origin of the loss of mirror symmetry in the band structure of the time-dependent superlattice, we construct perturbative solutions to the elastic wave functions. In the long-wavelength limit, propagation of longitudinal elastic waves in a one-dimensional medium perturbed by a spatio-temporal modulation of its stiffness, $C(x,t)$, obeys the following equation of motion:

$$\rho\frac{\partial^2 u(x,t)}{\partial t^2} = \frac{\partial}{\partial x}\left(C(x,t)\frac{\partial u(x,t)}{\partial x}\right) \tag{3.54}$$

In (3.54), $u(x,t)$ is the displacement field and ρ is the mass density of the medium. For the sake of analytical simplicity we choose a sinusoidal variation of the stiffness with position and time:

$$C(x,t) = C_0 + 2C_1 \sin\left(Kx + \Omega t\right) \tag{3.55}$$

where C_0 and C_1 are positive constants. $K = \frac{2\pi}{L}$ where L is the period of the stiffness modulation. Ω is a frequency associated with the velocity of the stiffness modulation, V. The quantities K and V are independent. The sign of Ω determines the direction of propagation of the modulation. In this representation, the maximum stiffness of the chalcogenide material is $C_{11}^{max} = C_0 + 2C_1$.

The periodicity of the modulated one-dimensional medium suggests that we should be seeking solutions of (3.54) in the form of Bloch waves:

$$u(x,t) = \sum_k \sum_g u(k,g,t)e^{i(k+g)x} \tag{3.56}$$

where $x \in [0,L]$. The wave number k is limited to the first Brillouin zone: $\left[\frac{-\pi}{L},\frac{\pi}{L}\right]$ and $g = \frac{2\pi}{L}m$ with m being a positive or negative integer. With this choice of form for the solution and inserting (3.55) into (3.54), the equation of propagation takes the form:

$$\frac{\partial^2 u(k+g,t)}{\partial t^2} + v_a^2(k+g)^2 u(k+g,t) = i\varepsilon\left\{f(k')u(k',t)e^{i\Omega t} + h\left(k''\right)u\left(k'',t\right)e^{-i\Omega t}\right\} \tag{3.57}$$

where $f(k) = Kk + k^2$, $h(k) = Kk - k^2$, $k' = k + g - K$ and $k'' = k + g + K$. In this equation, we have defined: $v_a^2 = \frac{C_0}{\rho}$ and $\varepsilon = \frac{C_1}{\rho}$. We solve this equation by using the multiple time scale perturbation method [38]. For the sake of analytical simplicity, we treat ε as a perturbation and write the displacement as a second order power series in the perturbation, namely:

$$\begin{aligned} u(k+g,\tau_0,\tau_1,\tau_2) &= u_0(k+g,\tau_0,\tau_1,\tau_2) + \varepsilon u_1(k+g,\tau_0,\tau_1,\tau_2) \\ &\quad + \varepsilon^2 u_2(k+g,\tau_0,\tau_1,\tau_2) \end{aligned} \tag{3.58}$$

In (3.58), u_i with $i = 0, 1, 2$ are displacement functions expressed to zeroth-order, first-order and second-order in the perturbation. We have also replaced the single time variable, t, by three variables representing different time scales: $\tau_0 = t$, $\tau_1 = \varepsilon t$, and $\tau_2 = \varepsilon^2 t = \varepsilon^2 \tau_0$. We can subsequently decompose (3.57) into three equations: one equation to zeroth-order in ε, one equation to first-order in ε and a third equation to second-order in ε. This decomposition is illustrated in Appendix 4. The zeroth-order equation represents propagation of an elastic wave in a homogeneous medium.

$$\frac{\partial^2 u_0(k+g,\tau_0,\tau_1,\tau_2)}{\partial \tau_0^2} + v_a^2(k+g)^2 u_0(k+g,\tau_0,\tau_1,\tau_2) = 0 \tag{3.59}$$

Its solution is taking the form of the Bloch wave:

$$u_0(k+g,\tau_0,\tau_1,\tau_2) = a_0(k+g,\tau_1,\tau_2)e^{i\omega_0(k+g)\tau_0} \tag{3.60}$$

To zeroth-order the dispersion relation takes the usual form: $\omega_0(k+g) = v_a|k+g|$. The first-order equation is used to solve for u_1.

$$\begin{aligned} &\frac{\partial^2 u_1(k+g,\tau_0,\tau_1,\tau_2)}{\partial \tau_0{}^2} + \omega_0^2(k+g)u_1(k+g,\tau_0,\tau_1,\tau_2) + 2\frac{\partial^2 u_0(k+g,\tau_0,\tau_1,\tau_2)}{\partial \tau_1 \partial \tau_0} = \\ &\quad i\left\{f(k')u_0(k',\tau_0,\tau_1,\tau_2)e^{i\Omega\tau_0} + h\left(k''\right)u_0\left(k'',\tau_0,\tau_1,\tau_2\right)e^{-i\Omega\tau_0}\right\} \end{aligned} \tag{3.61}$$

The third term in (3.61) is a secular term that is set to zero by assuming that the displacement, $u_0(k+g,\tau_0,\tau_2)$ is not a function of τ_1. Subsequently, we will assume that the displacement at all orders of expansion is independent of odd time scales. The solution to (3.61) is obtained in the form of the sum of homogeneous and particular solutions with split frequency:

$$u_1(k+g,\tau_0,\tau_2) = a_1(k+g,\tau_2)e^{i\omega_0(k+g)\tau_0} + i\frac{f(k')a_0(k',\tau_2)}{\omega_0^2(k+g) - (\omega_0(k')+\Omega)^2 + i\varphi}e^{i(\omega_0(k')+\Omega)\tau_0} + i\frac{h(k'')a_0(k'',\tau_2)}{\omega_0^2(k+g) - (\omega_0(k'')-\Omega)^2 + i\varphi}e^{i(\omega_0(k'')-\Omega)\tau_0} \tag{3.62}$$

We have introduced in the first-order solution given by (3.61) a small damping term $i\varphi$ to address the divergence of the two resonances that occur at $\omega_0^2(k+g) = (\omega_0(k')+\Omega)^2$ and $\omega_0^2(k+g) = (\omega_0(k'')-\Omega)^2$. We will later take the limit $\varphi \to 0$. The first term in the right-hand-side of (3.62) is solution of the homogeneous part of (3.7) and takes the same form as the zeroth-order solution of (3.60). The other two terms are particular solutions. They are equivalent to solutions for a driven harmonic oscillator. As seen in Fig. 3.9c, the particular solutions introduce additional dispersion curves in the band structure of the time-dependent superlattice obtained by shifting the zeroth-order band structure by $\pm\Omega$. These bands are reminiscent of Stokes and anti-Stokes bands in Brillouin scattering. The faint intensity of these bands reflects the non-resonant conditions for the amplitudes in (3.62). We also make the important observation that there is a phase difference of π between the first-order particular solution and the homogeneous (and zeroth-order) solutions of (3.62). This phase is due to the change in sign of the amplitude of the first-order displacement function as the wave number is varied across the resonance.

Finally, the second-order equation of motion is given by:

$$\frac{\partial^2 u_2(k+g,\tau_0,\tau_2)}{\partial \tau_0^2} + \omega_0^2(k+g)u_2(k+g,\tau_0,\tau_2) + 2\frac{\partial^2 u_0(k+g,\tau_0,\tau_2)}{\partial \tau_2 \partial \tau_0} = i\{f(k')u_1(k',\tau_0,\tau_2)e^{i\Omega\tau_0} + h(k'')u_1(k'',\tau_0,\tau_2)e^{-i\Omega\tau_0}\} \tag{3.63}$$

Inserting (3.62) into (3.63), leads to terms of the form $e^{i\omega_0(k+g)\tau_0}$ in the right-hand-side of the equation. These terms lead to secular behavior that can be cancelled by equating them to the third term in the left-hand-side of the equation:

$$2\frac{\partial^2 u_0(k+g,\tau_0,\tau_2)}{\partial \tau_2 \partial \tau_0} = -\left\{f(k')h(k+g)\left(\frac{1}{\omega_0^2(k') - (\omega_0(k+g)-\Omega)^2}\right) + h(k'')f(k+g)\left(\frac{1}{\omega_0^2(k'') - (\omega_0(k+g)+\Omega)^2}\right)\right\}a_0(k+g,\tau_2)e^{i\omega_0(k+g)\tau_0} \tag{3.64}$$

Introducing, $a_0(k+g,\tau_2)=\alpha_0(k+g)e^{i\gamma\tau_2}$, one may rewrite $u_0(k+g,\tau_0,\tau_2)$ as $u_0(k+g,\tau_0,\tau_2)=\alpha_0(k+g)e^{i\gamma\tau_2}e^{i\omega_0(k+g)\tau_0}=\alpha_0(k+g)e^{i\left[\omega_0(k+g)+\gamma\varepsilon^2\right]\tau_0}=\alpha_0(k+g)e^{i\omega_0^*(k+g)\tau_0}$. Then one obtains a correction to $\omega_0(k+g)$, leading to a frequency shift and damping. This frequency shift is most pronounced for values of the wave number leading to strong resonances in (3.62) and is given by:

$$\delta\omega_0(k+g)=\omega_0^*(k+g)-\omega_0(k+g)=\varepsilon^2(\gamma)_{pp}$$
$$=\frac{\varepsilon^2}{2\omega_0(k+g)}\left\{\begin{array}{l} f(k')h(k+g)\left(\frac{1}{\omega_0^2(k')-(\omega_0(k+g)-\Omega)^2}\right)_{pp} \\ +h(k'')f(k+g)\left(\frac{1}{\omega_0^2(k'')-(\omega_0(k+g)+\Omega)^2}\right)_{pp}\end{array}\right\} \tag{3.65}$$

The symbol $(\square)_{pp}$ in this expression represents Cauchy's principle part that results from taking the limit: $\varphi\to 0$. This frequency shift is the signature of the formation of hybridization band gaps between the zeroth-order and the first-order dispersion relations at the resonance wave numbers. Particular solutions of the (3.63) will also include terms in $e^{i(\omega_0(k'\,or\,k'')\pm 2\Omega)\tau_0}$. These terms introduce additional dispersion curves in the band structure of the time-dependent superlattice obtained by shifting the zeroth-order band structure by $\pm 2\Omega$ (see Fig. 3.9c). The denominators of the resonance conditions: $\omega_0^2(k')-(\omega_0(k+g)-\Omega)^2=0$ and $\omega_0^2(k'')-(\omega_0(k+g)+\Omega)^2=0$ determine the location of the formation the two hybridization gaps observed in Fig. 3.9c. These conditions predict hybridization gaps where the lowest first-order dispersion branch $(g=0)$ and second lowest branch $(g=\frac{2\pi}{L})$ intersect a first-order dispersion curve. The two gaps form only on one side (positive or negative side) of the first Brillouin zone depending on the sign of Ω (*i.e.*, the direction of propagation of the modulation of the stiffness). These two gaps occur at the same wave number: k_g. This leads to a band structure that does not possess mirror symmetry about the frequency axis as seen in Fig. 3.9c. The band structure now possesses a center of inversion, the origin, rather than a mirror plane.

3.4.3 Topology of Elastic Wave Functions

We can shed light on the non-conventional topology of the displacement Bloch function in the time-dependent superlattice by following a closed continuous paths in wave number space and determine the phase difference acquired by the amplitude of the wave function over the course of such a cycle. These paths are illustrated in Fig. 3.9b, d and c, e. In the case of the modulated system, we start from the origin of the band structure ($k=0$, point A) and follow the lowest $(g=0)$ zeroth-order branch by increasing k toward the first hybridization gap. The path takes us along the lowest branch of the gap and the wave function transitions from a state corresponding to a zeroth-order type wave ($e^{i\omega_0(k+g)\tau_0}$) to a wave having the characteristics of a first-order wave ($e^{i(\omega_0(k')-\Omega\ )\tau_0}$). As one crosses, k_g and transitions between these two types of solutions, the amplitude of the wave function changes sign and the wave function accumulates a geometric phase difference of π. One then reaches the edge of the Brillouin zone ($k=\frac{\pi}{L}$ or point B). Since the wave functions are Bloch waves, point B is equivalent by translational symmetry to point B′ located on the other edge of the Brillouin zone ($k=-\frac{\pi}{L}$). One then follows the first-order branch corresponding to a wave of the form ($e^{i\left(\omega_0\left(k''\right)-\Omega\right)\tau_0}$) back to the wavenumber $k=0$ (point C). At this stage we have closed a $\frac{2\pi}{L}$ loop in wave number space and the amplitude of the wave function has accumulated a geometric phase of π. Further increase in wave number takes us back along the first-order branch corresponding to the wave: $e^{i(\omega_0(k')-\Omega)\tau_0}$. One then reaches the top of the hybridization gap, again at k_g, and transitions back to the zeroth-order state $e^{i\omega_0(k+g)\tau_0}$. This transition accumulates an additional geometric phase difference of π. At the $k=\frac{\pi}{L}$ edge of the Brillouin zone, one has reached the point D. Point D is equivalent by translational periodicity to D′. We can close the continuous path by increasing k again toward the origin along the lowest dispersion branch of the zeroth-order wave. This action takes us back to the starting point A. This second stage of our continuous path corresponds to closing a second $\frac{2\pi}{L}$ loop in k space. For each complete loop, the displacement function accumulated a π geometric phase when one crosses the wave number k_g *i.e.*, when transitioning between zeroth-order and first-order wave functions at the gap. One therefore needs to complete two loops in k space (*i.e.*, a $4\pi/L$ rotation) to obtain a 2π geometric phase difference in the amplitude of the wave function. This behavior is characteristic of a non-trivial topology of k space whereby the wave function is supported on a wavenumber manifold that has the topology of a strip with a half twist (Fig. 3.9e). Note that here the twist in the manifold is not distributed along the entire length of the strip but is localized in k-space. The strip exhibits no phase difference along most of its length. The local twist leads to a π phase difference only near k_g which is associated with the narrow gap resulting from fully destructive interferences between first-order and zeroth-order waves. We can represent the evolution of the geometric phase of the wave function by following a closed path in k space on this strip and parallel transporting a tangent vector field. Starting at point A, and following a closed loop in k space, the upward-pointing tangent vector remains parallel until it approached the twist in the strip at k_g. The parallel transport condition imposed on the vector leads to a π inversion of the direction in which the vector points. The vectors remain parallel to each other through a full loop ($\frac{2\pi}{L}$ rotation) in k space reaching point C. One needs another full turn to go through the twist a second time and rotate the vector by π again. The vectors remain parallel until they close the continuous path and reach the point A. The vector has accumulated a 2 π phase difference along a $\frac{4\pi}{L}$ closed path. In contrast, we have also illustrated in Fig. 3.9d the manifold for a Bloch wave with conventional topology that corresponds to a time-independent superlattice. In this case the manifold does not possess a twist. The vector field is transported along a single closed path in k space without a change in phase. The amplitude of the wave function in this case does not depend on the wave number.

Through a series of numerical simulations, we now demonstrate the application of the concept of time-dependent modulation of elastic properties in achieving bulk wave propagation functionalities such non-reciprocity and immunity to scattering.

3.4.3.1 Non-reciprocity of Bulk Elastic Wave Propagation

We have seen that hybridization between a zeroth-order wave function and a first-order wave function occurs only on one side of the Brillouin zone. This hybridization forms a gap at the intersection of the zeroth-order and first-order dispersion branches. On the opposite side of the Brillouin zone, the zeroth-order and first-order branches cross and do not form a gap. The zeroth-order and first-order wave functions do not hybridize. This loss of mirror symmetry of the band structure leads to the existence of unidirectional bulk propagative modes within the frequency range of the hybridization gap. This asymmetry can then be exploited to achieve non-reciprocal wave propagation. To illustrate this phenomenon, we consider a finite system composed of a time-dependent superlattice sandwiched between two homogeneous media. Absorbing boundary conditions are imposed at the ends of the sandwich system. The medium to the left of the superlattice contains a source (S) of monochromatic elastic waves. A detector is located in the medium to the right of the superlattice (D).

The sandwich system is discretized and transmission is investigated using MD. The frequency of the incident elastic wave is chosen to be $f_0 = 70$ kHz which falls within the first band gap of the superlattice. For $V > 0$ the incident wave can excite first-order and higher-order type modes with weak intensities which will lead to some low level transmission. It can also excite one zeroth-order mode with positive group velocity which will lead to a more significant transmission at that frequency. The transmission coefficient in that case approaches 1. Excitation of the zeroth-order modes at the frequency f_0 leads through (3.61) to first-order modes at the frequencies $f_0 \pm \Omega/2\pi$ [see (3.63)]. These modes appear as small peaks in Fig. 3.10. For $V < 0$, the zeroth-order mode can be excited has a negative group velocity and will therefore not transmit elastic energy. Only higher-order modes with positive group velocity will transmit a small amount of energy leading to a very small transmission coefficient. The unidirectional propagation of elastic waves in the time-dependent superlattice is enabled by the asymmetric band structure and therefore the existence of a bulk zeroth-order propagative mode within the frequency range of a band gap. The change in sign of the group velocity of the propagative mode with the sign of the stiffness modulation velocity leads to the asymmetry in transmission coefficient. We note that the characteristics of the transmission coefficient are inverted when one launches an incident wave with a frequency, $f_1 = 105$ kHz, falling within the second band gap. This inversion results from the change in sign of the group velocity of the propagative mode associated with band folding.

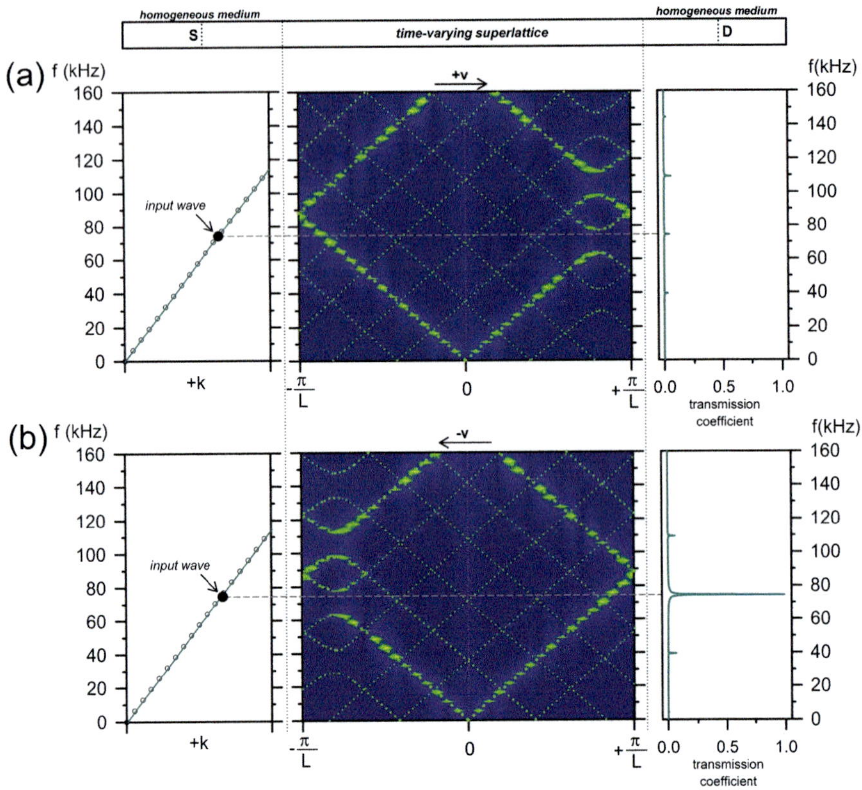

Fig. 3.10 One-way transport of bulk elastic waves—Illustration of the condition for transmission and transmission coefficient of a finite size time-dependent superlattice sandwiched between two homogeneous regions (inlet with sound source (S) and outlet with detector (D)) in the case (**a**) $V > 0$ and (**b**) $V < 0$. In both cases, the band structure of the homogeneous medium is represented on the *left* of the figure with the source emitting a monochromatic wave with frequency f_0. The plots on the *right* represent the transmission spectrum of the superlattice around the incident frequency

3.4.3.2 Demonstration of Topologically Backscattering-Immune Bulk States

One of the signatures of edge states in topological insulators is their robustness with respect to backscattering by defects. The same robustness exists for the bulk elastic waves in the time-dependent superlattice. This phenomenon is demonstrated by inserting a mass defect inside the elastic sandwich considered previously. Here the defect is constructed by changing the value of the masses of a region of the superlattice (Fig. 3.11a). We consider increasing levels of mismatch between the defect and the original superlattice. The defect mass is chosen to take the values $M = 2$, 3 and 4 m. The width of this region is taken to be equal to the period of the

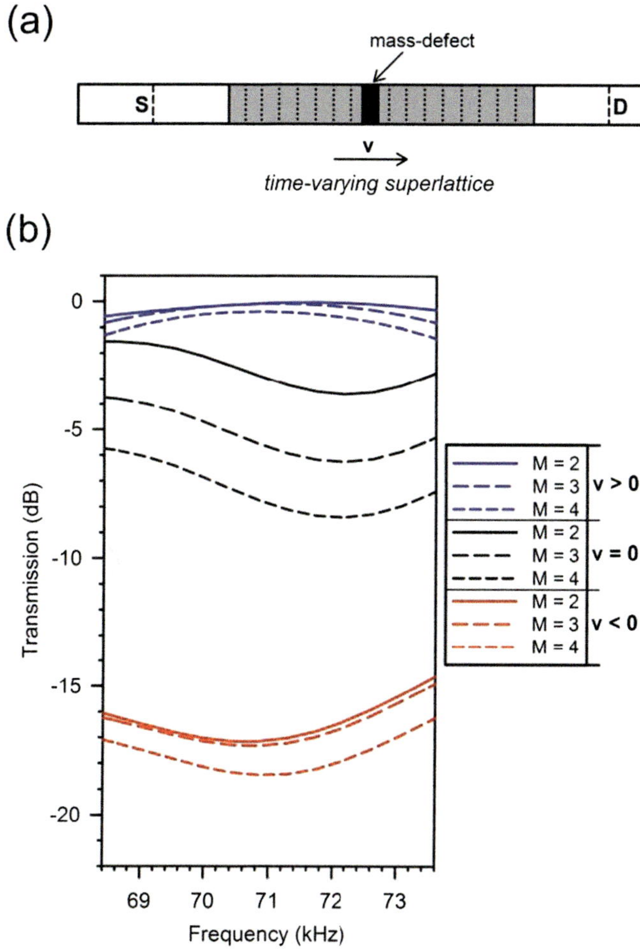

Fig. 3.11 Immunity to backscattering of bulk elastic waves—(**a**) Schematic representation of the sandwich system of Fig. 3.2 with a mass defect (see text for details). (**b**) Transmission through the defected superlattice. The incident wave has a frequency $f_o = 70$ kHz for which propagation is topologically constrained to be unidirectional. The *blue* and *red curves* correspond to time-dependent superlattices with opposite modulation velocities. The *black curves* represent transmission though a time-independent superlattice. *Solid, short-dashed* and *long-dashed curves* correspond to increasing levels of mass mismatch between the defect and the superlattice

modulation. The mass defect does not affect the spatio-temporal modulation of the stiffness.

The incident wave has the frequency f_0 whereby unidirectional propagation is attained. As can be seen in Fig. 3.11b, the transmission of the bulk wave through the time-independent superlattice is increasingly degraded by an increase in the mass mismatch *i.e.*, backscattering. However, the time-dependent superlattice exhibits no significant transmission for $V < 0$ as expected. For $V > 0$, the mass of the defect does not appear to have a significant effect on the transmission. Since the superlattice is unable in this case of supporting a defect induced backscattered wave, the elastic energy essentially propagates without scattering. The time-dependent superlattice demonstrates unambiguously immunity to backscattering.

3.5 Mixed Intrinsic and Extrinsic Topological Phononic Structure

In Sect. 3, we revealed the spinor character of the elastic wave function in the two-crystal system and one crystal coupled elastically to a rigid substrate. In Sect. 4, we applied an external stimulus in the form of a spatio-temporal modulation to break symmetry of a one-dimensional elastic wave guide taking the form of a harmonic chain. Here, we consider the effect of a spatio-temporal modulation on one crystal coupled to a substrate. We are particularly interested in understanding the effect of the external stimulus on the fermion-like elastic waves. More specifically, we are developing further the analogy between classical elastic waves and quantum phenomena. Mechanical analogues of electromagnetic and quantum phenomena have a long history. Maxwell in his seminal paper "A dynamical theory of the electromagnetic field" [39] sought an elastic model of electrical and magnetic phenomena and electromagnetic waves. Mechanical models of physical phenomena abound, including quantum mechanical behavior. For instance, the localization of ultrasound waves in two-dimensional [40] and three-dimensional [41] disordered media serve as mechanical analogues of Anderson localization of electrons. Tunneling of classical waves through phononic crystal barriers establishes a correspondence with its quantum counterpart [42, 43]. The motion of sound waves in convergent fluid flow exhibits the same properties of motion as electromagnetic waves in gravitational fields in space and time [21]. Mixed intrinsic-extrinsic topological phononic crystals have the potential of emulating the behavior of quantum particles in electromagnetic field. In Chap. 6, we will develop a scalar Quantum Field Theory that demonstrates the analogy between our simple one-dimensional elastic system described by a mechanical form of the Klein–Gordon equation subjected to a spatio-temporal modulation of its elastic properties and the one-dimensional Dirac equation in the presence of an electromagnetic field. More detailed examples of mechanical analogues will also be presented in Chap. 6.

3.5.1 One-Dimensional Mass-Spring Harmonic Crystal Grounded to a Substrate with Spatio-Temporal Modulation

The purpose of this section is to investigate the behavior of the field Ψ when the parameter α (spring stiffness, K_0) is subjected to a spatio-temporal modulation *i.e.*,

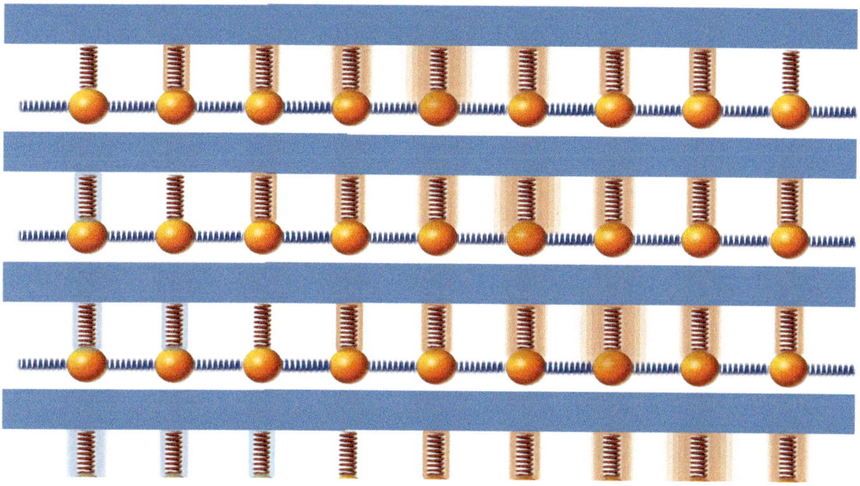

Fig. 3.12 Schematic illustration of the time evolution (*top* to *bottom*) of the harmonic crystal grounded to a substrate with spatial modulation of the side spring stiffness illustrated as a "pink glow" of varying width when $\alpha > \alpha_0$ and a "blue glow" when $\alpha < \alpha_0$

$\alpha = \alpha_0 + \alpha_1 2\sin(Kx + \Omega t)$ where α_0 and α_1 are constants (see Fig. 3.12). Here, $K = \frac{2\pi}{L}$ where L is the period of the modulation. Ω is the frequency modulation and its sign determines the direction of propagation of the modulation.

The question arises as to the effect of such a modulation on the state of the fermion-like phonons. The periodicity of the modulated one-dimensional medium suggests that we should be seeking solutions of (3.8a) in the form of Bloch waves: $\Psi(x,t) = \sum_k \sum_g \psi(k,g,t) e^{i(k+g)x}$ where $x \in [0, L]$. The wave number k is limited to the first Brillouin zone: $\left[\frac{-\pi}{L}, \frac{\pi}{L}\right]$ and $g = \frac{2\pi}{L} l$ with l being an integer. Choosing (3.6), we obtain the modulated Dirac-like equation in the Fourier domain:

$$\left[\boldsymbol{\sigma}_x \frac{\partial}{\partial t} + i\beta \boldsymbol{\sigma}_y (ik^*) - i\alpha_0 \boldsymbol{I}\right] \psi(k^*, t) - \alpha_1 \boldsymbol{I}\left[\psi(k^* + K, t) e^{i\Omega t} - \psi(k^* - K, t) e^{-i\Omega t}\right] = 0 \tag{3.66}$$

where $k^* = k + g$.

Consistent with Quantum Field Theory (QFT) approaches, we solve (3.66) using perturbation theory and in particular multiple time scale perturbation theory (see Appendix 4) up to second-order.

The parameter α_1 is treated as a perturbation ε. The wave function is written as a second-order power series in ε, namely:

$$\psi(k^*, \tau_0, \tau_1, \tau_2) = \psi^{(0)}(k^*, \tau_0, \tau_1, \tau_2) + \varepsilon \psi^{(1)}(k^*, \tau_0, \tau_1, \tau_2) + \varepsilon^2 \psi^{(2)}(k^*, \tau_0, \tau_1, \tau_2)$$

Here $\psi^{(j)}$ with $j = 0, 1, 2$ are wave functions expressed to zeroth, first and second-order. We have also replaced the single time variable, t, by three variables representing different time scales: $\tau_0 = t$, $\tau_1 = \varepsilon t$, and $\tau_2 = \varepsilon^2 t = \varepsilon^2 \tau_0$. Again, we can subsequently decompose (3.66) into equations to zeroth, first and second-order in ε. The zeroth-order equation consists of the Dirac-like equation in the absence of modulation:

$$\left[\boldsymbol{\sigma}_x \frac{\partial}{\partial \tau_0} + i\beta \boldsymbol{\sigma}_y (ik^*) - i\alpha_0 \boldsymbol{I}\right] \psi^{(0)}(k^*, \tau_0, \tau_1, \tau_2) = 0 \tag{3.67}$$

As seen in Sect. 3, its solutions take the form $\psi^{(0)}(k^*, \tau_0, \tau_1, \tau_2) = a^{(0)}(k^*, \tau_1, \tau_2)$ $e^{i\omega_0 \tau_0}$ with $a^{(0)}(k^*, \tau_1, \tau_2) = \begin{pmatrix} a_1^{(0)} \\ a_2^{(0)} \end{pmatrix} = a_0 \begin{pmatrix} \sqrt{\omega_0 + \beta k^*} \\ \sqrt{\omega_0 - \beta k^*} \end{pmatrix}$. $e^{i\omega_0 \tau_0}$ and $a^{(0)}$ represent the orbital and the spinor parts of the solution, respectively. We have the usual eigen values: $\omega_0^2 = \alpha^2 + \beta^2 (k^*)^2$. Inserting the zeroth-order solution into (3.66) expressed to first-order leads to

$$\left[\boldsymbol{\sigma}_x \frac{\partial}{\partial \tau_0} + i\beta \boldsymbol{\sigma}_y (ik^*) - i\alpha_0 \boldsymbol{I}\right] \psi^{(1)}(k^*, \tau_0, \tau_1, \tau_2) = -\sigma_x \frac{\partial}{\partial \tau_1} \psi^{(0)}(k^*, \tau_0, \tau_1, \tau_2)$$
$$+\boldsymbol{I}\left[\psi^{(0)}(k^* + K, \tau_0, \tau_1, \tau_2) e^{i\Omega\tau_0} - \psi^{(0)}(k^* - K, \tau_0, \tau_1, \tau_2) e^{-i\Omega\tau_0}\right] \tag{3.68}$$

The solutions of the first-order Dirac equation are the sum of solutions of the homogeneous equation and particular solutions. The homogenous solution is isomorphic to the zeroth-order solution, it will be corrected in a way similar to the zeroth-order solution as one accounts for higher and higher terms in the perturbation series. Under these conditions and to ensure that there are no secular terms (*i.e.*, terms that grow with time and that are incompatible with the assumption that $\psi^{(1)}$ must be a correction to $\psi^{(0)}$ in the particular solution of (3.68), the pre-factors of terms like $e^{i\omega_0 \tau_0}$ which lead to secular terms are forced to be zero. Subsequently, the derivative of the amplitudes $a^{(0)}(k^*, \tau_1, \tau_2)$ with respect to τ_1 must vanish and these amplitudes only depend on τ_2. The right-hand side of (3.68) reduces to the second term only. The particular solution of that simplified equation contains frequency shifted terms given by:

$$\psi_{1,P}^{(1)} = b_1 e^{i(\omega_0 + \Omega)\tau_0} + b_1' e^{i(\omega_0 - \Omega)\tau_0}$$
$$\psi_{2,P}^{(1)} = b_2 e^{i(\omega_0 + \Omega)\tau_0} + b_2' e^{i(\omega_0 - \Omega)\tau_0}$$

The coefficients b_1, b_1', b_2, b_2' are resonant terms:

$$b_1(k^*) = \frac{i}{\alpha_0} a_1^{(0)}(k^* + K) + \frac{1}{\alpha_0} b_2(k^*)[(\omega_0 + \Omega) + \beta k^*]$$

$$b_1'(k^*) = \frac{-i}{\alpha_0} a_1^{(0)}(k^* - K) + \frac{1}{\alpha_0} b_2'(k^*)[(\omega_0 - \Omega) + \beta k^*]$$

$$b_2(k^*) = -i \frac{\left\{ a_1^{(0)}(k^* + K)[(\omega_0 + \Omega) - \beta k^*] + \alpha_0 a_2^{(0)}(k^* + K) \right\}}{(\omega_0 + \Omega)^2 - (\beta k^*)^2 - \alpha_0^2}$$

$$b_2'(k^*) = +i \frac{\left\{ a_1^{(0)}(k^* - K)[(\omega_0 - \Omega) - \beta k^*] + \alpha_0 a_2^{(0)}(k^* - K) \right\}}{(\omega_0 - \Omega)^2 - (\beta k^*)^2 - \alpha_0^2}$$

In the preceding relations, we have defined: $\omega_0 + \Omega = \omega_0(k^* + K) + \Omega$ and $\omega_0 - \Omega = \omega_0(k^* - K) + \Omega$. We have also omitted the time dependencies for the sake of compactness.

To second order, the Dirac-like equation is written as:

$$\left[\boldsymbol{\sigma}_x \frac{\partial}{\partial \tau_0} + i\beta \boldsymbol{\sigma}_y (ik^*) - i\alpha_0 \boldsymbol{I} \right] \psi^{(2)}(k^*, \tau_0, \tau_2) =$$
$$-\boldsymbol{\sigma}_x \frac{\partial \psi^{(0)}(k^*, \tau_0, \tau_2)}{\partial \tau_2} + \boldsymbol{I}\left[\psi^{(1)}(k^* + K, \tau_0, \tau_2) e^{i\Omega\tau_0} - \psi^{(1)}(k^* - K, \tau_0, \tau_2) e^{-i\Omega\tau_0} \right] \tag{3.69}$$

The derivative $\frac{\partial \psi^{(0)}(k^*, \tau_0, \tau_2)}{\partial \tau_2}$ leads to secular terms. The homogeneous part of the first-order solution does not contribute secular terms but the particular solution does. Combining all secular terms and setting them to zero lead to the conditions:

$$\frac{\partial a_1^{(0)}(k^*)}{\partial \tau_2} = i a_1^{(0)}(k^*) G' + i a_2^{(0)}(k^*) F \tag{3.70a}$$

$$\frac{\partial a_2^{(0)}(k^*)}{\partial \tau_2} = i a_1^{(0)}(k^*) F + i a_2^{(0)}(k^*) G \tag{3.70b}$$

where we have defined:

$$F = \alpha_0 \left\{ \frac{1}{(\omega_0(k^*) - \Omega)^2 - \omega_0^2(k^* + K)} + \frac{1}{(\omega_0(k^*) + \Omega)^2 - \omega_0^2(k^* - K)} \right\}$$

$$G = \left\{ \frac{(\omega_0(k^*) - \Omega) + \beta(k^* + K)}{(\omega_0(k^*) - \Omega)^2 - \omega_0^2(k^* + K)} + \frac{(\omega_0(k^*) + \Omega) + \beta(k^* - K)}{(\omega_0(k^*) + \Omega)^2 - \omega_0^2(k^* - K)} \right\}$$

$$G' = \left\{ \frac{(\omega_0(k^*) - \Omega) - \beta(k^* + K)}{(\omega_0(k^*) - \Omega)^2 - \omega_0^2(k^* + K)} + \frac{(\omega_0(k^*) + \Omega) - \beta(k^* - K)}{(\omega_0(k^*) + \Omega)^2 - \omega_0^2(k^* - K)} \right\}$$

We note the asymmetry of these quantities. The terms G, G' and F diverge within the Brillouin zone of the modulated systems when the condition $(\omega_0(k^*) + \Omega)^2 - \omega_0^2(k^* - K) \to 0$ is satisfied but not when $(\omega_0(k^*) - \Omega)^2 - \omega_0^2(k^* + K) \to 0$. This asymmetry reflects a breaking of symmetry in wave number space due to the directionality of the modulation.

Equations (3.70a, 3.70b) impose second-order corrections onto the zeroth-order solution. We now seek solutions of (3.69) with spinor part $a^{(0)}(k^*, \tau_2)$ satisfying the second-order conditions given by (3.70a, 3.70b). These conditions can be reformulated as $\frac{\partial \tilde{a}}{\partial \tau_2} = iM\tilde{a}$ where the vector $\tilde{a} = \begin{pmatrix} a_1^{(0)} \\ a_2^{(0)} \end{pmatrix}$ and the matrix $M = \begin{pmatrix} G' & F \\ F & G \end{pmatrix}$. Solutions of this 2×2 system of first-order linear equations are easily obtained as: $\tilde{a} = C\tilde{e}\, e^{i\lambda\tau_2} + C'\tilde{e}' e^{i\lambda'\tau_2}$ where λ', $\tilde{e}$, and $\tilde{e}'$ are the eigen values and eigen vectors of the matrix M. The coefficients C and C' are determined by the boundary condition: $\lim_{\varepsilon\to 0}\tilde{a} = a_0 \begin{pmatrix} \sqrt{\omega_0 + \beta k^*} \\ \sqrt{\omega_0 - \beta k^*} \end{pmatrix}$. We find the eigen values $\lambda = \frac{G+G'}{2} + \sqrt{\left(\frac{G-G'}{2}\right)^2 + F^2}$ and $\lambda' = \frac{G+G'}{2} - \sqrt{\left(\frac{G-G'}{2}\right)^2 + F^2}$. The respective eigen vectors are $\tilde{e} = a_0\sqrt{\omega_0(k^*) + \beta k^*}F \begin{pmatrix} \frac{G-G'}{2F} + \sqrt{\left(\frac{G-G'}{2F}\right)^2 + 1} \\ 1 \end{pmatrix}$ and $\tilde{e}' = a_0\sqrt{\omega_0(k^*) + \beta k^*}F \begin{pmatrix} \frac{G-G'}{2F} - \sqrt{\left(\frac{G-G'}{2F}\right)^2 + 1} \\ 1 \end{pmatrix}$. The coefficients are given by

$$C = \frac{1}{F}\frac{1}{2\sqrt{\left(\frac{G-G'}{2F}\right)^2+1}}\left\{\frac{\sqrt{\omega_0(k^*)-\beta k^*}}{\sqrt{\omega_0(k^*)+\beta k^*}} - \frac{G-G'}{2F} + \sqrt{\left(\frac{G-G'}{2F}\right)^2 + 1}\right\}$$

and $C' = \frac{1}{F}\frac{1}{2\sqrt{\left(\frac{G-G'}{2F}\right)^2+1}}\left\{-\frac{\sqrt{\omega_0(k^*)-\beta k^*}}{\sqrt{\omega_0(k^*)+\beta k^*}} + \frac{G-G'}{2F} + \sqrt{\left(\frac{G-G'}{2F}\right)^2 + 1}\right\}$. We note that although the quantities G, G' and F may diverge the ratio $\frac{G-G'}{2F}$ remains finite. The directed spatio-temporal modulation impacts both the orbital part and the spinor part of the zeroth-order modes. The orbital part of the wave function is frequency shifted to $\omega_0 + \varepsilon^2\lambda$ and $\omega_0 + \varepsilon^2\lambda'$. Near the resonant condition: $(\omega_0(k^*) + \Omega)^2 - \omega_0^2(k^* - K) \to 0$, it is the eigen value $\lambda \to \frac{1}{(\omega_0(k^*)+\Omega)-\omega_0(k^*-K)}$ which diverges. This divergence is indicative of the formation of a gap in the dispersion relation as was seen in Sect. 4. The Eigen value $\lambda' \to \frac{1}{(\omega_0(k^*)+\Omega)+\omega_0(k^*-K)} \sim \frac{1}{2(\omega_0(k^*)+\Omega)}$ does not diverge. Since the frequency shift, $\varepsilon^2\lambda'$, is expected to be small compared to $\omega_0(k^*)$, the orbital term $e^{i\lambda\tau_2} \to 1$. Considering that the lowest frequency ω_0 is α_0, this condition would occur for all

k^*. Therefore, the term $e^{i\left(\omega_0+\varepsilon^2\lambda'\right)\tau_0}\sim e^{i\omega_0\tau_0}$ will essentially contribute to the band structure in a perturbative way similar to that of the uncorrected zeroth-order solution or homogeneous parts of the first- or second-order equations. The spinor part of the zeroth-order solution is also modified through the coupling between the orbital and "spin" part of the wave function as seen in the expressions for $\tilde{e}$ and $\tilde{e}'$. This coupling suggests an approach for the manipulation of the "spin" part of the elastic wave function by exciting the medium using a spatio-temporal modulation. Again, these alterations can be achieved by manipulating independently the magnitude of the modulation, α_1 as well as the spatio-temporal characteristics Ω and K.

The perturbative approach used here is showing the capacity of a spatio-temporal modulation to control the "spin-orbit" characteristics of elastic modes in a manner analogous to electromagnetic waves enabling the manipulation of the spin state of electrons [44]. However, the pertubative method is not able to give a complete picture of the effect of the modulation on the entire band structure of the elastic modes. For this, the vibrational properties of the mechanical system are also investigated numerically beyond perturbation theory. We calculate the phonon band structure of the modulated elastic Klein–Gordon equation since its eigen values are identical to those of the modulated Dirac-like equation. We use a one-dimensional chain that contains $N = 2400$ masses, $m = 4.361 \times 10^{-9}$ kg, with Born–Von Karman boundary conditions. The masses are equally spaced by $h = 0.1$ mm. The parameters $\kappa_1 = 0.018363$ kg m^2 s^{-2} and $\kappa_2 = 2295$ kg s^{-2}. The spatial modulation has a period $L = 100\ h$ and an angular frequency $\Omega = 1.934 \times 10^5$ rad/s. We have also chosen the magnitude of the modulation: $\alpha_1 = \frac{1}{10}\alpha_0$. As in Sect. 4.1, we apply MD with an integration time step is $dt = 1.624 \times 10^{-9}$ s. The dynamical trajectories generated by the MD simulation are analyzed within the framework of the Spectral Energy Density (SED) method (See Appendix 3) for generating the band structure. To ensure adequate sampling of the system's phase-space the SED calculations are averaged over four individual MD simulations, each simulation lasting 2^{20} time steps and starting from randomly generated initial conditions. We report in Fig. 3.13, the calculated band structure of the modulated system.

Figure 3.13 retains the essential features of the unperturbed band structure but for frequency shifted Bloch modes $\omega_0(k^*) \pm \Omega$ (Stokes and anti-Stokes modes due to Brillouin scattering) and two band gaps in the positive half of the Brillouin zone. The periodicity of the band structure in wave number space is retained but its symmetry is broken. The frequency shifted modes are illustrative of the first-order particular solutions. Second-order frequency shifted modes $\omega_0(k^*) \pm 2\Omega$ do not show in the figure due to their very weak amplitude. The two band gaps occur at the wave vector k_{gap} defined by the condition $\left(\omega_0\left(k_{gap}+g\right)+\Omega\right)^2-\omega_0^2\left(k_{gap}+g-K\right)=0$ for $g=0$ and $g=K$. It is the band folding due to the spatial modulation which enables overlap and hybridization between the frequency-shifted Bloch modes and the original Bloch modes of the lattice without the time dependency of the spatial modulation. The hybridization opens gaps in a band structure that has lost its mirror symmetry about the origin of the Brillouin zone. Considering

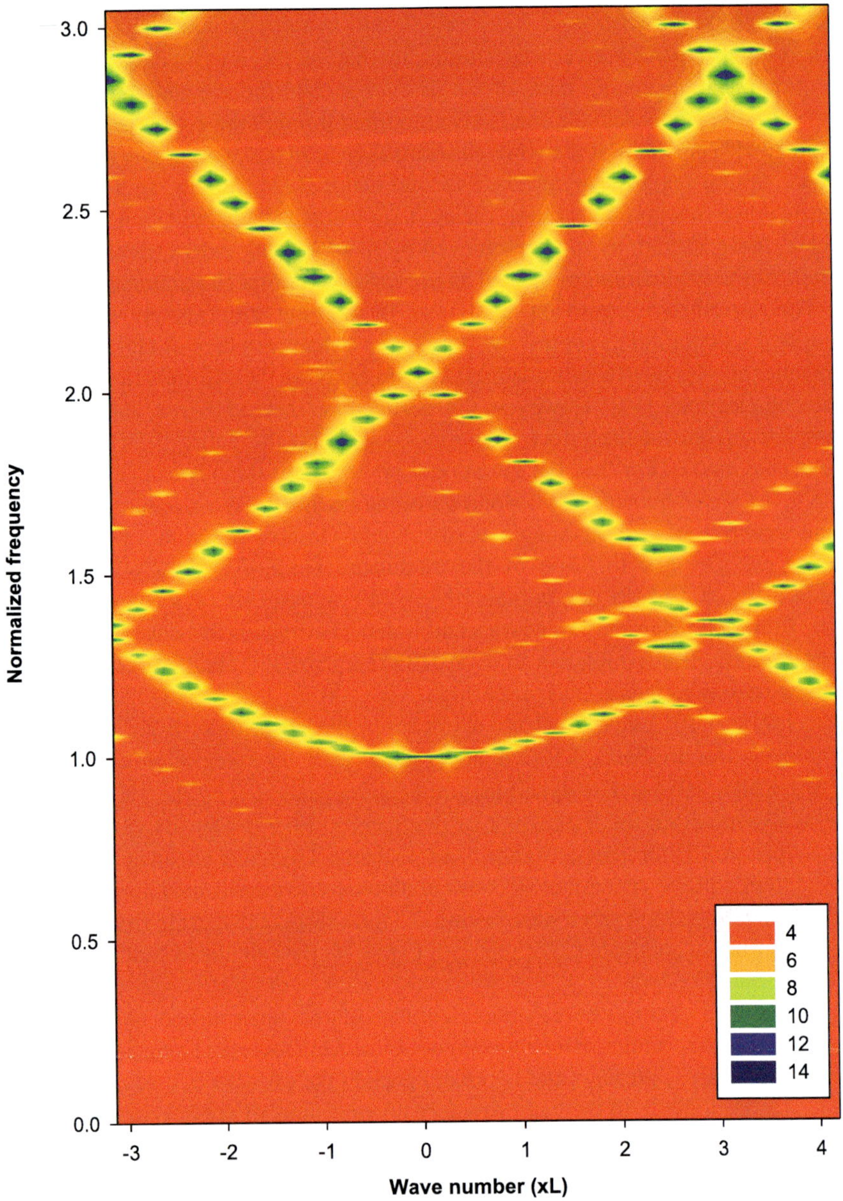

Fig. 3.13 Band structure of the mechanical model system of Fig. 3.2 calculated using the Spectral Energy Density (SED) method. The band structure is reported as a contour plot of the natural logarithm of the SED (*color bar*) versus normalized frequency and reduced wave number. The frequency is normalized to the lowest value of unperturbed band, namely 1.507×10^5 Hz. The *horizontal axis* is extended to the right beyond the first Brillouin zone $[-\pi,\pi]$ to highlight the asymmetry and therefore the modulation-induced symmetry breaking of the band structure. The brighter branches correspond to the usual zeroth-order type wave ($e^{i\omega_0(k+g)\tau_0}$). The fainter branches parallel to the brighter ones are characteristic of first-order waves ($e^{i(\omega_0(k+g)\pm\Omega)\tau_0}$)

the first gap and following a path in k space, starting at $k = 0$ at the bottom of the lowest branch, the wave function transitions from a state corresponding to a zeroth-order type wave, with orbital part ($e^{i\omega_0(k)\tau_0}$) and spinor part $\begin{pmatrix} \sqrt{\omega_0 + \beta k} \\ \sqrt{\omega_0 - \beta k} \end{pmatrix}$ to a wave having the characteristics of the first-order wave with orbital part ($e^{i(\omega_0(k^*)-\Omega\,)\tau_0}$) and spinor part $\begin{pmatrix} b_1' \\ b_2' \end{pmatrix}$. The control of the position of the gap through Ω and K enables strategies for tuning the spinor character of the elastic wave. The effect of these 'spin-orbit' manipulations of the elastic system can be measured by examining the transmission of plane waves.

It is also instructive to consider the symmetry of the Dirac-like equations in the presence of a spatio-temporal modulation to best understand its effect on the spinor character of the wave function. In the case of a modulation with a general phase φ, (3.8a, 3.8b) take the overall form:

$$\left[\boldsymbol{\sigma}_x \frac{\partial}{\partial t} + i\beta \boldsymbol{\sigma}_y \frac{\partial}{\partial x} - i\alpha_0 \boldsymbol{I} - i\alpha_1 \boldsymbol{I} \sin\left(Kx + \Omega t + \varphi\right)\right] \Psi = 0 \tag{3.71a}$$

$$\left[\boldsymbol{\sigma}_x \frac{\partial}{\partial t} + i\beta \boldsymbol{\sigma}_y \frac{\partial}{\partial x} + i\alpha_0 \boldsymbol{I} + i\alpha_1 \boldsymbol{I} \sin\left(Kx + \Omega t + \varphi\right)\right] \overline{\Psi} = 0 \tag{3.71b}$$

Applying the joint T-symmetry and parity symmetry to (3.71a) does not result in (3.71b) for all phases φ but a few special values. The modulated (3.71a, 3.71b) have lost the symmetry properties of the unmodulated Dirac equations (3.8a, 3.8b). The gap that formed at k_{gap} in Fig. 3.10 is therefore not a Dirac point. The transformations $T_{\substack{\omega \to \omega \\ k \to -k}}$ and $T_{\substack{\omega \to -\omega \\ k \to k}}$ do not apply near k_{gap}. The constraints imposed on the spinor component of the elastic wave function may be released in the vicinity of that wavenumber. This constraint was associated with fermion-like wave functions which have the character of quasistanding waves *i.e.*, being composed of forward and backward waves with a very specific proportion of their respective amplitudes. The release of the Dirac constraint associated with the impossibility for the medium to support forward propagating waves ($+k_{gap}$) but only backward propagating waves ($-k_{gap}$), may lead again to boson-like behavior with no restriction on the amplitude of the backward propagating waves.

3.6 Separable and Non-separable States in Elastic Structures

In this section, we consider the partitioning elastic systems into subsystems. We are particularly interested in the separability and non-separability of the wave functions into tensor products. More specifically, we show that separability is relative to the choice of the subsystems in which one partitions the system of interest. We consider three elastic systems: a system composed of non-interacting harmonic oscillators, the partitioning of a harmonic crystal into normal modes and the partitioning of the phononic structure composed of two coupled 1-D harmonic crystals into two uncoupled 1-D harmonic crystals attached to a rigid substrate.

3.6.1 Separability of Systems Composed of Harmonic Oscillators

Let us consider a single one dimensional harmonic oscillator. Its equation of motion is:

$$\frac{d^2u}{dt^2} + \beta^2 u = 0 \tag{3.72}$$

Where u is the displacement of the oscillator's mass. $\beta^2 = \frac{K}{m}$ with K and m being the stiffness and mass of the oscillator, respectively. We can write (3.72) in the form of first-order time derivatives:

$$\left(\frac{d}{dt} + i\beta\right)\left(-\frac{d}{dt} + i\beta\right)u = 0 \tag{3.73}$$

Equation (3.73) reflects the time-reversal symmetry of the oscillations. Equation (3.2) can therefore be written as a set of two ordinary differential equations:

$$\begin{cases} \dfrac{d\varphi_-}{dt} + i\beta\varphi_- = 0 \\ -\dfrac{d\varphi_+}{dt} + i\beta\varphi_+ = 0 \end{cases} \tag{3.74}$$

The solutions $\varphi_- = \varphi_-^0 e^{-i\beta t}$ and $\varphi_+ = \varphi_+^0 e^{+i\beta t}$ are oscillations that evolve along the negative and positive timelines, respectively. The solutions of (3.74) span the Hilbert space, H, of the functions, $e^{\pm i\beta t}$. These functions form an orthonormal basis for H, with the inner product $\int_{-\infty}^{\infty} dt e^{-i\beta t} e^{i\beta' t} = \delta(\beta - \beta')$. Here $\delta(\beta - \beta')$ is the Dirac delta function. Let us consider now a number N of independent harmonic oscillators evolving along the positive timeline. The equations of motion can be written as:

$$-\frac{d\varphi_+^n}{dt} + i\beta_n \varphi_+^n = 0 \text{ with } n = 1, \ldots, N \tag{3.75}$$

Each harmonic oscillator has solutions spanning the space, H.
We can write a single N-oscillator equation:

$$\left(-\frac{d}{dt} + i\beta_1 + i\beta_2 + \ldots + i\beta_N\right)\Phi = 0 \tag{3.76}$$

where the N-oscillator solution Φ can be written as a Cartesian product (actually a tensor product of scalars):

$$\Phi = \varphi_+^1 \varphi_+^2 \ldots \varphi_+^N \tag{3.77}$$

Inserting (3.77) into (3.76) gives:

$$-\varphi_+^2 \ldots \varphi_+^N \frac{d\varphi_+^1}{dt} - \varphi_+^1 \varphi_+^3 \ldots \varphi_+^N \frac{d\varphi_+^2}{dt} + \ldots + \varphi_+^1 \varphi_+^2 \ldots \varphi_+^{N-1} \frac{d\varphi_+^N}{dt} + i\varphi_+^2 \ldots \varphi_+^N \beta_1 \varphi_+^1$$
$$+ i\varphi_+^1 \varphi_+^3 \ldots \varphi_+^N \beta_2 \varphi_+^2 + \ldots + i\varphi_+^1 \varphi_+^2 \ldots \varphi_+^{N-1} \beta_N \varphi_+^N = 0$$

This equation can be factored as:

$$\varphi_+^2 \ldots \varphi_+^N \left[-\frac{d\varphi_+^1}{dt} + i\beta_1 \varphi_+^1\right] + \varphi_+^1 \varphi_+^3 \ldots \varphi_+^N \left[-\frac{d\varphi_+^2}{dt} + i\beta_2 \varphi_+^2\right]$$
$$+ \ldots + \varphi_+^1 \varphi_+^2 \ldots \varphi_+^{N-1} \left[-\frac{d\varphi_+^N}{dt} + i\beta_N \varphi_+^N\right] = 0 \tag{3.78}$$

For non-trivial $\varphi_+^n \neq 0$, the brackets in (3.78) must be equal to zero, thus recovering the set of N (3.75). The system of N non-interacting oscillators is separable. The state of one oscillator, "n" is defined by φ_+^n which spans a Hilbert space, H. The states of the N oscillators system span the N tensor product space $H \otimes H \otimes \ldots \otimes H$ and take the form of tensor products (here simple Cartesian products) of single harmonic oscillator states: $\Phi = \varphi_+^1 \otimes \varphi_+^2 \otimes \ldots \otimes \varphi_+^N = \varphi_+^1 \varphi_+^2 \ldots \varphi_+^N$.

3.6.2 Partitioning of a Harmonic Crystal into Normal Modes

The 1-D harmonic chain can be visualized as a set of coupled harmonic oscillators. This is a chain of oscillators, 1, ..., N, interacting via first nearest neighbor interactions. In the long wavelength limit equation the displacement $u(x, t)$ is given by:

$$\left(\frac{\partial^2}{\partial t^2} - \beta^2 \frac{\partial^2}{\partial x^2}\right) u(x, t) = 0 \tag{3.79}$$

Because of time reversal symmetry, (3.79) can be expressed in terms of first-order partial differential equations:

$$\left[\boldsymbol{\sigma}_x \frac{\partial}{\partial t} + i\beta \boldsymbol{\sigma}_y \frac{\partial}{\partial x}\right] \psi = 0 \tag{3.80}$$

where $\boldsymbol{\sigma}_x$ and $\boldsymbol{\sigma}_y$ are again the 2×2 Pauli matrices. $\psi = \begin{pmatrix} \varphi_- \\ \varphi_+ \end{pmatrix}$ is the 2×1 vector whose components represent plane waves propagating in opposite directions (or time reversed plane waves): $\varphi_- = \varphi_-^0 e^{i(\omega t + kx)}$ with $\omega = -\beta k$ (k is a wave number) and $\varphi_+ = \varphi_+^0 e^{i(\omega t + kx)}$ with $\omega = +\beta k$. The amplitudes $\varphi_\pm^0$ are arbitrary.

With a plane wave form for the solution of (3.80), $\varphi_\pm = \varphi_\pm^0 e^{ikx} \phi_\pm^k(t)$ we can write:

$$\begin{cases} \dfrac{\partial \phi_+^k}{\partial t} + i\beta k \phi_+^k = 0 \\ \dfrac{\partial \phi_-^k}{\partial t} - i\beta k \phi_-^k = 0 \end{cases} \tag{3.81}$$

Equations (3.81) are isomorphic to (3.74). The $\phi_\pm^k(t)$ span the space H. Subsequently, the solutions of (3.80) can be written as tensor products (Cartesian products) of the solutions of independent harmonic oscillators described by (3.75) and (3.76), provided one replaces the discrete index "n" by the continuous variable, "k" and defines: $\beta_k = \beta k$. The wave number k serves as a label for the different independent normal mode "harmonic oscillators". This is the usual normal mode decomposition. These normal modes are defined in an infinite tensor product space of single harmonic oscillator Hilbert space, namely $H \otimes H \otimes \ldots \otimes H \ldots$. In any linear elastic system, where we can perform a normal mode analysis, the states of the system are product states in the tensor product space of states of harmonic oscillators.

3.6.3 Uncoupled 1-D Harmonic Crystals

The question we pose now is: for a composite elastic system constituted of two coupled (linearly) subsystem, can we express the state of the composite system in terms of tensor products of the states of the uncoupled subsystems. For this, first let us focus on a composite system made of two uncoupled harmonic crystals. The equations of motions in terms of first-order differentials for crystals a and b take the form:

$$\left[\boldsymbol{\sigma}_x \frac{\partial}{\partial t} + i\beta\boldsymbol{\sigma}_y \frac{\partial}{\partial x_a}\right]\psi_a = 0 \tag{3.82a}$$

$$\left[\boldsymbol{\sigma}_x \frac{\partial}{\partial t} + i\beta\boldsymbol{\sigma}_y \frac{\partial}{\partial x_b}\right]\psi_b = 0 \tag{3.82b}$$

where $\psi_a = \begin{pmatrix} \psi_1^a(x_a,t) \\ \psi_2^a(x_a,t) \end{pmatrix}$ and $\psi_b = \begin{pmatrix} \psi_1^b(x_b,t) \\ \psi_2^b(x_b,t) \end{pmatrix}$. The states of the two uncoupled crystals span the individual Hilbert spaces H_a and H_b. The indices *1* and *2* play the role of + and – in Sects. 6.1 and 6.2.

We can construct a *single* two uncoupled crystals equation by the following procedure:

$$\left[\boldsymbol{\sigma}_x \otimes \boldsymbol{\sigma}_x \frac{\partial}{\partial t} + i\beta\boldsymbol{\sigma}_y \otimes \boldsymbol{\sigma}_x \frac{\partial}{\partial x_a} + i\beta\boldsymbol{\sigma}_x \otimes \boldsymbol{\sigma}_y \frac{\partial}{\partial x_b}\right]\Psi = 0 \tag{3.83}$$

where the 4×1 vector $\Psi = \begin{pmatrix} \Psi_1 \\ \Psi_2 \\ \Psi_3 \\ \Psi_4 \end{pmatrix}$.

Equation (3.83) can be written in matrix form:

$$\left[\begin{pmatrix} 0 & 0 & 0 & 1 \\ 0 & 0 & 1 & 0 \\ 0 & 1 & 0 & 0 \\ 1 & 0 & 0 & 0 \end{pmatrix}\frac{\partial}{\partial t} + \beta\begin{pmatrix} 0 & 0 & 0 & 1 \\ 0 & 0 & 1 & 0 \\ 0 & -1 & 0 & 0 \\ -1 & 0 & 0 & 0 \end{pmatrix}\frac{\partial}{\partial x_a} + \beta\begin{pmatrix} 0 & 0 & 0 & 1 \\ 0 & 0 & -1 & 0 \\ 0 & 1 & 0 & 0 \\ -1 & 0 & 0 & 0 \end{pmatrix}\frac{\partial}{\partial x_b}\right]\Psi = 0 \tag{3.84}$$

We now assume that the solution Ψ can be written as a tensor product of ψ_a and ψ_b:

$$\Psi = \begin{pmatrix} \psi_1^a\psi_1^b \\ \psi_1^a\psi_2^b \\ \psi_2^a\psi_1^b \\ \psi_2^a\psi_2^b \end{pmatrix} = \psi_a \otimes \psi_b \tag{3.85}$$

Inserting (3.85) into (3.84) leads to a set of four equations:

$$\begin{cases} \frac{\partial}{\partial t}(\psi_2^a\psi_2^b) + \beta\frac{\partial}{\partial x_a}(\psi_2^a\psi_2^b) + \beta\frac{\partial}{\partial x_b}(\psi_2^a\psi_2^b) = 0 \\ \frac{\partial}{\partial t}(\psi_2^a\psi_1^b) + \beta\frac{\partial}{\partial x_a}(\psi_2^a\psi_1^b) - \beta\frac{\partial}{\partial x_b}(\psi_2^a\psi_1^b) = 0 \\ \frac{\partial}{\partial t}(\psi_1^a\psi_2^b) - \beta\frac{\partial}{\partial x_a}(\psi_1^a\psi_2^b) + \beta\frac{\partial}{\partial x_b}(\psi_1^a\psi_2^b) = 0 \\ \frac{\partial}{\partial t}(\psi_1^a\psi_1^b) - \beta\frac{\partial}{\partial x_a}(\psi_1^a\psi_1^b) - \beta\frac{\partial}{\partial x_b}(\psi_1^a\psi_1^b) = 0 \end{cases} \tag{3.86}$$

Each one of these equations can now be factored:

$$\begin{cases} \psi_2^a\left[\frac{\partial\psi_2^b}{\partial t} + \beta\frac{\partial\psi_2^b}{\partial x_b}\right] + \psi_2^b\left[\frac{\partial\psi_2^a}{\partial t} + \beta\frac{\partial\psi_2^a}{\partial x_a}\right] = 0 \\ \psi_2^a\left[\frac{\partial\psi_1^b}{\partial t} - \beta\frac{\partial\psi_1^b}{\partial x_b}\right] + \psi_1^b\left[\frac{\partial\psi_2^a}{\partial t} + \beta\frac{\partial\psi_2^a}{\partial x_a}\right] = 0 \\ \psi_1^a\left[\frac{\partial\psi_2^b}{\partial t} + \beta\frac{\partial\psi_2^b}{\partial x_b}\right] + \psi_2^b\left[\frac{\partial\psi_1^a}{\partial t} - \beta\frac{\partial\psi_1^a}{\partial x_a}\right] = 0 \\ \psi_1^a\left[\frac{\partial\psi_1^b}{\partial t} - \beta\frac{\partial\psi_1^b}{\partial x_b}\right] + \psi_1^b\left[\frac{\partial\psi_1^a}{\partial t} - \beta\frac{\partial\psi_1^a}{\partial x_a}\right] = 0 \end{cases} \tag{3.87}$$

Since we are seeking non-trivial solutions $\psi_{1,2}^a \neq 0$ and $\psi_{1,2}^b \neq 0$, the brackets in (3.87) must be equal to zero. These equations reduce to (3.82a, 3.82b). The states of two uncoupled chains can be expressed as tensor products in the Hilbert space $H_a \otimes H_b$.

This concept could be expanded to more than two crystals. For instance for three crystals, a, b, and c, the Dirac-like dynamical equations are:

$$\left[\boldsymbol{\sigma}_x\frac{\partial}{\partial t} + i\beta\boldsymbol{\sigma}_y\frac{\partial}{\partial x_a}\right]\psi_a = 0$$

$$\left[\boldsymbol{\sigma}_x\frac{\partial}{\partial t} + i\beta\boldsymbol{\sigma}_y\frac{\partial}{\partial x_b}\right]\psi_b = 0$$

$$\left[\boldsymbol{\sigma}_x\frac{\partial}{\partial t} + i\beta\boldsymbol{\sigma}_y\frac{\partial}{\partial x_c}\right]\psi_c = 0$$

where $\psi_a = \begin{pmatrix}\psi_1^a(x_a,t)\\ \psi_2^a(x_a,t)\end{pmatrix}$, $\psi_b = \begin{pmatrix}\psi_1^b(x_b,t)\\ \psi_2^b(x_b,t)\end{pmatrix}$, and $\psi_c = \begin{pmatrix}\psi_1^c(x_c,t)\\ \psi_2^c(x_c,t)\end{pmatrix}$. The states of the three uncoupled chains span the individual Hilbert spaces H_a, H_b, and H_c.

We can construct a *single* three uncoupled crystals equation by the following procedure:

$$\left[\boldsymbol{\sigma}_x\otimes\boldsymbol{\sigma}_x\otimes\boldsymbol{\sigma}_x\frac{\partial}{\partial t} + i\beta\boldsymbol{\sigma}_y\otimes\boldsymbol{\sigma}_x\otimes\boldsymbol{\sigma}_x\frac{\partial}{\partial x_a} + i\beta\boldsymbol{\sigma}_x\otimes\boldsymbol{\sigma}_y\otimes\boldsymbol{\sigma}_x\frac{\partial}{\partial x_b} + i\beta\boldsymbol{\sigma}_x\otimes\boldsymbol{\sigma}_x\otimes\boldsymbol{\sigma}_y\frac{\partial}{\partial x_c}\right] \times \Psi = 0$$

where solutions are 8×1 vector $\Psi = \begin{pmatrix} \Psi_1 \\ \Psi_2 \\ \Psi_3 \\ \Psi_4 \\ \Psi_5 \\ \Psi_6 \\ \Psi_7 \\ \Psi_8 \end{pmatrix} = \psi_a \otimes \psi_b \otimes \psi_c = \begin{pmatrix} \psi_1^a\psi_1^b\psi_1^c \\ \psi_1^a\psi_2^b\psi_1^c \\ \psi_2^a\psi_1^b\psi_1^c \\ \psi_2^a\psi_2^b\psi_1^c \\ \psi_1^a\psi_1^b\psi_2^c \\ \psi_1^a\psi_2^b\psi_2^c \\ \psi_2^a\psi_1^b\psi_2^c \\ \psi_2^a\psi_2^b\psi_2^c \end{pmatrix}$.

It is straightforward but tedious to show that by introducing the tensor product solution into the single three uncoupled crystals equation, one recovers the three Dirac-like dynamical equations for the individual crystals. This can be generalized to any number of crystals, N. The complexity increases exponentially as 2^N.

3.6.4 Separability and Non-separability of the States of Coupled Harmonic Crystals

Let us now consider some arbitrary coupling between two crystals. For this we add to (3.84) an interaction operator, $\boldsymbol{\varepsilon}$:

$$\left[\begin{pmatrix} 0&0&0&1 \\ 0&0&1&0 \\ 0&1&0&0 \\ 1&0&0&0 \end{pmatrix}\frac{\partial}{\partial t}+\beta\begin{pmatrix} 0&0&0&1 \\ 0&0&1&0 \\ 0&-1&0&0 \\ -1&0&0&0 \end{pmatrix}\frac{\partial}{\partial x_a}+\beta\begin{pmatrix} 0&0&0&1 \\ 0&0&-1&0 \\ 0&1&0&0 \\ -1&0&0&0 \end{pmatrix}\frac{\partial}{\partial x_b}+\begin{pmatrix} \varepsilon_{11}&\varepsilon_{12}&\varepsilon_{13}&\varepsilon_{14} \\ \varepsilon_{21}&\varepsilon_{22}&\varepsilon_{23}&\varepsilon_{24} \\ \varepsilon_{31}&\varepsilon_{32}&\varepsilon_{33}&\varepsilon_{34} \\ \varepsilon_{41}&\varepsilon_{42}&\varepsilon_{43}&\varepsilon_{44} \end{pmatrix}\right]\Psi=0 \tag{3.88}$$

The components of the interaction operator depend on position within both crystals, that is $\varepsilon_{ij}(x_a, x_b)$. The interaction matrix couples points along the crystal a with points along the crystal b.

Assuming a solution in the form of a tensor product of states of uncoupled crystals [*i.e.*, (3.85)] and inserting such a solution in (3.88) leads to

$$\begin{cases} \frac{\partial}{\partial t}(\psi_2^a\psi_2^b)+\beta\frac{\partial}{\partial x_a}(\psi_2^a\psi_2^b)+\beta\frac{\partial}{\partial x_b}(\psi_2^a\psi_2^b)+\varepsilon_{11}\psi_1^a\psi_1^b+\varepsilon_{12}\psi_1^a\psi_2^b+\varepsilon_{13}\psi_2^a\psi_1^b+\varepsilon_{14}\psi_2^a\psi_2^b=0 \\ \frac{\partial}{\partial t}(\psi_2^a\psi_1^b)+\beta\frac{\partial}{\partial x_a}(\psi_2^a\psi_1^b)-\beta\frac{\partial}{\partial x_b}(\psi_2^a\psi_1^b)+\varepsilon_{21}\psi_1^a\psi_1^b+\varepsilon_{22}\psi_1^a\psi_2^b+\varepsilon_{23}\psi_2^a\psi_1^b+\varepsilon_{24}\psi_2^a\psi_2^b=0 \\ \frac{\partial}{\partial t}(\psi_1^a\psi_2^b)-\beta\frac{\partial}{\partial x_a}(\psi_1^a\psi_2^b)+\beta\frac{\partial}{\partial x_b}(\psi_1^a\psi_2^b)+\varepsilon_{31}\psi_1^a\psi_1^b+\varepsilon_{32}\psi_1^a\psi_2^b+\varepsilon_{33}\psi_2^a\psi_1^b+\varepsilon_{34}\psi_2^a\psi_2^b=0 \\ \frac{\partial}{\partial t}(\psi_1^a\psi_1^b)-\beta\frac{\partial}{\partial x_a}(\psi_1^a\psi_1^b)-\beta\frac{\partial}{\partial x_b}(\psi_1^a\psi_1^b)+\varepsilon_{41}\psi_1^a\psi_1^b+\varepsilon_{42}\psi_1^a\psi_2^b+\varepsilon_{43}\psi_2^a\psi_1^b+\varepsilon_{44}\psi_2^a\psi_2^b=0 \end{cases} \tag{3.89}$$

An attempt to factoring (3.89) into a symmetric form leads to:

$$\begin{cases} \psi_2^a\left[\frac{\partial \psi_2^b}{\partial t}+\beta\frac{\partial \psi_2^b}{\partial x_b}+\varepsilon_{13}\psi_1^b+\frac{1}{2}\varepsilon_{14}\psi_2^b\right]+\psi_2^b\left[\frac{\partial \psi_2^a}{\partial t}+\beta\frac{\partial \psi_2^a}{\partial x_a}+\varepsilon_{12}\psi_1^a+\frac{1}{2}\varepsilon_{14}\psi_2^a\right]+\varepsilon_{11}\psi_1^a\psi_1^b=0 \\ \psi_2^a\left[\frac{\partial \psi_1^b}{\partial t}-\beta\frac{\partial \psi_1^b}{\partial x_b}+\varepsilon_{24}\psi_2^b+\frac{1}{2}\varepsilon_{23}\psi_1^b\right]+\psi_1^b\left[\frac{\partial \psi_2^a}{\partial t}+\beta\frac{\partial \psi_2^a}{\partial x_a}+\varepsilon_{21}\psi_1^a+\frac{1}{2}\varepsilon_{23}\psi_2^a\right]+\varepsilon_{22}\psi_1^a\psi_2^b=0 \\ \psi_1^a\left[\frac{\partial \psi_2^b}{\partial t}+\beta\frac{\partial \psi_2^b}{\partial x_b}+\varepsilon_{31}\psi_1^b+\frac{1}{2}\varepsilon_{32}\psi_2^b\right]+\psi_2^b\left[\frac{\partial \psi_2^a}{\partial t}-\beta\frac{\partial \psi_2^a}{\partial x_a}+\varepsilon_{34}\psi_2^a+\frac{1}{2}\varepsilon_{32}\psi_1^a\right]+\varepsilon_{33}\psi_2^a\psi_1^b=0 \\ \psi_1^a\left[\frac{\partial \psi_1^b}{\partial t}-\beta\frac{\partial \psi_1^b}{\partial x_b}+\varepsilon_{42}\psi_2^b+\frac{1}{2}\varepsilon_{41}\psi_1^b\right]+\psi_1^b\left[\frac{\partial \psi_1^a}{\partial t}-\beta\frac{\partial \psi_1^a}{\partial x_a}+\varepsilon_{43}\psi_2^a+\frac{1}{2}\varepsilon_{41}\psi_1^a\right]+\varepsilon_{44}\psi_2^a\psi_2^b=0 \end{cases}$$

Upon inspection, one cannot conclude that later equation may not always lead to a separable form in terms of the states of two uncoupled crystals. This issue is addressed below. For the coupled crystals system we derive a Dirac-like form of the equations of motion which in the long wavelength limit was written in the form (3.3a, 3.3b): $\left(\left[\boldsymbol{A}\frac{\partial}{\partial t}+\beta\boldsymbol{B}\frac{\partial}{\partial x}\right]\pm i\frac{\alpha}{\sqrt{2}}\boldsymbol{C}\right)\Psi=0$ with

$$\boldsymbol{A}=\begin{pmatrix}0&0&1&0\\0&0&0&1\\1&0&0&0\\0&1&0&0\end{pmatrix},\boldsymbol{B}=\begin{pmatrix}0&0&0&1\\0&0&1&0\\0&-1&0&0\\-1&0&0&0\end{pmatrix},\boldsymbol{C}=\begin{pmatrix}1&-1&0&0\\-1&1&0&0\\0&0&1&-1\\0&0&-1&1\end{pmatrix}.$$

This equation can be reformulated into a more compact form by considering the equivalent representation:

$$\left[\boldsymbol{\sigma_x}\otimes\boldsymbol{\sigma_x}\frac{\partial}{\partial t}+i\beta\boldsymbol{\sigma_x}\otimes\boldsymbol{\sigma_y}\frac{\partial}{\partial x}\pm i\delta\boldsymbol{C}\right]\Psi=0 \tag{3.90}$$

with $\boldsymbol{C}=\begin{pmatrix}1&-1\\-1&1\end{pmatrix}\otimes\begin{pmatrix}1&0\\0&1\end{pmatrix}$ and $\delta=\frac{1}{\sqrt{2}}\alpha$. In matrix form and for the minus part of the $\pm$, (3.90) becomes:

$$\left[\begin{pmatrix}0&0&0&1\\0&0&1&0\\0&1&0&0\\1&0&0&0\end{pmatrix}\frac{\partial}{\partial t}+\beta\begin{pmatrix}0&0&0&1\\0&0&-1&0\\0&1&0&0\\-1&0&0&0\end{pmatrix}\frac{\partial}{\partial x}-i\delta\begin{pmatrix}1&0&-1&0\\0&1&0&-1\\-1&0&1&0\\0&-1&0&1\end{pmatrix}\right]\Psi=0 \tag{3.91}$$

We choose plane wave solutions, $\Psi_j=a_je^{+i\omega t}e^{+ikx}$. The functions $\begin{pmatrix}1\\0\\0\\0\end{pmatrix},\begin{pmatrix}0\\1\\0\\0\end{pmatrix},\begin{pmatrix}0\\0\\1\\0\end{pmatrix},\begin{pmatrix}0\\0\\0\\1\end{pmatrix}\times e^{i\omega t}e^{ikx}$ form the orthonormal basis for the solutions of the coupled system. Equation (3.91) transforms into the set of four linear equations:

$$\begin{cases} (\omega+\beta k)a_4 = \delta(a_1 - a_3) \\ (\omega-\beta k)a_3 = \delta(a_2 - a_4) \\ (\omega+\beta k)a_2 = -\delta(a_1 - a_3) \\ (\omega-\beta k)a_1 = -\delta(a_2 - a_4) \end{cases} \tag{3.92}$$

By inspection, we find that $a_1 = a_3$ is a solution. It leads to the eigen values: $\omega = \pm\beta k$ which yields in turn $a_2 = a_4$. The actual values of a_1 and a_2 are arbitrary. This solution corresponds to the symmetric mode we described earlier. However, now we see that if we establish the relations $\psi_1^a = a_1$, $\psi_2^a = a_3$, and $\psi_1^b = a_2$, $\psi_2^b = a_4$, as well as, use the interaction matrix

$$\begin{pmatrix} \varepsilon_{11} & \varepsilon_{12} & \varepsilon_{13} & \varepsilon_{14} \\ \varepsilon_{21} & \varepsilon_{22} & \varepsilon_{23} & \varepsilon_{24} \\ \varepsilon_{31} & \varepsilon_{32} & \varepsilon_{33} & \varepsilon_{34} \\ \varepsilon_{41} & \varepsilon_{42} & \varepsilon_{43} & \varepsilon_{44} \end{pmatrix} = -i\delta \begin{pmatrix} 1 & 0 & -1 & 0 \\ 0 & 1 & 0 & -1 \\ -1 & 0 & 1 & 0 \\ 0 & -1 & 0 & 1 \end{pmatrix}$$

Then the coupling terms in (3.89) become

$$\varepsilon_{11}\psi_1^a\psi_1^b + \varepsilon_{12}\psi_1^a\psi_2^b + \varepsilon_{13}\psi_2^a\psi_1^b + \varepsilon_{14}\psi_2^a\psi_2^b = -i\delta\psi_1^b\left(\psi_1^a - \psi_2^a\right) = 0$$
$$\varepsilon_{21}\psi_1^a\psi_1^b + \varepsilon_{22}\psi_1^a\psi_2^b + \varepsilon_{23}\psi_2^a\psi_1^b + \varepsilon_{24}\psi_2^a\psi_2^b = -i\delta\psi_2^b\left(\psi_1^a - \psi_2^a\right) = 0$$
$$\varepsilon_{31}\psi_1^a\psi_1^b + \varepsilon_{32}\psi_1^a\psi_2^b + \varepsilon_{33}\psi_2^a\psi_1^b + \varepsilon_{34}\psi_2^a\psi_2^b = i\delta\psi_1^b\left(\psi_1^a - \psi_2^a\right) = 0$$
$$\varepsilon_{41}\psi_1^a\psi_1^b + \varepsilon_{42}\psi_1^a\psi_2^b + \varepsilon_{43}\psi_2^a\psi_1^b + \varepsilon_{44}\psi_2^a\psi_2^b = i\delta\psi_2^b\left(\psi_1^a - \psi_2^a\right) = 0$$

Equation (3.89) then reduces to that of two uncoupled crystals. The states of the two crystal system with the dispersion relation $\omega = \pm\beta k$ are tensor products of the states of the independent crystals.

Another solution of (3.92) can be found when $a_1 = -a_3$ and $a_2 = -a_4$. In that case, the eigen values are $\omega^2 = \beta^2 k^2 + (2\delta)^2$. Inserting $2\delta = +\sqrt{\omega^2 - \beta^2 k^2}$ (the + sign is chosen because δ represents physically a stiffness), into (3.92) yields:

$$\begin{cases} (\omega+\beta k)a_4 = \sqrt{\omega^2 - \beta^2 k^2}\,a_1 \\ (\omega-\beta k)a_3 = \sqrt{\omega^2 - \beta^2 k^2}\,a_2 \\ (\omega+\beta k)a_2 = \sqrt{\omega^2 - \beta^2 k^2}\,a_3 \\ (\omega-\beta k)a_1 = \sqrt{\omega^2 - \beta^2 k^2}\,a_4 \end{cases} \tag{3.93}$$

Substituting, $\omega+\beta k = \sqrt{\omega+\beta k}\sqrt{\omega+\beta k} = \sqrt{+}\sqrt{+}$, $\omega-\beta k = \sqrt{\omega-\beta k}\sqrt{\omega-\beta k} = \sqrt{-}\sqrt{-}$, and $\sqrt{\omega^2-\beta^2k^2} = \sqrt{\omega+\beta k}\sqrt{\omega-\beta k} = \sqrt{+}\sqrt{-} = \sqrt{-}\sqrt{+}$ into (3.93) leads to

$$\begin{cases} \sqrt{+}\sqrt{+}a_4 = \sqrt{+}\sqrt{-}a_1 \\ \sqrt{-}\sqrt{-}a_3 = \sqrt{+}\sqrt{-}a_2 \\ \sqrt{+}\sqrt{+}a_2 = \sqrt{+}\sqrt{-}a_3 \\ \sqrt{-}\sqrt{-}a_1 = \sqrt{+}\sqrt{-}a_4 \end{cases} \tag{3.94}$$

A possible solution of the set of (3.94) is:

$$\begin{pmatrix} a_1 \\ a_2 \\ a_3 \\ a_4 \end{pmatrix} = a_0 \begin{pmatrix} \sqrt{+}\sqrt{+} \\ -\sqrt{+}\sqrt{-} \\ -\sqrt{+}\sqrt{+} \\ \sqrt{+}\sqrt{-} \end{pmatrix} \tag{3.95}$$

One may now wonder if the solution given by (3.95) can be expressed as a tensor product of two non-interacting subsystems each composed of a single harmonic crystal attached elastically to a rigid substrate. We have seen in previous subsections (see Table 3.1) that spinor states of one of these subsystems take the form $\begin{pmatrix} s_1\sqrt{\pm} \\ s_2\sqrt{\mp} \end{pmatrix}$ with s_1 and s_2 having the values +1 or -1. The tensor product of two subsystems a and b would therefore be expressed as tensor products:

$$\begin{pmatrix} s_1^a\sqrt{\pm} \\ s_2^a\sqrt{\mp} \end{pmatrix} \otimes \begin{pmatrix} s_1^b\sqrt{\pm} \\ s_2^b\sqrt{\mp} \end{pmatrix} = \begin{pmatrix} s_1^a s_1^b\sqrt{\pm}\sqrt{\pm} \\ s_1^a s_2^b\sqrt{\pm}\sqrt{\mp} \\ s_2^a s_1^b\sqrt{\mp}\sqrt{\pm} \\ s_2^a s_2^b\sqrt{\mp}\sqrt{\mp} \end{pmatrix} \tag{3.96}$$

By inspection, one sees that (3.95) cannot in general be written in the form of (3.96). However, a few specific cases it can be expressed as a tensor product.

For instance, when $k=0$, then $\omega=2\delta$ and $\sqrt{+}=\sqrt{-}=\sqrt{2\delta}$. Equation (3.95) becomes:

$$\begin{pmatrix} a_1 \\ a_2 \\ a_3 \\ a_4 \end{pmatrix} = a_0 2\delta \begin{pmatrix} 1 \\ -1 \\ -1 \\ 1 \end{pmatrix} = a_0 2\delta \begin{pmatrix} 1 \\ -1 \end{pmatrix} \otimes \begin{pmatrix} 1 \\ -1 \end{pmatrix} \tag{3.97}$$

For $\omega=2\delta>0$ (and $k=0$), the spinor states that are pertinent to (3.97) are therefore

$$\bar{\xi}_k = \begin{pmatrix} \sqrt{\omega-\beta kh} \\ -\sqrt{\omega+\beta kh} \end{pmatrix} e^{+ikx} e^{+i\omega_k t} \text{ or } \bar{\xi}_k = \begin{pmatrix} \sqrt{\omega+\beta kh} \\ -\sqrt{\omega-\beta kh} \end{pmatrix} e^{-ikx} e^{+i\omega_k t}.$$

Let us verify the observation of tensor product states further. We treat the case of two individual subsystems with wave functions:

where $\psi_a = \begin{pmatrix} \psi_1^a(x_a,t) \\ \psi_2^a(x_a,t) \end{pmatrix} = \begin{pmatrix} \xi_1 \\ \xi_2 \end{pmatrix} e^{i\omega_a t} e^{ik_a x}$ and $\psi_b = \begin{pmatrix} \psi_1^b(x_b,t) \\ \psi_2^b(x_b,t) \end{pmatrix}$ $= \begin{pmatrix} \zeta_1 \\ \zeta_2 \end{pmatrix} e^{i\omega_b t} e^{ik_b x}$. The functions $\begin{pmatrix} 1 \\ 0 \end{pmatrix} e^{i\omega_{a,b} t} e^{ik_{a,b} x}$ and $\begin{pmatrix} 0 \\ 1 \end{pmatrix} e^{i\omega_{a,b} t} e^{ik_{a,b} x}$ form the orthonormal basis of the Hilbert spaces, H_a and H_b. We can insert the functions ψ_a and ψ_b into the left-hand side of the first of (3.89) for illustration and we obtain:

$$LHS_1 = \{\xi_2\zeta_2[(\omega_a + \omega_b) + \beta(k_a + k_b)] - \delta[\xi_1\zeta_1 - \xi_2\zeta_1]\} i e^{i(\omega_a+\omega_b)t} e^{i(k_a+k_b)x}$$

In this calculation, we have used: $\varepsilon_{11}=1$, $\varepsilon_{12}=\varepsilon_{14}=0$, and $\varepsilon_{13}=-1$. At k=0, we have found (3.97) that $\xi_2\zeta_2 = a_4 = a_0 2\delta$, $\xi_1\zeta_1 = a_1 = a_0 2\delta$ and $\xi_2\zeta_1 = a_3 = -a_0 2\delta$. Imposing $\omega_a + \omega_b = \omega = 2\delta$ and $k_a + k_b = k = 0$, the term in parenthesis in LHS_1 reduces to:

$$\{a_0 2\delta[2\delta + 0] - \delta[a_0 2\delta - (-a_0 2\delta)]\} = 0$$

Here, the two non-interacting subsystems would have side springs with stiffness equal to δ^2. They would be in states $\omega_a = \omega_b = \delta$ and $k_a = k_b = 0$, and $\begin{pmatrix} \xi_1 \\ \xi_2 \end{pmatrix} \propto \begin{pmatrix} 1 \\ -1 \end{pmatrix}$ and $\begin{pmatrix} \zeta_1 \\ \zeta_2 \end{pmatrix} \propto \begin{pmatrix} 1 \\ -1 \end{pmatrix}$. The tensor product of (3.97) that represents the infinite wavelength standing wave supported by the coupled two-crystal system is that of one infinite wavelength standing wave in one subsystem and another infinite wavelength standing wave in the other. This constitutes the representation of the state of the coupled two-crystal system in the tensor product space $H_a \otimes H_b$ of two non-interacting subsystems composed of 1-D harmonic crystals elastically coupled to a rigid substrate.

When $\delta \to 0$ then $\omega \to \beta k$ and $\sqrt{+} \to \sqrt{2\beta k}$ and $\sqrt{-} \to 0$. Equation (3.95) reduces to

$$\begin{pmatrix} a_1 \\ a_2 \\ a_3 \\ a_4 \end{pmatrix} = a_0\sqrt{2\beta k}\begin{pmatrix} 1 \\ 0 \\ -1 \\ 0 \end{pmatrix} = a_0\sqrt{2\beta k}\begin{pmatrix} 1 \\ -1 \end{pmatrix} \otimes \begin{pmatrix} 1 \\ 0 \end{pmatrix} \tag{3.98}$$

The parenthesis in the LHS_1 of (3.89) reduces to

$$\{\xi_2\zeta_2[(\omega_a+\omega_b)+\beta(k_a+k_b)] - \delta[\xi_1\zeta_1 - \xi_2\zeta_1]\} = \{0[(\omega_a+\omega_b)+\beta(k_a+k_b)] - 2\delta\} \to 0.$$

We note that for $\delta \to 0$, the single subsystem state $\begin{pmatrix} 1 \\ -1 \end{pmatrix}$ actually consists of a zero frequency, infinite wavelength standing wave *i.e.*, a static state. The state $\begin{pmatrix} 1 \\ 0 \end{pmatrix}$ corresponds to a traveling wave in the other subsystem. The state of the coupled two-crystal system is the tensor product of a static state in one subsystem and one traveling wave in the other.

The two special cases: $k=0$ with $\omega=2\delta$ and $\delta\to 0$ with $\omega\to\beta k$, have been shown to lead to spinor states of the two coupled crystals that are tensor products of the spinor states of the two single crystal coupled to a substrate subsystems, however, in general the solution given by (3.95) does not lead to coupling terms in (3.89) that vanish. In general, the antisymmetric state of the coupled crystals is not a tensor product of states of the individual crystals but for a few special cases. In Chap. 6, we will demonstrate the importance of non-separable states in creating quantum computing analogues.

We note that given a multipartite system (*e.g.*, a linearly coupled two-crystal system), the way to subdivide it into subsystems is not unique [45]. For instance, the coupled two-crystal system can be readily decomposed into normal modes. The states of the coupled system span the tensor product space of harmonic oscillators since all its admissible states can be expressed as tensor products of the states of harmonic oscillators. However, it is also possible to express the states of the coupled system in the tensor product space of two subsystems, namely two non-interacting 1-D crystals elastically coupled to a substrate. While some of the states of the multipartite system can be written as tensor products of states of the subsystems, there exists many states that display correlations that do not allow such decomposition. These states exist in the tensor product space of the Hilbert spaces of the subsystems but are not expressible by a simple tensor product. These states can be referred as nonseparable and are analogous to "entangled" states. Therefore the notion of non-separability possesses some ambiguity determined by the relative selection of the decomposition of a multipartite system. The states of the coupled crystals are separable relative to a normal mode decomposition but appear non-separable relative to a decomposition in the tensor product Hilbert space of the two subsystems. The question arises as to the choice of a decomposition into subsystems. That choice may be motivated by some criterion such as the ability to perform observations and measurements [46]. For instance, when band structures and phonon spectra are measurable of interest, the multipartite system may be decomposed into harmonic oscillator subsystems which wave function is the tensor product of normal modes. However, wave transmission can also be a measureable (for a coupled system it would indicate the amount of forward versus backward propagating waves in each crystal of the coupled system), and one may choose subsystem for which transmission can also be measured (such as individual 1-D crystals coupled to a rigid substrate). In this case, the states of the coupled system are not separable and may not be expressible as tensor products of the subsystems.

Appendix 1: Berry Phase of the Spinor Amplitude for the Discrete Harmonic Crystal Grounded to a Rigid Substrate

In this appendix, we characterize the phase properties of the spinor amplitude given by (3.25a):

$$\xi = \begin{pmatrix} a_1 \\ a_2 \\ a_3 \\ a_4 \end{pmatrix} = \begin{pmatrix} ie^{-i\frac{kh}{4}}\left(\alpha - \beta\left(e^{i\frac{kh}{2}} - e^{-i\frac{kh}{2}}\right)\right) \\ \mp e^{+i\frac{kh}{4}}\sqrt{\alpha^2 - \beta^2\left(e^{i\frac{kh}{2}} - e^{-i\frac{kh}{2}}\right)^2} \\ \mp ie^{-i\frac{kh}{4}}\sqrt{\alpha^2 - \beta^2\left(e^{i\frac{kh}{2}} - e^{-i\frac{kh}{2}}\right)^2} \\ e^{+i\frac{kh}{4}}\left(\alpha - \beta\left(e^{i\frac{kh}{2}} - e^{-i\frac{kh}{2}}\right)\right) \end{pmatrix} \tag{3.99}$$

We need to determine the normalized spinor: $\hat{\xi} = \frac{\xi}{\xi^* \xi}$ and its complex conjugate $\hat{\xi}^* = \frac{\xi^*}{\xi^* \xi}$ with $\xi^* \xi = a_1^* a_1 + a_2^* a_2 + a_3^* a_3 + a_4^* a_4 = 4\left(\alpha^2 + 4\beta^2 \sin^2\frac{kh}{2}\right)$ since $a_i^* a_i = \alpha^2 + 4\beta^2 \sin^2\frac{kh}{2}$.

The Berry connection is given by:

$$BC(k) = -i\hat{\xi}^* \frac{\partial \hat{\xi}}{\partial k} \tag{3.100}$$

We find

$$\frac{\partial \hat{\xi}}{\partial k} = \frac{1}{2\sqrt{\alpha^2 + 4\beta^2 \sin^2\frac{kh}{2}}} \frac{ih}{4} \begin{pmatrix} a_1 g_1 \\ a_2 g_2 \\ a_3 g_3 \\ a_4 g_4 \end{pmatrix} - \beta^2 h \frac{\sin\frac{kh}{2}\cos\frac{kh}{2}}{\left(\alpha^2 + 4\beta^2 \sin^2\frac{kh}{2}\right)\sqrt{\left(\alpha^2 + 4\beta^2 \sin^2\frac{kh}{2}\right)}} \begin{pmatrix} a_1 \\ a_2 \\ a_3 \\ a_4 \end{pmatrix} \tag{3.101}$$

with

$$g_1 = 1 - 4\alpha\beta \frac{\cos\frac{kh}{2}}{\alpha^2 + 4\beta^2 \sin^2\frac{kh}{2}} - i4\beta^2 \frac{2\sin\frac{kh}{2}\cos\frac{kh}{2}}{\alpha^2 + 4\beta^2 \sin^2\frac{kh}{2}}$$

$$g_2 = -1 - i4\beta^2 \frac{2\sin\frac{kh}{2}\cos\frac{kh}{2}}{\alpha^2 + 4\beta^2 \sin^2\frac{kh}{2}}$$

$$g_3 = 1 - i4\beta^2 \frac{2\sin\frac{kh}{2}\cos\frac{kh}{2}}{\alpha^2 + 4\beta^2\sin^2\frac{kh}{2}}$$

$$g_4 = -1 - 4\alpha\beta \frac{\cos\frac{kh}{2}}{\alpha^2 + 4\beta^2\sin^2\frac{kh}{2}} - i4\beta^2 \frac{2\sin\frac{kh}{2}\cos\frac{kh}{2}}{\alpha^2 + 4\beta^2\sin^2\frac{kh}{2}}$$

We then get

$$\hat{\xi}^* \frac{\partial\hat{\xi}}{\partial k} = \frac{1}{4}\frac{ih}{4}(g_1 + g_2 + g_3 + g_4) - \beta^2 h \frac{2\sin\frac{kh}{2}\cos\frac{kh}{2}}{\left(\alpha^2 + 4\beta^2\sin^2\frac{kh}{2}\right)} \tag{3.102}$$

with $g_1 + g_2 + g_3 + g_4 = -2 \times 4\alpha\beta \frac{\cos\frac{kh}{2}}{\alpha^2 + 4\beta^2\sin^2\frac{kh}{2}} - 4i4\beta^2 \frac{2\sin\frac{kh}{2}\cos\frac{kh}{2}}{\alpha^2 + 4\beta^2\sin^2\frac{kh}{2}}$.

Equation (3.102) becomes

$$\hat{\xi}^* \frac{\partial\hat{\xi}}{\partial k} = -\frac{ih}{2}\alpha\beta \frac{\cos\frac{kh}{2}}{\alpha^2 + 4\beta^2\sin^2\frac{kh}{2}} \tag{3.103}$$

After some rearrangements, the Berry connection takes on the simple form:

$$BC(k) = -\frac{h}{2}\beta\cos\frac{kh}{2}\frac{\alpha}{4\beta^2\sin^2\frac{kh}{2} + \alpha^2} \tag{3.104}$$

The Berry phase accumulated as the wave number spans the Brillouin zone: $k \in \left[-\frac{\pi}{h}, \frac{\pi}{h}\right]$ is obtained by integration:

$$\eta = -\int_{-\frac{\pi}{2}}^{\frac{\pi}{2}} \beta\cos\theta \frac{\alpha}{4\beta^2\sin^2\theta + \alpha^2} d\theta \tag{3.105}$$

where we have made the change of variable $\theta = \frac{kh}{2}$.

We can easily evaluate the integral in the limit of small α. In that limit, we have:

$$\lim_{\alpha\to 0^+} \frac{\alpha}{4\beta^2\sin^2\theta + \alpha^2} = \pi\delta(2\beta\sin\theta) = \frac{\pi}{2}\frac{1}{\beta}\frac{1}{|\cos\theta|_{\theta=0}}\delta(\theta) \tag{3.106}$$

In (3.106), we are limited to $\theta = 0$ as the only pole of $\sin\theta$ within the Brillouin zone. Inserting (3.106) into (3.105) yields the Berry phase: $\eta = -\frac{\pi}{2}$. The spinor part of the amplitude of the wave function given by (3.25a, 3.25b) therefore contributes $-\frac{\pi}{2}$ to the phase as one spans the Brillouin zone.

Appendix 2: Rotational Modes in a Two-Dimensional Phononic Crystal

Phononic crystals (PCs) comprised of periodically arranged elastic scatterers of one material dispersed periodically throughout a different homogeneous matrix material can strongly affect the propagation of acoustic and/or elastic waves [47]. Several studies have considered the role rigid body rotations may play in modifying the bulk modes of propagation in the phononic structure [48–54]. Sainidou et al. [51] and Zhao et al. [52] revealed theoretically that rotary resonance modes can strongly interact with Bragg gaps to yield extremely wide absolute acoustic band gaps. Peng et al. [54] proposed a one-dimensional lumped model composed of finite-sized masses and massless springs to provide an understanding of the underlying physics behind rotary resonance in two-dimensional (2D) solid/solid PCs. The notion of modeling a PC from a continuum perspective (with additional degrees of freedom, namely rotation) relates to the work done by the Cosserat brothers over 100 years ago [55]. In 1909, the Cosserat brothers pioneered a continuum theory of elasticity that accounted for the rotational degrees of freedom of individual elements in addition to the standard translational degrees of freedom used in classical elasticity theory. In the Cosserat model, each material element has six degrees of freedom—three for translation and three for rotation. The theory introduces a coupled-stress tensor (a component arising from the coupling of rotational and shear waves) that fulfills the same role for torques as the stress tensor of classical elasticity plays for forces. Ultimately, Cosserat continuum elasticity theory predicts that rotational degrees of freedom (*e.g.*, rotational wave modes) can strongly modify the dispersion of shear waves [56]. Several studies have characterized rotational elastic waves in three-dimensional (3D) granular structures comprised of pre-compressed, regular arrangements of spherical elastic particles [57–59]. In these works, the Hertz–Mindlin contact model is used to represent the connection between the elements of the structure. The existence of transverse vibrations in the structures necessitates the consideration of rotation for the individual spherical particles. Rotational degrees of freedom in the structure showed individual rotational modes as well as coupled rotary/translational modes in the dispersion relations [57, 58].

In this appendix, we consider a PC composed of a square array of polystyrene (PS) cylinders with square cross section embedded in a homogeneous, elastic matrix of polydimethylsiloxane (PDMS). This combination of materials offers distinctive elastic band structures with modes corresponding to rotational waves. PS and PDMS are unique due to their sharp contrast in transverse speed of sound and low-density. The elastic parameters for PS and PDMS are as follows: $\rho_{PS} = 1050$ kg/m^3, $C_{L,\,PS} = 2350$ m/s, $C_{T,\,PS} = 1230$ m/s, $\rho_{PDMS} = 965$ kg/m^3, $C_{L,\,PDMS} = 1100$ m/s and $C_{T,\,PDMS} = 200$ m/s, where ρ, C_L and C_T denote density, longitudinal speed of sound and transverse speed of sound, respectively. The lattice constant of the PC is $a = 1$ cm. The length of the edge of the square inclusion is 0.6 cm for a filling fraction $ff = 0.36$ (see Fig. 3.14a for a schematic). The structural mechanics module of the commercial software package COMSOL Multiphysics is

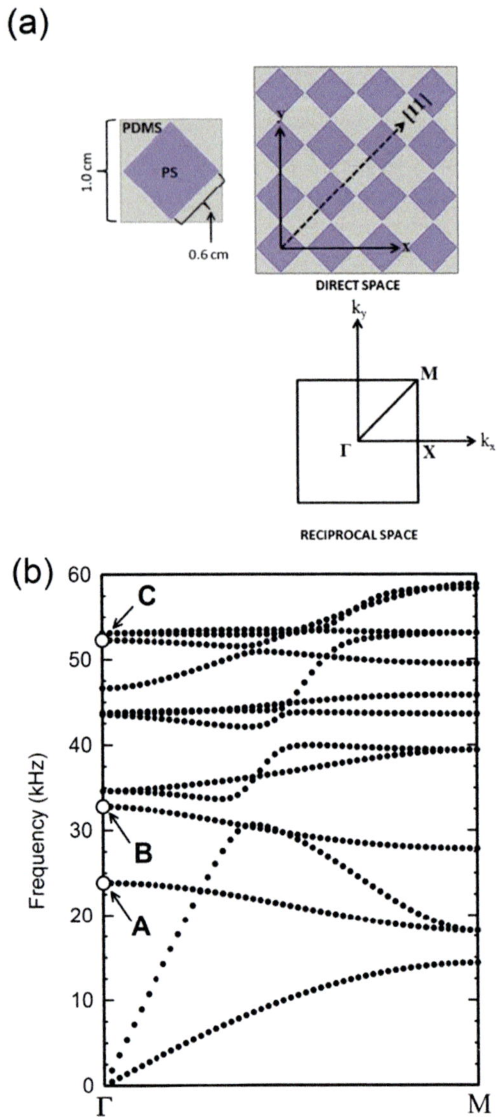

Fig. 3.14 (**a**) Schematic representation of the direct space and reciprocal space unit cells of the PC composed of a square lattice of PS inclusions (square cross-section) embedded in a PDMS matrix. (**b**) Phonon band structure (xy-modes) along ΓM-direction in reciprocal space

utilized to generate the phonon band structure of the PC and visualize displacement vector fields for specific Eigen modes.

The dispersion diagram along the ΓM-direction in reciprocal space ([11]-direction in direct space) is shown in Fig. 3.14b. Longitudinal and transverse acoustic branches stem from the Γ-point in the dispersion diagram. These bands span the width of the 1st Brillouin zone and fold back into the zone at the boundary (M-point), thus yielding multiple longitudinal and shear optical branches. Several hybridization gaps are observable between the longitudinal acoustic branch and higher frequency optical modes possessing similar symmetry, several of which possess mixed translation and rotational character.

Our specific interest in Fig. 3.14b is identifying rotational modes with pure rotational character. Modes "A", "B" and "C" in Fig. 3.14b at the Γ-point possess this characteristic. We illuminate these Eigen modes in Fig. 3.15 with finite-element calculations of displacement vector fields.

For mode A, along the [11]-direction, alternating regions of PS and PDMS exhibit out of phase rotations. This is also the case for mode C. Similar modes have been observed in a steel-epoxy solid/solid PC [54]. In contrast, mode B shows that the PS and PDMS regions are rotating in phase. The observation of a well-defined in-phase rotational mode at low frequency is enabled by the very low value of the transverse speed of sound in the PDMS compared to that of the PS inclusions. Indeed, we have verified that increasing $C_{T,\ PDMS}$ results in a shift of the mode "B" toward higher frequencies where this mode will hybridize with other modes. To elucidate the origin of the two types of rotational modes observed in the 2D PC, we use a 1D Cosserat-like micromechanics model. It was demonstrated in [54], that such a 1D model can reproduce the dispersion relations associated with rotational modes in a 2D PC composed of a square array of solid inclusions in a solid matrix.

Following Vasiliev et al. [60, 61] we now use a discrete linear 1D micromechanics model that includes longitudinal, shear and rotational degrees of freedom. This 1D discrete Cosserat-like lattice model consists of an infinite chain of square block elements (Cosserat elements) connected with multiple harmonic springs. Each element in the model is considered to have two translational degrees of freedom (displacement in the x and y directions) and one rotational degree of freedom (rotation about an axis perpendicular to the xy-plane). Fig. 3.16a and b show the repeatable unit cells for the monoblock and diblock Cosserat lattice models, respectively. Figure 3.16a shows periodicity (h) and Fig. 3.16b shows periodicity ($2h$). The diblock system mimics the behavior of alternating extended regions of PS and PDMS connected via elastic springs.

Three different harmonic springs (spring constants k_0, k_1, and k_2) connect different parts of the Cosserat elements. The Cosserat element in Fig. 3.16a has mass (m) and moment of inertia (I). The Cosserat elements that make-up the diblock unit cell have masses (m_1 and m_2) and inertial moments (I_1 and I_2). For Fig. 3.16a, the Cosserat element in the nth unit cell has x-displacement (u_n), y-displacement (v_n) and rotation component (φ_n). u_n and v_n represent displacements associated with longitudinal and transverse vibrations, respectively. The potential energy associated with the elastic connections of elements (n) and (n + 1) in the monoblock chain is written as follows:

$$\begin{aligned} E_{n,n+1} = {} & \frac{1}{2}K_0(u_{n+1}-u_n)^2 + \frac{1}{2}K_1\left[(v_{n+1}-v_n) + \frac{h}{2}(\varphi_{n+1}+\varphi_n)\right]^2 \\ & + \frac{1}{2}K_2(\varphi_{n+1}-\varphi_n)^2 \end{aligned} \tag{3.107}$$

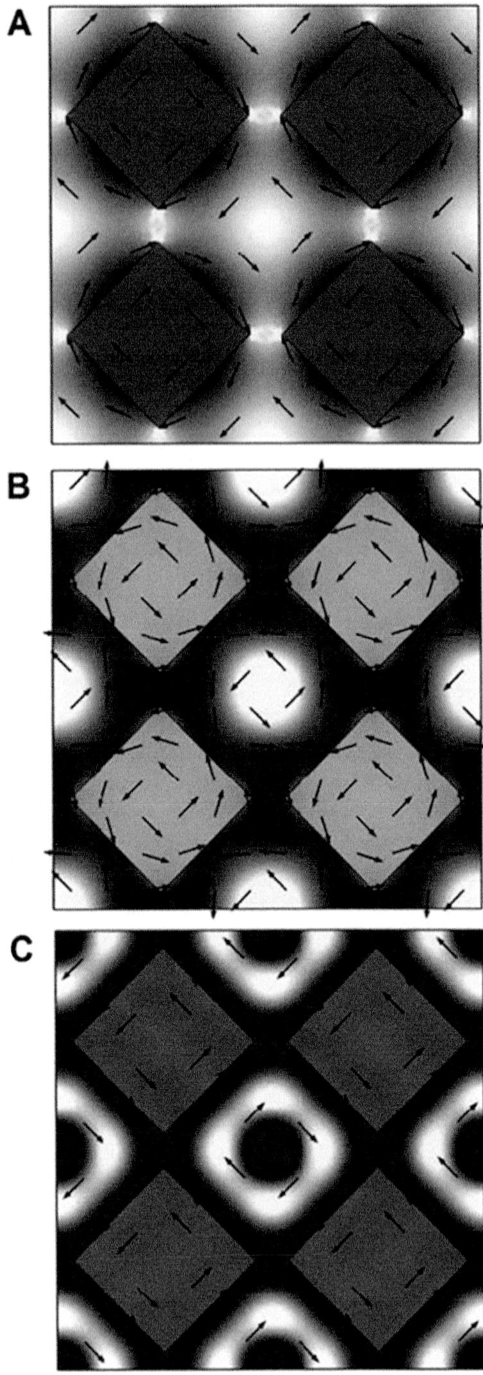

Fig. 3.15 Displacement vector field (*arrows*) of the PS/PDMS phononic crystal corresponding to Eigenmodes "A", "B" and "C" in Fig. 3.14b. The *grey color*-scale (not shown) corresponds to the z-component of the curl of the displacement field

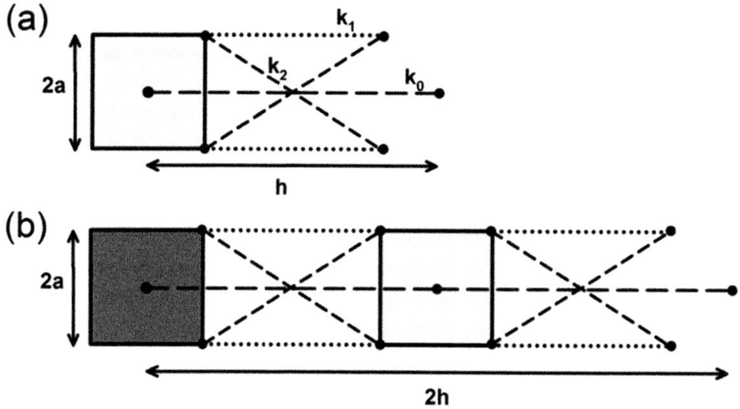

Fig. 3.16 Schematic illustration of the discrete Cosserat-like micromechanics model. (**a**) Unit cell of the monoblock model with Cosserat elements (blocks) connected by three types of harmonic springs (spring constants k_0, k_1, and k_2). Each Cosserat element possesses translational longitudinal (u), shear (v) and rotational (φ) degrees of freedom. (**b**) Unit cell of a diblock model

where $K_0 = \left(\frac{k_0}{h^2} + \frac{2k_1}{l^2} + \frac{2k_2 l^2}{l_d^4}\right)$, $K_1 = \left(\frac{2k_2 (2a)^2}{l_d^4}\right)$, $K_2 = \left(\frac{2a^2 k_1}{l^2}\right)$, $l = h - (2a)$, and $l_d = \sqrt{\left(l^2 + (2a)^2\right)}$. Accordingly, the equations of motion for the Cosserat element in the nth unit cell of the monoblock lattice are written as:

$$m\frac{d^2 u_n}{dt^2} = K_0(u_{n+1} - 2u_n + u_{n-1}) \tag{3.108}$$

$$m\frac{d^2 v_n}{dt^2} = K_1(v_{n+1} - 2v_n + v_{n-1}) + \frac{hK_1}{2}(\varphi_{n+1} - \varphi_{n-1}) \tag{3.109}$$

$$I\frac{d^2 \varphi_n}{dt^2} = K_2(\varphi_{n+1} - 2\varphi_n + \varphi_{n-1}) + \frac{hK_1}{2}(v_{n-1} - v_{n+1}) - \frac{h^2 K_1}{4}(\varphi_{n+1} + 2\varphi_n + \varphi_{n-1}) \tag{3.110}$$

Utilizing (3.107–3.110), we numerically generate phonon band structures for the monoblock and diblock systems by implementing a finite-difference scheme employing the Spectral Energy Density (SED) method (this method is detailed in Appendix 3). For the monoblock model we chose m = 25 g, a = 0.35 cm, h = 1.414 cm, k_0 = 55,000 N·m, k_1 = 9500 N·m and k_2 = 6500 N·m. The numerical values of these parameters are chosen to establish a semi-quantitative correspondence between the band structure of the 1-D block systems and that of the 2-D PC. In particular, we obtain semi-quantitative agreement in (a) the ordering of the bands (modes A,B,C), (b) the symmetry of the modes as well as (c) the order of magnitude of the frequencies of these modes. The diblock system is identical to the monoblock system except for the mass (and consequently moment of inertia) of the

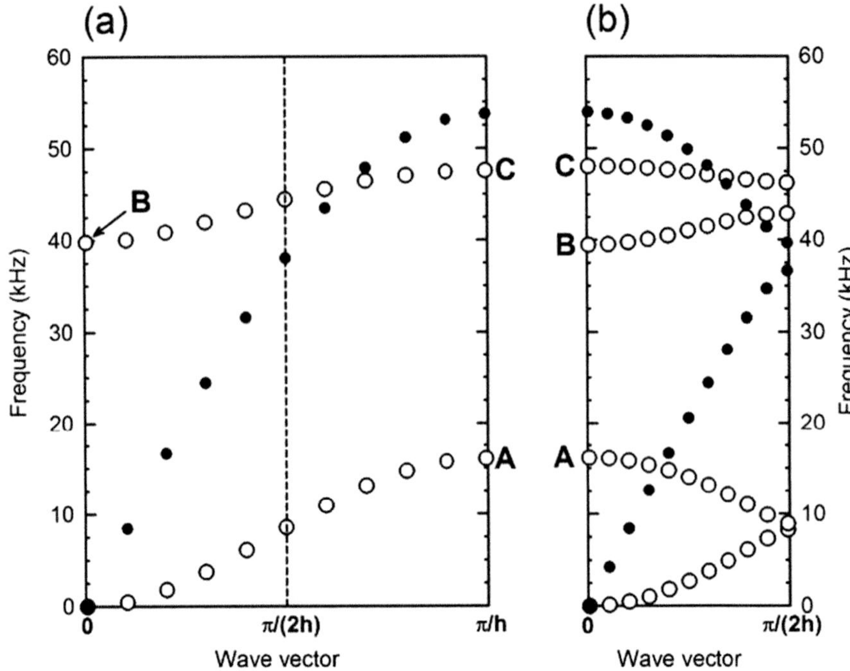

Fig. 3.17 Band structure of the monoblock (**a**) and of the diblock (**b**) Cosserat 1D discrete lattices. Modes "A" and "C" in (**a**) and (**b**) correspond to successive Cosserat elements rotating out-of-phase. Oppositely, mode "B" shows successive Cosserat elements rotating in-phase

Cosserat elements ($m_1 = 27$ g and $m_2 = 23$ g). Figures 3.17a and b show the dispersion diagrams for the monoblock and diblock systems, respectively.

The band structure of the monoblock lattice contains three bands corresponding to the three degrees of freedom (u, v, and φ). The bands rendered with unfilled circles support modes with rotational character [as shown by (3.109) and (3.110)] while the band comprised of solid circles corresponds to modes with translational character only [see (3.108)]. The points marked by "A", "B", and "C" correspond to modes with pure rotational character. The band structure of the diblock system (Fig. 3.17b) exhibits the characteristic band folding associated with the doubling of the period. We have retained on that figure the location of the monoblock rotational modes (eigenmodes "A", "B", and "C"). At the origin, the rotational mode marked by "B" is characterized by each block rotating in phase with its neighbor. Modes "A" and "C" correspond to rotational waves whereby alternating blocks rotate with a π phase shift. These modes are isomorphic to those observed in the more complex band structure of the 2D PC. Mode "B" is the same for the monoblock and diblock systems; this mode can be investigated within the formalism of Dirac and possesses spinor characteristics.

Appendix 3: Molecular Dynamics (MD) and Spectral Energy Density (SED) Methods

MD is a simulation technique for computing the thermodynamic as well as kinetic properties of a classical many-body system [62]. Classical MD methods consist of solving numerically Newton's equations of motion of a collection of N interacting massive particles or atoms. The most critical component of a MD simulation is the interatomic potential from which interatomic forces may be derived. The equation of motion of each individual atom is solved numerically in time to obtain the trajectories of the system, namely, the time evolution of the positions and momenta of every particle. In some systems, the computational task of solving the equations of motion scales at best linearly with the number of particles, N, and more generally as N^2. Periodic boundary conditions (PBC) are often used to reduce the computational problem size. PBC consist of repeating periodically in all directions of space a "small" simulation cell. One allows interaction between the N atoms within the simulation cell but also between atoms inside the simulation cell and atoms in the periodically repeated "image" cells. Interactions are cut-off to less than half the minimum characteristic length of the simulation cell to avoid spurious effects such as interaction of an atom with its own image. This method effectively reduces the effects that may be associated with surfaces in a finite size system. However, while trying to mimic the behavior of an infinite system, the simulated system still possesses the characteristics of a finite system. For instance, the finiteness of an MD system with PBC leads to a discretization of the phonon modes and a suppression of the modes with wavelength longer than the simulation cell length. This is easily seen by considering a one-dimensional (1D) monoatomic system composed of N atoms interacting via a nearest neighbor harmonic (or anharmonic) potential. In this case, imposing PBC leads to atom N interacting with atom 1 thus forming a ring. Modes with wavelengths exceeding the length $L=Na$, where a is the interatomic spacing, are not compatible with the constraint of the ring geometry and cannot be supported by that structure.

To initiate a simulation, every mass in the MD simulation cell is randomly displaced from its equilibrium position. The maximum value in which a mass can be displaced is constrained such that instabilities do not emerge in the potential energy function. For atomic systems, the random velocities may be scaled such that the total kinetic energy relates to the temperature of the system according to the equipartition theorem. For post-processing spectral energy density (SED) calculations, velocity data is collected for each mass in the simulation cell over the entire simulation time.

The SED method is a technique for predicting phonon dispersion relations and lifetimes from the atomic velocities of the particles in a crystal generated by classical MD [63]. The SED method offers a comprehensive description of phonon properties because individual phonon modes can be isolated for analysis and is computationally affordable for the systems that have been examined in this chapter. Formally, the expression for SED is written as follows:

$$\Phi\left(\vec{k},\omega\right)=\frac{1}{4\pi\tau_0 N}\sum\nolimits_{\alpha}\sum\nolimits_{b}^{B}m_b\left|\int_0^{\tau_0}\sum\nolimits_{n_{x,y,z}}^{N}v_\alpha\begin{pmatrix}n_{x,y,z}\\ b\end{pmatrix};t\right)\times e^{\left(i\vec{k}\cdot\vec{r_0}-i\omega t\right)}dt\right|^2 \tag{3.111}$$

where τ_0 represents the length of time over which velocity data is collected from a given MD simulation, N is the total number of unit cells represented in the MD simulation and $v_\alpha\left(\begin{matrix}n_{x,y,z}\\ b\end{matrix};t\right)$ represents the velocity of atom b (of mass m_b in unit cell $n_{x,y,z}$) in the α-direction. For a specified wave vector ($\vec{k}$), the spectrum relating SED to frequency is found by adding the square of the absolute value of the Fourier transform of the discrete temporal signal $f(t)=\sum\nolimits_{n_{x,y,z}}^{N_T}v_\alpha\left(\begin{matrix}n_{x,y,z}\\ b\end{matrix};t\right)\times e^{\left(i\vec{k}\cdot\vec{r_0}\right)}$ for every $[\alpha, b]$ pair. A SED value represents the average kinetic energy per unit cell as a function of wave vector and frequency. A peak in the spectrum relating SED to frequency signifies a vibrational eigenmode for wave vector ($\vec{k}$). The shape of the frequency spread for eigen mode ($\vec{k}$) is represented with the Lorentzian function:

$$\Phi\left(\vec{k},\omega\right)=\frac{I}{1+\left[(\omega-\omega_c)/\gamma\right]^2} \tag{3.112}$$

where I is the peak magnitude, ω_c is the frequency at the center of the peak and γ is the half-width at half-maximum. The lifetime for phonon mode ($\vec{k}$) is defined as $\tau=1/2\gamma$ [63]. Non-degenerate wave vector modes are dependent on the size of the MD simulation cell and are written as follows: $k_i=2\pi n_i/aN_i$, where a is the lattice constant, N_i is the total number of unit cells in the i-direction, and n_i is an integer ranging from $-N_i+1$ to N_i.

Larkin et al. [64] have recently corrected the expression (3.111). This corrected expression takes the form:

$$\Phi\left(\vec{k},\omega\right)=\sum\nolimits_{\nu}^{3B}A_o\begin{pmatrix}\vec{k}\\ \nu\end{pmatrix}\frac{\Gamma\begin{pmatrix}\vec{k}\\ \nu\end{pmatrix}/\pi}{\left[\omega_o\begin{pmatrix}\vec{k}\\ \nu\end{pmatrix}-\omega\right]^2+\Gamma^2\begin{pmatrix}\vec{k}\\ \nu\end{pmatrix}} \tag{3.113}$$

which is a superposition of 3B Lorentzian functions with centers at $\omega_o\begin{pmatrix}\vec{k}\\ \nu\end{pmatrix}$ and linewidths $\Gamma\begin{pmatrix}\vec{k}\\ \nu\end{pmatrix}$ (one for each polarization, ν). $A_o\begin{pmatrix}\vec{k}\\ \nu\end{pmatrix}$ are phonon mode-dependent constants. The kinetic energy normal mode coordinate is:

$$\dot{q}\begin{pmatrix}\vec{k}\\ \nu\end{pmatrix};t\Big)=\sum\nolimits_{\alpha=1}^{3}\sum\nolimits_{b=1}^{B}\sum\nolimits_{n=1}^{N_T}\sqrt{\frac{m_b}{N_T}}v_\alpha\begin{pmatrix}n\\ b\end{pmatrix};t\Big)e^*\begin{pmatrix}\vec{k} & b\\ \nu & \alpha\end{pmatrix}exp\left[i\ \vec{k}\cdot\vec{r_0}\right] \tag{3.114}$$

where $e^*\begin{pmatrix} \vec{k} & b \\ \nu & \alpha \end{pmatrix}$ are the components of the time independent phonon mode eigenvector (generated from quasi-harmonic Lattice Dynamics calculations), B is the number of atoms in the unit cell, m_b is the mass of the bth atom in the unit cell and $\vec{r_0}$ is the equilibrium position vector of the nth unit cell. There are N_T total unit cells and $v_\alpha\begin{pmatrix} n \\ b \end{pmatrix}; t$ is the α component of the velocity of the bth atom in the nth unit cell at time t. Given a set of atomic velocities from MD simulation and the phonon mode eigenvector, $\Phi\left(\vec{k}, \omega\right)$ can be calculated using

$$\Phi\left(\vec{k}, \omega\right) = 2\sum_{\nu}^{3B} \lim_{\tau_o \to \infty} \frac{1}{2\tau_o} \left| \frac{1}{\sqrt{2\pi}} \int_0^{\tau_o} \dot{q}\begin{pmatrix} \vec{k} \\ \nu \end{pmatrix}; t \Big) exp(-i\omega t)dt \right|^2 \tag{3.115}$$

and then fit using (3.113) to extract the phonon properties $\omega_o\begin{pmatrix} \vec{k} \\ \nu \end{pmatrix}$ and $\Gamma\begin{pmatrix} \vec{k} \\ \nu \end{pmatrix}$. The phonon lifetime, $\tau\begin{pmatrix} \vec{k} \\ \nu \end{pmatrix}$ is defined as $\left[2\Gamma\begin{pmatrix} \vec{k} \\ \nu \end{pmatrix}\right]^{-1}$. In practice, τ_o should be much larger than the longest phonon lifetime and the continuous Fourier transform in (3.115) is performed using a discrete Fast Fourier transform (FFT).

Spectral Energy Density Code

This Fortran 90 code is based on 3.111 used to calculate the band structure of a simple one-dimensional (1-D) harmonic crystal. All physical and geometric parameters of the 1-D crystal model are normalized. The code utilizes the FFT routine FOUR1 from reference [65]. Figure 3.18 illustrates output from this code.

```
program Spectral_Energy_Density
  implicit none
  integer :: i, ic, it, ineg, itemp, ip, im
  integer :: n0, ig, jf, j
  integer, parameter :: nt=2**20
  integer, parameter :: nt2=2**19
  integer, parameter :: fmax=2**10
  integer :: inave, NAVE
  double precision ::  ao
  integer :: N, NUCs, NM
  double precision :: amp
  double precision :: ur, up, um, dt, beta
  double precision :: k0, pi, coskr, sinkr, freq, kvec, mi, accr,acci
  double precision, dimension(:), allocatable :: u, ur1
  double precision, dimension(:), allocatable :: vi, v, vp1
```

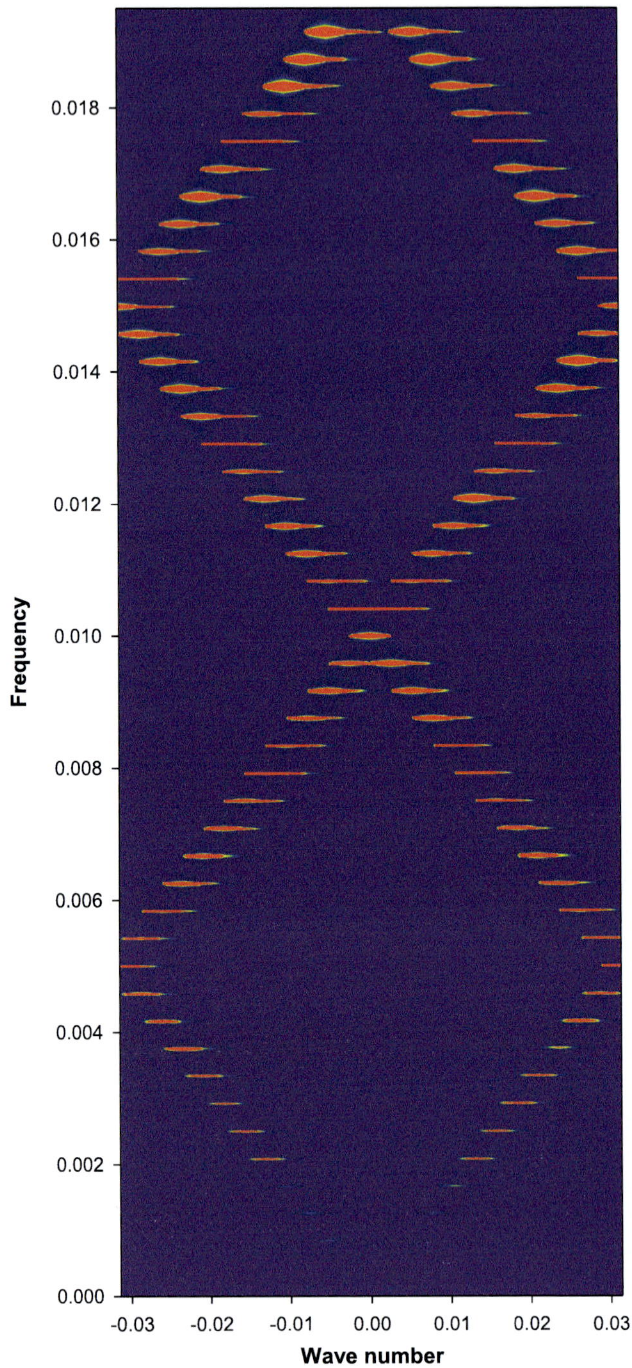

Fig. 3.18 Elastic band structure of a 1-D harmonic crystal composed of 2400 masses calculated with the SED code above. The band structure is plotted as a contour map of the SED in the wave number and frequency plane. The wavenumber axis is discretized in 25 points. The folded bands result from the supercell representation of the crystal (a supercell contains 100 masses). The SED

```
  double precision, dimension(:), allocatable :: r, k
  double precision, dimension(fmax) :: SED, SEDavg
  real, dimension(2*nt2,100) :: data
  real, dimension(2*nt2) :: datatemp
  double precision, dimension(:), allocatable :: tempREAL, tempIMAG
  integer, dimension(:), allocatable :: id123
  double precision, dimension(:), allocatable :: modi
  real :: N1

  N = 2400            !! number of masses N = 2400
  NUCs = 24                     !! number of unit cells with 100
masses
  NM=100 !! number of masses per unit cell
  NAVE=10!!number of dynamical simulation over which one averages the
SED
  allocate(tempREAL(NM))
  allocate(tempIMAG(NM))
  allocate(id123(NM))
  allocate(modi(NM))

  open(unit=2, file='band_struct.dat', status='old') !! output file

  dt=0.1d0     !! time step
  mi=1.d0      !! mass
  ao=100.d0  !! length of one unit cell with 100 masses
  beta=1.d0  !!spring constant divided by mass
  pi=4.0d0*atan(1.0d0)

  allocate(r(NUCs))
  allocate(u(N))
  allocate(ur1(N))
  allocate(vi(N))
  allocate(v(N))
  allocate(vp1(N))

  !!================= Start SED parameters ===================
  !!=============== loop over wave vectors ===================
  !! n0: [-12,12]
  !! Brillouin zone divided into 25 discrete wave numbers
   do n0=-12,12
```

Fig. 3.18 (continued) is averaged over 10 dynamical simulation of the crystal. Elastic modes appear as elongated spots; the width of the spots decreases with increasing number of dynamical simulations used for averaging

```
  k0 = ((2.0*pi)/ao)*((1.0d0*n0)/(1.0d0*NUCs))
  do i=1,NUCs
     r(i)=(0.50d0*ao)+((i-1)*ao)   !!coordinates of the unit cells
  end do
  !!================== End SED parameters ======================
!! Single mass-spring chain : 1-D harmonic crystal (this section is
system
!!dependent)
  amp = 0.1d0    !!maximum initial random displacement of masses

    do  inave=1,NAVE

          u(1:N)=0.0d0
          do i=1,N
               call random_number(N1)
               u(i) = (N1-0.5)*amp
          end do

        v(1:N)=0.0d0
        ig=0

      do it=1,nt

          do i=1,N
               ip = i+1      !! periodic boundary conditions
               if (i.eq.N) ip = 1
               im = i-1
               if (i.eq.1) im = N

               ur = u(i)
               up = u(ip)
               um = u(im)

               accr =beta*((up-ur)-(ur-um))
               !! accelleration
                       vp1(i)=v(i)+dt*accr
                       ur1(i)=u(i)+dt*vp1(i)
          end do
 !! vi : velocities used for the SED analysis
               do i=1,N
               vi(i)=vp1(i)
               enddo
 !! time-shift displacement
               do i=1,N
               u(i) = ur1(i)
```

```
                v(i)=vp1(i)
                enddo

!! eliminating the early part of the dynamics before nt2, ig is the
new time index
          if (it.GT.nt2) then
                ig=ig+1
                      do i=1,NM
                      tempREAL(i)=0.0d0
                      tempIMAG(i)=0.0d0
                      enddo

                do i=1,NUCs
                         do j=1,NM
                         id123(j)=j+((i-1)*NM)  !!indexing the
masses
                         enddo

                     coskr = dcos(k0*r(i))
                     sinkr = dsin(k0*r(i))

                             do j=1,NM
                             tempREAL(j)=tempREAL(j)+vi(id123
(j))*coskr

                             tempIMAG(j)=tempIMAG(j)+vi(id123
(j))*sinkr
                             end do
                end do

                      do ic=1,NM
                data(2*ig-1,ic)=tempREAL(ic)
                data(2*ig,ic)=tempIMAG(ic)
                      enddo
          end if

      end do
        do ic=1,NM
       do itemp=1,2*nt2
       datatemp(itemp)=data(itemp,ic)
       enddo
      call four1(datatemp,nt2,1)
       do itemp=1,2*nt2
       data(itemp,i)=datatemp(itemp)
       enddo
       enddo
```

```
      do jf=1,fmax
         SED(jf)=0.0d0
         do ic=1,NM
            modi(ic)=(mi/NUCs)*((data(2*jf-1,ic))**2+(data(2*jf,
ic))**2)

           SED(jf)=SED(jf)+modi(ic)
      end do
        end do

      do jf=1,fmax
      SEDavg(jf)=SEDavg(jf)+SED(jf)
      enddo
    enddo

      do jf=1,fmax
      SEDavg(jf)=SEDavg(jf)/(1.d0*NAVE)
      enddo

   do jf=1,fmax
      kvec = k0
      freq = (jf-1)/(nt2*dt)
      write(2,'(3e20.8)'), kvec, freq, SEDavg(jf)
   end do
 end do
  close(unit=2)

end program Spectral_Energy_Density

!! FFT subroutine
SUBROUTINE four1(data,nn,isign)
   implicit none
   INTEGER, intent(IN) :: isign,nn
   REAL, intent(IN OUT) :: data(2*nn)
   INTEGER              :: i,istep,j,m,mmax,n
   REAL                 :: tempi,tempr
   real (kind(0D0))     :: theta,wi,wpi,wpr,wr,wtemp
   n=2*nn
   j=1
   do i=1,n,2
      if(j.gt.i)then
         tempr=data(j)
         tempi=data(j+1)
         data(j)=data(i)
         data(j+1)=data(i+1)
```

```
          data(i)=tempr
          data(i+1)=tempi
       end if
       m=n/2
1      if ((m.ge.2).and.(j.gt.m)) then
          j=j-m
          m=m/2
          goto 1
       end if
      j=j+m
  end do
  mmax=2
2 if (n.gt.mmax) then
       istep=2*mmax
       theta=6.28318530717959d0/(isign*mmax)
       wpr=-2.d0*sin(0.5d0*theta)**2
       wpi=sin(theta)
       wr=1.d0
       wi=0.d0
       do m=1,mmax,2
          do i=m,n,istep
             j=i+mmax
             tempr=sngl(wr)*data(j)-sngl(wi)*data(j+1)
             tempi=sngl(wr)*data(j+1)+sngl(wi)*data(j)
             data(j)=data(i)-tempr
             data(j+1)=data(i+1)-tempi
             data(i)=data(i)+tempr
             data(i+1)=data(i+1)+tempi
          end do
          wtemp=wr
          wr=wr*wpr-wi*wpi+wr
          wi=wi*wpr+wtemp*wpi+wi
       end do
       mmax=istep
       goto 2
   endif
   return
end SUBROUTINE four1
```

Appendix 4: Mutiple Time-Scale Perturbation Theory

Nowadays, the multiple time scale perturbation theory (MTSPT) for differential equations is a very popular method to approximate solutions of weakly nonlinear differential equations. Several implementations of this method were proposed in various fields of mathematics, mechanics, and physics [66–72]. Moreover, Khoo and Wang [73] have shown that the MTSPT is a reliable theoretical tool for studying the lattice dynamics of an anharmonic crystal. More recently, Swinteck et al. [74] applied successfully the MTSPT, as described in [73], for solving propagation equations in a quadratically nonlinear monoatomic chain of infinite extent (we use this example here). Consequently, we have used the MTSPT for solving a variety of nonlinear phononic problems throughout this chapter as well as other Chapters.

We illustrate the method in the case of a one-dimensional continuous elastic wave equation (displacement u) with some weak non-linear force $\varepsilon f(u)$:

$$\frac{\partial^2 u}{\partial t^2} - \beta^2 \frac{\partial^2 u}{\partial x^2} + \varepsilon f(u) = 0 \tag{3.116}$$

We treat ε as a perturbation and write the displacement as a second order power series in the perturbation, namely

$$\begin{aligned} u(k+g, \tau_0, \tau_1, \tau_2) &= u_0(k+g, \tau_0, \tau_1, \tau_2) + \varepsilon u_1(k+g, \tau_0, \tau_1, \tau_2) \\ &\quad + \varepsilon^2 u_2(k+g, \tau_0, \tau_1, \tau_2) + \ldots \end{aligned} \tag{3.117}$$

In (3.117), u_i with $i = 0, 1, 2$ are displacement functions expressed to zeroth-order, first-order and second-order in the perturbation. We have also replaced the single time variable, t, in (3.116) by three variables representing different time scales: $\tau_0 = t$, $\tau_1 = \varepsilon t$, and $\tau_2 = \varepsilon^2 t = \varepsilon^2 \tau_0$. We can subsequently decompose (3.116) into three equations: one equation to zeroth-order in ε, one equation to first-order in ε and a third equation to second-order in ε.

For this we calculate:

$$\frac{\partial u}{\partial t} = \frac{\partial u}{\partial \tau_0}\frac{\partial \tau_0}{\partial t} + \frac{\partial u}{\partial \tau_1}\frac{\partial \tau_1}{\partial t} + \frac{\partial u}{\partial \tau_2}\frac{\partial \tau_2}{\partial t} + \ldots = \frac{\partial u}{\partial \tau_0} + \varepsilon \frac{\partial u}{\partial \tau_1} + \varepsilon^2 \frac{\partial u}{\partial \tau_2} + \ldots \tag{3.118}$$

Inserting (3.117) into (3.118) yields the expansion for the first derivative of u with respect to time:

$$\frac{\partial u}{\partial t} = \frac{\partial u_0}{\partial \tau_0} + \varepsilon \left(\frac{\partial u_1}{\partial \tau_0} + \frac{\partial u_0}{\partial \tau_1} \right) + \varepsilon^2 \left(\frac{\partial u_2}{\partial \tau_0} + \frac{\partial u_1}{\partial \tau_1} + \frac{\partial u_0}{\partial \tau_2} \right) + \ldots \tag{3.119}$$

The second derivative is obtained from:

$$\frac{\partial^2 u}{\partial t^2} = \frac{\partial}{\partial \tau_0}\left(\frac{\partial u}{\partial t}\right) + \varepsilon \frac{\partial}{\partial \tau_1}\left(\frac{\partial u}{\partial t}\right) + \varepsilon^2 \frac{\partial}{\partial \tau_2}\left(\frac{\partial u}{\partial t}\right) + \dots \tag{3.120}$$

Inserting (3.119) into (3.120) and retaining only terms up to 2nd order yields:

$$\frac{\partial^2 u}{\partial t^2} = \frac{\partial^2 u_0}{\partial \tau_0^2} + \varepsilon\left(\frac{\partial^2 u_1}{\partial \tau_0^2} + 2\frac{\partial^2 u_0}{\partial \tau_1 \partial \tau_0}\right) + \varepsilon^2\left(\frac{\partial^2 u_2}{\partial \tau_0^2} + 2\frac{\partial^2 u_1}{\partial \tau_1 \partial \tau_0} + 2\frac{\partial^2 u_0}{\partial \tau_2 \partial \tau_0} + \frac{\partial^2 u_0}{\partial \tau_1^2}\right) \tag{3.121}$$

The second derivative with respect to position making up the second term in (3.116) is expanded in the form:

$$\frac{\partial^2 u}{\partial x^2} = \frac{\partial^2 u_0}{\partial x^2} + \varepsilon \frac{\partial^2 u_1}{\partial x^2} + \varepsilon^2 \frac{\partial^2 u_2}{\partial x^2} + \dots \tag{3.122}$$

The expansion of the third term in (3.116) depends on its functional form. For illustrative purposes, we use a quadratic function and expand it to first order (the third term is already multiplied by ε:

$$f(u) = u^2 = \left(u_0 + \varepsilon u_1 + \varepsilon^2 u_2 + \dots\right)^2 = u_0^2 + \varepsilon(u_0 u_1 + u_1 u_0) + \dots \tag{3.123}$$

Combining (3.121), (3.122) and (3.123) into (3.116) and separating the terms by orders in ε, gives:

$$\begin{aligned} &\left\{\frac{\partial^2 u_0}{\partial \tau_0^2} - \beta^2 \frac{\partial^2 u_0}{\partial x^2}\right\} + \varepsilon\left\{\frac{\partial^2 u_1}{\partial \tau_0^2} + 2\frac{\partial^2 u_0}{\partial \tau_1 \partial \tau_0} - \beta^2 \frac{\partial^2 u_1}{\partial x^2} + u_0^2\right\} \\ &+\varepsilon^2\left\{\frac{\partial^2 u_2}{\partial \tau_0^2} + 2\frac{\partial^2 u_1}{\partial \tau_1 \partial \tau_0} + 2\frac{\partial^2 u_0}{\partial \tau_2 \partial \tau_0} + \frac{\partial^2 u_0}{\partial \tau_1^2} - \beta^2 \frac{\partial^2 u_2}{\partial x^2} + (u_0 u_1 + u_1 u_0)\right\} = 0 \end{aligned} \tag{3.124}$$

The set of coupled equations to solve are therefore:

To zeroth order: $\dfrac{\partial^2 u_0}{\partial \tau_0^2} - \beta^2 \dfrac{\partial^2 u_0}{\partial x^2} = 0$

To first order: $\dfrac{\partial^2 u_1}{\partial \tau_0^2} + 2\dfrac{\partial^2 u_0}{\partial \tau_1 \partial \tau_0} - \beta^2 \dfrac{\partial^2 u_1}{\partial x^2} + u_0^2 = 0$

To second order: $\dfrac{\partial^2 u_2}{\partial \tau_0^2} + 2\dfrac{\partial^2 u_1}{\partial \tau_1 \partial \tau_0} + 2\dfrac{\partial^2 u_0}{\partial \tau_2 \partial \tau_0} + \dfrac{\partial^2 u_0}{\partial \tau_1^2} - \beta^2 \dfrac{\partial^2 u_2}{\partial x^2} + (u_0 u_1 + u_1 u_0) = 0$

Finally, in Chap. 4, we will utilize the multiple time scale perturbation theory up to fourth order in ε, we therefore give below the expansion of $\frac{\partial^2 u}{\partial t^2}$ up to fourth order:

$$\text{To zeroth order:} \quad \frac{\partial^2 u_0}{\partial \tau_0^2}$$

$$\text{To first order:} \quad \frac{\partial^2 u_1}{\partial \tau_0^2} + 2\frac{\partial^2 u_0}{\partial \tau_1 \partial \tau_0}$$

$$\text{To second order:} \quad \frac{\partial^2 u_2}{\partial \tau_0^2} + 2\frac{\partial^2 u_1}{\partial \tau_1 \partial \tau_0} + 2\frac{\partial^2 u_0}{\partial \tau_2 \partial \tau_0} + \frac{\partial^2 u_0}{\partial \tau_1^2}$$

$$\text{To third order:} \quad \frac{\partial^2 u_3}{\partial \tau_0^2} + 2\frac{\partial^2 u_0}{\partial \tau_1 \partial \tau_2} + 2\frac{\partial^2 u_0}{\partial \tau_3 \partial \tau_0} + 2\frac{\partial^2 u_1}{\partial \tau_0 \partial \tau_2} + 2\frac{\partial^2 u_2}{\partial \tau_1 \partial \tau_0} + \frac{\partial^2 u_1}{\partial \tau_1^2}$$

$$\text{To fourth order:} \quad \frac{\partial^2 u_4}{\partial \tau_0^2} + 2\frac{\partial^2 u_3}{\partial \tau_0 \partial \tau_1} + 2\frac{\partial^2 u_2}{\partial \tau_0 \partial \tau_2} + 2\frac{\partial^2 u_1}{\partial \tau_0 \partial \tau_3} + 2\frac{\partial^2 u_1}{\partial \tau_2 \partial \tau_1}$$

$$+2\frac{\partial^2 u_0}{\partial \tau_4 \partial \tau_0} + 2\frac{\partial^2 u_0}{\partial \tau_1 \partial \tau_3} + \frac{\partial^2 u_2}{\partial \tau_1^2} + \frac{\partial^2 u_0}{\partial \tau_2^2}$$

References

1. M.Z. Hasan, C.L. Kane, Colloquium: topological insulators. Rev. Mod. Phys. **82**, 3045 (2010)
2. A.B. Khanikaev, S.H. Mousavi, W.-K. Tse, M. Kargarian, A.H. MacDonald, G. Shvets, Photonic topological insulators. Nat. Mat. **12**, 233 (2013)
3. F.D.M. Haldane, S. Raghu, Possible realization of directional optical waveguides in photonic crystals with broken time-reversal symmetry. Phys. Rev. Lett. **100**, 013904 (2008)
4. P.A. Deymier, K. Runge, N. Swinteck, K. Muralidharan, Torsional topology and fermion-like behavior of elastic waves in phononic structures. C. R. Acad. Sci. Méc. **343**, 700–711 (2015)
5. P.A. Deymier, K. Runge, N. Swinteck, K. Muralidharan, Rotational modes in a phononic crystal with fermion-like behaviour. J. Appl. Phys. **115**, 163510 (2014)
6. E. Prodan, C. Prodan, Topological phonon modes and their role in dynamic instability of microtubules. Phys. Rev. Lett. **103**, 248101 (2009)
7. C.L. Kane, T.C. Lubensky, Topological boundary modes in isostatic lattices. Nat. Phys. **10**, 39 (2013)
8. S. Mousavi, A.B. Khanikaev, Z. Wang, Topologically protected elastic waves in phononic metamaterials. Nat. Commun. **6**, 8682 (2015)
9. B.G. Chen, N. Upadhyaya, V. Vitelli, Nonlinear conduction via solitons in a topological mechanical insulator. Proc. Natl. Acad. Sci. U.S.A. **111**, 13004 (2014)
10. R. Süsstrunk, S.D. Huber, Observation of phononic helical edge states in a mechanical topological insulator. Science **349**, 47 (2015)
11. M. Xiao, G. Ma, Z. Yang, P. Sheng, Z.Q. Zhang, C.T. Chan, Geometric phase and band inversion in periodic acoustic systems. Nat. Phys. **11**, 240 (2015)
12. J. Paulose, B.G. Chen, V. Vitelli, Topological modes bound to dislocations in mechanical metamaterials. Nat. Phys. **11**, 153 (2015)

13. N. Berg, K. Joel, M. Koolyk, E. Prodan, Topological phonon modes in filamentary structures. Phys. Rev. E **83**, 021913 (2011)
14. R.K. Pal, M. Schaeffer, M. Ruzzene, Helical edge states and topological phase transitions in phononic systems using bi-layered lattices. J. Appl. Phys. **119**, 084305 (2016)
15. A.B. Khanikaev, R. Fleury, S.H. Mousavi, A. Alù, Topologically robust sound propagation in an angular-momentum-biased graphene-like resonator lattice. Nat. Commun. **6**, 8260 (2016)
16. G. Salerno, T. Ozawa, H.M. Price, I. Carusotto, Floquet topological system based on frequency-modulated classical coupled harmonic oscillators. arXiv:1510.04697 (2015)
17. J. Paulose, A.S. Meeussen, V. Vitelli, Selective buckling via states of self-stress in topological metamaterials. Proc. Natl. Acad. Sci. U.S.A. **112**, 7639 (2015)
18. L.M. Nash, D. Kleckner, A. Read, V. Vitelli, A.M. Turner, W.T.M. Irvine, Topological mechanics of gyroscopic metamaterials. ArXiv:150403362 Cond-Mat (2015)
19. P. Wang, L. Lu, K. Bertoldi, Topological phononic crystals with one-way elastic edge waves. Phys. Rev. Lett. **115**, 104302 (2015)
20. Z. Yang, F. Gao, X. Shi, X. Lin, Z. Gao, Y. Chong, B. Zhang, Topological acoustics. Phys. Rev. Lett. **114**, 114301 (2015)
21. W.G. Unruh, Experimental black-hole evaporation? Phys. Rev. Lett. **46**, 1351 (1981)
22. M. Visser, C. Molina-París, Acoustic geometry for general relativistic barotropic irrotational fluid flow. New J. Phys. **12**, 095014 (2010)
23. N. Bilic, Relativistic acoustics geometry. Class. Quantum Grav. **16**, 3953–3964 (1999)
24. R. Fleury, D.L. Sounas, C.F. Sieck, M.R. Haberman, A. Alu, Sound isolation and giant linear nonreciprocity in a compact acoustic circulator. Science **343**, 516 (2014)
25. L.C. Garcia de Andrade, Non-Riemannian geometry of vortex acoustics. Phys. Rev. D. **064004**, 70 (2004)
26. H.W. Hoogstraten, B. Kaper, Propagation of sound waves in a moving medium. J. Eng. Math. **5**, 295–305 (1971)
27. O.A. Godin, An exact wave equation for sound in inhomogeneous, moving, and non-stationary fluids. OCEANS, 1–5 (2011)
28. Q. Wang, Y. Yang, X. Ni, Y.-L. Xu, X.-C. Sun, Z.-G. Chen, L. Feng, X.-P. Liu, M.-H. Lu, Y.-F. Chen, Acoustic asymmetric transmission based on time-dependent dynamical scattering. Nat. Sci. Rep. **5**, 10880 (2015)
29. M. Stone, Acoustic energy and momentum in a moving medium. Phys. Rev. E **62**, 1341 (2000)
30. N. Swinteck, S. Matsuo, K. Runge, J.O. Vasseur, P. Lucas, P.A. Deymier, Bulk elastic waves with unidirectional backscattering-immune topological states in a time-dependent superlattice. J. Appl. Phys. **118**, 063103 (2015)
31. A. Vasiliev, A. Miroshnichenko, M. Ruzzene, A discrete model and analysis of one-dimensional deformations in a structural interface with micro-rotations. Mech. Res. Commun. **37**, 225 (2010)
32. A. Vasiliev, A. Miroshnichenko, M. Ruzzene, Multifield model for Cosserat media. J. Mech. Mater. **3**, 1365 (2008)
33. A.G. Webster, Acoustical impedance and the theory of horns and of the phonograph. Proc. Natl. Acad. Sci. U.S.A. **5**, 275 (1919)
34. B.J. Forbes, Acoustical Klein–Gordon equation: a time-independent perturbation analysis. Phys. Rev. Lett. **93**, 054301 (2004)
35. B.J. Forbes, E. Roy Pike, D.B. Sharp, The acoustical Klein–Gordon equation: the wave-mechanical step and barrier potential functions. J. Acoust. Soc. Am. **114**, 1291 (2003)
36. J. Gump, I. Finckler, H. Xia, R. Sooryakumar, W.J. Bresser, P. Boolchand, Phys. Rev. Lett. **92**, 245501 (2004)
37. J.A. Thomas, J.E. Turney, R.M. Iutzi, C.H. Amon, A.J.H. McGaughey, Predicting phonon dispersion relations and lifetimes from the spectral energy density. Phys. Rev. B **81**, 081411 (2010)
38. R.H.G. Helleman, E.W. Montroll, On a nonlinear perturbation theory without secular terms: I. Classical coupled anharmonic oscillators. Physica **74**, 22–74 (1974)

39. J.C. Maxwell, A dynamical theory of the electromagnetic field. Philos. Trans. R. Soc. Lond. **155**, 459 (1865)
40. R.L. Weaver, Anderson localization of ultrasound. Wave Motion **12**, 129 (1990)
41. H.F. Hu, A. Strybulevych, J.H. Page, S.E. Skipetrov, B.A. Van Tiggelen, Localization of ultrasound in a three-dimensional elastic network. Nat. Phys. **4**, 945 (2008)
42. F. Van der Biest, A. Sukhovich, A. Tourin, J.H. Page, B.A. van Tiggelen, Z. Liu, M. Fink, Resonant tunneling of acoustic waves through a double barrier consisting of two phononic crystals. Europhys. Lett. **71**, 63 (2005)
43. S. Yang, J.H. Page, L. Zhengyou, M.L. Cowan, C.T. Chan, P. Sheng, Ultrasound tunneling through 3D phononic crystals. Phys. Rev. Lett. **88**, 104301 (2002)
44. J.A. Gupta, R. Knobel, N. Samarth, D.D. Awschalom, Ultrafast manipulation of electron spin coherence. Science **292**, 2458 (2001)
45. P. Zanardi, Virtual quantum systems. Phys. Rev. Lett. **87**, 077901 (2001)
46. P. Zanardi, D.A. Lidar, S. Lloyd, Quantum tensor product structure are observable dependent. Phys. Rev. Lett. **92**, 060402 (2004)
47. P. A. Deymier (ed.), *Acoustic Metamaterials and Phononic Crystals*, Springer Series in Solid State Sciences 173 (Springer, Heidelberg, 2013)
48. F. Liu, Y. Lai, X. Huang, C.T. Chan, Dirac cones at $\vec{k} = 0$ in phononic crystals. Phys. Rev. B **84**, 224113 (2011)
49. G. Wang, X. Wen, J. Wen, L. Shao, Y. Liu, Two-dimensional locally resonant phononic crystals with binary structures. Phys. Rev. Lett. **93**, 154302 (2004)
50. Y. Lai, Y. Wu, P. Sheng, Z.Q. Zhang, Hybrid elastic solids. Nat. Mater. **10**, 620 (2011)
51. R. Sainidou, N. Stefanou, A. Modinos, Formation of absolute frequency gaps in three-dimensional solid phononic crystals. Phys. Rev. B **66**, 212301 (2002)
52. H. Zhao, Y. Liu, G. Wang, J. Wen, D. Yu, X. Han, X. Wen, Resonance modes and gap formation in a two-dimensional solid phononic crystal. Phys. Rev. B **72**, 012301 (2005)
53. K. Maslov, V.K. Kinra, B.K. Henderson, Lattice resonances of a planar array of spherical inclusions: an experimental study. Mech. Mater. **31**, 175 (1999)
54. P. Peng, J. Mei, Y. Wu, Lumped model for rotational modes in phononic crystals. Phys. Rev. B **86**, 134304 (2012)
55. E. Cosserat, F. Cosserat, *Théorie des Corps Déformables* (Hermann et Fils, Paris, 1909)
56. Lakes R, in Continuum Models for Materials with Microstructure, H. Muhlhaus (Wiley, New York, 1995).
57. A. Merkel, V. Tournat, V. Gusev, Dispersion of elastic waves in three-dimensional noncohesive granular phononic crystals: properties of rotational modes. Phys. Rev. E **82**, 031305 (2010)
58. V. Tournat, I. Pérez-Arjona, A. Merkel, V. Sanchez-Morcillo, V. Gusev, Elastic waves in phononic monolayer granular membranes. New J. Phys. **13**, 073042 (2011)
59. A. Merkel, V. Tournat, V. Gusev, Experimental evidence of rotational elastic waves in granular phononic crystals. Phys. Rev. Lett. **107**, 225502 (2011)
60. A. Vasiliev, A. Miroshnichenko, M. Ruzzene, A discrete model and analysis of one-dimensional deformations in a structural interface with micro-rotations. Mech. Res. Commun. **37**, 225 (2010)
61. A. Vasiliev, A. Miroshnichenko, M. Ruzzene, Multifield model for Cosserat media. J. Mech. Mater. **3**, 1365 (2008)
62. J.M. Haile, *Molecular Dynamics Simulation: Elementary Methods* (Wiley Inter-Science, 1992)
63. J.A. Thomas, J.E. Turney, R.M. Iutzi, C.H. Amon, A.J.H. McGaughey, Predicting phonon dispersion relations and lifetimes from the spectral energy density. Phys. Rev. B **81**, 081411 (2010)
64. J.M. Larkin, J.E. Turney, A.D. Massicotte, C.H. Amon, A.J.H. McGaughey, Comparison and evaluation of spectral energy methods for predicting phonon properties. J. Comput. Theor. Nanosci. **11**, 249–256 (2014)

65. W.H. Press, S.A. Teukolsky, W.T. Vetterling, B.P. Flannery, *Numerical Recipes in Fortran 77*, 2nd edn. (Cambridge University Press, Cambridge, 1992)
66. C.M. Bender, S.A. Orszag, *Advanced Mathematical Methods for Scientists and Engineers I, Asymptotic Methods and Perturbation Theory* (Springer, New York, 1999)
67. J. Kevorkian, J.D. Cole, *Scale and singular perturbation methods* (Springer, New York, 1996)
68. M. Belhaq, R.L. Clerc, C. Hartmann, Multiple scales methods for finding invariant solutions of two dimensional maps and application. Mech. Res. Commun. **15**, 361 (1988)
69. A. Maccari, A perturbation method for nonlinear two dimensional maps. Nonlinear Dyn. **19**, 295 (1999)
70. W.T. van Horssen, M.C. ter Brake, On the multiple scales perturbation method for difference equations. Nonlinear Dyn. **55**, 401 (2009)
71. R.H.G. Helleman, E.W. Montroll, On a nonlinear perturbation theory without secular terms. Physica **74**, 22 (1974)
72. P.S. Lee, Y.C. Lee, C.T. Chang, Multiple-time-scale analysis of spontaneous radiation processes. I. One- and two-particle systems. Phys. Rev. A **8**, 1722 (1973)
73. I.C. Khoo, Y.K. Wang, Multiple time scale analysis of an anharmonic crystal. J. Math. Phys. **17**, 222 (1976)
74. N. Swinteck, K. Muralidharan, P.A. Deymier, Phonon scattering in one-dimensional anharmonic crystals and superlattices: analytical and numerical study. ASME J. Vib. Acoust. **135**, 041016 (2013)

Chapter 4
Coherence

4.1 Introduction

Interactions in nonlinear elastic media cause multiple scattering and resonances of sound waves, which lead to the loss of phase coherence and acoustic wave degradation through amplitude reduction. However, recently, extraordinary modes of phonon transport have been demonstrated. For instance, granular materials are highly nonlinear prototypical phononic structures that have been extensively studied whereby strong localization of modes such as solitary waves enforce long-range coherent energy propagation [1–6]. High frequency phonons have been shown to propagate over long distance coherently even at room temperature. The occurrence of ballistic transport over distances of many microns, with significant contributions to thermal conductivity [7–9] and coherent propagation through phononic materials with numerous interfaces [10], have been reported. Consequently, sound-supporting media, phononic structures and acoustic metamaterials offer a broader palette of nonlinear responses with possibility of control of the coherence of phonon propagation. These responses extend over a range of nonlinear types, strengths and orders. These include: (a) geometrical nonlinearity associated with Hertzian contact in granular media [1–6]; (b) intrinsic nonlinearity of the constituent materials and components [11–14]; nonlinear rotational degrees of freedom in composite structures [15]; (c) hysteretic nonlinearity [16]; and (d) open system nonlinearity from exchanging matter or energy with an external reservoir [17]. Furthermore, the strength and order of nonlinearity can be selectively amplified in composite media comprising linear and nonlinear constituents [18]. In this Chapter, we present a number of simple models of nonlinear media to illustrate some notions related to nonlinear waves, coherence, and decoherence.

The simplest model of a nonlinear medium consists of a one-dimensional crystal of discrete springs and masses with quadratic nonlinearity. This model includes three-phonon scattering processes that break time-reversal symmetry by introducing temporal damping. The scattering process introduces phonon correlations.

P. Deymier, K. Runge, *Sound Topology, Duality, Coherence and Wave-Mixing*, Springer Series in Solid-State Sciences 188, DOI 10.1007/978-3-319-62380-1_4

A nonlinear spring-mass model composed to two coupled crystals is then used to shed further light on the controllability of three-phonon scattering phenomena by tuning the nonlinearity. We demonstrate coupling between symmetric and antisymmetric phonons through the nonlinearity. Multi-phonon scattering processes beyond three-phonon scattering in one-dimensional anharmonic biological mineralized tissues (bone-like hydrated collagen-hydroxy-apatite (HAP) superlattice) are then investigated. The nonlinearity in the mechanical response of the collagen–water system enables a variety of multi-phonon scattering processes resulting in an increase in the number of channels for the dissipation of elastic waves and therefore for the dissipation of mechanical energy. This latter model offers an entryway to studying time-reversal symmetry breaking of calcium wave propagation in chains of endothelial cells with nonlinear reaction dynamics. The rich dynamics of physical, chemical, and biological systems containing bi-stable elements have been studied extensively. The model of nonlinear calcium waves in biological systems serve here as a metaphor for the propagation of mechanical waves in a variety of bi-stable mechanical and elastic systems in which wave propagation may exhibit propagation failure or non-reciprocity [19, 20]. Finally, we consider the nonlinear interactions between elastic waves and multiple spatio-temporal modulations of stiffness in a one-dimensional waveguide. This is an extension of the one-dimensional elastic model exhibiting non-reciprocal wave propagation due to extrinsic application of a single sinusoidal spatio-temporal modulation of its stiffness. This model introduces the notion of non-conventional band structure design through the interaction between Fourier components of general periodic spatial pattern of symmetry breaking spatio-temporal modulations.

These simple models are analyzed employing a variety of analytical and numerical methods that may be useful to a number of other nonlinear problems involving wave propagation. The analytical methods include multiple-time-scale perturbation theory (to second and fourth-order), Homotopy Perturbation Method, Lippman-Schwinger propagator theory for time dependent processes, and Green's function approaches. The numerical methods employed here include molecular dynamics (MD), Spectral Energy Density methods (SED), and Finite Difference Time Domain (FDTD) method.

4.2 The Anharmonic One-Dimensional Monatomic Crystal

4.2.1 Multiple-Time Scale Perturbation Theory of the Monatomic Anharmonic Crystal

In a harmonic crystal, the vibrational modes are not interacting, however anharmonic lattice dynamics methods have been applied to introduce phonon interactions in three-dimensional crystals as perturbations to the harmonic solution [21–23]. Anharmonic forces lead to mode-dependent frequency shifts and introduce finite phonon lifetimes (*i.e.*, line-widths). In this section, we consider the anharmonic one-dimensional monatomic crystal as a simple model to shed light on the effect of nonlinear interatomic forces on the vibrational modes that this medium can support. Amplitude-dependent self-interaction of a wave in a monatomic and diatomic chain of masses and springs with nonlinear cubic forces has been studied [24]. It was shown that the dispersion curves undergo frequency shifts dependent on the amplitude of the wave. The interaction between two different waves in a nonlinear monoatomic chain results in the formation of different dispersion branches that are amplitude and frequency dependent [25]. Here, we employ a second-order perturbation theory based on the multiple time scale analysis [26, 27] (see Appendix 4, Chap. 3) and provide a detailed derivation of the anharmonic modes.

A schematic illustration of the 1D monatomic crystal is shown in Fig. 4.1a. The potential energy function detailing the interaction between neighboring masses in the 1D crystal is shown in Fig. 4.1b. The parameter (ε), characterizes the strength of nonlinearity in the springs connecting the masses. As ε increases in magnitude a region of instability emerges in the potential energy function.

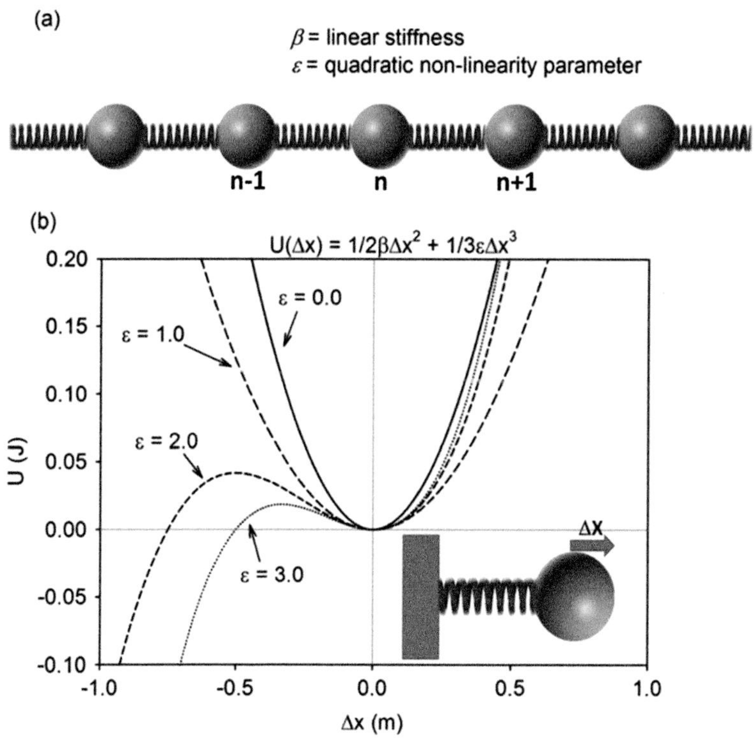

Fig. 4.1 (**a**) Schematic representation of 1D crystal with linear stiffness β and quadratic nonlinearity parameter ε. (**b**) The potential energy function describing the 1D crystal

The equation of motion for the quadratic nonlinear monatomic chain is represented by (4.1):

$$m\frac{d^2u_n(t)}{dt^2} = \beta(u_{n+1} - 2u_n + u_{n-1}) + \varepsilon\left[(u_{n+1} - u_n)^2 - (u_n - u_{n-1})^2\right] \quad (4.1)$$

where m is mass, $u_n(t)$ is the displacement from equilibrium of the nth mass, β is linear stiffness and ε is a small parameter characterizing quadratic nonlinearity. In multiple time scale perturbation theory, the time variable (t) is replaced by a collection of variables $\tau = (\tau_0, \tau_1, \tau_2)$ whereby: $\tau_0 = t$, $\tau_1 = \varepsilon t$, $\tau_2 = \varepsilon^2 t$. Under this condition, (4.1) becomes:

$$\frac{d^2u_n(\tau_0,\tau_1,\tau_2)}{dt^2} = {\omega_n}^2(u_{n+1} - 2u_n + u_{n-1}) + \frac{\varepsilon}{m}\left[(u_{n+1} - u_n)^2 - (u_n - u_{n-1})^2\right] \quad (4.2)$$

where $\omega_n = \sqrt{\frac{\beta}{m}}$. The dependent variable in (4.2), $u_n(\tau)$, is expressed as an asymptotic expansion at multiple time scales:

$$u_n(\tau) = u_n^{(0)}(\tau) + \varepsilon u_n^{(1)}(\tau) + \varepsilon^2\, u_n^{(2)}(\tau) + \textit{higher order terms} \quad (4.3)$$

With this (4.2) is decomposed into equations for each order of expansion of ε, namely, the following set of equations:

$$O(\varepsilon^0)\text{:}\ \frac{\partial^2 u_n^{(0)}}{\partial{\tau_0}^2} = {\omega_n}^2\left(u_{n+1}^{(0)} - 2u_n^{(0)} + u_{n-1}^{(0)}\right)$$

$$O(\varepsilon^1)\text{:}\ \frac{\partial^2 u_n^{(1)}}{\partial{\tau_0}^2} + 2\frac{\partial^2 u_n^{(0)}}{\partial\tau_0\partial\tau_1} = {\omega_n}^2\left(u_{n+1}^{(1)} - 2u_n^{(1)} + u_{n-1}^{(1)}\right)$$
$$+\frac{1}{m}\left[u_{n+1}^{(0)}u_{n+1}^{(0)} - 2u_{n+1}^{(0)}u_n^{(0)} + 2u_{n-1}^{(0)}u_n^{(0)} - u_{n-1}^{(0)}u_{n-1}^{(0)}\right]$$

$$O(\varepsilon^2)\text{:}\ \frac{\partial^2 u_n^{(2)}}{\partial{\tau_0}^2} + 2\frac{\partial^2 u_n^{(1)}}{\partial\tau_0\partial\tau_1} + 2\frac{\partial^2 u_n^{(0)}}{\partial\tau_0\partial\tau_2} + \frac{\partial^2 u_n^{(0)}}{\partial{\tau_1}^2} = {\omega_n}^2\left(u_{n+1}^{(2)} - 2u_n^{(2)} + u_{n-1}^{(2)}\right)$$
$$+\frac{2}{m}\left[u_{n+1}^{(1)}u_{n+1}^{(0)} - u_{n+1}^{(1)}u_n^{(0)} - u_{n+1}^{(0)}u_n^{(1)} + u_{n-1}^{(1)}u_n^{(0)} + u_{n-1}^{(0)}u_n^{(1)} - u_{n-1}^{(1)}u_{n-1}^{(0)}\right]$$

4.2.2 Self-Interaction

We first address the self-interaction of a vibrational mode, that is, the effect of the lattice deformation on itself. To solve the ε^0-equation, a general solution of the following form is proposed:

$$u_{n,G}^{(0)}(\tau_0,\tau_1,\tau_2)=A_0(\tau_1,\tau_2)e^{ikna}e^{-i\omega_0\tau_0}+\bar{A}_0(\tau_1,\tau_2)e^{-ikna}e^{i\omega_0\tau_0},\tag{4.4}$$

where

$$A_0(\tau_1,\tau_2)=\alpha(\tau_1,\tau_2)e^{-i\varphi(\tau_1,\tau_2)}$$
$$\bar{A}_0(\tau_1,\tau_2)=\alpha(\tau_1,\tau_2)e^{i\varphi(\tau_1,\tau_2)}$$

$A_0(\tau_1,\tau_2)$ is a complex quantity that permits slow time evolution of amplitude and phase and $\alpha(\tau_1,\tau_2)$ and $\varphi(\tau_1,\tau_2)$ are real-valued functions. Inserting (4.4) into the ε^0-order equation yields the well-known dispersion relation for the harmonic system equation (4.5):

$$\omega_0{}^2=\omega_n{}^2\left(2-e^{ika}-e^{-ika}\right)=\frac{\beta}{m}[2-2\cos(ka)]\tag{4.5}$$

Equation (4.4) is now utilized in the ε^1-order equation to resolve the general solution for $u_n^{(1)}$. The ε^1-order equation is written as follows:

$$\frac{\partial^2u_n^{(1)}}{\partial\tau_0{}^2}+\omega_n{}^2\left(2u_n^{(1)}-u_{n+1}^{(1)}-u_{n-1}^{(1)}\right)=$$
$$2i\omega_0\left[\frac{\partial A_0}{\partial\tau_1}e^{ikna}e^{-i\omega_0\tau_0}-\frac{\partial\bar{A}_0}{\partial\tau_1}e^{-ikna}e^{i\omega_0\tau_0}\right]$$
$$+\frac{1}{m}\left[\left(e^{i2ka}-2e^{ika}+2e^{-ika}-e^{-i2ka}\right)\left(A_0A_0e^{i2kna}e^{-i2\omega_0\tau_0}-\bar{A}_0\bar{A}_0e^{-i2kna}e^{i2\omega_0\tau_0}\right)\right]$$

It is assumed that the solution to the homogeneous equation of the ε^1-order equation takes similar form to the general solution of the ε^0-order equation. Under this assumption, terms on the right-hand side of the ε^1-order equation with functional form $e^{i\omega_0\tau_0}$ or $e^{-i\omega_0\tau_0}$ contribute to secular behavior. These terms are eliminated by setting them equal to zero. Accordingly, A_0 and $\bar{A}_0$ are considered to be independent functions of τ_1. This modifies the form of the general solution to the ε^0-equation:

$$u_{n,G}^{(0)}(\tau_0,\tau_2)=A_0(\tau_2)e^{ikna}e^{-i\omega_0\tau_0}+\bar{A}_0(\tau_2)e^{-ikna}e^{i\omega_0\tau_0},\tag{4.6}$$

where

$$A_0(\tau_2)=\alpha(\tau_2)e^{-i\varphi(\tau_2)}$$
$$\bar{A}_0(\tau_2)=\alpha(\tau_2)e^{i\varphi(\tau_2)}$$

The homogeneous solution to the ε^1-order equation takes the following form:

$$u_{n,H}^{(1)}(\tau_0,\tau_2) = B_0(\tau_2)e^{ikna}e^{-i\omega_0\tau_0} + \bar{B}_0(\tau_2)e^{-ikna}e^{i\omega_0\tau_0} \tag{4.7}$$

The particular solution to the ε^1-order equation is of the form:

$$u_{n,P}^{(1)}(\tau_0,\tau_2) = C_0(\tau_2)e^{i2kna}e^{-i2\omega_0\tau_0} + \bar{C}_0(\tau_2)e^{-i2kna}e^{i2\omega_0\tau_0} \tag{4.8}$$

Inserting (4.8) into the ε^1-order equation and relating like terms reveals relationships for the exponential pre-factors $C_0(\tau_2)$ and $\bar{C}_0(\tau_2)$. Equation (4.8) becomes:

$$\begin{aligned} u_{n,P}^{(1)}(\tau_0,\tau_2) &= \frac{2i(\sin(2ka) - 2\sin(ka))}{\beta((2-2\cos(2ka)) - 4(2-2\cos(ka)))} \\ &\times \left[A_0{}^2 e^{i2kna}e^{-i2\omega_0\tau_0} - \bar{A}_0{}^2 e^{-i2kna}e^{i2\omega_0\tau_0}\right] \end{aligned}$$

The general solution to the ε^1-order equation is a sum of the homogeneous ($u_{n,H}^{(1)}$) and particular solutions ($u_{n,P}^{(1)}$):

$$\begin{aligned} u_{n,G}^{(1)}(\tau_0,\tau_2) &= B_0 e^{ikna}e^{-i\omega_0\tau_0} + \bar{B}_0 e^{-ikna}e^{i\omega_0\tau_0} \\ &+ \frac{2i(\sin(2ka) - 2\sin(ka))}{\beta((2-2\cos(2ka)) - 4(2-2\cos(ka)))}\left[A_0A_0 e^{i2kna}e^{-i2\omega_0\tau_0}\right. \\ &\left.-\bar{A}_0\bar{A}_0 e^{-i2kna}e^{i2\omega_0\tau_0}\right] \end{aligned}$$

The values for B_0 and $\bar{B}_0$ are found from initial conditions. With the general solutions to the ε^0-equation and the ε^1-equation, the ε^2-order equation is developed. Inserting $u_{n,G}^{(0)}$ and $u_{n,G}^{(1)}$ into the ε^2-order equation, utilizing the expressions for $A_0(\tau_2)$ and $\bar{A}_0(\tau_2)$, and noting that $u_n^{(0)}$ and $u_n^{(1)}$ are independent functions of τ_1, the ε^2-order equation is written as:

$$\begin{aligned} &\frac{\partial^2 u_n^{(2)}}{\partial \tau_0{}^2} + \omega_n{}^2\left(2u_n^{(2)} - u_{n+1}^{(2)} - u_{n-1}^{(2)}\right) = \\ &e^{ikna}e^{-i\omega_0\tau_0}\left(2\omega_0\alpha\frac{\partial\varphi}{\partial\tau_2}e^{-i\varphi} + 2i\omega_0 e^{-i\varphi}\frac{\partial\alpha}{\partial\tau_2}\right) + e^{-ikna}e^{i\omega_0\tau_0}\left(2\omega_0\alpha\frac{\partial\varphi}{\partial\tau_2}e^{i\varphi} - 2i\omega_0 e^{i\varphi}\frac{\partial\alpha}{\partial\tau_2}\right) \\ &\quad + \frac{2}{m}\Big\{\Big[\left(e^{i2ka} - 2e^{ika} + 2e^{-ika} - e^{-i2ka}\right)\left(A_0B_0e^{i2kna}e^{-i2\omega_0\tau_0}\right. \\ &\quad \left.- \bar{A}_0\bar{B}_0e^{-i2kna}e^{i2\omega_0\tau_0}\right)\Big] \\ &\quad + \Big[\left(e^{i3ka} - e^{i2ka} - e^{ika} + e^{-ika} + e^{-i2ka} - e^{-i3ka}\right)\left(A_0C_0e^{i3kna}e^{-i3\omega_0\tau_0}\right. \\ &\quad \left.- \bar{A}_0\bar{C}_0e^{-i3kna}e^{i3\omega_0\tau_0}\right)\Big] \\ &\quad + \Big[\left(e^{i2ka} - 2e^{ika} + 2e^{-ika} - e^{-i2ka}\right)\left(A_0\bar{C}_0e^{-ikna}e^{i\omega_0\tau_0} - \bar{A}_0C_0e^{ikna}e^{-i\omega_0\tau_0}\right)\Big]\Big\} \end{aligned}$$

The homogeneous solution to the ε^2-order equation is similar in form to the general solution of the ε^0-equation and the homogeneous solution of the ε^1-equation. Accordingly, terms on the right-hand side of the ε^2-order equation with functional form $e^{i\omega_0\tau_0}$ or $e^{-i\omega_0\tau_0}$ contribute to secular behavior and must be eliminated. Setting exponential pre-factors equal to zero yields the following relationships for $\alpha(\tau_2)$ and $\varphi(\tau_2)$:

$$\alpha(\tau_2) = \alpha_0 \tag{4.9}$$

$$\varphi(\tau_2) = -\frac{\alpha^2}{\omega_0\beta m} \cdot \frac{4(\sin(2ka) - 2\sin(ka))^2}{(2 - 2\cos(2ka)) - 4(2 - 2\cos(ka))}\tau_2 + \varphi_0 \tag{4.10}$$

where α_0 and φ_0 are constants determined from initial plane wave conditions. The general solution to the ε^0-equation (4.6) is considered again with (4.9) and (4.10) substituted in for expressions A_0 and $\bar{A}_0$. Here, the constant φ_0 can be set equal to zero without loss of generality.

$$u_{n,G}^{(0)}(\tau_0, \tau_2) = \alpha_0 e^{i\left(kna - \left(\omega_0 - \varepsilon^2 \frac{\alpha^2}{\omega_0\beta m} \cdot \frac{4(\sin(2ka) - 2\sin(ka))^2}{(2-2\cos(2ka)) - 4(2-2\cos(ka))}\right)\tau_0\right)} + \alpha_0 e^{-i\left(kna - \left(\omega_0 - \varepsilon^2 \frac{\alpha^2}{\omega_0\beta m} \cdot \frac{4(\sin(2ka) - 2\sin(ka))^2}{(2-2\cos(2ka)) - 4(2-2\cos(ka))}\right)\tau_0\right)}$$

This 0th order result in the asymptotic expansion of u_n shows that the harmonic dispersion curve is shifted by a quantity with quadratic dependence on the strength of the nonlinearity parameter ε.

4.2.3 *Three-Wave Interactions*

Now we consider the interaction between three waves with different wave vectors and frequencies. The analysis begins with the equation of motion (4.1) from the single wave dispersion analysis. The displacement of the nth mass is represented by a superposition of wave modes each with a unique, time and wave vector dependent amplitude factor (4.11):

$$u_n(t) = \Sigma_k A(k, t) e^{ikna} \tag{4.11}$$

We use a discrete summation over the wave numbers instead of an integral over a continuum of wave vectors. This is done to help the reader to conceptualize the interactions between specific phonons and to facilitate the comparison with the MD models presented subsequently. Indeed, MD simulations are limited to finite size systems for which the phonon modes do not form a continuum but a discrete set of possible wave vectors. Inserting (4.11) into the equation of motion for the 1D monatomic crystal yields a modified equation of motion (4.12).

$$m\sum_k \frac{d^2A(k,t)}{dt^2}e^{ikna} =$$

$$-4\beta\Sigma_k A(k,t)e^{ikna}\sin^2\left(\frac{ka}{2}\right) + \varepsilon\left[\Sigma_{k'}\Sigma_{k''}A(k',t)A(k'',t)e^{i(k'+k'')na}f(k',k'')\right] \quad (4.12)$$

where $f(k',k'') = -8i\sin\left(\frac{k'a}{2}\right)\sin\left(\frac{k''a}{2}\right)\sin\left(\frac{(k'+k'')a}{2}\right)$. Equation (4.12) is multiplied by e^{-ik^*na} and a summation over all n masses is imposed. This procedure selects the mode k^* as a reference wave vector. With $\omega_n^2 = 4\beta/m$, (4.12) becomes

$$\frac{d^2A(k^*,t)}{dt^2} + \omega_n^2\sin^2\left(\frac{k^*a}{2}\right)A(k^*,t) = \frac{\varepsilon}{m}\Sigma_{k'}\Sigma_{k''}A(k',t)A(k'',t)f(k',k'')\delta_{k'+k'',k^*} \quad (4.13)$$

$\delta_{k'+k'',\,k^*}$ imposes the wave vector conservation rule $k^* = k' + k'' + mG$ where m is an integer and G is a reciprocal lattice vector of the periodic structure. We do not label G in the delta function for the sake of simplicity of the notation. For $m = 0$, one has the so-called normal three phonon scattering process. The case of $m \neq 0$ corresponds to umklapp processes where $k' + k''$ is located outside the first Brillouin zone. In (4.13), the variable τ is introduced, where $\tau = \omega_n t$. Single time variables (τ) are replaced by a collection of variables $\tau = (\tau_0, \tau_1, \tau_2)$ whereby: $\tau_0 = \tau$, $\tau_1 = \varepsilon\tau$, $\tau_2 = \varepsilon^2\tau$. Additionally, $A(k^*, \tau)$ is replaced by an asymptotic expansion whereby:

$$A(k^*,\tau) = A_0(k^*,\tau) + \varepsilon A_1(k^*,\tau) + \varepsilon^2 A_2(k^*,\tau)$$
$$A(k^*,\tau_0,\tau_1,\tau_2) = A_0(k^*,\tau_0,\tau_1,\tau_2) + \varepsilon A_1(k^*,\tau_0,\tau_1,\tau_2) + \varepsilon^2 A_2(k^*,\tau_0,\tau_1,\tau_2)$$

With these considerations, (4.13) is separated into expressions at order ε^0, ε^1 and ε^2:

$$O(\varepsilon^0):\ \frac{\partial^2 A_0(k^*,\tau)}{\partial\tau_0^2} + \sin^2\left(\frac{k^*a}{2}\right)A_0(k^*,\tau) = 0$$

$$O(\varepsilon^1):\ \frac{\partial^2 A_1(k^*,\tau)}{\partial\tau_0^2} + 2\frac{\partial^2 A_0(k^*,\tau)}{\partial\tau_1\partial\tau_0} + \sin^2\left(\frac{k^*a}{2}\right)A_1(k^*,\tau)$$
$$= \frac{1}{m\omega_n^2}\Sigma_{k'}\Sigma_{k''}f(k',k'')\delta_{k'+k'',k^*}\left[A_0(k',\tau)A_0(k'',\tau)\right]$$

$$O(\varepsilon^2):\ \frac{\partial^2 A_2(k^*,\tau)}{\partial\tau_0^2} + 2\frac{\partial^2 A_1(k^*,\tau)}{\partial\tau_1\partial\tau_0} + 2\frac{\partial^2 A_0(k^*,\tau)}{\partial\tau_2\partial\tau_0} + \frac{\partial^2 A_0(k^*,\tau)}{\partial\tau_1^2}$$
$$+ \sin^2\left(\frac{k^*a}{2}\right)A_2(k^*,\tau)$$
$$= \frac{1}{m\omega_n^2}\Sigma_{k'}\Sigma_{k''}f(k',k'')\delta_{k'+k'',k^*}\left[A_0(k',\tau)A_1(k'',\tau) + A_1(k',\tau)A_0(k'',\tau)\right]$$

To solve the ε^0-equation, a general solution of the following form is proposed:

$$A_0(k^*,\tau_0,\tau_1,\tau_2)=a_0(k^*,\tau_1,\tau_2)e^{i\omega_0^*\tau_0}+\bar{a}_0(k^*,\tau_1,\tau_2)e^{-i\omega_0^*\tau_0} \tag{4.14}$$

Inserting (4.14) into the ε^0-equation offers the expected relationship between ω_0^* and k^*: $\omega_0^{*2}=\sin^2\left(\frac{k^*a}{2}\right)$. Inserting (4.14) into the ε^1-equation offers an expression to solve for $A_1(k^*,\tau)$. After rearranging and utilizing the following definitions:

$$A_0(k',\tau_0,\tau_1,\tau_2)=a_0(k',\tau_1,\tau_2)e^{i\omega_0'\tau_0}+\bar{a}_0(k',\tau_1,\tau_2)e^{-i\omega_0'\tau_0}$$
$$A_0(k'',\tau_0,\tau_1,\tau_2)=a_0(k'',\tau_1,\tau_2)e^{i\omega_0''\tau_0}+\bar{a}_0(k',\tau_1,\tau_2)e^{-i\omega_0''\tau_0}$$

The ε^1-equation becomes:

$$\begin{aligned}\frac{\partial^2A_1(k^*,\tau)}{\partial\tau_0^2}&+\omega_0^{*2}A_1(k^*,\tau)\\&=-2i\omega_0^*\left[\frac{\partial a_0^*}{\partial\tau_1}e^{i\omega_0^*\tau_0}-\frac{\partial\bar{a}_0^*}{\partial\tau_1}e^{-i\omega_0^*\tau_0}\right]\\&+\frac{1}{m\omega_n^2}\sum_{k'}\sum_{k''}f(k',k'')\delta_{k'+k'',k^*}\left[a_0'a_0''e^{i\left(\omega_0'+\omega_0''\right)\tau_0}+a_0'\bar{a}_0''e^{i\left(\omega_0'-\omega_0''\right)\tau_0}\right.\\&\left.+\bar{a}_0'a_0''e^{-i\left(\omega_0'-\omega_0''\right)\tau_0}+\bar{a}_0'\bar{a}_0''e^{-i\left(\omega_0'+\omega_0''\right)\tau_0}\right]\end{aligned}$$

where terms like a_0^*, a_0', $a_0''\ldots etc.$ are compact representations of $a_0(k^*,\tau_1,\tau_2)$, $a_0(k',\tau_1,\tau_2)$, $a_0(k'',\tau_1,\tau_2)\ldots.etc.$ A homogeneous solution to the ε^1-equation is proposed:

$$A_{1,H}(k^*,\tau_0,\tau_2)=a_1(k^*,\tau_2)e^{i\omega_0^*\tau_0}+\bar{a}_1(k^*,\tau_2)e^{-i\omega_0^*\tau_0}=a_1^*e^{i\omega_0^*\tau_0}+\bar{a}_1^*e^{-i\omega_0^*\tau_0} \tag{4.15}$$

The forcing terms on the right-hand side of the ε^1-equation with functional form $e^{i\omega_0^*\tau_0}$ or $e^{-i\omega_0^*\tau_0}$ contribute to secular behavior. These terms must be eliminated such that the final representation of $A(k^*,\tau)$ is well behaved (*e.g.*, contains no terms that temporally grow without bound). These terms are set to zero by making a_0 and $\bar{a}_0$ functions of k^* and τ_2 only. With this stipulation, an appropriate form of the particular solution to the ε^1-equation is:

$$\begin{aligned}A_{1,P}(k^*,\tau)=&\frac{1}{m\omega_n^2}\Sigma_{k'}\Sigma_{k''}f(k',k'')\delta_{k'+k'',\ k^*}\left[b_1e^{i\left(\omega_0'+\omega_0''\right)\tau_0}+\bar{b}_1e^{-i\left(\omega_0'+\omega_0''\right)\tau_0}\right.\\&\left.+c_1e^{i\left(\omega_0'-\omega_0''\right)\tau_0}+\bar{c}_1e^{-i\left(\omega_0'-\omega_0''\right)\tau_0}\right]\end{aligned} \tag{4.16}$$

The exponential pre-factors $b_1, \bar{b}_1, c_1, \bar{c}_1$ have dependency on $k', k'', \tau_2, \omega_0^*$, ω_0', ω_0''. Substituting (4.16) into the ε^1-equation and relating like terms reveals the exponential pre-factors: $b_1, \bar{b}_1, c_1, \bar{c}_1$

$$b_1 = \frac{a_0(k', \tau_2)a_0(k'', \tau_2)}{\omega_0^{*2} - (\omega_0' + \omega_0'')^2}; \ \bar{b}_1 = \frac{\bar{a}_0(k', \tau_2)\bar{a}_0(k'', \tau_2)}{\omega_0^{*2} - (\omega_0' + \omega_0'')^2}$$

$$c_1 = \frac{a_0(k', \tau_2)\bar{a}_0(k'', \tau_2)}{\omega_0^{*2} - (\omega_0' - \omega_0'')^2}; \ \bar{c}_1 = \frac{\bar{a}_0(k', \tau_2)a_0(k'', \tau_2)}{\omega_0^{*2} - (\omega_0' - \omega_0'')^2}$$

In the long wavelength limit, angular frequency has nearly linear dependence on wave vector. In considering the stipulated wave vector relationship inside the double summation in (4.16), ($k' + k'' = k^*$), it is conceivable that $\omega_0(k') + \omega_0(k'') = \omega_0(k^*)$ or $\omega_0(k') - \omega_0(k'') = \omega_0(k^*)$. In this instance, the denominator terms in the expressions for $b_1, \bar{b}_1, c_1, \bar{c}_1$ will go to zero. To avoid this complication, following the procedure stipulated by Khoo et al. [27], a small imaginary part φ is introduced in the denominator. At the final step of the calculation the limit will be taken as $\varphi \to 0$. The general solution to the ε^1-equation is a sum of the homogeneous and particular solutions:

$$\begin{aligned} A_1(k^*, \tau_0, \tau_2) &= a_1^* e^{i\omega_0^* \tau_0} + \bar{a}_1^* e^{-i\omega_0^* \tau_0} \\ &\quad + \frac{1}{m\omega_n^2} \Sigma_{k'} \Sigma_{k''} f(k', k'') \delta_{k'+k'', k^*} \left[\frac{a_0' a_0''}{g_1^*} e^{i(\omega_0' + \omega_0'')\tau_0} + \frac{\bar{a}_0' \bar{a}_0''}{g_1^*} e^{-i(\omega_0' + \omega_0'')\tau_0} \right. \\ &\quad \left. + \frac{a_0' \bar{a}_0''}{g_2^*} e^{i(\omega_0' - \omega_0'')\tau_0} + \frac{\bar{a}_0' a_0''}{g_2^*} e^{-i(\omega_0' - \omega_0'')\tau_0} \right], \end{aligned} \tag{4.17}$$

where $g_1^* = \omega_0^{*2} - (\omega_0' + \omega_0'')^2 + i\varphi; g_2^* = \omega_0^{*2} - (\omega_0' - \omega_0'')^2 + i\varphi$

The ε^2-equation is reduced to the following expressions because $A_0(k^*, \tau)$ and $A_1(k^*, \tau)$ are independent of τ_1:

$$\begin{aligned} &\frac{\partial^2 A_2(k^*, \tau_0, \tau_1, \tau_2)}{\partial \tau_0^2} + \omega_0^{*2} A_2(k^*, \tau_0, \tau_1, \tau_2) \\ &\quad = -2i\omega_0^* \frac{\partial a_0(k^*, \tau_2)}{\partial \tau_2} e^{i\omega_0^* \tau_0} + 2i\omega_0^* \frac{\partial \bar{a}_0(k^*, \tau_2)}{\partial \tau_2} e^{-i\omega_0^* \tau_0} \\ &\quad + \frac{1}{m\omega_n^2} \sum_{k'} \sum_{k''} f(k', k'') \delta_{k'+k'', k^*} [A_0(k', \tau_0, \tau_2) A_1(k'', \tau_0, \tau_2) \\ &\quad + A_1(k', \tau_0, \tau_2) A_0(k'', \tau_0, \tau_2)] \end{aligned}$$

As before, the solution to the homogeneous equation of the ε^2-equation is of the form:

$$A_{2,H}(k^*, \tau_0, \tau_2) = a_2(k^*, \tau_2) e^{i\omega_0^* \tau_0} + \bar{a}_2(k^*, \tau_2) e^{-i\omega_0^* \tau_0}$$

Terms on the right-hand side of the ε^2-equation with functional form $e^{i\omega_0^*\tau_0}$ or $e^{-i\omega_0^*\tau_0}$ contribute to secular behavior. Using (4.14) and (4.17) to develop the right-hand side of the ε^2-equation gives (4.18).

$$
\begin{aligned}
&\frac{\partial^2 A_2(k^*,\tau_0,\tau_1,\tau_2)}{\partial\tau_0^2}+\omega_0^{*2}A_2(k^*,\tau_0,\tau_1,\tau_2)\\
&=-2i\omega_0^*\frac{\partial a_0(k^*,\tau_2)}{\partial\tau_2}e^{i\omega_0^*\tau_0}+2i\omega_0^*\frac{\partial\bar{a}_0(k^*,\tau_2)}{\partial\tau_2}e^{-i\omega_0^*\tau_0}\\
&+\frac{1}{m\omega_n{}^2}\sum_{k'}\sum_{k''}f(k',k'')\,\delta_{k'+k'',k^*}\left(a_0'a_1''e^{i(\omega_0'+\omega_0'')\tau_0}+a_0'\bar{a}_1''e^{i(\omega_0'-\omega_0'')\tau_0}\right.\\
&\qquad\left.+\bar{a}_0'a_1''e^{-i(\omega_0'-\omega_0'')\tau_0}+\bar{a}_0'\bar{a}_1''e^{-i(\omega_0'+\omega_0'')\tau_0}\right)\\
&+\frac{1}{m\omega_n{}^2}\sum_{k'}\sum_{k''}f(k',k'')\,\delta_{k'+k'',k^*}\left(a_0''a_1'e^{i(\omega_0''+\omega_0')\tau_0}+a_0''\bar{a}_1'e^{i(\omega_0''-\omega_0')\tau_0}\right.\\
&\left.+\bar{a}_0''a_1'e^{-i(\omega_0''-\omega_0')\tau_0}+\bar{a}_0''\bar{a}_1'e^{-i(\omega_0''+\omega_0')\tau_0}\right)\\
&+\frac{1}{m\omega_n^2}\Sigma_{k'}\Sigma_{k''}f(k',k'')\delta_{k'+k'',k^*}\left\{\left[\frac{1}{m\omega_n^2}\Sigma_{k_1}\Sigma_{k_2}f(k_1,k_2)\delta_{k_1+k_2,k''}\right.\right.\\
&\times\left[\frac{a_0'a_0^{(1)}a_0^{(2)}}{g_1''}e^{i\left(\omega_0^{(1)}+\omega_0^{(2)}+\omega_0'\right)\tau_0}+\frac{a_0'\bar{a}_0^{(1)}\bar{a}_0^{(2)}}{g_1''}e^{-i\left(\omega_0^{(1)}+\omega_0^{(2)}-\omega_0'\right)\tau_0}\right.\\
&+\frac{a_0'a_0^{(1)}\bar{a}_0^{(2)}}{g_2''}e^{i\left(\omega_0^{(1)}-\omega_0^{(2)}+\omega_0'\right)\tau_0}+\frac{a_0'\bar{a}_0^{(1)}a_0^{(2)}}{g_2''}e^{-i\left(\omega_0^{(1)}-\omega_0^{(2)}-\omega_0'\right)\tau_0}\\
&+\frac{\bar{a}_0'a_0^{(1)}a_0^{(2)}}{g_1''}e^{i\left(\omega_0^{(1)}+\omega_0^{(2)}-\omega_0'\right)\tau_0}+\frac{\bar{a}_0'\bar{a}_0^{(1)}\bar{a}_0^{(2)}}{g_1''}e^{-i\left(\omega_0^{(1)}+\omega_0^{(2)}+\omega_0'\right)\tau_0}\\
&\left.\left.+\frac{\bar{a}_0'a_0^{(1)}\bar{a}_0^{(2)}}{g_2''}e^{i\left(\omega_0^{(1)}-\omega_0^{(2)}-\omega_0'\right)\tau_0}+\frac{\bar{a}_0'\bar{a}_0^{(1)}a_0^{(2)}}{g_2''}e^{-i\left(\omega_0^{(1)}-\omega_0^{(2)}+\omega_0'\right)\tau_0}\right]\right]\\
&+\left[\frac{1}{m\omega_n^2}\Sigma_{k_1}\Sigma_{k_2}f(k_1,k_2)\delta_{k_1+k_2,k'}\left[\frac{a_0''a_0^{(1)}a_0^{(2)}}{g_1'}e^{i\left(\omega_0^{(1)}+\omega_0^{(2)}+\omega_0''\right)\tau_0}\right.\right.\\
&+\frac{a_0''\bar{a}_0^{(1)}\bar{a}_0^{(2)}}{g_1'}e^{-i\left(\omega_0^{(1)}+\omega_0^{(2)}-\omega_0''\right)\tau_0}+\frac{a_0''a_0^{(1)}\bar{a}_0^{(2)}}{g_2'}e^{i\left(\omega_0^{(1)}-\omega_0^{(2)}+\omega_0''\right)\tau_0}\\
&+\frac{a_0''\bar{a}_0^{(1)}a_0^{(2)}}{g_2'}e^{-i\left(\omega_0^{(1)}-\omega_0^{(2)}-\omega_0''\right)\tau_0}+\frac{\bar{a}_0''a_0^{(1)}a_0^{(2)}}{g_1'}e^{i\left(\omega_0^{(1)}+\omega_0^{(2)}-\omega_0''\right)\tau_0}\\
&+\frac{\bar{a}_0''\bar{a}_0^{(1)}\bar{a}_0^{(2)}}{g_1'}e^{-i\left(\omega_0^{(1)}+\omega_0^{(2)}+\omega_0''\right)\tau_0}+\frac{\bar{a}_0''a_0^{(1)}\bar{a}_0^{(2)}}{g_2'}e^{i\left(\omega_0^{(1)}-\omega_0^{(2)}-\omega_0''\right)\tau_0}\\
&\left.\left.\left.+\frac{\bar{a}_0''a_0^{(1)}a_0^{(2)}}{g_2'}e^{-i\left(\omega_0^{(1)}-\omega_0^{(2)}+\omega_0''\right)\tau_0}\right]\right]\right\}
\end{aligned}
\tag{4.18}
$$

There is notable similarity between the terms on the right-hand side of the ε^1-equation that was solved to yield (4.17) and the third and fourth terms on the right-hand side of (4.18). These terms are treated with the same procedure as that used for the ε^1-equation. Accordingly, they will not contribute to secular terms. The objective is to identify terms in the ε^2-equation with $e^{i\omega_0^* \tau_0}$ or $e^{-i\omega_0^* \tau_0}$ dependency. This will be done by systematically evaluating all wave vector pairs $\{k_1, k_2\}$ that satisfy the wave vector constraints stipulated by (4.18). Specifically,

$$\delta_{k'+k'',\ k^*}\ \delta_{k_1+k_2,\ k''} \rightarrow k' + k_1 + k_2 = k^*$$
$$\delta_{k'+k'',\ k^*}\ \delta_{k_1+k_2,\ k'} \rightarrow k'' + k_1 + k_2 = k^*.$$

If a certain pair of wave vectors satisfies the above mentioned wave vector constraints, then an analysis will be carried through to see if these wave vectors give rise to terms with $e^{i\omega_0^* \tau_0}$ or $e^{-i\omega_0^* \tau_0}$ dependence. As before, terms with $e^{i\omega_0^* \tau_0}$ or $e^{-i\omega_0^* \tau_0}$ dependence will be removed. In (4.18), inside the summation over k', k'', there are two summations over k_1, k_2. For the first summation over k_1, k_2, two conditions must be met: (1) $k' + k'' = k^*$ and (2) $k_1 + k_2 = k''$. The only possible combinations for k_1, k_2 that give wave vector relationships that are compatible with $\delta_{k'+k'',\ k^*}$ are shown as Condition A and Condition B:

$$\text{Condition A: } k_1 = -k', k_2 = k^*, \text{and} - k' + k^* = k''$$
$$\text{Condition B: } k_1 = k^*, k_2 = -k', \text{and}\ k^* - k' = k''$$

Now that wave vector constraints are satisfied, an analysis is carried out to see if any terms with $e^{i\omega_0^* \tau_0}$ or $e^{-i\omega_0^* \tau_0}$ dependence arise in the first summation over k_1, k_2. The following frequency relationships are present in the first summation over k_1, k_2 in (4.18):

$$\text{(i).}\ \ \omega_0^{(1)} + \omega_0^{(2)} + \omega_0'$$
$$\text{(ii).}\ \ \omega_0^{(1)} + \omega_0^{(2)} - \omega_0'$$
$$\text{(iii).}\ \ \omega_0^{(1)} - \omega_0^{(2)} + \omega_0'$$
$$\text{(iv).}\ \ \omega_0^{(1)} - \omega_0^{(2)} - \omega_0'$$
$$\text{Condition A: } k_1 = -k' \rightarrow \omega_0^{(1)} = \omega_0' \text{ and } k_2 = k^* \rightarrow \omega_0^{(2)} = \omega_0^*$$

Applying Condition A to these frequency relationships show two relationships that offer terms with $e^{i\omega_0^* \tau_0}$ or $e^{-i\omega_0^* \tau_0}$ dependence:

$$\text{(i).}\ \ \omega_0^{(1)} + \omega_0^{(2)} + \omega_0' \rightarrow \omega_0' + \omega_0^* + \omega_0' = \omega_0^* + 2\omega_0'$$
$$\text{(ii).}\ \ \omega_0^{(1)} + \omega_0^{(2)} - \omega_0' \rightarrow \omega_0' + \omega_0^* - \omega_0' = \omega_0^*$$
$$\text{(iii).}\ \ \omega_0^{(1)} - \omega_0^{(2)} + \omega_0' \rightarrow \omega_0' - \omega_0^* + \omega_0' = -\omega_0^* + 2\omega_0'$$

$$\text{(iv)}.\ \omega_0^{(1)} - \omega_0^{(2)} - \omega'_0 \rightarrow \omega'_0 - \omega_0^* - \omega'_0 = -\omega_0^*$$

As a result, with Condition A, the following terms in the first summation over k_1, k_2 contribute to secular terms:

$$\frac{a'_0\bar{a}_0^{(1)}\bar{a}_0^{(2)}}{g''_1}e^{-i\left(\omega_0^{(1)}+\omega_0^{(2)}-\omega'_0\right)\tau_0} = \frac{a'_0\bar{a}'_0\bar{a}_0^*}{g''_1}e^{-i\left(\omega'_0+\omega_0^*-\omega'_0\right)\tau_0} = \frac{a'_0\bar{a}'_0\bar{a}_0^*}{g''_1}e^{-i\left(\omega_0^*\right)\tau_0}$$

$$\frac{\bar{a}'_0a_0^{(1)}a_0^{(2)}}{g''_1}e^{i\left(\omega_0^{(1)}+\omega_0^{(2)}-\omega'_0\right)\tau_0} = \frac{\bar{a}'_0a'_0a_0^*}{g''_1}e^{i\left(\omega'_0+\omega_0^*-\omega'_0\right)\tau_0} = \frac{\bar{a}'_0a'_0a_0^*}{g''_1}e^{i\left(\omega_0^*\right)\tau_0}$$

$$\frac{a'_0\bar{a}_0^{(1)}a_0^{(2)}}{g''_2}e^{-i\left(\omega_0^{(1)}-\omega_0^{(2)}-\omega'_0\right)\tau_0} = \frac{a'_0\bar{a}'_0a_0^*}{g''_2}e^{-i\left(\omega'_0-\omega_0^*-\omega'_0\right)\tau_0} = \frac{a'_0\bar{a}_0^{(1)}a_0^{(2)}}{g''_2}e^{i\left(\omega_0^*\right)\tau_0}$$

$$\frac{\bar{a}'_0a_0^{(1)}\bar{a}_0^{(2)}}{g''_2}e^{i\left(\omega_0^{(1)}-\omega_0^{(2)}-\omega'_0\right)\tau_0} = \frac{\bar{a}'_0a'_0\bar{a}_0^*}{g''_2}e^{i\left(\omega'_0-\omega_0^*-\omega'_0\right)\tau_0} = \frac{\bar{a}'_0a_0^{(1)}\bar{a}_0^{(2)}}{g''_2}e^{-i\left(\omega_0^*\right)\tau_0}$$

Recall condition B: $k_1 = k^* \rightarrow \omega_0^{(1)} = \omega_0^*$, $k_2 = -k' \rightarrow \omega_0^{(2)} = \omega'_0$, applying Condition B to these frequency relationships show two different relationships that offer terms with $e^{i\omega_0^*\tau_0}$ or $e^{-i\omega_0^*\tau_0}$ dependence.

$$\text{(i)}.\ \omega_0^{(1)} + \omega_0^{(2)} + \omega'_0 \rightarrow \omega_0^* + \omega'_0 + \omega'_0 = \omega_0^* + 2\omega'_0$$

$$\text{(ii)}.\ \omega_0^{(1)} + \omega_0^{(2)} - \omega'_0 \rightarrow \omega_0^* + \omega'_0 - \omega'_0 = \omega_0^*$$

$$\text{(iii)}.\ \omega_0^{(1)} - \omega_0^{(2)} + \omega'_0 \rightarrow \omega_0^* - \omega'_0 + \omega'_0 = \omega_0^*$$

$$\text{(iv)}.\ \omega_0^{(1)} - \omega_0^{(2)} - \omega'_0 \rightarrow \omega_0^* - \omega'_0 - \omega'_0 = \omega_0^* - 2\omega'_0$$

As a result, with Condition B, the following terms in the first summation over k_1, k_2 contribute to secular terms:

$$\frac{a'_0\bar{a}_0^{(1)}\bar{a}_0^{(2)}}{g''_1}e^{-i\left(\omega_0^{(1)}+\omega_0^{(2)}-\omega'_0\right)\tau_0} = \frac{a'_0\bar{a}_0^*\bar{a}'_0}{g''_1}e^{-i\left(\omega_0^*+\omega'_0-\omega'_0\right)\tau_0} = \frac{a'_0\bar{a}_0^*\bar{a}'_0}{g''_1}e^{-i\left(\omega_0^*\right)\tau_0}$$

$$\frac{\bar{a}'_0a_0^{(1)}a_0^{(2)}}{g''_1}e^{i\left(\omega_0^{(1)}+\omega_0^{(2)}-\omega'_0\right)\tau_0} = \frac{\bar{a}'_0a_0^*a'_0}{g''_1}e^{i\left(\omega_0^*+\omega'_0-\omega'_0\right)\tau_0} = \frac{\bar{a}'_0a_0^*a'_0}{g''_1}e^{i\left(\omega_0^*\right)\tau_0}$$

$$\frac{a'_0a_0^{(1)}\bar{a}_0^{(2)}}{g''_2}e^{i\left(\omega_0^{(1)}-\omega_0^{(2)}+\omega'_0\right)\tau_0} = \frac{a'_0a_0^*\bar{a}'_0}{g''_2}e^{i\left(\omega_0^*-\omega'_0+\omega'_0\right)\tau_0} = \frac{a'_0a_0^*\bar{a}'_0}{g''_2}e^{i\left(\omega_0^*\right)\tau_0}$$

$$\frac{\bar{a}'_0\bar{a}_0^{(1)}a_0^{(2)}}{g''_2}e^{-i\left(\omega_0^{(1)}-\omega_0^{(2)}+\omega'_0\right)\tau_0} = \frac{\bar{a}'_0\bar{a}_0^*a'_0}{g''_2}e^{-i\left(\omega_0^*-\omega'_0+\omega'_0\right)\tau_0} = \frac{\bar{a}'_0\bar{a}_0^*a'_0}{g''_2}e^{-i\left(\omega_0^*\right)\tau_0}$$

For the second summation over k_1, k_2, two conditions must be met, (1) $k' + k'' = k^*$ and (2) $k_1 + k_2 = k'$. The only possible combinations for k_1, k_2 that give wave vector relationships that are compatible with $\delta_{k'+k'',\ k^*}$ are shown as Conditions C and D:

$$\text{Condition C: } k_1 = -k'', k_2 = k^*, \text{ and } -k'' + k^* = k'$$
$$\text{Condition D: } k_1 = k^*, k_2 = -k'', \text{ and } k^* - k'' = k'$$

Now that wave vector constraints are satisfied, an analysis is carried out to see if any terms with $e^{i\omega_0^* \tau_0}$ or $e^{-i\omega_0^* \tau_0}$ dependency arise in the second summation over k_1, k_2. The following frequency relationships are present in the second summation over k_1, k_2 in (4.18):

$$\text{(i).}\ \omega_0^{(1)} + \omega_0^{(2)} + \omega_0''$$
$$\text{(ii).}\ \omega_0^{(1)} + \omega_0^{(2)} - \omega_0''$$
$$\text{(iii).}\ \omega_0^{(1)} - \omega_0^{(2)} + \omega_0''$$
$$\text{(iv).}\ \omega_0^{(1)} - \omega_0^{(2)} - \omega_0''$$

$$\text{Condition C: } k_1 = -k'' \rightarrow \omega_0^{(1)} = \omega_0'' \text{ and } k_2 = k^* \rightarrow \omega_0^{(2)} = \omega_0^*$$

Applying Condition C to these frequency relationships show two relationships that offer terms with $e^{i\omega_0^* \tau_0}$ or $e^{-i\omega_0^* \tau_0}$ dependence:

$$\text{(v).}\ \omega_0^{(1)} + \omega_0^{(2)} + \omega_0'' \rightarrow \omega_0'' + \omega_0^* + \omega_0'' = \omega_0^* + 2\omega_0''$$
$$\text{(vi).}\ \omega_0^{(1)} + \omega_0^{(2)} - \omega_0'' \rightarrow \omega_0'' + \omega_0^* - \omega_0'' = \omega_0^*$$
$$\text{(vii).}\ \omega_0^{(1)} - \omega_0^{(2)} + \omega_0'' \rightarrow \omega_0'' - \omega_0^* + \omega_0'' = -\omega_0^* + 2\omega_0''$$
$$\text{(viii).}\ \omega_0^{(1)} - \omega_0^{(2)} - \omega_0'' \rightarrow \omega_0'' - \omega_0^* - \omega_0'' = -\omega_0^*$$

As a result, with Condition C, the following terms in the second summation over k_1, k_2 contribute to secular terms:

$$\frac{a_0''\bar{a}_0^{(1)}\bar{a}_0^{(2)}}{g_1'} e^{-i\left(\omega_0^{(1)}+\omega_0^{(2)}-\omega_0''\right)\tau_0} = \frac{a_0''\bar{a}_0''\bar{a}_0^*}{g_1'} e^{-i\left(\omega_0''+\omega_0^*-\omega_0''\right)\tau_0} = \frac{a_0''\bar{a}_0''\bar{a}_0^*}{g_1'} e^{-i\left(\omega_0^*\right)\tau_0}$$

$$\frac{\bar{a}_0''a_0^{(1)}a_0^{(2)}}{g_1'} e^{i\left(\omega_0^{(1)}+\omega_0^{(2)}-\omega_0''\right)\tau_0} = \frac{\bar{a}_0''a_0''a_0^*}{g_1'} e^{i\left(\omega_0''+\omega_0^*-\omega_0''\right)\tau_0} = \frac{\bar{a}_0''a_0^*a_0''}{g_1'} e^{i\left(\omega_0^*\right)\tau_0}$$

$$\frac{a_0''\bar{a}_0^{(1)}a_0^{(2)}}{g_2'} e^{-i\left(\omega_0^{(1)}-\omega_0^{(2)}-\omega_0''\right)\tau_0} = \frac{a_0''\bar{a}_0''a_0^*}{g_2'} e^{-i\left(\omega_0''-\omega_0^*-\omega_0''\right)\tau_0} = \frac{a_0''\bar{a}_0''a_0^*}{g_2'} e^{i\left(\omega_0^*\right)\tau_0}$$

$$\frac{\bar{a}_0''a_0^{(1)}\bar{a}_0^{(2)}}{g_2'}e^{i\left(\omega_0^{(1)}-\omega_0^{(2)}-\omega_0''\right)\tau_0}=\frac{\bar{a}_0''a_0''\bar{a}_0^{*}}{g_2'}e^{i\left(\omega_0''-\omega_0^{*}-\omega_0''\right)\tau_0}=\frac{\bar{a}_0''a_0''\bar{a}_0^{*}}{g_2'}e^{-i\left(\omega_0^{*}\right)\tau_0}$$

Condition D: $k_1=k^*\rightarrow\omega_0^{(1)}=\omega_0^*$ and $k_2=-k''\rightarrow\omega_0^{(2)}=\omega_0''$

Applying Condition D to these frequency relationships show two different relationships that offer terms with $e^{i\omega_0^*\tau_0}$ or $e^{-i\omega_0^*\tau_0}$ dependence:

$$(\mathrm{v}).\ \omega_0^{(1)}+\omega_0^{(2)}+\omega_0''\rightarrow\omega_0^*+\omega_0''+\omega_0''=\omega_0^*+2\omega_0''$$
$$(\mathrm{vi}).\ \omega_0^{(1)}+\omega_0^{(2)}-\omega_0''\rightarrow\omega_0^*+\omega_0''-\omega_0''=\omega_0^*$$
$$(\mathrm{vii}).\ \omega_0^{(1)}-\omega_0^{(2)}+\omega_0''\rightarrow\omega_0^*-\omega_0''+\omega_0''=\omega_0^*$$
$$(\mathrm{viii}).\ \omega_0^{(1)}-\omega_0^{(2)}-\omega_0''\rightarrow\omega_0^*-\omega_0''-\omega_0''=\omega_0^*-2\omega_0''$$

As a result, with Condition D, the following terms in the second summation over k_1,k_2 contribute to secular terms:

$$\frac{a_0''\bar{a}_0^{(1)}\bar{a}_0^{(2)}}{g_1'}e^{-i\left(\omega_0^{(1)}+\omega_0^{(2)}-\omega_0''\right)\tau_0}=\frac{a_0''\bar{a}_0^{*}\bar{a}_0''}{g_1'}e^{-i\left(\omega_0^{*}+\omega_0''-\omega_0''\right)\tau_0}=\frac{a_0''\bar{a}_0^{*}\bar{a}_0''}{g_1'}e^{-i\left(\omega_0^{*}\right)\tau_0}$$

$$\frac{\bar{a}_0''a_0^{(1)}a_0^{(2)}}{g_1'}e^{i\left(\omega_0^{(1)}+\omega_0^{(2)}-\omega_0''\right)\tau_0}=\frac{\bar{a}_0''a_0^{*}a_0''}{g_1'}e^{i\left(\omega_0^{*}+\omega_0''-\omega_0''\right)\tau_0}=\frac{\bar{a}_0''a_0^{*}a_0''}{g_1'}e^{i\left(\omega_0^{*}\right)\tau_0}$$

$$\frac{a_0''a_0^{(1)}\bar{a}_0^{(2)}}{g_2'}e^{i\left(\omega_0^{(1)}-\omega_0^{(2)}+\omega_0''\right)\tau_0}=\frac{a_0''a_0^{*}\bar{a}_0''}{g_2'}e^{i\left(\omega_0^{*}-\omega_0''+\omega_0''\right)\tau_0}=\frac{a_0''a_0^{*}\bar{a}_0''}{g_2'}e^{i\left(\omega_0^{*}\right)\tau_0}$$

$$\frac{\bar{a}_0''\bar{a}_0^{(1)}a_0^{(2)}}{g_2'}e^{-i\left(\omega_0^{(1)}-\omega_0^{(2)}+\omega_0''\right)\tau_0}=\frac{\bar{a}_0''\bar{a}_0^{*}a_0''}{g_2'}e^{-i\left(\omega_0^{*}-\omega_0''+\omega_0''\right)\tau_0}=\frac{\bar{a}_0''\bar{a}_0^{*}a_0''}{g_2'}e^{-i\left(\omega_0^{*}\right)\tau_0}$$

Assuming that terms $[a_0(k'),\bar{a}_0(k'),a_0(-k'),\bar{a}_0(-k'),\ldots,etc.]$ in (4.18) behave as follows:

$$a_0(k',\tau_2)=a_0(k',0)e^{i\beta(k')\tau_2}$$
$$\bar{a}_0(k',\tau_2)=\bar{a}_0(k',0)e^{-i\beta(k')\tau_2}$$
$$a_0(-k',\tau_2)=a_0(-k',0)e^{i\beta(-k')\tau_2}$$
$$\bar{a}_0(-k',\tau_2)=\bar{a}_0(-k',0)e^{-i\beta(-k')\tau_2}$$
$$\vdots$$

etc.

Additionally,

$$a_0(k',0) = a_0(-k',0)$$
$$\bar{a}_0(k',0) = \bar{a}_0(-k',0)$$
$$\beta(k') = \beta(-k')$$

Equation (4.18) can be rewritten in the form of (4.19):

$$\begin{aligned}
&\frac{\partial^2 A_2(k^*,\tau_0,\tau_1,\tau_2)}{\partial\tau_0^2} + \omega_0^{*2} A_2(k^*,\tau_0,\tau_1,\tau_2) = \left\{-2i\omega_0^* \frac{\partial a_0(k^*,\tau_2)}{\partial\tau_2}\right.\\
&\quad + a_0(k^*,0)e^{i\beta(k^*)\tau_2}\left(\frac{1}{m\omega_n^2}\right)^2 \sum_{k'}\sum_{k''} f(k',k'')\delta_{k'+k'',k^*}\\
&\quad \times\left[2f(-k',k^*)\delta_{-k'+k^*,\,k''}a_0(k',0)\bar{a}_0(k',0)\left[\frac{1}{g_1''}+\frac{1}{g_2''}\right]\right.\\
&\quad \left.\left.+2f(-k'',k^*)\delta_{-k''+k^*,\,k'}a_0(k'',0)\bar{a}_0(k'',0)\left[\frac{1}{g_1'}+\frac{1}{g_2'}\right]\right]\right\}\\
&\quad \times e^{i\omega_0^*\tau_0} + \left\{2i\omega_0^* \frac{\partial \bar{a}_0(k^*,\tau_2)}{\partial\tau_2} + \bar{a}_0(k^*,0)e^{-i\beta(k^*)\tau_2}\right.\\
&\quad \times\left(\frac{1}{m\omega_n^2}\right)^2 \sum_{k'}\sum_{k''} f(k',k'')\delta_{k'+k'',k^*}\\
&\quad \times\left[2f(-k',k^*)\delta_{-k'+k^*,\,k''}a_0(k',0)\bar{a}_0(k',0)\left[\frac{1}{g_1''}+\frac{1}{g_2''}\right]\right.\\
&\quad \left.\left.+2f(-k'',k^*)\delta_{-k''+k^*,\,k'}a_0(k'',0)\bar{a}_0(k'',0)\left[\frac{1}{g_1'}+\frac{1}{g_2'}\right]\right]\right\}e^{-i\omega_0^*\tau_0}\\
&\quad + \textit{other terms which will not give } e^{i\omega_0^*\tau_0} \textit{ or } e^{-i\omega_0^*\tau_0} \textit{ dependence}
\end{aligned} \tag{4.19}$$

The terms in front of $e^{i\omega_0^*\tau_0}$ and $e^{-i\omega_0^*\tau_0}$ are set to zero. This is shown by (4.20) and (4.21):

$$\begin{aligned}
&a_0(k^*,0)e^{i\beta(k^*)\tau_2}\left(\tfrac{1}{m\omega_n^2}\right)^2 \sum_{k'}\sum_{k''} f(k',k'')\\
&\qquad \times\delta_{k'+k'',k^*}\left[2f(-k',k^*)\delta_{-k'+k^*,\,k''}a_0(k',0)\bar{a}_0(k',0)\left[\frac{1}{g_1''}+\frac{1}{g_2''}\right]\right.\\
&\qquad \left.+2f(-k'',k^*)\delta_{-k''+k^*,\,k'}a_0(k'',0)\bar{a}_0(k'',0)\left[\frac{1}{g_1'}+\frac{1}{g_2'}\right]\right]\\
&\qquad = -2\omega_0^*\beta(k^*)a_0(k^*,0)e^{i\beta(k^*)\tau_2}
\end{aligned} \tag{4.20}$$

$$\begin{aligned}
&\bar{a}_0(k^*,0)e^{-i\beta(k^*)\tau_2}\left(\tfrac{1}{m\omega_n^2}\right)^2 \sum_{k'}\sum_{k''} f(k',k'')\\
&\qquad \times\delta_{k'+k'',k^*}\left[2f(-k',k^*)\delta_{-k'+k^*,k''}a_0(k',0)\bar{a}_0(k',0)\left[\frac{1}{g_1''}+\frac{1}{g_2''}\right]\right.
\end{aligned}$$

$$+2f(-k'',k^*)\delta_{-k''+k^*,k'}a_0(k'',0)\bar{a}_0(k'',0)\left[\frac{1}{g_1'}+\frac{1}{g_2'}\right]\Bigg]$$
$$= -2\omega_0^*\beta(k^*)\bar{a}_0(k^*,0)e^{-i\beta(k^*)\tau_2} \tag{4.21}$$

From (4.20) and (4.21), the same expression for β^* results (4.22):

$$\beta(k^*) = -\frac{1}{2\omega_0^*}\left(\frac{1}{m\omega_n^2}\right)^2\sum_{k'}\sum_{k''}f(k',k'')$$
$$\times\delta_{k'+k'',k^*}\Bigg[2f(-k',k^*)\delta_{-k'+k^*,k''}a_0(k',0)\bar{a}_0(k',0)\left[\frac{1}{g_1''}+\frac{1}{g_2''}\right]$$
$$+2f(-k'',k^*)\delta_{-k''+k^*,k'}a_0(k'',0)\bar{a}_0(k'',0)\left[\frac{1}{g_1'}+\frac{1}{g_2'}\right]\Bigg] \tag{4.22}$$

Recall that φ appears in the terms containing g_1'', g_2'', g_1', g_2'. The limit of (4.22) is taken as $\varphi \to 0$ and the following definition is utilized [17]:

$$\lim_{\theta\to 0}\frac{1}{(x\pm i\theta)} = \left(\frac{1}{x}\right)_{pp} \mp i\pi\delta(x)$$

where pp denotes the principle part. The real and imaginary parts of (4.22) are shown as (4.23) and (4.24), respectively.

$$Re(\beta^*) = \Delta_{k^*} = \frac{-64}{\omega_0^*}\left(\frac{1}{m\omega_n^2}\right)^2\sum_{k'}\sum_{k''}\sin^2\left(\frac{k'a}{2}\right)\sin^2\left(\frac{k''a}{2}\right)\sin^2\left(\frac{k^*a}{2}\right)$$
$$\times\Bigg\{a_0'\bar{a}_0'\left[\left(\frac{1}{\omega_0''^2-(\omega_0^*+\omega_0')^2}\right)_{pp}+\left(\frac{1}{\omega_0''^2-(\omega_0^*-\omega_0')^2}\right)_{pp}\right]$$
$$+a_0''\bar{a}_0''\left[\left(\frac{1}{\omega_0'^2-(\omega_0^*+\omega_0'')^2}\right)_{pp}+\left(\frac{1}{\omega_0'^2-(\omega_0^*-\omega_0'')^2}\right)_{pp}\right]\Bigg\} \tag{4.23}$$

$$Im(\beta^*) = \Gamma_{k^*} =$$
$$\frac{32\pi}{\omega_0^*}\left(\frac{1}{m\omega_n^2}\right)^2\sum_{k'}\sum_{k''}\sin^2\left(\frac{k'a}{2}\right)\sin^2\left(\frac{k''a}{2}\right)\sin^2\left(\frac{k^*a}{2}\right)$$
$$\times\Bigg\{\delta(\omega_0^*+\omega_0'+\omega_0'')\left[\frac{a_0'\bar{a}_0'}{\omega_0''}+\frac{a_0''\bar{a}_0''}{\omega_0'}\right]-\delta(\omega_0^*+\omega_0'-\omega_0'')\left[\frac{a_0'\bar{a}_0'}{\omega_0''}-\frac{a_0''\bar{a}_0''}{\omega_0'}\right]$$
$$+\delta(\omega_0^*-\omega_0'+\omega_0'')\left[\frac{a_0'\bar{a}_0'}{\omega_0''}-\frac{a_0''\bar{a}_0''}{\omega_0'}\right]-\delta(\omega_0^*-\omega_0'-\omega_0'')\left[\frac{a_0'\bar{a}_0'}{\omega_0''}+\frac{a_0''\bar{a}_0''}{\omega_0'}\right]\Bigg\} \tag{4.24}$$

In the above expressions for the real and imaginary parts of β^*:

$$d_0'\bar{a}_0' = a_0(k',0)\bar{a}_0(k',0)$$
$$a_0''\bar{a}_0'' = a_0(k'',0)\bar{a}_0(k'',0)$$

From here, the general solution to the ε^0-equation (4.14) is considered with the new found results for $a_0(k^*,\tau_2)$ and $\bar{a}_0(k^*,\tau_2)$:

$$a_0(k^*,\tau_2) = a_0(k^*,0)e^{i\beta(k^*)\tau_2}$$
$$\bar{a}_0(k^*,\tau_2) = \bar{a}_0(k^*,0)e^{-i\beta(k^*)\tau_2}$$

Equation (4.14) is written as $A_0(k^*,\tau_0,\tau_2) = a_0(k^*,\tau_2)e^{i\omega_0^*\tau_0} + \bar{a}_0(k^*,\tau_2)e^{-i\omega_0^*\tau_0}$.

Again, using the new $a_0(k^*,\tau_2)$ and $\bar{a}_0(k^*,\tau_2)$ one arrives at the following expression:

$$A_0(k^*,\tau_0,\tau_2) = a_0(k^*,0)e^{i\left(\omega_0^*\tau_0+\beta(k^*)\tau_2\right)} + \bar{a}_0(k^*,0)e^{-i\left(\omega_0^*\tau_0+\beta(k^*)\tau_2\right)}$$

Writing the above expression strictly in terms of τ_0, where $\tau_2 = \varepsilon^2\tau_0$, gives the following representation for $A_0(k^*,\tau_0)$:

$$A_0(k^*,\tau_0) = a_0(k^*,0)e^{i\left(\omega_0^*\tau_0+\varepsilon^2\beta(k^*)\tau_0\right)} + \bar{a}_0(k^*,0)e^{-i\left(\omega_0^*\tau_0+\varepsilon^2\beta(k^*)\tau_0\right)}$$

β^* is expressed in terms of its real and imaginary parts to yield the final representation for $A_0(k^*,\tau_0)$:

$$\beta(k^*) = \Delta_{k^*} + i\Gamma_{k^*}$$
$$A_0(k^*,\tau_0) = a_0(k^*,0)e^{i\left(\omega_0^*\tau_0+\varepsilon^2\left(\Delta_{k^*}+i\Gamma_{k^*}\right)\tau_0\right)} + \bar{a}_0(k^*,0)e^{-i\left(\omega_0^*\tau_0+\varepsilon^2\left(\Delta_{k^*}+i\Gamma_{k^*}\right)\tau_0\right)}$$
$$\begin{aligned} A_0(k^*,\tau_0) = {} & a_0(k^*,0)e^{i\left(\left(\omega_0^*+\varepsilon^2\Delta_{k^*}\right)\tau_0\right)}e^{-\varepsilon^2\Gamma_{k^*}\tau_0} \\ & + \bar{a}_0(k^*,0)e^{-i\left(\left(\omega_0^*+\varepsilon^2\Delta_{k^*}\right)\tau_0\right)}e^{\varepsilon^2\Gamma_{k^*}\tau_0} \end{aligned} \quad (4.25)$$

Three-wave interaction leads therefore to an additional frequency shift proportional to the square of the strength of the nonlinearity. Moreover, three wave interaction leads to a damping of each wave, that is, a finite lifetime. This result is the classical mechanics equivalent of that reported within the framework of quantum mechanics [21–23, 27]. Three wave interaction *i.e.*, three-phonon scattering process breaks time reversal symmetry of the elastic waves. Indeed, changing the sign of time: $t \to -t$ transforms a damped wave into an amplified wave.

4.2.4 Molecular Dynamics (MD) Simulation and Spectral Energy Density (SED) Approaches

For the present discussion, the equation of motion (4.1) for a simple model system is integrated by MD techniques with PBC using the velocity Verlet algorithm under the microcanonical ensemble (constant energy) (see Appendix 3, Chap. 3). This scheme ensures energy is conserved within 0.5%. Harmonic MD simulations of the 1D monatomic crystal utilize $\beta = 1.0$ N/m and $\varepsilon = 0.0$ N/m^2 whereas anharmonic simulations utilize $\beta = 1.0$ N/m and $\varepsilon = [0.9\text{–}3.7]$ N/m^2. The 1D crystal consists of a chain of 1.0 kg masses spaced periodically 1.0 meter apart. These parameters can be easily scaled down to represent an atomic system. To initiate a simulation, every mass in the MD simulation cell is randomly displaced from its equilibrium position. The maximum value in which a mass can be displaced is constrained such that instabilities do not emerge in the potential energy function. MD simulations are run for 2^{21} time steps with a timestep of 0.01 s. For post-processing Spectral Energy Density (SED) calculations, velocity data is collected for each mass in the simulation cell over the entire simulation time (See Appendix 3, Chap. 3).

The band structure generated by the SED method is shown for the 1D harmonic monatomic crystal (Fig. 4.2). Figure 4.2 shows contours of constant SED over the wave vector-frequency plane.

There are 101 discrete, non-degenerate wave vectors resolved between the center of the irreducible Brillouin zone and the zone edge at $k = \pi/a$. In the band structure, there is a nearly linear region that accounts for the propagative characteristics of long wavelength excitations in the 1D harmonic crystal. At larger wave vector values, a departure from the linear behavior is apparent and the phase velocity of propagative phonon modes is markedly different from the group velocity. This is similar to the expected dispersion behavior of the infinite monatomic harmonic crystal. At the edge of the irreducible Brillouin zone, a SED-frequency

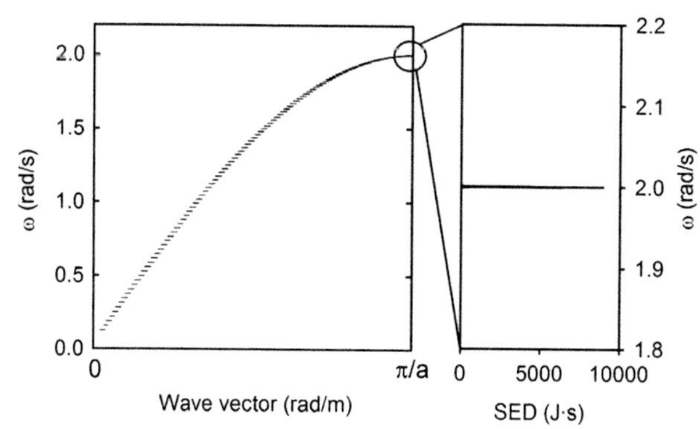

Fig. 4.2 (*Left*) Band structure of 1D harmonic monatomic crystal. (*Right*) SED-frequency plot showing wave vector mode $k = \pi/a$

plot is reported. A peak in the spectrum shows this vibrational mode contributing significantly to the average kinetic energy per unit cell. A Lorentzian function is fit to this peak and shows a finite value (γ) for full width at half-maximum (FWHM) because the fast Fourier transform scheme used in the SED calculation involves a signal sampled over a finite time window. This value for FWHM is subsequently used as a lower bound for the error on lifetime estimated with the SED method. This error amounts to one interval in the discrete frequency scale. The band structure of the harmonic system is highlighted in the long wavelength regime; Fig. 4.3 zooms in on a region of the dispersion curve near $\mathrm{k} = \pi/10a$.

In Fig. 4.3 on the right, four SED-frequency plots are shown (plots a–d). Each plot represents a different MD simulation of the 1D harmonic monatomic crystal. Each MD simulation begins with a random starting configuration for atomic displacements in the 1D crystal. It is observable from these four plots that for a given wave vector, the SED takes on different values. This is due to the fact that in a harmonic crystal, energy contained within a particular mode cannot be passed to other modes of vibration. This highlights the sensitivity of the vibrational modes of the harmonic crystal on the initial configuration. Consequently, to obtain a non-biased band structure, multiple MD simulations must be run such that an average can be taken of the different SED values for each discrete, non-degenerate wave vector mode. An average of plots (a–d) is shown on the left of Fig. 4.3 with the color of the contours signifying SED intensity. A Lorentzian function is fit to each of the peaks in the left hand figure and shows the same value for FWHM as that calculated in Fig. 4.2. For comparison, the band structure of the

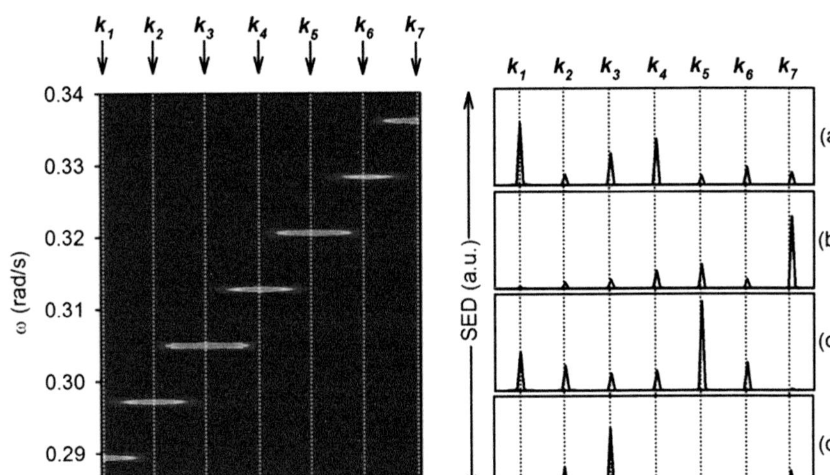

Fig. 4.3 (*Left*) Band structure (constant SED contours) for 1D harmonic monatomic crystal near $\mathrm{k} = \pi/10a$. (*Right*) SED-frequency plots for four MD simulations differing in their initial random configurations

1D anharmonic monatomic crystal near $k = \pi/10a$ is shown in Fig. 4.4. Here the parameter characterizing the degree of anharmonicity in the 1D crystal is $\varepsilon = 3.0$ (see Fig. 4.1b).

Similarly to Fig. 4.3, the four plots on the right of Fig. 4.4 represent SED-frequency plots generated from four different MD simulations. The SED intensity for a given mode varies from simulation to simulation, which indicates that energy does not easily exchange between modes of vibration in the 1D anharmonic crystal. In contrast, though, there are some peaks in the SED-frequency spectra that show slightly larger values for FWHM. However, it is critical that averages be taken for SED data extracted from several MD simulations such that an accurate quantification of phonon lifetime can be realized. The contour map on the left of Fig. 4.4 represents an average over plots (a–d). Lorentzian functions are fit to the peaks in this figure. The FWHM for all peaks is found to be comparable to the harmonic case. With a random initial displacement of the masses of at most 10% of the lattice spacing "a", the total energy of the anharmonic system is only 1.3% higher than that of the harmonic system. Under this condition, the system can be considered to be weakly anharmonic and second order perturbation theory is applicable. Considering the final expression for $A_0(k^*, \tau_0)$ in (4.25), which represents the 0th order term in the asymptotic expansion of $A(k^*, \tau)$ describing three-wave interactions, Γ_{k^*} (4.24) corresponds to a decay constant for mode k^*. FWHM calculations of peaks in SED-frequency spectra embody Γ_{k^*}. In the long wavelength regime, Γ_{k^*} is small because of squared sinusoidal terms inside the double summation over k' and k''. Accordingly, one should not expect large values for FWHM in the

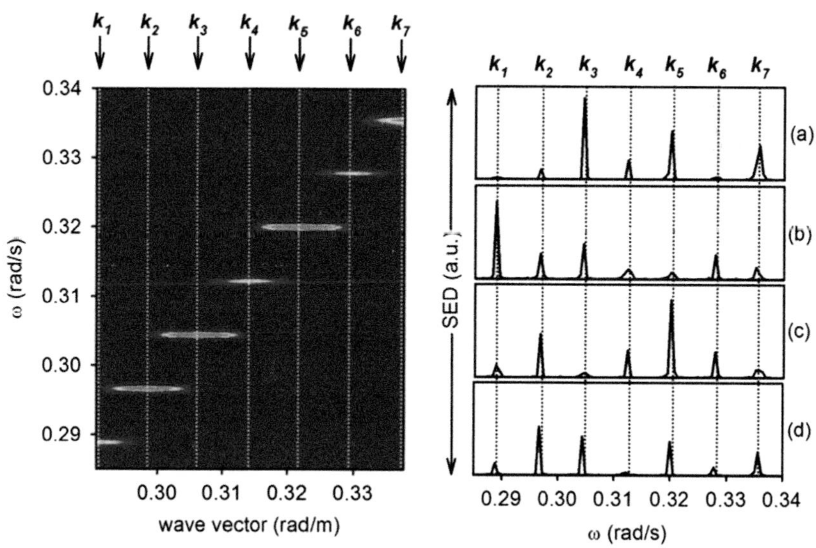

Fig. 4.4 (*Left*) Band structure (constant SED contours) for 1D anharmonic monatomic crystal near $k = \pi/10a$. (*Right*) SED-frequency plots for four MD simulations differing in their initial random configurations

long wavelength limit. The complete band structure for the 1D anharmonic monatomic crystal is shown in Fig. 4.5. The band structure is generated from SED averages taken from four MD simulations.

In Fig. 4.5, it seems that each non-degenerate wave vector is associated with multiple eigen frequencies due to the fact that multiple peaks appear in the SED. At the edge of the irreducible Brillouin zone, an intense central peak is seen along with multiple, less intense symmetrical satellite peaks. These satellite peaks, emerge when the anharmonicity of the system is adequately sampled (*i.e.*, large amplitudes of vibration). Equation (4.17) is utilized to explain the appearance of these satellite peaks. This equation represents the 1st order term in the asymptotic expansion of $A(k^*, \tau)$ describing three-wave interactions. Inside the double summation over (k', k'') in (4.17), conservation of wave vectors is imposed: $\delta_{k'+k'',\, k^*} \rightarrow k' + k'' = k^*$. If the mode of interest is $k^* = \pi/a$, then conservation of wave vector can be satisfied by adding non-degenerate wave vector pairs that yield k^*. With N = 400, non-degenerate wave vectors are limited to the following: $k_i = \frac{n_i}{400} \cdot \frac{2\pi}{a}$. If only wave vectors contained between the center of the irreducible Brillouin zone and the zone edge are considered, then n_i ranges from 0 to 200. As a first example, to satisfy wave vector conservation, consider two wave vectors: (1) the first non-degenerate wave vector before the zone edge at $(k = \pi/a)$ and (2) the first non-degenerate wave vector after the center of the irreducible Brillouin zone at $(k = 0)$. This pair of wave vectors is shown as Case I and satisfies wave vector conservation: (Case I) $k' = \frac{199}{400} \cdot \frac{2\pi}{a}$, $k'' = \frac{1}{400} \cdot \frac{2\pi}{a}$, $k^* = \frac{200}{400} \cdot \frac{2\pi}{a}$. As a second example, consider (1) the second non-degenerate wave vector before the zone edge at $(k = \pi/a)$ and (2) the second non-degenerate wave vector after the center of the irreducible Brillouin zone at $(k = 0)$. This pair of wave vectors is defined as Case II and satisfies wave vector conservation: (Case II) $k' = \frac{198}{400} \cdot \frac{2\pi}{a}$, $k'' = \frac{2}{400} \cdot \frac{2\pi}{a}$, $k^* = \frac{200}{400} \cdot \frac{2\pi}{a}$. We note that both cases do not conserve frequency. In both cases, since the dispersion relation for the 1D anharmonic monatomic crystal is not strictly linear, the

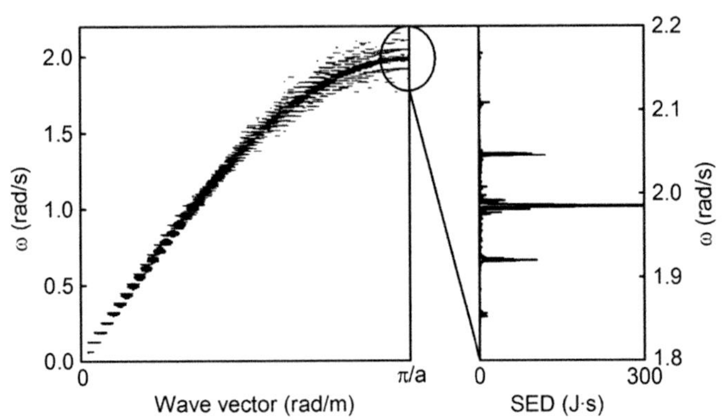

Fig. 4.5 (*Left*) Band structure for 1D anharmonic monatomic crystal. (*Right*) SED-frequency plot showing wave vector mode $k = \pi/a$. $\varepsilon = 3.0$ and initial random displacement does not exceed 10% of a

frequency of mode k' plus (or minus) the frequency of mode k'' will not exactly equal the frequency of mode k^*. Instead, the addition (or subtraction) of the frequencies associated with modes k' and k'' will be slightly greater than (or less than) the frequency of mode k^*. This forces the denominator of the pre-exponential factors in (4.17) to become small, thereby contributing to a large value of $A_1(k^*, \tau_0, \tau_2)$. The presence of non-zero $A_1(k^*, \tau_0, \tau_2)$ indicates that discrete, near-resonance modes are initiated for short wavelength phonons (k') interacting with long wavelength phonons (k''). On the left of Fig. 4.6, we show non-degenerate wave vector modes k' and k'' corresponding to Case I (top) and Case II (bottom). On the right, Fig. 4.6 shows the modes at $k^* = \pi/a$.

In this image, the satellite peaks coincide with discrete, near-resonance modes. The central peak frequencies of modes k' and k'' add (or subtract) to yield satellite peaks to the central peak for $k^* = \pi/a$. The primary satellite peaks at 1.999 and 1.969 rad/s come from Case I. The secondary satellite peaks at 2.015 and 1.951 rad/s come from Case II. Tertiary, quaternary and other higher order satellite peaks exist and are revealed if the scale on the right hand SED plot is adjusted. The magnitude of the satellite peaks depends upon the "distance" from the central peak at $k = \pi/a$ in accordance with their near resonant character. This distance depends upon the size of the MD simulation. For an MD simulation with N = 100 atoms, there are 51 discrete, non-degenerate wave vector modes available between the center of the irreducible Brillouin zone and the zone edge. For N = 1000 atoms, there are 501 available modes. The resolution in wave vector space is finer for larger MD systems as is the resolution in frequency space. Higher frequency resolution results in smaller spacing between satellite peaks. This is shown in Fig. 4.7. As the number of atoms (N) increases, the satellite peaks congregate around the central peak and increase in relative amplitude. In the limit of an infinite system all satellite peaks merge into the central peak.

For a phonon mode to decay, wave vector and frequency conservation rules must be satisfied. For short wavelength phonon modes, these constraints are pathologically difficult to satisfy because the monatomic dispersion curve is not linear. The central frequency peaks in Fig. 4.7 represent the resonance mode of wave vector $k = \pi/a$. The satellite peaks in Fig. 4.7 represent frequency non-conserving near-resonance modes spawned from nonlinear wave interactions between short wavelength phonons and long wavelength phonons. The lifetime of phonon mode $k = \pi/a$ comes from fitting a Lorentzian function to the central peak. As Fig. 4.7 shows, the FWHM for phonon mode $k = \pi/a$ is rather insensitive to the number of atoms in the MD simulation cell. It is found that the FWHM for $k = \pi/a$ is the same order of magnitude as the error estimate found from the harmonic case in Fig. 4.2. As a result, the lifetime of high frequency phonon modes in the anharmonic monatomic crystal is inherently long because wave vector and frequency conservation constraints cannot be satisfied.

In comparing the anharmonic band structure with the harmonic band structure at $(k = \pi/a)$, there is an obvious shift in frequency of the central peak. The perturbation analysis of the single wave dispersion has shown that the anharmonic dispersion curve is shifted in frequency (with respect to the harmonic dispersion curve) by a

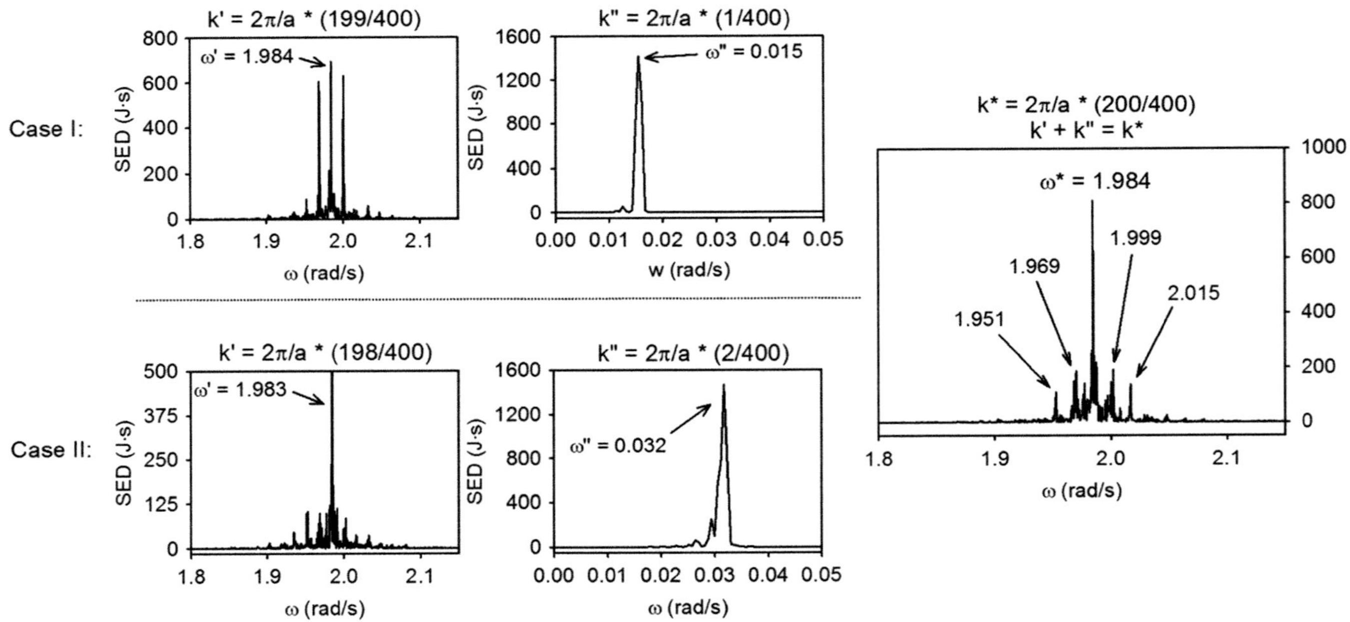

Fig. 4.6 (*Top, Left*) SED-frequency plots for wave vector modes k' and k'' corresponding to Case I. (*Bottom, Left*) SED-frequency plots for wave vector modes k' and k'' corresponding to Case II. (*Right*) SED-frequency plot corresponding to $k^* = \pi/a$. For Cases I and II, wave vectors k' and k'' satisfy wave vector conservation for mode k^*. The frequencies of modes k' and k'' add (or subtract) to yield near-resonance peaks near $\omega^*(k^*)$. Notice the frequency scale difference between short and long wavelength modes

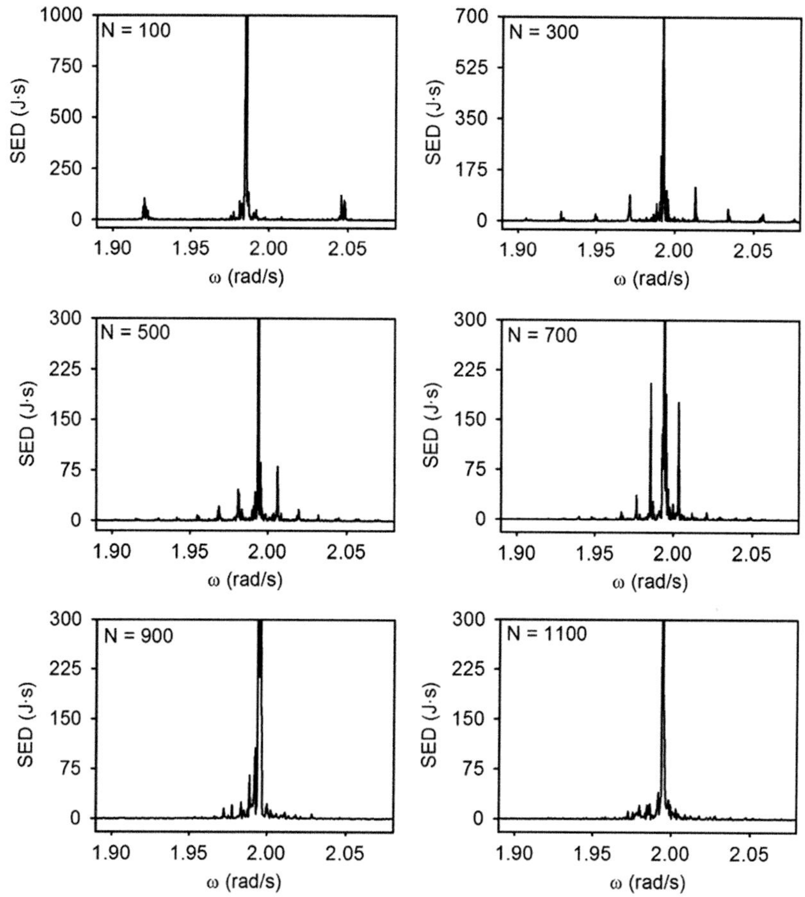

Fig. 4.7 SED-frequency plots for 1D anharmonic monatomic crystal at $k = \pi/a$ for MD systems of varying sizes. The parameter characterizing the degree of anharmonicity in the 1D crystal is $\varepsilon = 3.0$

quantity that has quadratic dependence on the strength of the nonlinearity parameter ε. Figure 4.8 shows a plot mapping the frequency shift relative to the harmonic system for several values of ε for a MD simulation cell consisting of N = 200 atoms. In Fig. 4.8, three different curves are rendered. Each curve represents a different magnitude for the initial random displacement imposed upon the masses in the 1D crystal as a percentage of the lattice spacing. The magnitude of the initial displacement controls the amplitude of the phonon modes. For triangles, the maximum value a mass can be displaced is 10% of the lattice spacing. For squares and circles, displacement values are 5% and 2%, respectively. Quadratic dependence is observed for values of ε ranging from 0.0 to 3.7. Beyond $\varepsilon = 3.7$, the potential energy function becomes completely unstable.

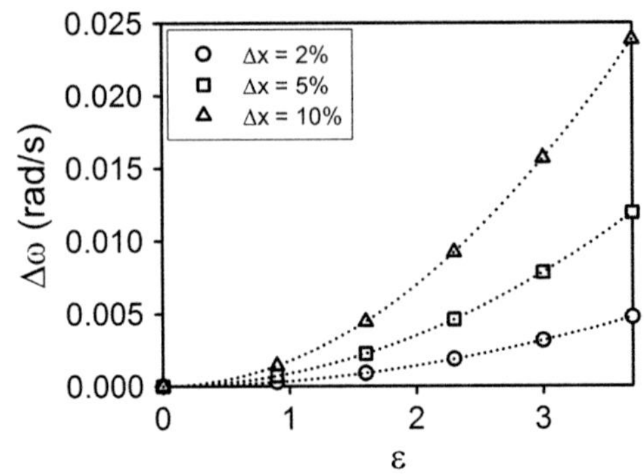

Fig. 4.8 Frequency shift evaluated at $k = \pi/a$ for the 1D anharmonic crystal relative to harmonic case. *Symbols* represent different magnitudes for the maximum initial random displacement imposed upon the masses in the 1D crystal as a percentage of the lattice spacing. *Circle*, *square* and *triangle symbols* represent small, intermediate and large initial displacements, respectively

Analysis of the weakly anharmonic 1D monatomic crystal has shown that the lifetime of phonon modes is not significantly affected by nonlinear interaction forces because it is pathologically difficult in satisfying the conditions for frequency and wave vector conservation. On the contrary, there exist conditions between short wavelength phonons and long wavelength phonons whereby near-resonance peaks emerge in plots of SED-frequency spectra. Satellite peaks materialize when the anharmonicity of the system is adequately sampled. Lastly, nonlinear interaction forces lead to amplitude dependent frequency shifts.

4.3 Nonlinear Phonon Modes in Second-Order Anharmonic Coupled Monatomic Chains

Coupled nonlinear and linear mechanical systems have recently received attention for their application in targeted energy transfer, whereby undesirable mechanical energy is directed irreversibly from a linear system to a nonlinear system [28]. Targeted energy transfer involves nonlinear resonances between the linear and nonlinear systems. So far linear-nonlinear coupled systems that have been investigated focusing on systems with a small number of degrees of freedom [29–31] or linear systems with small number of degrees of freedom coupled to a nonlinear attachments or absorber [32, 33]. Systems with multiple degrees of freedom have been addressed by considering finite linear chains with nonlinear end attachments [34, 35]. In this context, the model system presented in this section is a coupled linear-nonlinear discrete system with an infinite number of degrees of freedom. More specifically, we investigate the elastic vibrational propagative modes supported by a model of two coupled discrete infinite one-dimensional mass-spring chains with second-order nonlinearity. We contrast the behavior of coupled linear-nonlinear chains to that of the linear-linear and nonlinear-nonlinear two-chain models. We have seen in Chap. 3 that the phonon band structure of the linear two-chain model possesses an asymmetric and a symmetric band. We show here that irreversible energy transfer is achieved between the symmetric and the asymmetric bands of the linear-nonlinear coupled chain model. This transfer is associated with an inter-band resonance resulting from the quadratic nonlinearity. The nonlinear-nonlinear two-chain model does not possess such a resonance. We have also identified three-phonon processes that lead to the creation of a third band in the phonon band structure of the linear-nonlinear two-chain model. This nonlinear band bridges the symmetric and asymmetric bands. From the point of view of phononic systems that are isomorphic to the present nonlinear two-chain model, this model has also relevance to the study of a variety of low-dimensionality phononic structures or nanostructures such as elastic modes in coupled granular chains [36]. The study of nonlinear two-chain models is also relevant to the study of phonons in two-leg ladders [37]. Nonlinear vibrational modes in nonlinear discrete two-chain models have also been used to describe the thermal properties of DNA [38].

4.3.1 Model Systems

For the sake of generality, we consider the case of two parallel infinite chains of lattice parameter a, made of identical atoms of mass m. The two chains are coupled together with a spring of linear stiffness β'. Each atomic site "n" in chain 1 (2) is characterized by a displacement, u_n (v_n). We assume that the model is limited to first nearest neighbor interactions (see Fig. 4.9).

The two parallel chains are supposed to be anharmonic and the interaction forces between two adjacent masses n and n + 1 are taken as the sum of a linear and a second-order terms:

$$F^{(1)}_{n,n+1} = \beta(u_{n+1} - u_n) + \delta\varepsilon(u_{n+1} - u_n)^2, \tag{4.26}$$

in chain 1 and

$$F^{(2)}_{n,n+1} = \beta(v_{n+1} - v_n) + \varepsilon(v_{n+1} - v_n)^2, \tag{4.27}$$

in chain 2. In (4.26) and (4.27), the parameter ε characterizes the strength of the nonlinearity in the springs connecting the masses constituting the two chains and the parameter δ that can take the values 0 or 1, is used to switch on and off the nonlinearity of chain 1 with respect to that of chain 2. If $\delta = 0$, chain 1 is harmonic with springs of linear stiffness β while chain 2 remains anharmonic. This case is named subsequently the linear-nonlinear two chain model. When $\delta = 1$, both chain possess the same level of anharmonicity, this is denoted as the nonlinear-nonlinear

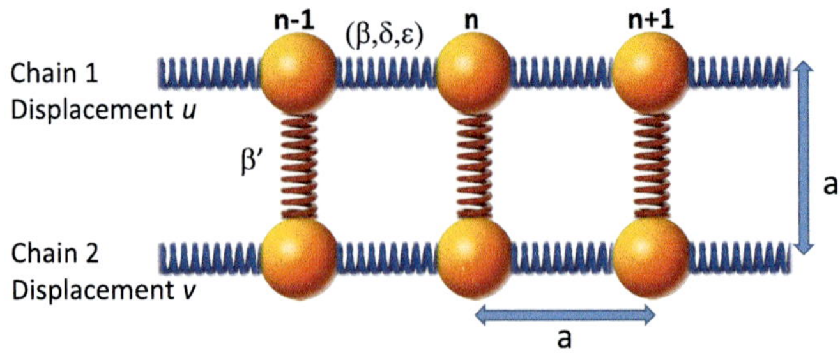

Fig. 4.9 Schematic illustration of the two coupled infinite atomic chains made of identical atoms of mass m with a lattice parameter a. β would correspond to the stiffness of the spring linking atoms in the two horizontal chains if these springs were linear. A degree of nonlinearity can be introduced inside the model via parameters δ and ε (see text for definition). β′ is the stiffness of the linear spring connecting together atoms "n" in the two horizontal chains

system. Consequently the equations of motion for atomic sites "n" in chains 1 and 2 are given as

$$\begin{cases} \dfrac{\partial^2 u_n}{\partial t^2} - \dfrac{\Omega^2}{4}(u_{n+1} - 2u_n + u_{n-1}) - \Omega'^2(v_n - u_n) \\ \qquad - \dfrac{\delta\varepsilon}{m}\left[(u_{n+1} - u_n)^2 - (u_n - u_{n-1})^2\right] = 0 \\ \dfrac{\partial^2 v_n}{\partial t^2} - \dfrac{\Omega^2}{4}(v_{n+1} - 2v_n + v_{n-1}) - \Omega'^2(u_n - v_n) \\ \qquad - \dfrac{\varepsilon}{m}\left[(v_{n+1} - v_n)^2 - (v_n - v_{n-1})^2\right] = 0 \end{cases} \tag{4.28}$$

where $\Omega^2 = 4\beta/m$ and $\Omega'^2 = \beta'/m$. The third terms in the left-hand side of these equations are associated with the linear springs connecting the chains.

4.3.2 Multiple Time Scale Perturbation Theory

We use multiple time scale perturbation theory (MTSPT) for solving (4.28) (see Appendix 4, Chap. 3). For the sake of analytical simplicity, we treat ε as a perturbation and write the displacements u_n and v_n as a second order power series in the perturbation, namely:

$$u_n(\tau_0, \tau_1, \tau_2) = u_n^{(0)}(\tau_0, \tau_1, \tau_2) + \varepsilon u_n^{(1)}(\tau_0, \tau_1, \tau_2) + \varepsilon^2 u_n^{(2)}(\tau_0, \tau_1, \tau_2), \tag{4.29}$$

$$v_n(\tau_0, \tau_1, \tau_2) = v_n^{(0)}(\tau_0, \tau_1, \tau_2) + \varepsilon v_n^{(1)}(\tau_0, \tau_1, \tau_2) + \varepsilon^2 v_n^{(2)}(\tau_0, \tau_1, \tau_2). \tag{4.30}$$

In (4.29) and (4.30), $u_n^{(i)}$ and $v_n^{(i)}$ with $i = 0, 1, 2$ are displacement functions expressed to zeroth-order, first-order and second-order in the perturbation. We have also replaced the single time variable, t, by three variables representing different time scales: $\tau_0 = t$, $\tau_1 = \varepsilon t = \varepsilon\tau_0$, and $\tau_2 = \varepsilon^2 t = \varepsilon^2\tau_0$. We can subsequently decompose (4.29) and (4.30) into three pairs of equations: one to zeroth-order in ε, one to first-order in ε and a third pair of equations to second-order in ε:

$$O(\varepsilon^0): \left\{ \frac{\partial^2 u_n^{(0)}}{\partial\tau_0^2} - \frac{\Omega^2}{4}\left(u_{n+1}^{(0)} - 2u_n^{(0)} + u_{n-1}^{(0)}\right) - \Omega'^2\left(v_n^{(0)} - u_n^{(0)}\right) = 0 \right. \quad (4.31a)$$

$$O(\varepsilon^0): \left\{ \frac{\partial^2 v_n^{(0)}}{\partial\tau_0^2} - \frac{\Omega^2}{4}\left(v_{n+1}^{(0)} - 2v_n^{(0)} + v_{n-1}^{(0)}\right) + \Omega'^2\left(v_n^{(0)} - u_n^{(0)}\right) = 0 \right. \quad (4.31b)$$

$$O(\varepsilon^1): \left\{ \begin{aligned} &\frac{\partial^2 u_n^{(1)}}{\partial\tau_0^2} - \frac{\Omega^2}{4}\left(u_{n+1}^{(1)} - 2u_n^{(1)} + u_{n-1}^{(1)}\right) - \Omega'^2\left(v_n^{(1)} - u_n^{(1)}\right) \\ &= -2\frac{\partial^2 u_n^{(0)}}{\partial\tau_0\partial\tau_1} + \frac{\delta}{m}\left[\left(u_{n+1}^{(0)} - u_n^{(0)}\right)^2 - \left(u_n^{(0)} - u_{n-1}^{(0)}\right)^2\right] \end{aligned} \right. \quad (4.32a)$$

$$O(\varepsilon^1): \left\{ \begin{aligned} &\frac{\partial^2 v_n^{(1)}}{\partial\tau_0^2} - \frac{\Omega^2}{4}\left(v_{n+1}^{(1)} - 2v_n^{(1)} + v_{n-1}^{(1)}\right) + \Omega'^2\left(v_n^{(1)} - u_n^{(1)}\right) \\ &= -2\frac{\partial^2 v_n^{(0)}}{\partial\tau_0\partial\tau_1} + \frac{1}{m}\left[\left(v_{n+1}^{(0)} - v_n^{(0)}\right)^2 - \left(v_n^{(0)} - v_{n-1}^{(0)}\right)^2\right] \end{aligned} \right. \quad (4.32b)$$

$$O(\varepsilon^2): \left\{ \begin{aligned} &\frac{\partial^2 u_n^{(2)}}{\partial\tau_0^2} - \frac{\Omega^2}{4}\left(u_{n+1}^{(2)} - 2u_n^{(2)} + u_{n-1}^{(2)}\right) - \Omega'^2\left(v_n^{(2)} - u_n^{(2)}\right) \\ &= -2\frac{\partial^2 u_n^{(1)}}{\partial\tau_0\partial\tau_1} - 2\frac{\partial^2 u_n^{(0)}}{\partial\tau_0\partial\tau_2} - 2\frac{\partial^2 u_n^{(0)}}{\partial\tau_1^2} + \frac{2\delta}{m}\left[\begin{array}{l}\left(u_{n+1}^{(0)} - u_n^{(0)}\right)\left(u_{n+1}^{(1)} - u_n^{(1)}\right) \\ -\left(u_n^{(0)} - u_{n-1}^{(0)}\right)\left(u_n^{(1)} - u_{n-1}^{(1)}\right)\end{array}\right] \end{aligned} \right.$$
(4.33a)

$$O(\varepsilon^2): \left\{ \begin{aligned} &\frac{\partial^2 v_n^{(2)}}{\partial\tau_0^2} - \frac{\Omega^2}{4}\left(v_{n+1}^{(2)} - 2v_n^{(2)} + v_{n-1}^{(2)}\right) + \Omega'^2\left(v_n^{(2)} - u_n^{(2)}\right) \\ &= -2\frac{\partial^2 v_n^{(1)}}{\partial\tau_0\partial\tau_1} - 2\frac{\partial^2 v_n^{(0)}}{\partial\tau_0\partial\tau_2} - 2\frac{\partial^2 v_n^{(0)}}{\partial\tau_1^2} + \frac{2}{m}\left[\begin{array}{l}\left(v_{n+1}^{(0)} - v_n^{(0)}\right)\left(v_{n+1}^{(1)} - v_n^{(1)}\right) \\ -\left(v_n^{(0)} - v_{n-1}^{(0)}\right)\left(v_n^{(1)} - v_{n-1}^{(1)}\right)\end{array}\right] \end{aligned} \right.$$
(4.33b)

In the course of the calculations, general solutions to (4.31a, 4.31b) are first determined and then inserted into (4.32a, 4.32b), to resolve for $u_n^{(1)}$and $v_n^{(1)}$. Subsequently, solutions to (4.31a, 4.31b) and (4.32a, 4.32b) allow us to solve (4.33a, 4.33b).

4.3.3 Self-Interaction

We first address the self interaction of a vibrational mode (the effect of the lattice deformation on itself).

4.3.3.1 Zeroth Order Solutions

One notes that (4.31a, 4.31b) are strictly equivalent to the equations of motion when chains 1 and 2 are linear. The solutions of these equations can be written in the usual form

$$\begin{cases} u_n^{(0)} = A_0 e^{ikna} e^{-i\omega\tau_0} \\ v_n^{(0)} = B_0 e^{ikna} e^{-i\omega\tau_0} \end{cases} \tag{4.34}$$

where A_0 and B_0 are constants to be determined with the initial conditions, k is the wave-number of the propagation mode, ω, its angular frequency and a, the lattice parameter. Inserting these solutions into (4.31a, 4.31b) leads to the following matrix equation:

$$\begin{pmatrix} \omega^2 - \Omega^2 sin^2\left(\frac{ka}{2}\right) - \Omega'^2 & \Omega'^2 \\ \Omega'^2 & \omega^2 - \Omega^2 sin^2\left(\frac{ka}{2}\right) - \Omega'^2 \end{pmatrix} \begin{pmatrix} A_0 \\ B_0 \end{pmatrix} = \begin{pmatrix} 0 \\ 0 \end{pmatrix}. \tag{4.35}$$

Equation (4.35) admits two non-trivial solutions when the determinant of the matrix vanishes. These solutions are

$$\omega = \omega_0^- = \Omega \left| sin\left(\frac{ka}{2}\right) \right| \tag{4.36}$$

and

$$\omega = \omega_0^+ = \sqrt{\Omega^2 sin^2\left(\frac{ka}{2}\right) + 2\Omega'^2}. \tag{4.37}$$

These dispersion relations are periodic in wave number, k, with a period of $\frac{2\pi}{a}$ and are represented graphically in the band structure of Fig. 4.10.

The lower band passing by $k = 0$ in Fig. 4.10 corresponds to the lower frequency solution *i.e.*, ω_0^-. In this case the two chains vibrate in phase ($A_0 = B_0$) and do not interact with each other resulting in the dispersion relation of an isolated atomic chain. The upper band in Fig. 4.10 represents the higher frequency solution *i.e.*, ω_0^+

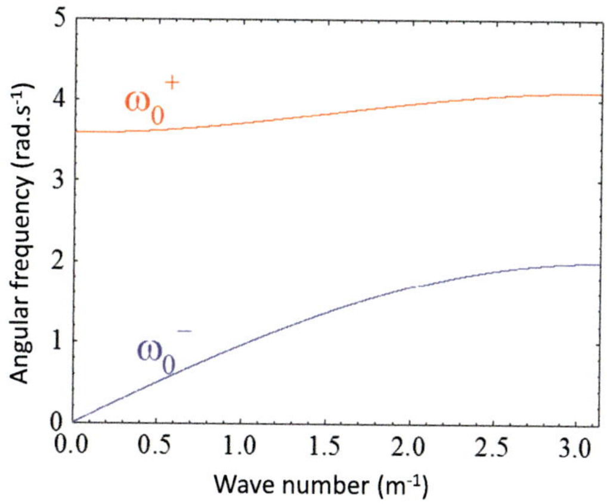

Fig. 4.10 Band structure associated with the zeroth order solutions. Parameters a, m, β and β/ (see text for definitions) were chosen to be a = 1 m, m = 1 kg, $\beta = 1\,\text{N}\cdot\text{m}^{-1}$, $\beta\prime = 6.4\,\text{N}\cdot\text{m}^{-1}$

where the two chains vibrate out of phase ($A_0 = -B_0$). These are the well-known symmetric and antisymmetric modes.

One may write a general solution of (4.31a, 4.31b) as a superposition of modes corresponding to the lower and upper branches of the band structure:

$$\begin{aligned} u_{n,G}^{(0)} &= A_0^+(\tau_1,\tau_2)e^{ikna}e^{-i\omega_0^+\tau_0} + \overline{A_0^+}(\tau_1,\tau_2)e^{-ikna}e^{i\omega_0^+\tau_0} \\ &\quad + A_0^-(\tau_1,\tau_2)e^{ikna}e^{-i\omega_0^-\tau_0} + \overline{A_0^-}(\tau_1,\tau_2)e^{-ikna}e^{+i\omega_0^-\tau_0} \end{aligned} \tag{4.38}$$

$$\begin{aligned} v_{n,G}^{(0)} &= -A_0^+(\tau_1,\tau_2)e^{ikna}e^{-i\omega_0^+\tau_0} - \overline{A_0^+}(\tau_1,\tau_2)e^{-ikna}e^{i\omega_0^+\tau_0} \\ &\quad + A_0^-(\tau_1,\tau_2)e^{ikna}e^{-i\omega_0^-\tau_0} + \overline{A_0^-}(\tau_1,\tau_2)e^{-ikna}e^{+i\omega_0^-\tau_0} \end{aligned} \tag{4.39}$$

where

$$\begin{aligned} A_0^+(\tau_1,\tau_2) &= \alpha_0^+(\tau_1,\tau_2)e^{-i\varphi_0^+(\tau_1,\tau_2)} \\ A_0^-(\tau_1,\tau_2) &= \alpha_0^-(\tau_1,\tau_2)e^{-i\varphi_0^-(\tau_1,\tau_2)}. \end{aligned} \tag{4.40}$$

$A_0^\pm(\tau_1,\tau_2)$ are complex quantities that permit slow time evolution of amplitude and phase and $\alpha_0^\pm(\tau_1,\tau_2)$ and $\varphi_0^\pm(\tau_1,\tau_2)$ are real-valued functions. In (4.38) and (4.39), $\bar{X}$ means the complex conjuguate of X.

4.3.3.2 First Order Solutions

Equations (4.38) and (4.39) are now inserted in the ε^1-order equation (4.32a, 4.32b) to obtain the general solutions for $u_n^{(1)}$ and $v_n^{(1)}$. However, one observes that the right-hand side of (4.32a) only depends on $u_n^{(0)}$ and includes a second order derivative of $u_n^{(0)}$ with respect to τ_0 and τ_1. Substituting (4.38) in (4.32a), the right-hand side of (4.32a) exhibits terms of the forms $i\omega_0^+\frac{\partial A_0^+(\tau_1,\tau_2)}{\partial\tau_1}e^{-i\omega_0^+\tau_0}$, $i\omega_0^-\frac{\partial A_0^-(\tau_1,\tau_2)}{\partial\tau_1}$

$e^{-i\omega_0^- \tau_0}$ and their complex conjugates. It is assumed that the solution to the homogeneous equation of (4.32a) takes similar form to the general solution of (4.31a, 4.31b). Under this assumption and to ensure that there are no secular terms (*i.e.*, terms that grow with time and that are incompatible with the assumption that $u_n^{(1)}$ must be a correction to $u_n^{(0)}$) in the particular solution of (4.32a), the pre-factors of $e^{\pm i\omega_0^+ \tau_0}$ and $e^{\pm i\omega_0^- \tau_0}$ in the right-hand side of (4.32a) are forced to be zero. Subsequently, the derivatives of the amplitudes A_0^+ and A_0^- with respect to τ_1 must vanish and these amplitudes only depend on τ_2. The same result should be obtained by considering the first order equation in ε for displacement v_n *i.e.*, (4.32b). The general solutions of (4.32a, 4.32b) are the sums of homogeneous and particular solutions. Equations (4.32a, 4.32b) have homogeneous solutions that look like those of (4.38) and (4.39) where the amplitudes $A_0^{\mp}(\tau_2)$ are replaced by $A_1^{\mp}(\tau_2)$. The second terms in the right-hand sides of (4.32a, 4.32b) constitute forcing functions at the frequencies $2\omega_0^{\pm}$ and $\left(\omega_0^+ + \omega_0\right)$. We seek particular solutions in the form

$$u_{n,P}^{(1)} = D_1 e^{2ikna} e^{-2i\omega_0^+ \tau_0} + D'_1 e^{-2ikna} e^{2i\omega_0^+ \tau_0} + E_1 e^{2ikna} e^{-2i\omega_0^- \tau_0} + E'_1 e^{-2ikna} e^{2i\omega_0^- \tau_0} + F_1 e^{2ikna} e^{-i\left(\omega_0^+ + \omega_0^-\right)\tau_0} + F'_1 e^{-2ikna} e^{i\left(\omega_0^+ + \omega_0^-\right)\tau_0}, \tag{4.41}$$

$$v_{n,P}^{(1)} = I_1 e^{2ikna} e^{-2i\omega_0^+ \tau_0} + I'_1 e^{-2ikna} e^{2i\omega_0^+ \tau_0} + J_1 e^{2ikna} e^{-2i\omega_0^- \tau_0} + J'_1 e^{-2ikna} e^{2i\omega_0^- \tau_0} + K_1 e^{2ikna} e^{-i\left(\omega_0^+ + \omega_0^-\right)\tau_0} + K'_1 e^{-2ikna} e^{i\left(\omega_0^+ + \omega_0^-\right)\tau_0}. \tag{4.42}$$

We find

$$D_1 = \frac{-2i(sin\,(2ka) - 2sin\,(ka))}{m}.\Phi_1\left(2\omega_0^+\right).{A_0^+}^2 \tag{4.43}$$

$$D'_1 = \frac{2i(sin\,(2ka) - 2sin\,(ka))}{m}.\Phi_1\left(2\omega_0^+\right).\overline{A_0^+}^2 \tag{4.44}$$

$$E_1 = \frac{-2i(sin\,(2ka) - 2sin\,(ka))}{m}.\Phi_1\left(2\omega_0^-\right).{A_0^-}^2 \tag{4.45}$$

$$E'_1 = \frac{2i(sin\,(2ka) - 2sin\,(ka))}{m}.\Phi_1\left(2\omega_0^-\right).\overline{A_0^-}^2 \tag{4.46}$$

$$F_1 = \frac{-4i(sin\,(2ka) - 2sin\,(ka))}{m}.\Phi_3\left(\omega_0^+ + \omega_0^-\right).A_0^+.A_0^- \tag{4.47}$$

$$F'_1 = \frac{4i(sin\,(2ka) - 2sin\,(ka))}{m}.\Phi_3\left(\omega_0^+ + \omega_0^-\right).\overline{A_0^+}.\overline{A_0^-} \tag{4.48}$$

$$I_1 = \frac{-2i(sin\,(2ka) - 2sin\,(ka))}{m}.\Phi_2\left(2\omega_0^+\right).A_0^{+2} \tag{4.49}$$

$$I'_1 = \frac{2i(sin\,(2ka) - 2sin\,(ka))}{m}.\Phi_2\left(2\omega_0^+\right).\overline{A_0^+}^2 \tag{4.50}$$

$$J_1 = \frac{-2i(sin\,(2ka) - 2sin\,(ka))}{m}.\Phi_2\left(2\omega_0^-\right).A_0^{-2} \tag{4.51}$$

$$J'_1 = \frac{2i(sin\,(2ka) - 2sin\,(ka))}{m}.\Phi_2\left(2\omega_0^-\right).\overline{A_0^-}^2 \tag{4.52}$$

$$K_1 = \frac{4i(sin\,(2ka) - 2sin\,(ka))}{m}.\Phi_4\left(\omega_0^+ + \omega_0^-\right).A_0^+A_0^- \tag{4.53}$$

$$K'_1 = \frac{-4i(sin\,(2ka) - 2sin\,(ka))}{m}.\Phi_4\left(\omega_0^+ + \omega_0^-\right).\overline{A_0^+}.\overline{A_0^-} \tag{4.54}$$

where

$$\Phi_1(\omega) = \frac{\delta\left(\omega^2 - \Omega^2\sin^2(\mathrm{ka}) - \Omega'^2\right) - \Omega'^2}{\left(\omega^2 - \Omega^2\sin^2(\mathrm{ka}) - 2\Omega'^2\right)\left(\omega^2 - \Omega^2\sin^2(\mathrm{ka})\right)} \tag{4.55}$$

$$\Phi_2(\omega) = \frac{\left(\omega^2 - \Omega^2\sin^2(\mathrm{ka}) - \Omega'^2\right) - \delta\Omega'^2}{\left(\omega^2 - \Omega^2\sin^2(\mathrm{ka}) - 2\Omega'^2\right)\left(\omega^2 - \Omega^2\sin^2(\mathrm{ka})\right)} \tag{4.56}$$

$$\Phi_3(\omega) = \frac{\delta\left(\omega^2 - \Omega^2\sin^2(\mathrm{ka}) - \Omega'^2\right) + \Omega'^2}{\left(\omega^2 - \Omega^2\sin^2(\mathrm{ka}) - 2\Omega'^2\right)\left(\omega^2 - \Omega^2\sin^2(\mathrm{ka})\right)} \tag{4.57}$$

$$\Phi_4(\omega) = \frac{\left(\omega^2 - \Omega^2\sin^2(\mathrm{ka}) - \Omega'^2\right) + \delta\Omega'^2}{\left(\omega^2 - \Omega^2\sin^2(\mathrm{ka}) - 2\Omega'^2\right)\left(\omega^2 - \Omega^2\sin^2(\mathrm{ka})\right)}. \tag{4.58}$$

The functions Φ_1, Φ_2, Φ_3, and Φ_4 may diverge and contain information about the resonant processes that underlie the nonlinear modes of our system.

4.3.3.3 Second Order Solutions

The first order solutions are now inserted into (4.33a, 4.33b). Since zeroth and first order solutions are independent of τ_1, the first and third terms in the right-hand side of (4.33a, 4.33b) vanish. Some of the terms in the square brackets, lead to forcing functions in $e^{\pm i\omega_0^{\pm}}$. This results in secular terms that need to be canceled by the second term in the right-hand side of (4.33a, 4.33b). After several algebraic steps, we find that α_0^+ and α_0^- are real and that $\varphi_0^-(\tau_2)$ and $\varphi_0^+(\tau_2)$ are linear functions of τ_2. Equation (4.33a) leads to corrections to the frequencies of the zeroth order lower and upper branches:

$$\omega^+ = \omega_0^+ + \frac{4\delta(\sin(2ka) - 2\sin(ka))^2}{\omega_0^+ m^2}\varepsilon^2\left[\alpha_0^{+^2}\Phi_1(2\omega_0^+) + 2\alpha_0^{-^2}\Phi_3(\omega_0^+ + \omega_0^-)\right] \tag{4.59}$$

$$\omega^- = \omega_0^- + \frac{4\delta(\sin(2ka) - 2\sin(ka))^2}{\omega_0^- m^2}\varepsilon^2\left[\alpha_0^{-^2}\Phi_1(2\omega_0^-) + 2\alpha_0^{+^2}\Phi_3(\omega_0^+ + \omega_0^-)\right] \tag{4.60}$$

and (4.33b) leads to the corrections:

$$\omega^+ = \omega_0^+ + \frac{4(\sin(2ka) - 2\sin(ka))^2}{\omega_0^+ m^2}\varepsilon^2\left[\alpha_0^{+^2}\Phi_2(2\omega_0^+) + 2\alpha_0^{-^2}\Phi_4(\omega_0^+ + \omega_0^-)\right] \tag{4.61}$$

$$\omega^- = \omega_0^- + \frac{4(\sin(2ka) - 2\sin(ka))^2}{\omega_0^- m^2}\varepsilon^2\left[\alpha_0^{-^2}\Phi_2(2\omega_0^-) + 2\alpha_0^{+^2}\Phi_4(\omega_0^+ + \omega_0^-)\right] \tag{4.62}$$

In the case of the nonlinear-nonlinear two chain model, if $\delta=1$, $\Phi_1 = \Phi_2$, and $\Phi_3 = \Phi_4$ then (4.59) and (4.60) are identical to (4.61) and (4.62). These functions do not exhibit divergences that are the signature of resonances. In the case of the linear-nonlinear model, $\delta = 0$, the set of (4.59) and (4.60) does not lead to corrections to the band structure, however (4.61) and (4.62) do result in second-order frequency shifts. The functions $\Phi_2(2\omega_0^+)$ and $\Phi_4(\omega_0^+ + \omega_0^-)$ do not diverge. A resonance appears due to the divergence of the function $\Phi_2(2\omega_0^-)$ at some non-vanishing specific value of the wave number $k = k_c$. Therefore, the self-interaction results in a resonance for $k = k_c$. The condition of resonance is

$$4\omega_0^{-^2}(k_c) - \Omega^2\sin^2(k_c a) - 2\Omega'^2 = 0 \tag{4.63}$$

which is equivalent to

$$4\omega_0^{-^2}(k_c) - \omega_0^{+^2}(2k_c) = 0. \tag{4.64}$$

From (4.63) and (4.36), one obtains k_c as

$$k_c = \pm \frac{2}{a} arcsin\left[\left(\frac{\Omega'^2}{2\Omega^2}\right)^{\frac{1}{4}}\right], \tag{4.65}$$

and one notes that k_c depends on the ratio $\frac{\Omega'}{\Omega}$.

In Fig. 4.11, we plot the corrected frequencies in the case $\delta = 0$ (the linear-nonlinear system). Resonance occurs between two modes in the lower band and one mode in the upper band. We note that if $k_c > 0$ and $2k_c$ lies outside the first Brillouin zone of the system, then the upper branch phonon associated with the resonance will be located at $\frac{2\pi}{a} - 2k_c$. If $k_c < 0$ and $2k_c$ is outside the first Brillouin zone, then the upper branch resonant phonon occurs at $2k_c - \frac{2\pi}{a}$. Figure 4.11 illustrates the effect of the magnitude of the spring constant coupling the chains. Even though these coupling springs are linear, they influence the location of the self-interaction resonance. As the strength of the coupling spring constant increases the value of the critical wavenumber, k_c, in the positive half of the first Brillouin zone varies from the origin to $\frac{\pi}{a}$. We also note that as the wavenumber approaches k_c from below, the frequency of the lower branch dips dramatically. The frequency ω^- increases as one approaches k_c from above. It can be noted that in the case of identical nonlinear chains (δ=1), the left-hand side of (4.63) appears both in the numerator and the denominator of functions Φ_1 and Φ_2. Therefore, the singularity disappears and no resonance effect is observed in the band structure.

In summary, when one chain is linear and the other is nonlinear, the two-chain model exhibits a nonlinear resonance near a critical wave number due to mode self-interaction. The existence of such resonance is clearly indicated in (4.62). This resonance occurs between two modes in the lower band (passing by $k = 0$) and one mode in the upper band in the band structure of the two coupled chains.

4.3.3.4 Energy Transfer in the Linear-Nonlinear Two-Chain Model

We now investigate the possibility of energy transfer in the linear-nonlinear two-chain model. The total energy is expressed as the sum of the quadratic kinetic energy (T) and quadratic potential energy of each chain (U), the quadratic energy associated with the coupling between chains (U_c) and a source of nonlinearity taking the form of a cubic term for one of the chains proportional to $\varepsilon(U_{NL})$:

$$\begin{aligned} E = T + U + U_c + U_{NL} = \frac{1}{2} m\Sigma_n\left[\dot{u}_n\bar{\dot{u}}_n + \dot{v}_n\bar{\dot{v}}_n\right] + \frac{1}{2}\beta\Sigma_n(u_{n+1} - u_n)(\bar{u}_{n+1} - \bar{u}_n) \\ + \frac{1}{2}\beta\Sigma_n(v_{n+1} - v_n)(\bar{v}_{n+1} - \bar{v}_n) + \frac{1}{2}\beta'\Sigma_n(u_n - v_n)\times(\bar{u}_n - \bar{v}_n) \\ + \frac{1}{3}\varepsilon\Sigma_n|v_{n+1} - v_n|^2(v_{n+1} - v_n) \end{aligned} \tag{4.66}$$

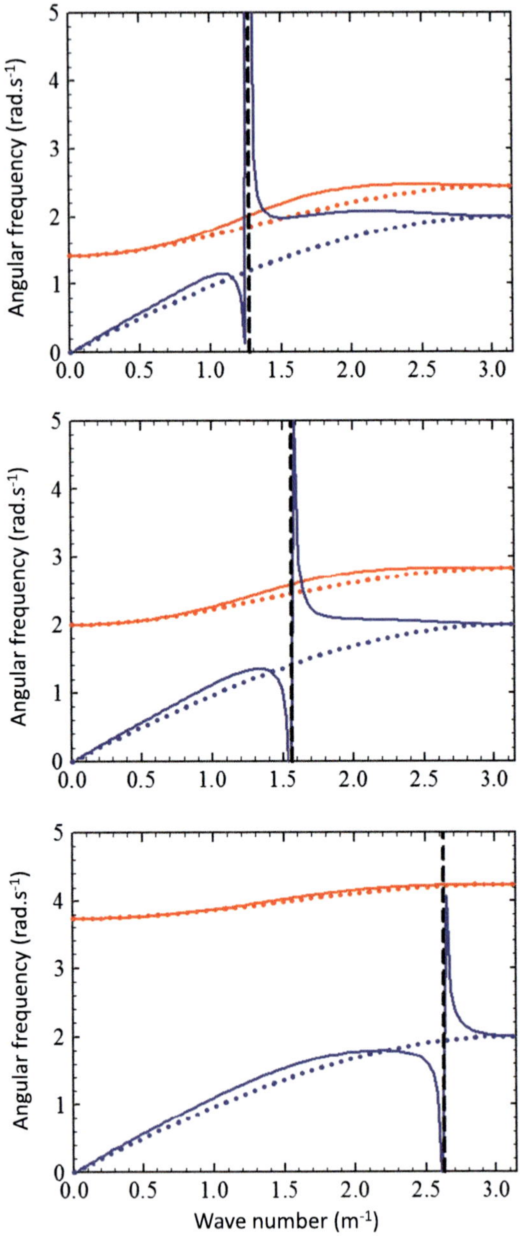

Fig. 4.11 $\omega^{+}(k)$ and $\omega^{-}(k)$ as functions of wavenumber for three values of the spring constant β′, namely 1 N·m^{-1} (*top*), 2 N·m^{-1} (*middle*), and 7 N·m^{-1} (*bottom*). The *dotted lines* correspond to the linear-linear two chain model. The system parameters are a = 1 m, m = 1 kg, β = 1 N·m^{-1}, ε = 1.6 N·m^{-2}, $\alpha_0^{+} = \alpha_0^{-} = 0.17$ m. The *black dashed line* indicates the value of the critical wave number: $k_c = 1.27$ m^{-1} (*top*), 1.57 m^{-1} (*middle*), 2.63 m^{-1} (*bottom*)

While the calculation of the total energy is out of reach, we can estimate the contribution of the resonant zeroth order displacements to the energy of the system. For this, we insert the polynomial series given by (4.29) and (4.30) into (4.66) and neglect all terms in ε, ε^2 and higher powers of ε which include first- and higher-order displacements. We obtain an expression that involves only the zeroth order solutions: $u_n^{(0)}(t)$ and $v_n^{(0)}(t)$. However to account for the nonlinear behavior of the model, we employ a superposition of zeroth order displacements in the lower and upper dispersion branches with the second-order corrected frequency of (4.61) and (4.62):

$$u_n^{(0)}(t) = \Sigma_k\Big\{C^+(k)e^{ikna}e^{-i\omega^+t} + \overline{C^+}(k)e^{-ikna}e^{i\omega^+t} + C^-(k)e^{ikna}e^{-i\omega^-t} + \overline{C^-}(k)e^{-ikna}e^{+i\omega^-t}\Big\} \tag{4.67}$$

$$v_n^{(0)}(t) = \Sigma_k\Big\{-C^+(k)e^{ikna}e^{-i\omega^+t} - \overline{C^+}(k)e^{-ikna}e^{i\omega^+t} + C^-(k)e^{ikna}e^{-i\omega^-t} + \overline{C^-}(k)e^{-ikna}e^{+i\omega^-t}\Big\}. \tag{4.68}$$

In this form the effect of the self-interaction is embedded in the frequencies $\omega^+(k)$ and $\omega^-(k)$.

After numerous algebraic steps we obtain:

$$T = \Sigma_k T(k) = \Sigma_k m(\omega^-)^2|C^-(k)|^2(2-2\cos(2\omega^-t)) + m(\omega^+)^2|C^+(k)|^2(2-2\cos(2\omega^+t)), \tag{4.69}$$

$$U = \Sigma_k U(k) = \Sigma_k \beta 4\left(\sin\frac{ka}{2}\right)^2\Big[|C^-(k)|^2(2+2\cos(2\omega^-t)) + |C^+(k)|^2(2+2\cos(2\omega^+t))\Big] \tag{4.70}$$

and

$$U_c = \Sigma_k U_c(k) = \Sigma_k 2\beta'|C^+(k)|^2(2+2\cos(2\omega^+t) \tag{4.71}$$

and

$$U_{NL} = \frac{1}{3}\varepsilon\Sigma_n\left|v_{n+1}^{(0)} - v_n^{(0)}\right|^2\left(v_{n+1}^{(0)} - v_n^{(0)}\right). \tag{4.72}$$

We verify that when $\omega^\pm = \omega_0^\pm$ and $\varepsilon = 0$, *i.e.*, in the limit of the linear system, the energy may be written as:

$$E_0 = E_0^- + E_0^+ \tag{4.73}$$

with the contribution from the linear lower and upper branches to the energy given by

$$E_0^- = \Sigma_k E_0^-(k) = \Sigma_k\, 16\beta |C^-(k)|^2 sin^2\left(\frac{ka}{2}\right) \quad (4.74)$$

$$E_0^+ = \Sigma_k E_0^+(k) = \Sigma_k |C^+(k)|^2 \left[16\beta sin^2\left(\frac{ka}{2}\right) + 4\beta'\right] \quad (4.75)$$

The energy is independent of time as expected.

Equation (4.69) includes the nonlinear resonance through the terms $(\omega^-)^2$ and $(\omega^+)^2$ in the kinetic energy. Although the nonlinear energy U_{NL} is first order in ε and the kinetic energy incorporates only second order powers of ε, the resonant form of the frequency in (4.69) enables us to neglect the non-resonant zeroth-order nonlinear term given by (4.72). Under this condition, for a given wavenumber, the contribution of the zeroth-order solution with nonlinear correction due to self-interaction is simplified to:

$$E(k) = T(k) + U(k) + U_c(k) = E^-(k) + E^+(k). \quad (4.76)$$

We regroup the terms into the respective contributions from the lower and upper branches $E^-(k)$ and $E^+(k)$. These contributions are functions of time, hence we take the time average of (4.51). We plot in Fig. 4.12, the normalized functions

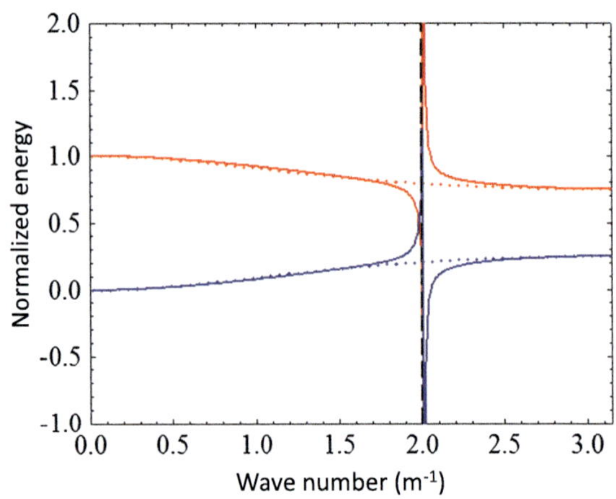

Fig. 4.12 Contributions of the *lower* and *upper* dispersion branches to the energy of the linear two-chain model (*dashed lines*), $E_0^-(k)$ (*blue*), $E_0^+(k)$ (*red*) and the linear-nonlinear model (*solid lines*), $E^-(k)$ (*blue*), $E^+(k)$ (*red*) normalized to their respective total energy, $E_0(k)$ and $E(k)$. The system parameters are a = 1 m, m = 1 kg, β = 1 N·m^{-1}, β′ = 4 N·m^{-1}, ε = 1.6 N·m^{-2}, and $\alpha_0^+ = \alpha_0^- = 0.17$ m. The *black dashed line* indicates the value of the critical wave number: $k_c = 2.0$ m^{-1}

$E_0^-(k)$, $E_0^+(k)$, $E^-(k)$ and $E^+(k)$. Approaching the critical wave number, k_c, from below, and keeping total energy constant, energy is transferred from the upper branch to the lower branch. The inverse phenomenon is observed when approaching k_c from above. We note that in the case of the nonlinear/nonlinear two chain model, this inter-band transfer of energy resulting from the self-interaction resonant phenomenon would not occur. One needs the coupling between a linear chain and a nonlinear chain to achieve the selective exchange of energy between the symmetric (lower band) and antisymmetric (upper band) modes of the two-chain system.

4.3.4 *Three Wave Interactions*

Now we investigate the interactions between three phonons with general wave numbers k, k', and k^*. We focus on the linear-nonlinear two chains model (δ=0) for which three phonons resonant processes have been identified through the self-interaction analysis. Specifically we are interested in identifying nonlinear modes which arise from three phonon interactions. For this, we need to insert a superposition of zeroth order solutions into (4.32a, 4.32b). Again, we assume that all solutions are independent of τ_1. The time derivatives on the right-hand side of (4.32a, 4.32b) vanish. For the linear-nonlinear two chains system, (4.32a, 4.32b) take the simpler form

$$\begin{cases} \dfrac{\partial^2 u_n^{(1)}}{\partial \tau_0^2} - \dfrac{\Omega^2}{4}\left(u_{n+1}^{(1)} - 2u_n^{(1)} + u_{n-1}^{(1)}\right) - \Omega'^2\left(v_n^{(1)} - u_n^{(1)}\right) \\ \qquad = 0 \\ \dfrac{\partial^2 v_n^{(1)}}{\partial \tau_0^2} - \dfrac{\Omega^2}{4}\left(v_{n+1}^{(1)} - 2v_n^{(1)} + v_{n-1}^{(1)}\right) + \Omega'^2\left(v_n^{(1)} - u_n^{(1)}\right) \\ \qquad = \dfrac{1}{m}\left[\left(v_{n+1}^{(0)} - v_n^{(0)}\right)^2 - \left(v_n^{(0)} - v_{n-1}^{(0)}\right)^2\right] \end{cases}. \tag{4.77}$$

Using the superposition of zeroth order solutions:

$$v_{n,G}^{(0)} = \Sigma_k\Big\{-A_0^+(\tau_2)e^{ikna}e^{-i\omega_0^+\tau_0} - \overline{A_0^+}(\tau_2)e^{-ikna}e^{i\omega_0^+\tau_0} + A_0^-(\tau_2)e^{ikna}e^{-i\omega_0^-\tau_0}$$
$$+\overline{A_0^-}(\tau_2)e^{-ikna}e^{+i\omega_0^-\tau_0}\Big\} \tag{4.78}$$

in the right-hand side of the equation above, produces intra-band and inter-band wave mixing. In light of the origin of the self-interaction resonances, we pay particular attention to the forcing terms corresponding to the mixing of two modes in the lower band. These are terms proportional to $e^{\pm i\left(\omega_0^-(k)+\omega_0^-(k')\right)\tau_0}$. We choose first-order solutions in the form

$$u_n^{(1)} = \Sigma_k A_1(k, \tau_0, \tau_2)e^{ikna} \tag{4.79}$$

$$v_n^{(1)} = \Sigma_k B_1(k, \tau_0, \tau_2)e^{ikna} \tag{4.80}$$

Equation (4.77) are multiplied by e^{ik^*na}, and a summation over all n masses is imposed. Equations (4.77) reduce to

$$\frac{\partial^2 A_1(k^*,\tau_0,\tau_2)}{\partial\tau_0^2}+\Omega^2 A_1(k^*,\tau_0,\tau_2)sin^2\left(\frac{k^*a}{2}\right)-\Omega'^2[B_1(k^*,\tau_0,\tau_2)-A_1(k^*,\tau_0,\tau_2)]=0 \tag{4.81}$$

$$\frac{\partial^2 B_1(k^*,\tau_0,\tau_2)}{\partial\tau_0^2}+\Omega^2 B_1(k^*,\tau_0,\tau_2)sin^2\left(\frac{k^*a}{2}\right)-\Omega'^2[A_1(k^*,\tau_0,\tau_2)-B_1(k^*,\tau_0,\tau_2)]$$
$$=\frac{1}{m}\Sigma_k\Sigma_{k'}A_0^-(k)A_0^-(k')e^{-i\left(\omega_0^-(k)+\omega_0^-(k')\right)\tau_0}\delta_{k^*,k+k'}f(k,k')+other\ terms. \tag{4.82}$$

The function f is given by $f(k,k')=\left(e^{ika}-1\right)\left(e^{ik'a}-1\right)-\left(e^{-ika}-1\right)\left(e^{-ik'a}-1\right)$.

The homogeneous solutions of (4.81) and (4.82) are similar to the zeroth-order solutions. Nonlinear effects are sought in the particular solutions by considering the first-order particular amplitudes in (4.79) and (4.80) to follow the forcing term in the right-hand side of (4.82). These are:

$$A_{1,p}(k^*,\tau_0,\tau_2)=\Sigma_k\Sigma_{k'}U_1(k,k',\tau_2)e^{-i\left(\omega_0^-(k)+\omega_0^-(k')\right)\tau_0}\delta_{k^*,k+k'}f(k,k'), \tag{4.83}$$

$$B_{1,p}(k^*,\tau_0,\tau_2)=\Sigma_k\Sigma_{k'}V_1(k,k',\tau_2)e^{-i\left(\omega_0^-(k)+\omega_0^-(k')\right)\tau_0}\delta_{k^*,k+k'}f(k,k'). \tag{4.84}$$

Combining (4.81) and (4.82), as well as, (4.83) and (4.84) results in the following amplitudes of the particular solutions:

$$U_1(k,k')=\frac{1}{m}\frac{A_0^-(k)A_0^-(k')\Omega'^2}{\left[-\left(\omega_0^-(k)+\omega_0^-(k')\right)^2+\Omega^2\sin^2\left(\frac{k^*a}{2}\right)+\Omega'^2\right]^2-\Omega'^4} \tag{4.85}$$

$$V_1(k,k')=\frac{1}{m}\frac{A_0^-(k)A_0^-(k')\left[-\left(\omega_0^-(k)+\omega_0^-(k')\right)^2+\Omega^2\sin^2\left(\frac{k^*a}{2}\right)+\Omega'^2\right]}{\left[-\left(\omega_0^-(k)+\omega_0^-(k')\right)^2+\Omega^2\sin^2\left(\frac{k^*a}{2}\right)+\Omega'^2\right]^2-\Omega'^4} \tag{4.86}$$

with $k^*=k+k'$. The denominators of (4.85) and (4.86) diverge when:

$$\left(\omega_0^-(k)+\omega_0^-(k')\right)^2-\Omega^2\sin^2\left(\frac{(k+k')a}{2}\right)-2\Omega'^2=0. \tag{4.87}$$

This condition for resonance can be simplified to:

$$\left(\omega_0^-(k) + \omega_0^-(k')\right)^2 - \left(\omega_0^+(k+k')\right)^2 = 0. \tag{4.88}$$

Equation (4.88) generalizes (4.64) beyond self-interaction, indeed the resonant condition of (4.88) is satisfied by considering the self-interaction $k = k' = k_c$. A similar condition also exists for wavenumbers in the negative sector of the Brillouin zone. Equation (4.88) describes a three phonon resonant process whereby, two phonons in the lower branch of the band structure of the two-chain system interact to form a phonon in the upper branch. We may also consider first-order solutions in the vicinity of the resonance. For instance, taking $k' = k_c$ with $k \neq k_c$ will lead to first-order solutions which frequency $\omega_0^-(k) + \omega_0^-(k')$ does not lie on the upper branch of the band structure. The dispersion curve of these nonlinear modes is given by

$$\omega_0^{NL}(k + k_c) = \omega_0^-(k) + \omega_0^-(k_c) \text{ when } 0 < k < \frac{\pi}{a} \tag{4.89}$$

and is displayed in Fig. 4.13. This dispersion curve starts from the linear lower mode at $k_\ell = k_c$ and crosses the upper linear mode at $k_u = \frac{2\pi}{a} - 2k_c$. The branch folds when $k + k_c > \frac{\pi}{a}$. The amplitude of these nonlinear modes given by (4.85) and (4.86) will decrease as k deviates away from the critical wave number. Thus, if $k + k_c > \frac{\pi}{a}$, no mode will exhibit higher amplitude when $k = k_u$ at the crossing of nonlinear mode and linear upper mode branches.

4.3.5 Numerical Results

We employ the method of molecular dynamics (MD) (See Appendix 3C) to calculate the distribution of phonons under conditions of thermal equilibrium in the linear-nonlinear coupled chain system. This approach enables us to investigate this system beyond the limit of perturbation theory. We consider a system composed of two chains, each containing N = 800 masses. The parameters of the system are: $a = 1$ m, $m = 1$ kg, $\beta = 1$ N·m^{-1} and $\varepsilon = 1.6$ N·m^{-2}. Periodic boundary conditions are employed to mimic the behavior of an infinite system. A random distribution of mass displacements is used as initial condition and trajectories of the masses are numerically integrated under constant energy conditions. The velocity of each mass is recorded during the whole length of the simulation which amounts to 2^{21} integration steps ($\Delta t = 2.5$ ms). These velocities are then utilized to calculate the Spectral Energy Density (SED) of phonons in the space (k,ω). In Fig. 4.14, we report the SED results for three values of the elastic constant of the spring coupling the two chains, $\beta' = 5.4$, 6.4, and 7.4 N·m^{-1}. The vibrational modes with the highest SED values (orange color) describe quite well the two bands constituting the band structure of the linear system *i.e.*, $\omega_0^-(k)$ and $\omega_0^+(k)$ (see Fig. 4.10). The signature of the self-interaction resonance appears in the presence of a folded nonlinear

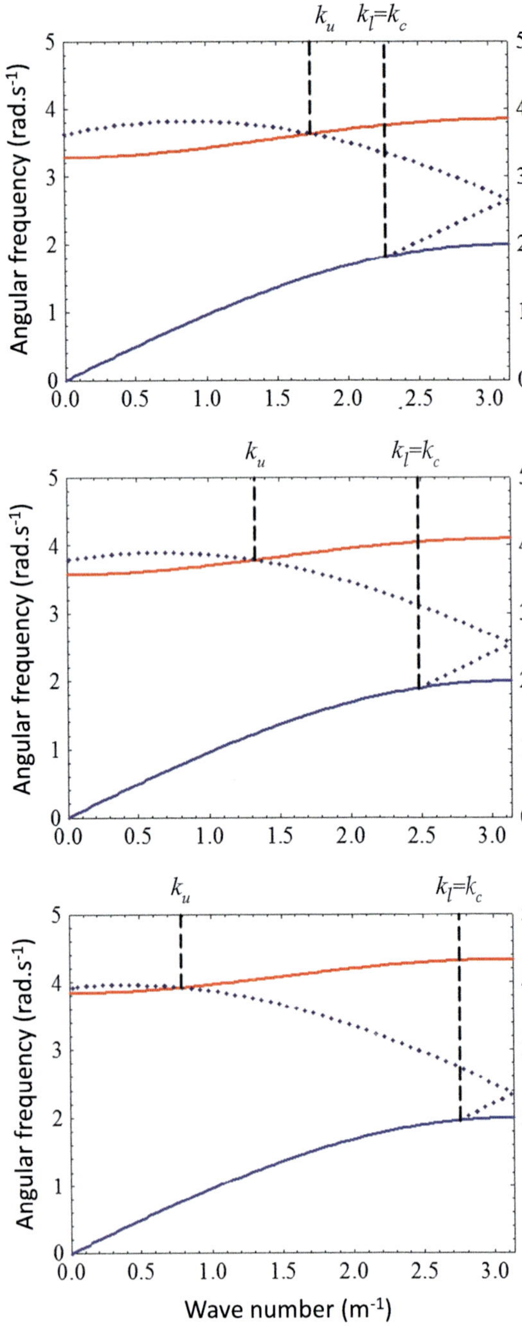

Fig. 4.13 $\omega_0^-(k)$ (*blue solid line*), $\omega_0^+(k)$ (*red solid line*) and $\omega_0^{NL}(k+k_c) = \omega_0^-(k_c) + \omega_0^-(k)$ (*dotted line*) as functions of wave number for three values of the spring constant β', namely 5.4 N·m^{-1} (*top*), 6.4 N·m^{-1} (*middle*) and 7.4 N·m^{-1} (*bottom*). The system parameters are a = 1 m, m = 1 kg and β = 1 N·m^{-1}. $k_\ell = k_c$ and $k_u = \frac{2\pi}{a} - 2k_c$ are the wave numbers at which the nonlinear branch (*dotted line*) intersects the *lower* and the *upper* linear bands. The *black dashed lines* indicate the values of the wave numbers: k_ℓ= 2.27 m^{-1} and k_u= 1.74 m^{-1} (*top*), k_ℓ= 2.48 m^{-1} and k_u= 1.32 m^{-1} (*middle*), k_ℓ= 2.75 m^{-1} and k_u= 0.78 m^{-1} (*bottom*)

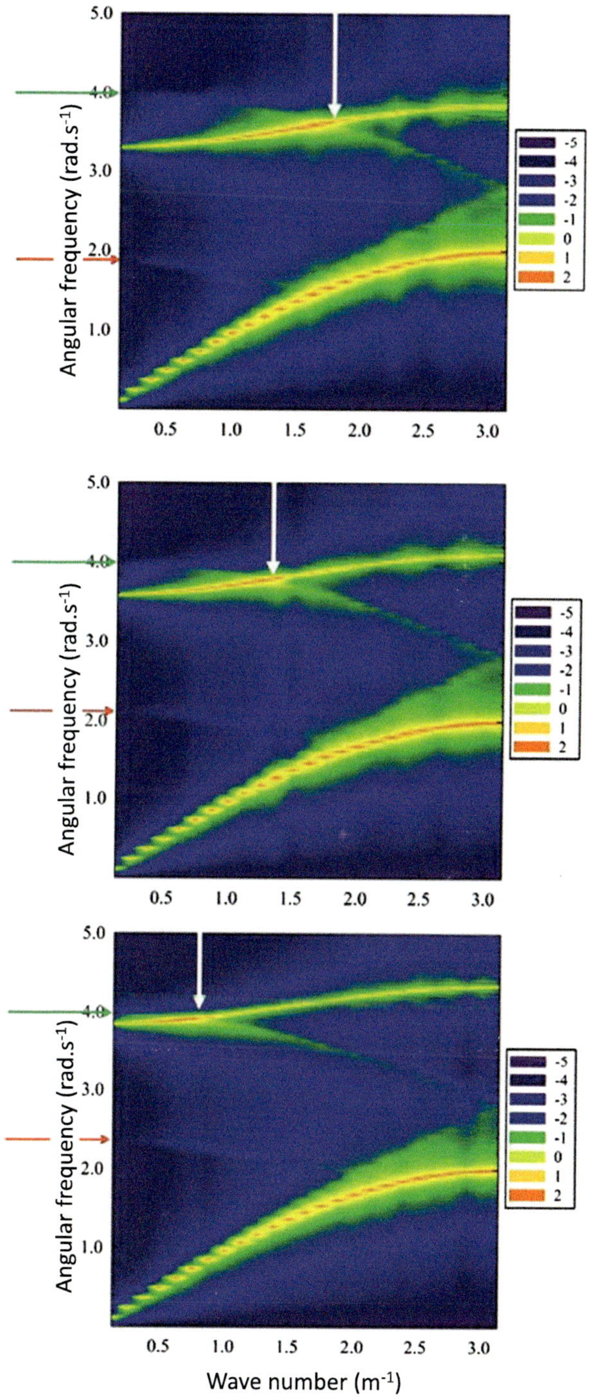

Fig. 4.14 Spectral energy densities (SEDs in J.s) calculated from the velocities of the atoms in the linear-nonlinear two-chain model for three values of the coupling elastic constant $\beta' = 5.4\ \text{N}\cdot\text{m}^{-1}$ (*top*), $6.4\ \text{N}\cdot\text{m}^{-1}$ (*middle*), and $7.4\ \text{N}\cdot\text{m}^{-1}$ (*bottom*). The system parameters are $a = 1$ m, $m = 1$ kg, $\beta = 1\ \text{N}\cdot\text{m}^{-1}$, $\varepsilon = 1.6\ \text{N}\cdot\text{m}^{-2}$, and $\alpha_0^+ = \alpha_0^- = 0.17\cdot\text{m}$. The *color* scale corresponds to $\log_{10}$(SED). The *white arrow* indicates the value of k_u^{SED}: $k_u^{SED} \approx 1.7\ m^{-1}$ (*top*), $k_u^{SED} \approx 1.3\ m^{-1}$ (*middle*), $k_u^{SED} \approx 0.7\ m^{-1}$ (*bottom*). The meaning of the *red* and *green arrows* is given in the text

dispersion branch (dark green color in Fig. 4.14) that connects the upper and the lower dispersion curves. Indeed, this additional branch disappears in the SED calculation of the band structure of the nonlinear-nonlinear two chains system (not reported) where self-interaction resonance does not occur. One notes that the nonlinear band calculated with the SED method has almost the same characteristics as those of the first-order modes in the vicinity of the resonant three-phonon process determined in Sect. 3.4 (see Fig. 4.13). Especially this band after folding at the border of the irreducible Brillouin zone (at $k = \pi/a$) intersects the upper linear band for a wave number k_u^{SED} very close to k_u determined analytically in Sect. 3.4. As predicted analytically k_u^{SED} shifts toward the origin of the band structure as β' increases. Beyond the intersection point other nonlinear modes (shown with the green arrow in Fig. 4.14) with much lower amplitude occur, as expected in Fig. 4.13. Again, the nonlinear band is formed of modes that have wave vector $k + k_c$ and frequencies $\omega_0^-(k) + \omega_0^-(k_c)$. Nevertheless in Fig. 4.14, additional modes appear, especially around the lower linear band. In this range of frequency, it seems that each non-degenerate wave vector is associated with multiple eigen frequencies and the number of these eigen frequency increases when the wave number approaches the edge of the irreducible Brillouin zone. A similar phenomenon was already observed in the framework of the three-phonon scattering processes in the one-dimensional anharmonic monatomic crystals (Sect. 2). Equations (4.85) and (4.86) of Sect. 3.4 are utilized to explain the appearance of these additional modes. These equations represent the first order term in the asymptotic expansion of $A_{1,p}(k^*, \tau_0, \tau_2)$ and $B_{1,p}(k^*, \tau_0, \tau_2)$ describing three-wave interactions. Inside the double summation over (k′, k″) in (4.83) and (4.84), conservation of wave vectors is imposed: $\delta_{k^*, k'+k''} \to k^* = k' + k''$. If the mode of interest is $k^* = \pi/a$, then conservation of wave vector can be satisfied by adding non-degenerate wave vector pairs that yield k^*. However such a combination of wave vectors does not conserve frequency. Indeed, since the dispersion relation for 1D anharmonic monatomic chain is not strictly linear, the frequency of mode k' plus (or minus) the frequency of mode k" will not exactly equal the frequency of mode $k^* = k' + k''$. Instead, the addition (or subtraction) of the frequencies associated with modes k' and k'' will be slightly greater than (or less than) the frequency of mode k^*. This forces the denominator of the pre-exponential factors in (4.83) and (4.84) to become small, thereby contributing to a large value of U_1 and V_1. These additional modes combine only modes of the lower branch and are nearly independent of β′ as it can be observed in Fig. 4.14. As a matter of results, the effect of β′ on the intersection point between the nonlinear band and the lower linear band (named as k_ℓ in Sect. 3.4) cannot be clearly observed because the classical three phonon interactions in the lower band overwhelm the three phonons inter-band nonlinear interaction. For the same reason, it is difficult to define exactly the frequency at which the nonlinear band is folded at the edge of the irreducible Brillouin zone *i.e.*, for $k = \pi/a$. Finally, the red arrows in Fig. 4.14 show vibrational modes starting at the center of the Brillouin zone and exhibiting a negative dispersion. Angular frequencies of these modes slightly depend on β′. The SED of these modes is at least ten times lower than that of the nonlinear modes connecting the lower and upper linear bands and this very low value of the SED indicates that they might correspond to even higher order nonlinear modes.

In summary, the linear-nonlinear two-chain model exhibits nonlinear modes within the spectral gap separating the lower asymmetric and upper symmetric branches of the phonon band structure. These modes result from the interaction between a phonon belonging to the nonlinear branch and two phonons lying on the lower branch. When stimulated at a frequency corresponding to a nonlinear mode, this linear-nonlinear system may therefore be used to split a phonon into a pair of phonons. The two phonons in the lower branch are correlated and have a combined frequency and wave number equal to the frequency and wavenumber of the nonlinear phonon. The amplitude of the two emitted correlated phonons depends on their spectral proximity to a critical wavenumber, k_c corresponding to a self-interaction resonance. These nonlinear modes may serve as a source of two non-separable phonons provided the split phonons are phase-matched in the frequency domain. Such two photon sources in second-order nonlinear photonic materials or even three photon sources in third-order nonlinear materials have been described [39, 40]. The present model illustrates an avenue to achieve similar type of phenomena with phonons instead of photons.

Furthermore, it is the interaction between waves in nonlinear media that can lead to a highly complex behavior such as non-separability of modes. Non-separable modal interactions can be used to create superposition of multiple wave states in the Hilbert space tensor product of the Hilbert spaces of the individual waves that cannot be written as tensor products of the individual waves. Nonlinear, non-separable superposition of phonons may then be used to form states in a Hilbert space displaying exponential complexity. The key to creating exponentially complex superposition of states lies in the controllable selectivity of order (2nd, 3rd, higher order, etc.), and magnitude of the nonlinearity. Such controllability is illustrated in Sects. 3 and 4 and will be used to explore quantum computing analogues in Chap. 6.

4.4 Multi-Phonon Scattering Processes in One-Dimensional Anharmonic Biological Superlattices

Mineralized biological tissues, such as bone and tooth, are hierarchical composite structures composed of a stiff hydroxyapatite (HAP) mineral phase, a compliant proteinaceous collagen phase, and water. At the nanoscale, bone and teeth are constituted of a periodic assembly of alternating regions of collagen and HAP in a hydrated environment [41]. This periodic composite structure, forming a one dimensional (1-D) superlattice, is believed to be responsible for the remarkable strength and toughness of these biological materials [42–44]. At the micron scale, mineralized tissues exhibit a large network of interconnected porosity, tubules in dentin and canaliculi and lacunae in bone, which allow for the transfer of nutrients, waste, and water throughout the tissue [45, 46]. This porosity allows bone and teeth to remain in equilibrium with the water, maintaining hydration of the tissues. Water molecules exhibit a variety of different interactions with the HAP and collagen including the formation of water-bridges within the collagen helix, filling channels within the HAP, and surface hydration of the collagen and HAP phases [47–49]. Three-point bend and notch testing indicate that hydration has a significant impact on the mechanical properties of mineralized tissues resulting in increased elastic moduli as well as decreased toughness and loss of plastic behavior [50–52]. Hydration increases the nonlinear behavior of collagen as well as its elastic modulus while increasing its toughness [53, 54]. The model considered in this section is therefore a one dimensional phononic crystal composed of alternate collagen and hydroxyapatite layers. This superlattice will serve as metaphor for mineralized tissues present in bones and teeth. The collagen is treated as an open system elastic medium which water content can vary depending on the level of stress applied. The open system nature of the collagen-water system leads to a nonlinear stress-strain response. The nonlinearity in the mechanical response of the collagen-water system enables us to investigate a variety of multi-phonon scattering processes.

4.4.1 *Model of Nonlinear Hydrated Collagen*

To address the problem of the mechanical behavior of bone material in the presence of water, we develop the chemo-mechanical equations of states of materials that can adsorb fluids under stress based on the work of Larche and Cahn [55, 56]. The total internal energy of the material is obtained as an integral of an internal Helmholtz energy density f':

$$E = \int_{V'} f'\, dV' \tag{4.90}$$

where the energy density is given by:

$$f' = f'\left(T, \varepsilon, \ldots, c'_i, \ldots\right) \tag{4.91}$$

The prime indicates that all densities are relative to the reference state for measuring strain. T, ε, and c'_I are the temperature, strain and molar density of chemical constituent "I". We consider K variable chemical species in the chosen materials. The differential form of (4.91) is written:

$$\begin{aligned} df'\left(T, \varepsilon_{ij}, c'_I, \ldots\right) = s'\left(T, \varepsilon_{ij}, c'_I, \ldots\right)ds' + \sigma_{ij}\left(T, \varepsilon_{ij}, c'_I, \ldots\right) \\ + \sum M_{I,K}\left(T, \varepsilon_{ij}, c'_I, \ldots\right)dc'_I \end{aligned} \tag{4.92}$$

The functions: $s'(T, \varepsilon_{ij}, c'_I, \ldots)$, $\sigma_{ij}(T, \varepsilon_{ij}, c'_I \ldots)$ and $M_{I,K}(T, \varepsilon_{ij}, c'_I, \ldots)$ are the density of entropy, stress and diffusion potential equations of state. The diffusion potential is used when considering a substitutional solid solution that constrains the molar densities according to: $c'_1 + \ldots + c'_I + \ldots + c'_K = c'_0$ where c'_0 is the density of substitutional sites of the different chemical species. Introducing mole fractions $X_I = \frac{c'_I}{c'_0}$ the diffusion potentials are therefore defined as:

$$\left(\frac{\partial e'}{\partial X_I}\right)_{T,\varepsilon,X_{I\neq K}} = c'_0 M_{I,K} \text{ with } I = 1, K - 1 \tag{4.93}$$

By choosing K as a dependent chemical specie, one may treat the problem with only K-1, independent variables. Let us now simplify the problem to a binary solution *i.e.*, $I=1$ and $K=2$. Note that the substitutional solid solution representation is equivalent to an interstitial solid solution if 1 represents the interstitial specie (*e.g.*, water) and 2 the interstitial sites (*i.e.*, available sites for water in a collagen matrix). Denoting by X the composition in specie 1 and using 2 as the dependent specie, the diffusion potential becomes:

$$M_{12} = \frac{1}{c'_0}\left(\frac{\partial f'}{\partial X_1}\right)_{T,\varepsilon} \tag{4.94}$$

It is convenient to define the free energy density Φ' by the Legendre transformation where strain is replaced by stress as variable.

$$\Phi' = f' - \sigma_{ij}\varepsilon_{ij} \tag{4.95}$$

The differential form of the density Φ' is:

$$d\Phi' = -\varepsilon_{ij} d\sigma_{ij} - s' dT + c'_0 M_{12} dX_1 \tag{4.96}$$

From this relation we deduce the following Maxwell relation

$$-c'_0\left(\frac{\partial M_{12}}{\partial \sigma_{ij}}\right)_{T,X_1} = \left(\frac{\partial \varepsilon_{ij}}{\partial X_1}\right)_{T,\sigma_{kl}} \tag{4.97}$$

In the case of a binary solution which elastic properties depend on the composition, the right hand side of (4.97) has to be written as:

$$\left(\frac{\partial \varepsilon_{ij}}{\partial X_1}\right)_{T,\sigma_{kl}} = \frac{\partial \varepsilon^c_{ij}}{\partial X_1} + \frac{\partial S_{ijkl}}{\partial X_1}\sigma_{kl} \tag{4.98}$$

where we have dropped the subscript in the differentials for the sake of simplifying the notation. ε^c_{ij} are the components of the chemical strain and S_{ijkl} are the components of the compliance tensor. The chemical strain is stress free and is only associated with the expansion or contraction of the material upon a change in composition. To simplify the notation, we take $\eta_{ij} = \partial\varepsilon_{ij}^c/\partial x$ where the linear coefficients η_{ij} are the components of the chemical expansion coefficient tensor. The simplest relationship between the change in composition $(X - X_0)$ and the chemical strain is therefore:

$$\varepsilon_{ij}^c = (X - X_0)\eta_{ij}\delta_{ij} \tag{4.99}$$

δ_{ij} in (4.99) is the Kroenecker symbol and X_0 is the composition of the reference state for measuring strain. To obtain the second term in (4.98), we have used Hooke's law:

$$s_{ij} = C_{ijkl}\left(e_{kl} - e_{kl}^c\right) \tag{4.100}$$

or:

$$\varepsilon_{ij} - \varepsilon_{ij}^c = S_{ijkl}\sigma_{kl} \tag{4.101}$$

C_{ijkl} are the components of the stiffness tensor and the quantity $\varepsilon_{ij}^m = \varepsilon_{ij} - \varepsilon_{ij}^c$ is the mechanical strain.

Inserting (4.98) into (4.97) and after integration, the diffusion potential becomes:

$$c_0'M_{12} = -\frac{\partial\varepsilon_{ij}^c}{\partial X}\sigma_{ij} - \frac{1}{2}\frac{\partial S_{ijkl}}{\partial X}\sigma_{kl}\sigma_{ij} + \varphi(X) \tag{4.102}$$

where $\varphi(X)$ is some unknown function of composition. This unknown function is eliminated by choosing a hydrostatic state of pressure, P, as reference state. Inserting (4.99) into (4.102), the diffusion potential equation of state for the binary solution is now given as:

$$M_{12} = \mu_1(T,P,X_2) - \mu_2(T,P,X_1) - V_0'\left[\eta_{ij}\sigma_{ij} - \eta_{kk}P - \frac{1}{2}\frac{\partial S_{ijkl}}{\partial X_1}\sigma_{kl}\sigma_{ij} + \frac{1}{2}P^2\frac{\partial S_{ijkl}}{\partial X_1}\delta_{kl}\delta_{ij}\right] \tag{4.103}$$

where $V_0' = 1/c_0'$.

In (4.103), we have introduced the chemical potentials of species 1 and 2. Under hydrostatic pressure, the diffusion potential is nothing but a difference in chemical potential. The chemical potentials are defined as: $\mu_i(T,P,X_i) = \frac{1}{c_0'}\left(\frac{\partial f'}{\partial X_I}\right)_{T,\varepsilon,X_{j\neq i}}$

The condition for chemical equilibrium of the binary solid solution in contact with a binary fluid solution is determined by the conservation of the diffusion potential,

$$\begin{aligned} M_{12}(\sigma, x) &= \mu_1 - \mu_2 \\ \mu_1^F &= \mu_1 \\ \mu_2^F &= \mu_2 \end{aligned} \tag{4.104}$$

We now make the temperature dependency implicit and drop, T, from the equations. Subtracting the diffusion potentials of a stresses solid solution, M_{12} and a solid solution under hydrostatic pressure, $\bar{M}_{12}$ yields:

$$M_{12} - \bar{M}_{12} = \mu_1^F - \overline{\mu}_1^F - \left(\mu_2^F - \overline{\mu}_2^F\right) \tag{4.105}$$

$\overline{\mu}_2^F \mu_1^F, \overline{\mu}_1^F$, and μ_2^F are the chemical potentials of the species 1 and 2 in the fluid when the solid is subjected to a stress or to a hydrostatic pressure only, respectively. Provided that the fluid behaves like a chemical reservoir, this is the case if we consider the fluid to be a reservoir of water only, the difference on the right hand side of (4.105) becomes identically equal to 0 and the diffusion potentials of the stressed and unstressed system are equal. Note that the diffusion potential of the stressed solid is evaluated at the composition X while the hydrostatic diffusion potential is the difference in chemical potential of the species 1 and 2 at equilibrium hydrostatic composition X^0. The diffusion potential of the stressed system can be calculated from:

$$M_{12} = \mu_1\left(P, X_l^0\right) - \mu_2\left(P, X_2^0\right) \tag{4.106}$$

Equation (4.106) is sufficient to solve for the change in composition, $X - X^0$, of the solid to maintain equilibrium under stress with the fluid reservoir. For this, we use the equation of state (4.103).

$$\mu_1\left(T, P, X_l^0\right) - \mu_2\left(T, P, X_2^0 = 1 - X_l^0\right) - \mu_1(T, P, X_1) - \mu_2(T, P, X_2 = 1 - X_1)$$
$$- V'_0\left[\eta_{ij}\sigma_{ij} - \eta_{kk}P - \frac{1}{2}\frac{\partial S_{ijkl}}{\partial X_1}\sigma_{kl}\sigma_{ij} + \frac{1}{2}P^2\frac{\partial S_{ijkl}}{\partial X_1}\delta_{kl}\delta_{ij}\right] \tag{4.107}$$

It is clear that (4.107) leads to a composition which is a nonlinear function of stress. Then, inserted into equation of states (4.99) and (4.101), the stress-strain relation becomes nonlinear. To simplify the problem, we assume that the collagen-water binary obeys the prototypical regular solution model. The molar free energy of mixing:

$$f_m = RT(X_1 \ln X_1 + (1 - X_1)\ln(1 - X_1)) + \Omega(1 - X_1)X_1 \tag{4.108}$$

Ω represents the interaction energy between the species 1 (water) and 2 (water sites in collagen). In this case the difference in chemical potential is given by:

$$\mu_1(T,P,X_1) - \mu_2(T,P,X_2 = 1 - X_1) = RT\ln\frac{X_1}{1-X_1} + \Omega(1-2X_1) \quad (4.109)$$

We now assume that the equilibrium mole fraction of water in collagen in absence of stress is $X_1^0 = 0.5$. Molecular dynamics simulations of the interaction between collagen and water indicate that the occupancy of water sites, both internal and external to the collagen helix, is approximately 40–60% [48, 49]. Under this condition, in absence of stress, the difference of chemical potential given by (4.109) is zero. We further assume that the elastic coefficients of collagen are independent of water content. This is justified based on the fact that the speed of sound and the density of polymer-based media and water are very similar. Finally, reducing the problem to a 1-D one, we rewrite (4.109) in the form:

$$RT\ln\frac{X_1}{1-X_1} + \Omega(1-2X_1) = V'_0\eta\sigma \quad (4.110)$$

To obtain (4.110) we have also neglected the hydrostatic pressure P compared to the stress. The chemo-mechanical problem of a collagen solid matrix in equilibrium with a water reservoir then can be solved by eliminating composition between (4.110) and (4.111):

$$\varepsilon = \frac{1}{Y}\sigma + \eta\left(X_1 - X_1^0\right) \quad (4.111)$$

Equation (4.111) is the 1-D version of (4.101) combined with (4.99). There Y is Young's modulus of the collagen with the stress-free water content, $X_1^0 = 0.5$.

Bowman et al. have measured the stress versus strain relationship in demineralized bovine bone under uniaxial tension and open conditions (*i.e.*, in equilibrium with water) [57]. The nonlinear stress-strain curve for tensile stress is extracted from Bowman et al. The part of the curve for the negative values of the strain is the symmetrical of the positive part, as it is in the equations above. Figure 4.15 shows a fit to this data obtained by eliminating composition between (4.110) and (4.111). We found that best fit is achieved for the following conditions: $\eta = 0.12$, $Y = 7.75 \times 10^8$ Pa, $\Omega = 5448.9$ J/mole and $V'_0 = 4.56 \times 10^{-4}$ m^3/mole. These values are in agreement with values of $\eta = 0.09$ and Y = 0.5 – 1 GPa, measured for type 1 collagen [58–60].

From the fit, the Young's modulus $Y(\varepsilon)$ and the energy $E(\varepsilon)$ are written as Taylor series up to the 8th order.

$$Y(\varepsilon) = a_0 + a_2\varepsilon^2 + a_4\varepsilon^4 + a_6\varepsilon^6 + a_8\varepsilon^8 \quad (4.112)$$

and

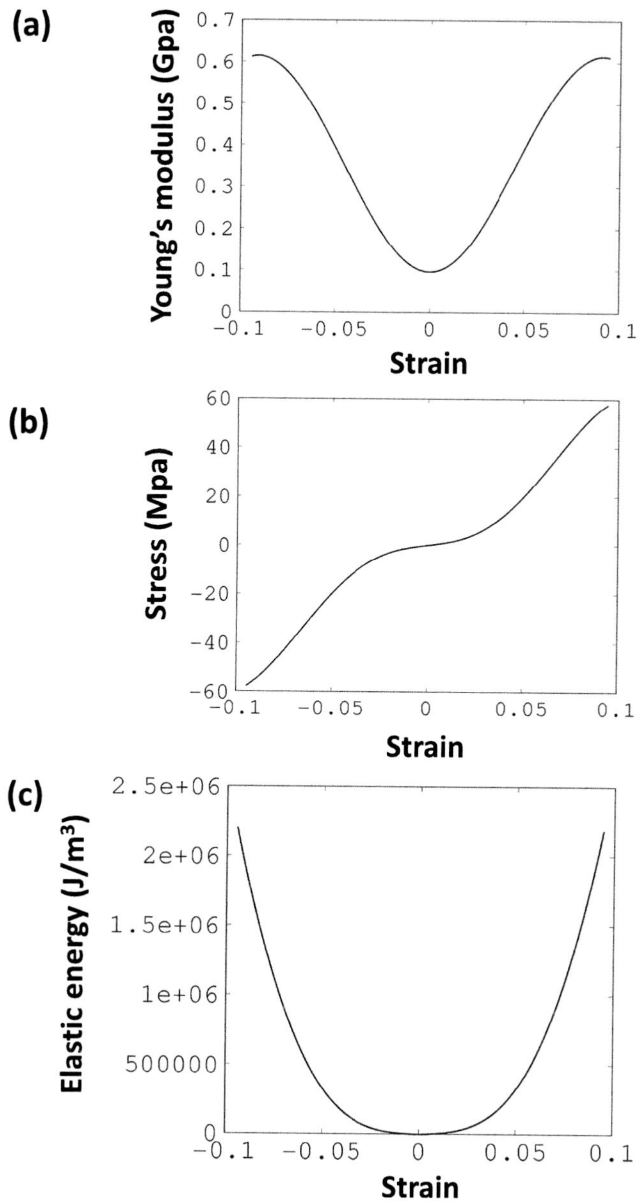

Fig. 4.15 (**a**) Young's modulus versus strain for open collagen/water system. The curve is a Taylor expansion up to eighth order that fits the experimental data. (**b**) Stress versus strain for collagen in the presence of water. This is obtained by multiplying the Young's modulus by the strain: $\sigma = Y(\varepsilon)\varepsilon$. (**c**) Energy versus strain for collagen in the presence of water. The curve is obtained by integrating the stress versus strain curve

$$E(\varepsilon) = \frac{a_0}{2}\varepsilon^2 + \frac{a_2}{4}\varepsilon^4 + \frac{a_4}{6}\varepsilon^6 + \frac{a_6}{8}\varepsilon^8 \quad (4.113)$$

With: $a_0 = 9.565\times10^7$ Pa, $a_2 = 1.543\times10^{11}$ Pa, $a_4 = -1.571\times10^{13}$ Pa, $a_6 = 6.958\times10^{14}$ Pa and $a_8 = -1.650\times10^{16}$ Pa.

The Young's modulus, stress and elastic energy versus the strain obtained with this Taylor series are represented in Fig. 4.15.

4.4.2 Model and Simulation Methods

We now consider the 1-D model illustrated in Fig. 4.16 as representative of the nanoscale, periodic assembly of alternating regions of collagen and HAP in bone and teeth. The finite super lattice is composed of N = 2560 masses of materials A and B connected by nonlinear springs. Material A is chemically inert and represents HAP. Material B represents the open collagen/water system *i.e.*, collagen that can adsorb water from a reservoir.

The behavior of elastic modes in this model is simulated using a combination of Spectral Energy Density method (see Appendix 3, Chap. 3) and Finite Difference Time Domain (FDTD) method [1]. In its original version, this method solves the springs and masses equation by discretizing time and space (x-axis) and by replacing derivatives by finite differences. We extend this approach to include chemo-mechanical effects by using the nonlinear stress-strain relation described in Sect. 4.1. The nonlinear spring constant depends on the water content in accordance with the nonlinear stress-strain relation of the collagen/water open system. Owing to the 1-D nature of the system, we assume that lateral diffusion of water into material B (collagen) is much faster than the acoustic wave and that material B is always at equilibrium with respect to chemical composition. Under this condition, the 1-D mass-spring equation is given by:

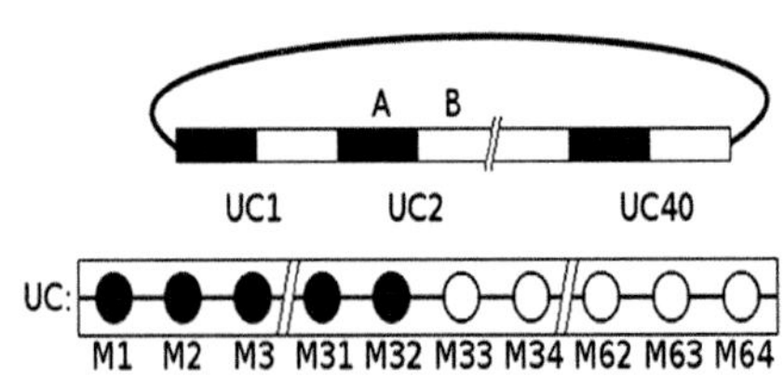

Fig. 4.16 Schematic representation of the simulated HAP/collagen 1-D periodic system. The unit cells (UC) are heterogeneous media composed 32 masses of a material (A) mimicking HAP that does not change composition when subjected to a stress and 32 masses of material B. B is a collagen-based material that can interact with a chemical reservoir of water in which it is embedded. Each mass is connected to its two neighbors by two springs. A spring is linear (nonlinear) if the mass on its *left* is composed of material A (B). Periodic Boundary Conditions are imposed at the end of the system, which contains 40 UCs

$$m_i \frac{\partial^2 u_i}{\partial t^2} = f_i + f_{i-1} \quad \text{with} \quad m_i = \rho_i \times dx^3 \tag{4.114}$$

where t is time, ρ_i is the mass density, $u(t)$ is the displacement. The displacement is related to the mechanical strain by $\varepsilon = \frac{\partial u}{\partial x}$ and σ is the stress. In 1-D, we describe the stress-strain relation by:

$$\sigma = Y(\varepsilon)\varepsilon \tag{4.115}$$

or the force-displacement relation by:

$$f_i = k_i du_i = \rho_i c_{L,i}^2 \Delta x du_i \quad \text{with } k_i = \rho_i c_{L,i}^2 \Delta x \text{ and } c_{L,i} = \frac{Y(\varepsilon)_i}{\rho_i} \tag{4.116}$$

Equation (4.114) is solved discretely and takes the form:

$$\rho_i \Delta x^3 \frac{\partial^2 u_i}{\partial t^2} = \left[\rho_i c_{L,i}^2 (u_{i+1} - u_i) - \rho_{i-1} c_{L,i-1}^2 (u_i - u_{i-1})\right] \Delta x \tag{4.117}$$

In (4.117) we assume that the mass density is independent of composition. This is the case in material A. Since the mass density of water and collagen are similar, the mass densities of discrete points in material B are also taken to be constant.

The structure is composed of 40 unit cells each of them composed of 64 masses (32 of them composing the material A and the 32 others composing the material B). Each mass is connected to its two neighbors by two springs. A spring is linear (nonlinear) if the mass on its left is composed of material A (B). The nonlinear spring constant is calculated using (4.116) where $Y(\varepsilon)$ is the Taylor expansion of the Young's modulus described in Sect. 4.1.

We employ a spatial grid with mesh size $\Delta x = 9.32 \times 10^{-13}$ m. The time integration step, Δt, is given by $\Delta t = \Delta x/(30c^{(A)})$ where $c^{(A)}$ is the speed of sound in medium A. The total number of time steps used in our calculations is 2^{22} time steps. Periodic boundary conditions are applied at the free ends of the homogeneous regions to simulate an infinite superlattice. The thickness of the segments of material A and B are designated by L_A and L_B. For the sake of simplicity, we take: $L_A = L_B = 33.5$ nm. Hence the length of a unit cell is a = 67 nm. An initial random displacement is applied to each of the 2560 masses which imparts an initial potential energy to the structure. Then, the system is free to evolve during the total number of time steps (2^{22}) and the speed of each mass is recorded during the last 2^{21} time steps.

We use the SED method to calculate the presence of acoustic modes in the crystal for different wave vectors. For each mass, its speed is projected on a considered wave number and a Fourier transform provides the frequency distribution of the energy in the solid. This energy is averaged over the 64 masses of each unit cell. This operation is repeated 200 times for which the results are averaged, in order to allow a sufficient variety of initial energy distribution.

The whole operation is repeated over 21 wave numbers ranging from 0 to π/a by step of 0.05 π/a. The values used for the initial displacement of the masses are dx/45 and dx/150. The wave number interval $[0, \pi/a]$ constitutes the first Brillouin zone of the phononic periodic superlattice.

In what follows, the density and sound velocity in the elastic HAP (material A) are: $\rho^{(A)} = 3.16\times10^3$ kg/m^3 and $c^{(A)} = 4357.45$ m/s (elastic modulus of 60×10^9 Pa). The density of medium B is $\rho^{(B)} = 1.3\times10^3$ kg/m^3. The time step is therefore approximately $\Delta t = 3.2\times10^{-14}$ s. This time is significantly larger than the time for diffusion through the model system. If we approximate the 1-D system by a wire with a cross section $l = 0.5$ nm and if we consider a diffusion coefficient for water in collagen of $D = 5\times10^{-5}$ cm^2/s, the time for diffusion across the wire is on the order of $t = l^2/D = 5\times10^{-15}$ s. If we compare this time to Δt, we can assume that the water content of the system is at equilibrium at all time. This assumption is even more valid when considering low frequency modes.

As for the convergence of the method, the SED-FDTD calculation was performed using 16, 32 and 64 masses per unit cell in the limit of small displacements. We observed that the band structure obtained with 16 masses per unit cell almost matched the linear results below 10 GHz but not above this frequency. The band structure obtained with 64 masses per unit cell matched almost perfectly the linear band diagram shown in Fig. 4.17. The band structure obtained with 32 masses per unit cell was an intermediary result, with a match better than that for 16 but worse than that for 64 masses per unit cell. Hence, the value of 64 masses per unit cell was chosen for all the SED-FDTD calculations.

The band structure of the linear superlattice is obtained from (2.96) in Appendix 1, Chap. 2:

$$\cos kL = \cos k_A L_A \cos k_B L_B - \frac{1}{2}\left(F + \frac{1}{F}\right) \sin k_A L_A \sin k_B L_B$$

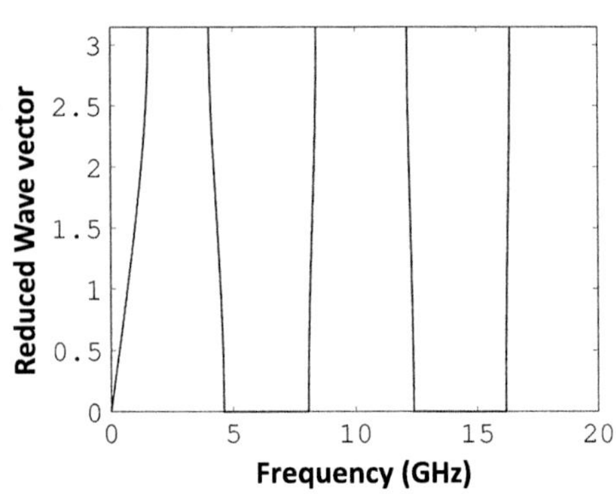

Fig. 4.17 Band structure of the AB superlattice. The wave vector is expressed in reduced units of 1/L where L is the period of the HAP/collagen superlattice

where $F = \frac{k_A \rho_A c_A^2}{k_B \rho_B c_B^2}$. The model structure of our mineralized biological tissue is a superlattice with the period L $= L_A + L_B = 67$ nm. We also have $k_A = \frac{\omega}{c_A}$ and $k_B = \frac{\omega}{c_B}$. The elastic band structure of the AB superlattice with fixed water content of B, $X_1^0 = 0.5$, can be calculated readily and is illustrated in Fig. 4.17. This is the band structure of the elastic superlattice, that is, when the collagen/water system keeps a constant composition and its corresponding Young's modulus remains constant. In this case, the superlattice exhibits elastic passing bands separated by wide band gaps. Elastic waves will propagate through the HAP/collagen superlattice only when their frequency will fall into the range of the passing bands. Waves with frequency in the band gap will not propagate and be only reflected by the superlattice.

The Spectral Energy Density of the structure for an initial displacement of dx/150 (low amplitude) is provided in Fig. 4.18a and that for an initial displacement of dx/45 (large amplitude) is provided in Fig. 4.18b. At low amplitude, the spectrum shows very thin peaks at specific frequencies corresponding to the vibrational modes that are found when the transmission is solved analytically (Fig. 4.17). The system behaves according to linear elasticity: it shows a harmonic behavior. In this regime, the elastic coefficient (Young's modulus) of the collagen (B) region retains its value at small strain. For the whole frequency range (0–20 GHz), an excellent agreement between this nonlinear result and the linear result is obtained.

At large amplitude, the system becomes strongly nonlinear and the peaks broaden significantly. A filling of the band gaps is observed, as shown on Fig. 4.19. This band gap filling is more visible at higher frequencies (10–20 GHz), which allows us to think that as frequency increases, the band gaps will be more and more filled. At higher frequencies (hundreds of GHz), this would mean that the energy is dissipated as thermal phonons.

The peak broadening and frequency shift are interpreted as follows. If we limit the Young's modulus function of strain to its constant term, which is exactly the linear case, then the Taylor series expansion of the elastic energy function is limited to its second order term, which appears in Fig. 4.20.

In this case, the calculated spectrum only contains the "primary" frequencies which are that of a linear system. This result is shown in Fig. 4.19. Now, if we limit the Young's modulus function of strain to its terms up to second order, the energy function contains a term of second order and a term of fourth order. As shown on Fig. 4.19, the simulated system behaves as a nonlinear system with the frequency shift and band gap filling described above. If we limit the Young's modulus function to its terms up to fourth (sixth) order, the energy function contains terms of second, fourth, and sixth (eighth) order. Taking those terms into account (see Fig. 4.19) does not change the behavior of the system, which proves that the second order term of the Young's modulus (fourth order term of the energy) plays the major role for the nonlinear behavior and the terms of higher order do not play a

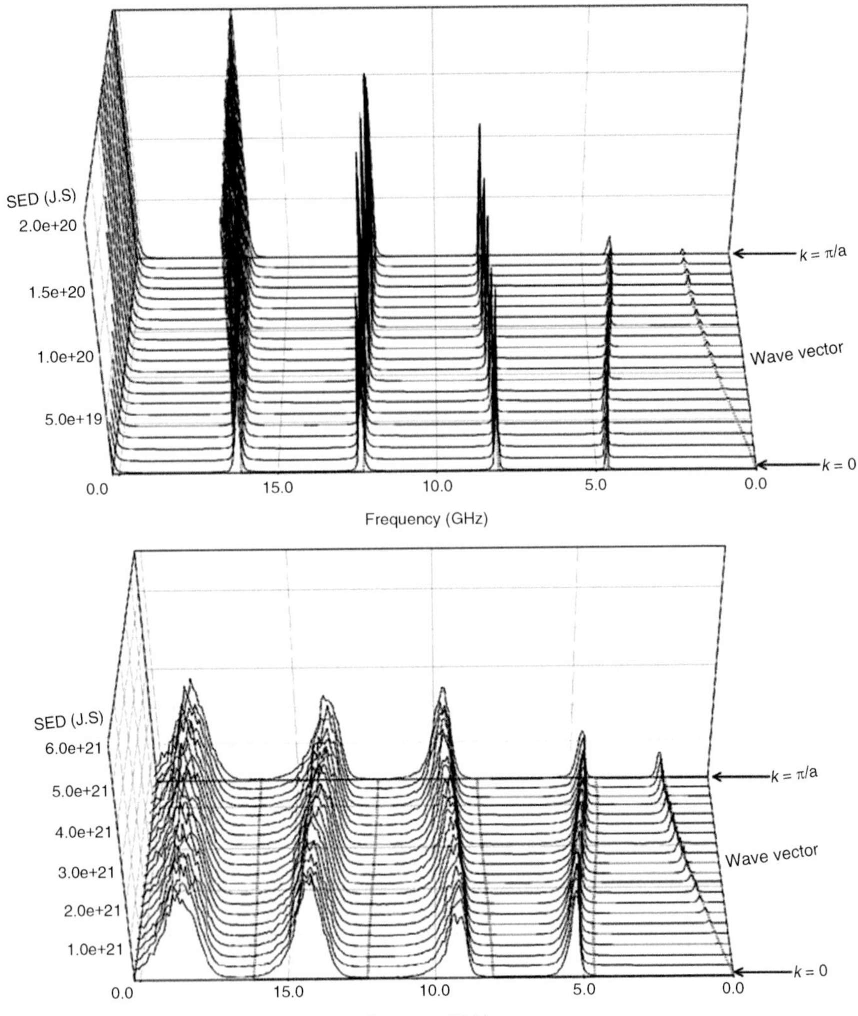

Fig. 4.18 *Solid black lines*: Spectral Energy Density of the phononic crystal structure with an initial displacement of (**a**) dx/150 and (**b**) dx/45. Gray dashes: linear band structure of the superlattice shown in Fig. 4.17

significant role. Hence, we can assert that this nonlinear behavior associated with a fourth order function in strain elastic energy is mainly due to four-wave (four-phonon) interactions. Indeed, considering that the dynamic strain is a linear superposition of plane waves with different wave number and frequencies and raising this superposition to the fourth power to estimate the energy, leads to four-wave interactions. These scattering interactions would conserve momentum and

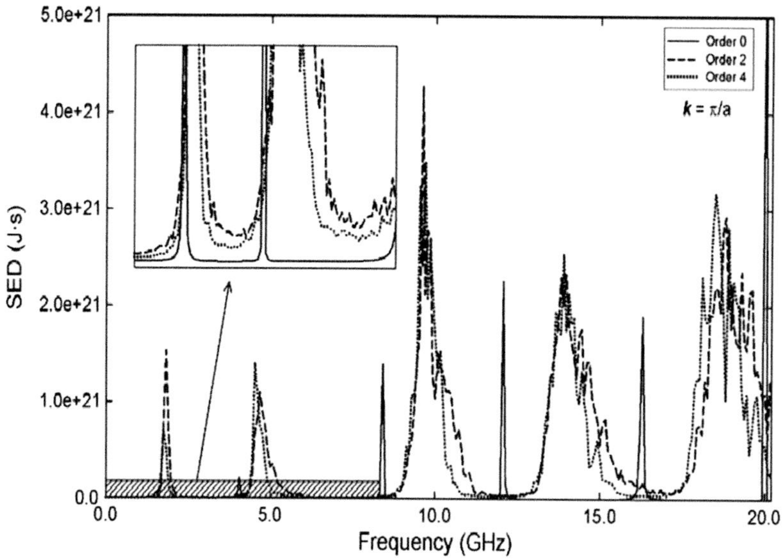

Fig. 4.19 Spectral Energy Density of the phononic crystal structure with an initial displacement of dx/45 for wave vector π/a. *Gray solid line*: with the Young's modulus truncated at the zeroth order (constant value, linear case); *dashes*: with the Young's modulus truncated at second order; *dots*: with the Young's modulus truncated at fourth order. Curves with the Young's modulus truncated at sixth and eighth order are not included since they do not carry significant changes compared to fourth order. *Insert*: highlight of the band gap filling in the frequency range between 0 and 10 GHz

frequency and may involve a variety of processes such as the splitting of a single phonon into three others, the scattering of two phonons forming two others, etc. Those new phonons interactions offer a number of channels for the dispersion of mechanical energy.

We can relate the behavior of the nonlinear superlattice to the behavior of a multiple well mass spring system. A multiple well system is a system for which the representation of the elastic energy as a function of strain is not parabolic but is a superposition of an infinity of parabolas. Each parabola corresponds to a single value of the spring constant. Hence, representing a non-quadratic energy map by a multiplicity of parabolic wells is equivalent to introducing a continuum of values of spring constants. At high amplitudes, waves will sample wider ranges of strain values, effectively visiting the multiplicity of energy wells. The band structure will show a continuity of modes above the "primary" frequencies of the linear system. Those bands broaden and finally fill the gaps. The passing bands of this system will also be shifted to higher frequencies. In our case, the reason why the spring constants increase as a function of the magnitude of the strain is that the model we chose for Young's modulus is symmetrical and has only even orders in its

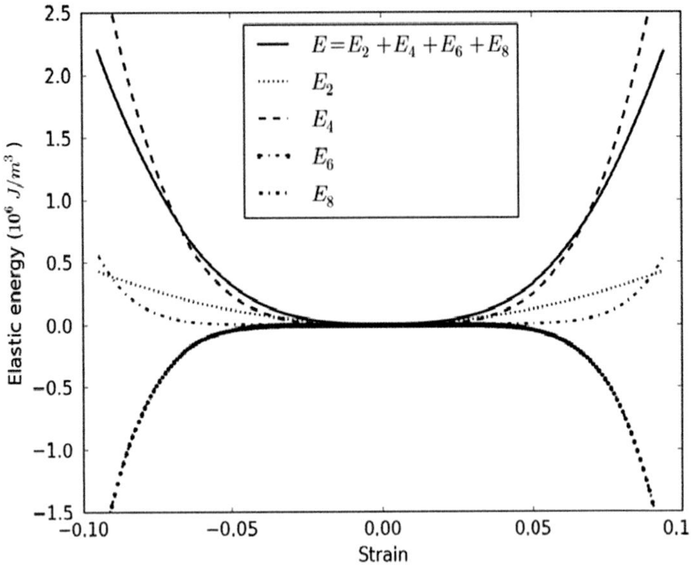

Fig. 4.20 Elastic energy versus strain for the collagen/water open system. The *solid line* is the same as that of Fig. 4.15, namely the integral of the Taylor series decomposition of the stress as a function of strain. The *dotted* and *dashed lines* are the terms of second, fourth, sixth, and eighth order of the Taylor series. The term of sixth order is particularly interesting as it is gives a concave down curve, characteristic of an unstable system

Taylor series expansion. If an asymmetrical function was used for Young's modulus, its Taylor series expansion would contain odd order terms. In that case, nonlinear modes with frequencies less than the "primary" frequencies would appear. We would also observe a shift of the passing bands to lower frequencies.

In summary, in the linear limit, the band structure of the linear collagen-apatite superlattice is composed of passing bands separated by band gaps that forbid the propagation of elastic waves over some ranges of frequency. The nonlinearity of the collagen layers in the superlattice give rise to multi-wave (phonon) scattering processes that lead to partial band gap filling. Multi-phonon scattering processes constitute ways of opening new channels for the dissipation of mechanical energy. This mechanism for mechanical energy dissipation is the direct consequence of the hydration of the collagen. The risk of dentin and bone fracture has been shown to increase with age. During aging, there is also a marked decrease in the level of interconnected porosity in both tissues [61, 62]. This filling of the porosity is associated with decreased fluid flow and therefore decreased hydration of the tissues. It has previously been theorized that this decrease in hydration may in part be responsible for the increased fragility of bone and dentin with age; however, the associated mechanism remained unclear [63].

4.5 Composite Media with Controllable Nonlinearity

We have seen in the previous three sections that the nature of phonon-phonon scattering processes depends on the characteristics (*i.e.*, order) of the nonlinearity. It is therefore interesting to consider the possibility of constructing an effective medium by mixing nonlinear elastic constituents to control the magnitude and order of the nonlinearity [18]. In particular, we treat the case of a simple one-dimensional mass-spring model with two types of nonlinear spring, 1 and 2. The system is illustrated in Fig. 4.21.

The model systems is constituted of N_1 and N_2 spring elements with nonlinear spring constants K_1 and K_2. The total number of element in the chain is $N = N_1 + N_2$. We assume that the system is supporting some wave excitation with long wavelength $\lambda \gg L$. In this limit, the stress is equal at each element and can be written as:

$$\sigma = \sigma_1 = K_1 X_1 \quad (4.118a)$$

$$\sigma = \sigma_2 = K_2 X_2 \quad (4.118b)$$

We rewrite the nonlinear constitutive equations for each element in the form:

$$\sigma_1(\varepsilon_1) = M_1(1 + f_1(\varepsilon_1))\varepsilon_1 \quad (4.119a)$$

$$\sigma_2(\varepsilon_2) = M_2(1 + f_2(\varepsilon_2))\varepsilon_2 \quad (4.119b)$$

Where M_1 and M_2 are independent of strain, which is at each element i, $\varepsilon_i = \frac{X_i}{L}$. The functions f_1 and f_2 are nonlinear and satisfy $f_1(\varepsilon) \approx f_2(\varepsilon) \ll 1$. We treat the case of soft elements, 2, and hard elements, 1, namely:

$$\frac{M_1}{M_2} = Q \gg 1 \quad (4.120)$$

According to (4.118a, 4.118b), the ratio of elongations is determined by the ratio of spring constants:

$$\frac{X_2}{X_1} = \frac{K_1}{K_2} = \frac{M_1}{M_2} = Q \quad (4.121)$$

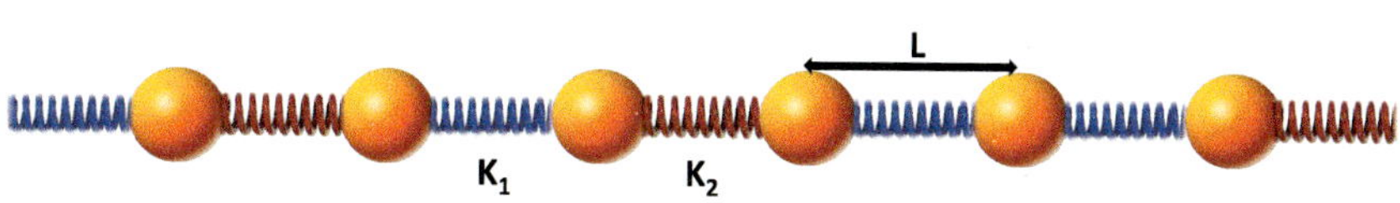

Fig. 4.21 Mass-spring chain system composed of nonlinear spring elements with spring constants K_1 and K_2. The inter-mass spacing is L

We seek now an expression for the nonlinear effective constitutive relation $\sigma(\varepsilon)$ of the composite system. For this we first calculate the total elongation of a segment of the chain of length NL:

$$X_T = N_1 X_1 + N_2 X_2 \tag{4.122}$$

The total strain of the segment is therefore:

$$\varepsilon = \frac{X_T}{NL} = (1-c)\varepsilon_1 + c\varepsilon_2 \tag{4.123}$$

where we have introduced a concentration of soft springs 2: $c = \frac{N_2}{N}$. In terms of elongations and strain of spring 1, (4.123) becomes:

$$\varepsilon = (1-c)\frac{X_1}{L} + c\frac{X_2}{L} = (1 + c(Q-1))\frac{X_1}{L} = (1 + c(Q-1))\varepsilon_1 \tag{4.124}$$

The average elastic energy density per spring stored in the segment of the chain is given by:

$$W = (1-c)W_1 + cW_2 \tag{4.125}$$

In (4.125), the elastic energy of spring i is expressed as:

$$W_i = \int_0^{\varepsilon_i} \sigma_i(\zeta)d\zeta = \int_0^{\varepsilon_i} M_i(1 + f_i(\zeta))\zeta d\zeta \tag{4.126}$$

Rewriting (4.126) in terms of elongations gives:

$$W_i = \int_0^{X_i} M_i\left(1 + f_i\left(\frac{\xi}{L}\right)\right)\frac{\xi}{L}\frac{1}{L}d\xi \tag{4.127}$$

The average elastic energy is therefore:

$$W = (1-c)\int_0^{X_1} M_1\left(1 + f_1\left(\frac{\xi}{L}\right)\right)\frac{\xi}{L}\frac{1}{L}d\xi + c\int_0^{X_2} M_2\left(1 + f_2\left(\frac{\xi}{L}\right)\right)\frac{\xi}{L}\frac{1}{L}d\xi \tag{4.128}$$

The stress in the segment of the chain is:

$$\sigma = \frac{dW}{d\varepsilon} = \frac{dW}{dX_1}\frac{dX_1}{d\varepsilon} \tag{4.129}$$

In (4.129), $\frac{dX_1}{d\varepsilon} = \frac{L}{(1 + c(Q - 1))}$ by virtue of (4.124). Using (4.128), we calculate:

$$\frac{dW}{dX_1} = (1 - c)\frac{M_1}{L}\left(1 + f_1\left(\frac{X_1}{L}\right)\right)\frac{X_1}{L} + c\frac{M_2}{L}\left(1 + f_2\left(\frac{QX_1}{L}\right)\right)\frac{QX_1}{L}\frac{dX_2}{dX_1}$$

or

$$\frac{dW}{dX_1} = (1 - c)\frac{M_1}{L}\left(1 + f_1\left(\frac{X_1}{L}\right)\right)\frac{X_1}{L} + c\frac{M_2}{L}\left(1 + f_2\left(\frac{QX_1}{L}\right)\right)\frac{QX_1}{L}Q \qquad (4.130)$$

With $M_2Q = M_1$ and (4.130), the stress takes the form:

$$\sigma = \frac{M_1}{L}\left\{(1 - c)\left(1 + f_1\left(\frac{X_1}{L}\right)\right)\frac{X_1}{L} + c\left(1 + f_2\left(\frac{QX_1}{L}\right)\right)\frac{QX_1}{L}\right\}\frac{L}{(1 + c(Q - 1))}$$

or

$$\sigma = \frac{M_1}{L}\left\{(1 - c)\left(1 + f_1\left(\frac{X_1}{L}\right)\right)\frac{X_1}{L} + c\left(1 + f_2\left(\frac{QX_1}{L}\right)\right)\frac{QX_1}{L}\right\}\frac{L}{(1 + c(Q - 1))}$$

$$\sigma = M_1\varepsilon_1\left\{1 + \frac{(1 - c)}{(1 + c(Q - 1))}f_1(\varepsilon_1) + \frac{cQ}{(1 + c(Q - 1))}f_2(Q\varepsilon_1)\right\} \qquad (4.131)$$

In the linear case when $f_1 = 0$ and $f_2 = 0$, we obtain $\sigma = M_1\varepsilon_1 = \frac{M_1}{1 + c(Q - 1)}\varepsilon$. $M_{eff} = \frac{M_1}{1 + c(Q - 1)}$ is the linear effective elastic coefficient.

We remember that $Q \gg 1$, so in the limit of very low concentration $c \ll 1/Q$, then $\varepsilon_1 \to \varepsilon$ and $\varepsilon_2 \to Qc$. The strain of springs of type 2 takes on very large values leading to strong nonlinear effects as seen in (4.132):

$$\sigma = M_1\varepsilon\{1 + f_1(\varepsilon) + cQf_2(Q\varepsilon)\} \qquad (4.132)$$

Let us now consider the case of high concentrations. For this, we now define the functions of f_1 and f_2 to be the sum of a linear term and a quadratic term. We chose:

$$f_i(\varepsilon) = \gamma_i\varepsilon + \delta_i\varepsilon^2 \qquad (4.133)$$

Since the linear term in the functions relates to a quadratic nonlinearity in energy and the quadratic term in the function arises from a cubic term in energy we will denote γ_i and δ_i quadratic and cubic terms.

We insert (4.133) into (4.131) which after algebraic manipulation becomes:

$$\sigma = M_1 A\varepsilon\left(1 + \gamma_1 \varepsilon B + \delta_1 \varepsilon^2 C\right) \tag{4.134}$$

with: $$A = \frac{1}{(1 + c(Q-1))};$$

and the second order and third order nonlinear parameters

$$B = \frac{1 - c + c\dfrac{\gamma_2}{\gamma_1}Q^2}{(1 + c(Q-1))^2} \tag{4.135a}$$

$$C = \frac{1 - c + c\dfrac{\delta_2}{\delta_1}Q^3}{(1 + c(Q-1))^3} \tag{4.135b}$$

We are now seeking the optimal concentration which maximizes the nonlinear parameters B and C. We now define a generalized nonlinear parameter:

$$X_n = \frac{1 - c + c\dfrac{x_2}{x_1}Q^n}{(1 + c(Q-1))^n} \tag{4.136}$$

When n = 2, $x_i = \gamma_i$ and $X_2 = B$. When n = 3, $x_i = \delta_i$ and $X_3 = C$. The derivative $\dfrac{dX_n}{dc} = 0$ when

$$\left(\frac{x_2}{x_1}Q^n - 1\right)(1 + c(Q-1)) - n(Q-1)\left(1 + c\left(\frac{x_2}{x_1}Q^n - 1\right)\right) = 0 \tag{4.137}$$

This condition gives an optimal concentration:

$$c_o(n) = \frac{1}{(n-1)(Q-1)} - \frac{n}{(n-1)\left(\dfrac{x_2}{x_1}Q^n - 1\right)} \approx \frac{1}{(n-1)Q} \text{ when } Q \gg 1 \tag{4.138}$$

The optimal concentration for the second order parameter is $c_o(2) \approx 1/Q$. The optimal concentration for the third-order parameter is $c_o(3) \approx 1/2Q$. These optimal concentrations are different which suggest that one might be able to tune the concentration in such a way that one selectively optimizes the second or third order parameters.

We note that at the optimum concentrations, the linear parameter is $A(c_o(n)) = \dfrac{(n-1)Q}{nQ-1} \approx \dfrac{n-1}{n}$. This parameter takes on the value 1/2 for $c_o(2)$ or 2/3 $c_o(3)$. One can conclude that the linear parameter does not vary significantly.

On the other hand, inserting the approximation $c_o(n) \approx \dfrac{1}{(n-1)Q}$ (4.138) into (4.136), yields:

$$X_n(c_o(n)) \approx \frac{x_2}{x_1} \frac{(n-1)^{n-1} Q^{n-1}}{n^n} \tag{4.139}$$

At the optimal concentration for quadratic nonlinearity, we get:

$$B(c_o(2)) \approx \frac{\gamma_2}{\gamma_1} \frac{Q}{4} \tag{4.140a}$$

$$C(c_o(2)) \approx \frac{\delta_2}{\delta_1} \frac{Q^2}{4} \tag{4.140b}$$

At the optimal concentration for cubic nonlinearity, the parameters are:

$$B(c_o(3)) \approx \frac{\gamma_2}{\gamma_1} \frac{2Q}{9} \tag{4.141a}$$

$$C(c_o(2)) \approx \frac{\delta_2}{\delta_1} \frac{4Q^2}{27} \tag{4.141b}$$

It is therefore possible to increase the cubic nonlinearity by a factor of 100 (*i.e.*, $\frac{4}{27}Q^2 = 100$) if one chooses: $Q \approx 26$ and $c_0 = 1/52$. In this case, $B \approx \frac{52}{9}\frac{\gamma_2}{\gamma_1}$ and the quadratic nonlinearity is only increased by a factor of 5.8. This concentration favors the cubic nonlinearity.

This simple example is therefore illustrating the possibility of tuning the nonlinear response of a composite medium constituted of nonlinear constitutive materials by varying the type of constitutive materials as well as their concentrations. The effective nonlinear response may be tuned with respect to magnitude and order of nonlinearity.

4.6 Calcium Waves in Chains of Endothelial Cells with Nonlinear Reaction Dynamics: A Metaphor for Wave Propagation in Bi-stable Media

The dynamics of physical, chemical, and biological systems containing bi-stable elements is very rich. In this section, we use a model of nonlinear calcium waves in biological systems to serve as a metaphor for the propagation of waves in other bi-stable media such as mechanical and elastic systems. We present a Green's function-based perturbative approach to solving the nonlinear reaction-diffusion problem in chains of endothelial cells. We focus on a single component (Ca2+), piecewise nonlinear model of endoplasmic calcium dynamics and trans-membrane diffusion. We verify analytically and then numerically that this approach leads to the known transition from propagation of calcium front to failure of propagation when the diffusion rate is varied relative to the reaction rates. Propagation in this bi-stable medium is also shown to be non-reciprocal.

4.6.1 Background for Calcium Waves in Endothelial Cells

From tumorigenesis, to tissue engineering to blood vessels, questions concerning the interactions between living cells and their environment/architecture have received increasing attention [64–72]. One aspect particularly relevant to these questions is the emerging behavior of a multicellular architecture in which cell-level functions, such as intracellular pathways, integrate with the architecture through cell-to-cell interactions. For example, downstream and upstream signal conduction between endothelial cells along the walls of vessels is playing an important role in circulatory function of formed vasculatures, vascular network remodeling, vasculogenesis and neovascularisation [73]. Central to this problem is that cellular networks inherently combine dynamical and structural complexity. Some mathematical approaches have tackled the problem by setting aside the dynamics of the network nodes (cells) and emphasizing the complexity of the network architecture [74]. On the other hand, Othmer and Scriven [75, 76] developed, following Turing's pioneering mathematical treatise of morphogenesis [77], a linear approach in which the information about the underlying network topology, through a connectivity matrix, is decoupled from that of the intracellular reaction pathway dynamics thus incorporating complexity at the level of the cell and the network. However, Plahte [78, 79] argued that linear analysis of reaction-dynamics problems must be complemented by nonlinear analysis to effectively explain pattern formation. Of particular interest in the study of systems with nonlinear reaction dynamics is the phenomenon of propagation failure, where the wave speed is zero. In a model of a chain of nearest-neighbor, diffusively-coupled, over-damped oscillators the propagation failure and the nature of the wave front depends on the nonlinearity of the dynamics of the model [80]. Similar behavior has been observed in a discrete version of Nagumo equation modeling excitable cells with a cubic dynamics [81]. There, propagation of a wave front occurs when the strength of the diffusion coupling is large enough. Similar results have been obtained by Kladko et al. [82]. Propagation failure of traveling waves in two-dimensional discrete lattice with nearest neighbor diffusive-coupling and bi-stable scalar ODE at each site has demonstrated the angular dependency of the zero speed conditions through the detuning parameter of the nonlinearity [83].

At the level of the individual cell, Ca^{2+} signals rely on the intake or release of Ca^{2+} ions from intracellular stores such as the endoplasmic reticulum (ER). The physiological state of the cytoplasm of a cell determines the nature of the dynamics of calcium release and intake [84]. Within endothelial cells of arterioles, a calcium based signaling pathway exists that contains a two-component negative feedback loop. This loop occurs between Ca^{2+} and IP_3. This inositol phospholipid signaling pathway is started by an extracellular signal molecule that activates a transmembrane G-protein coupled receptor which in turn activates phospholipase C-β. Phospholipase C-β cleaves intracellular membrane bound phospho inositol 4,5-bisphosphate [PI(4,5)P_2] causing the cytoplasmic release of IP_3. Cytoplasmic IP_3 can bind and open IP_3 gated Ca^{2+} channels in the endoplasmic reticulum leading to increased cytoplasmic Ca^{2+} concentration. IP_3 concentration is degraded by phosphorylation via Ca^{2+} regulated kinase. The cytoplasmic inositol 1,4,5-triphosphate 3-kinases (IP_3Ks) are a group of calcium-regulated inositol polyphosphate kinases that convert IP_3 into inositol 1,3,4,5-tetrakisphosphate. This later specie is inactive as a Ca^{2+} release inducer, thus reducing intracellular Ca^{2+} concentration. The overall effect of this signaling cascade is that of a two component (Ca^{2+} and IP_3) negative feedback loop. Consequently, the physiological state of a cell may be bi-stable with two resting states for calcium concentration; low basal and high corresponding to replete ER and empty ER, respectively, and separated by an intermediate unstable threshold concentration. The bi-stable state may lead to traveling fronts in spatially extended systems. The physiological state may also be that of an excitable cytoplasm, and may be considered to be a variant on the bi-stable state with the possibility of returning to the low basal concentration beyond a high cytosol concentration threshold. Excitable cytoplasm may produce pulse waves in spatially extended systems. There exist a large variety of mathematical models of calcium dynamics and calcium waves [84–87]. However, since the objective of the present work is to demonstrate the feasibility of a Green-function based approach to solving diffusion-reaction problems with nonlinear reaction dynamics, for the sake of mathematical tractability, we assume an effective nonlinear intracellular reaction dynamics involving only Ca^{2+}. For this we utilize a simple piecewise-linear model of the nonlinear Ca^{2+} reaction dynamics. This model mimics the Ca^{2+} depletion of the cytoplasm and repletion separated by a threshold concentration. Such piece-wise linear models have been shown to possess the essential features of nonlinear biological dynamical systems [88, 89].

In this section, we develop a Green's function-based theory of complex multicellular architectures with nonlinear reaction dynamics. This approach is enabled by the convergence of three areas. The first area is the development of a theory of complex multicellular architectures with linear reaction dynamics that provides solutions that can be used as starting approximations for perturbative methods to be applied to architectures with nonlinear dynamics. This approach is based on Green's functions (see Chap. 1). The second area relates to the so-called adiabatic switching formalism used in statistical mechanics for the calculation of the free energy of a system of interest by connecting it through a reversible path to a system with a known free energy. An effective Hamiltonian is constructed as a linear combination of the Hamiltonians of the two systems using a continuous variable [90, 91]. In fact, this approach is similar to He's Homotopy Perturbation Method (HPM) [92]. Here, an operator is separated into its linear and nonlinear parts. An embedding parameter that can vary between 0 and 1 is used to construct an operator that can transition between the linear operator and the nonlinear one. The solution to the switching operator is assumed to be expandable into a power series of the embedding parameters. Upon taking the limit of the embedding parameter toward 1 one obtains an approximate solution to the original nonlinear operator. The HPM has been shown to be able to effectively solve strongly nonlinear problems including nonlinear parabolic differential equations [93]. The final area is the use of Lippmann–Schwinger propagator theory for time dependent processes that enables us to write the Green's function of a perturbed system in terms of a series of integral terms involving the product of Green's function of the unperturbed system and of the perturbation operator. These perturbation expansions of the Green's function have been summarized in the form of Feynman diagrams which have found applications in both high energy and condensed matter physics. [94, 95]. We detail these approaches in the subsections below.

4.6.2 Linear and Nonlinear Reaction Dynamics

We address the problem of linear and nonlinear reaction dynamics to establish the mathematical foundations in terms of Green's functions that will enable us to derive an expression for a time propagator. Let us first start with a simple linear reaction problem where the kinetics of the reaction is described by the following equation:

$$\frac{du}{dt} = -Ku \tag{4.142}$$

where K is the reaction rate (taking positive or negative values), and u some composition variable. The solution to that equation is obtained by integration over time and takes the form:

$$u(t) = u(t')e^{-K(t-t')} \tag{4.143}$$

Equation (4.142) can also be solved by using Laplace transforms. It becomes

$$\omega\tilde{u} - u(0) = -K\tilde{u} \tag{4.144}$$

ω is the variable conjugated with time in the transform. The Laplace transform of the composition is obtained as:

$$\tilde{u} = u(0)\frac{1}{\omega + K} \tag{4.145}$$

The inverse Laplace transform of (4.145) gives the solution of (4.143) where t' is taken as the origin of time. Finally, one may recast (4.142) in terms of a time dependent Green's function or propagator, $G(t - t')$, in the form:

$$LG(t - t') = \delta(t - t') \tag{4.146}$$

where the operator L is defined as:

$$L = \frac{d}{dt} + K \tag{4.147}$$

The solution to (4.146) is given by:

$$G(t - t') = H(t - t')e^{-K(t-t')} \tag{4.148}$$

In (4.148), $H(t - t')$ is the Heaviside function with

$$H(t - t') = \begin{cases} 0 \text{ if } t \leq t' \\ 1 \text{ if } t > t' \end{cases} \tag{4.149}$$

The solution for the composition is recovered by the time integration of the product of the propagator and some initial condition $u(0)$. For instance using, $t = 0$ as origin of time and $u(0) = u_0\delta(t)$ with δ being the usual delta function, the time dependence of the composition is obtained as:

$$u(t) = \int_{-\infty}^{\infty} H(t - t')e^{-K(t-t')}u_0\delta(t')dt' = u_0e^{-Kt}.$$

Let us now consider a single component two-states nonlinear reaction problem that can be modeled via a two-segment piecewise linear function. The differential equation which solution is the time evolution of the composition, $u(t)$, is written as:

$$\frac{\partial u}{\partial t} + (1 - H(u - u_c))K_1u + H(u - u_c)K_2u = 0 \tag{4.150}$$

Here K_1 and K_2 are the reaction rates of the linear segments. u_c is the composition at which the system can switch between state 1 (reaction rate K_1) and state 2 (rate K_2). H is the Heaviside function or any other function that might describe the transition from one reaction rate to another upon change in composition.

He's Homotopy Perturbation Method (HPM) starts with the differential equation

$$A(u) = 0 \tag{4.151}$$

where A is a general differential operator than can be divided into a linear part, L and a nonlinear part N. The differential equation then becomes:

$$L(u) + N(u) = 0 \tag{4.152}$$

If the solution of the linear differential equation $L(u) = 0$ is u_L, then one rewrites the general differential equation in the form:

$$L(u) + (1 - p)L(u_L) + pN(u) = 0 \tag{4.153}$$

where p is an embedding parameter. The parameter p varies between 0 and 1 and continuously links the linear problem to the nonlinear one of interest. When $p = 0$, the problem reduces to the linear one with solution u_L. In the case $p = 1$, the differential equation becomes the original one. Assuming that the solution of (4.153) can be written in the form of a power series in p, in the limit of $p \to 1$, one recovers the solution to the nonlinear problem of interest. We illustrate the extension of He's HPM with propagator's theory to solve a nonlinear reaction differential equation. Replacing H in (4.150) by an embedding parameter p, (4.150) becomes isomorphic to (4.153). When $p = 0$, and $p = 1$, (4.150) becomes a set of two linear equations extending over the complete interval of values accessible to u:

$$\begin{aligned} \frac{\partial u}{\partial t} + K_1 u = 0 \\ \frac{\partial u}{\partial t} + K_2 u = 0 \end{aligned} \tag{4.154}$$

Introducing the Green's function formalism, (4.154) take the form:

$$\left(\frac{\partial}{\partial t} + K_i\right) G_i^0(t - t') = \delta(t - t') \text{ with } i = 1, 2 \tag{4.155}$$

with solutions

$$\begin{aligned} G_1^0(t - t') = H(t - t')e^{-K_1(t-t')} \\ G_2^0(t - t') = H(t - t')e^{-K_2(t-t')} \end{aligned} \tag{4.156}$$

Equation (4.150) also becomes:

$$\left(\frac{\partial}{\partial t} + K_1 + p(K_2 - K_1)\right) G(t - t') = \delta(t - t') \tag{4.157}$$

where the third term in the parenthesis can be thought of as a perturbation operator

$$V = p(K_1 - K_2) \tag{4.158}$$

To progress further, let us make for now the assumption that the perturbation V is independent of time, *i.e.*, we make the eikonal approximation for slow varying perturbations to calculate time integrals. We may take the value of the embedding parameter to be that corresponding to the initial composition, $p(u(0))$.

Considering a system described by a Green's function, G_1^0, the Green's function, G, for this same system but perturbed by the operator, V, satisfies the Lippmann–Schwinger equation:

$$\begin{aligned} G(t-t') = {} & G_1^0(t-t') - \int_{-\infty}^{+\infty} dt'' G_1^0(t-t'') V G_1^0(t''-t') \\ & + \int_{-\infty}^{+\infty} dt'' \int_{-\infty}^{+\infty} dt''' G_1^0(t-t'') V G_1^0(t''-t''') V G_1^0(t'''-t') - \ldots \end{aligned} \tag{4.159}$$

Inserting (4.156) into (4.159) and performing the appropriate integrations for times before or after switching, one obtains:

$$G(t-t') = H(t-t')e^{-K_1(t-t')}\left(1 - V(t-t') + \frac{1}{2}V^2(t-t')^2 - \frac{1}{6}V^3(t-t')^3 \ldots\right) \tag{4.160}$$

Equation (4.160) is a power series in V and therefore p. One notes that when $p=0$ ($H(u-u_c)=0$ or $u<u_c$) the solution to the problem is that of the differential equation with linear dynamics and rate constant K_1. The embedding parameter allows the transition from the state 1 to the state 2 (solution of the differential equation with reaction constant K_2). The parenthesis in (4.160) converges to $e^{-V(t-t')} = e^{-p(K_2-K_1)(t-t')}$ and in the limit $p=1$ ($H(u\text{-}u_c)=1$ or $u>u_c$), G converges to G_2^0. The solution for the propagator in the case of a constant perturbation (p) takes the form:

$$G(t-t') = H(t-t')e^{-K_1(t-t')}e^{-p(K_2-K_1)(t-t')} \tag{4.161}$$

Again, in performing the integrals in (4.159), we have considered that the perturbation operator V and therefore the embedding parameter p are independent of time. However, it is clear in (4.150) that the embedding parameter represents a transition function of composition which is time dependent. Consequently, the Green's function or propagator of the perturbed system given by (4.161) cannot be used as is to represent the solution of the nonlinear dynamics problem with time dependent embedding parameter. We therefore use an iterative process to solve the nonlinear dynamics of the system. Let us consider a small time difference, $\delta t = t - t'$, with known composition at t', u(t′), we assume that $p(u(t))$ takes the constant value $p(u(t'))$ during that interval of time. This enables us to propagate the composition using (4.161) to the later time t. At that time, the new composition is used to determine the new value of the embedding parameter and propagates the composition further in time. Equation (4.161) can therefore be used iteratively as a propagator over only small time intervals δt. The size of the time interval depends on the nature of the function describing the composition dependence of the embedding parameter.

The solution of (4.150) can be also obtained by performing a Laplace transform assuming that p is independent of time over some time interval $[0,t]$ taking the value $p(u(0))$:

$$(\omega + k_1 + p(K_2 - K_1))\tilde{u} = u(0) \tag{4.162}$$

The Laplace transform of the composition is:

$$\tilde{u} = u(0)\frac{1}{\omega + K_1 + p(K_2 - K_1)} \tag{4.163}$$

which leads to the sought solution when doing the inverse Laplace transform:

$$u(t) = u(0)e^{-K_1 t}e^{-p(K_2 - K_1)t} \tag{4.164}$$

This solution is obtained by assuming that p is constant. Again, to account for the actual nonlinear reaction dynamics where the embedding parameter is composition (and time) dependent, one may use (4.164) in an iterative process over short time intervals to solve for the time dependence of the composition over a series of short time intervals starting from some known initial condition. The approach developed here will now be used to derive an analytical expression for the propagator of the reaction-diffusion problem of a chain of cells with nonlinear reaction dynamics.

4.6.3 *Reaction-Diffusion Problem with Linear or Nonlinear Reaction Dynamics*

The objective of this subsection is to add spatial degrees of freedom to the approach described in Sect. 6.2 and derive an expression for the reaction-diffusion propagator with nonlinear reaction dynamics. We first tackle the problem of linear reaction kinetics in conjunction with diffusion. We consider a one dimensional chain of cells with diffusion limited to nearest neighbors.

The linear reaction-diffusion problem is described by the equation:

$$\frac{du_n}{dt} = W(u_{n+1} - 2u_n + u_{n-1}) - Ku_n \tag{4.165}$$

The first term in the right-hand side of the equation represents the diffusion process between adjacent cells via the discrete Laplacian operator. From a biological point of view trans-membrane diffusion may take place via gap junctions. The second term represents the linear reaction dynamics of cell "n". W relates to the nearest neighbor calcium transfer rate. K is the reaction rate.

We can Laplace transform (4.165) which becomes:

$$\omega\tilde{u}_n - u_n(0) = W(\tilde{u}_{n+1} - 2\tilde{u}_n + \tilde{u}_{n-1}) - K\tilde{u}_n \tag{4.166}$$

In (4.166), $u_n(0)$ is the initial concentration profile along the chain. If we choose $u_n(0) = \delta_{nm}$ where δ_{nm} is the usual Kroenecker symbol as initial condition, that is a unit initial concentration at the cell "m", then (4.166) may be rewritten into the form:

$$W\tilde{u}_{n+1} - (2W + \omega + K)\tilde{u}_n + W\tilde{u}_{n-1} = -\delta_{nm} \tag{4.167}$$

The composition $\tilde{u}_n$ in (4.167) can therefore be thought as a spatial Green's function, $\tilde{D}_{n,m}$, that is the response of the system at location "n" to a delta stimulus at location "m." Equation (4.167) can be rewritten in matrix form:

$$W\begin{pmatrix} \cdots & 0 & 1 & -2\xi & 1 & 0 & 0 & \cdots \\ \cdots & 0 & 0 & 1 & -2\xi & 1 & 0 & \cdots \\ \cdots & 0 & 0 & 0 & 1 & -2\xi & 1 & 0 \end{pmatrix}\begin{pmatrix} \cdot \\ \tilde{D}_{n-1,m} \\ \tilde{D}_{n,m} \\ \tilde{D}_{n+1,m} \\ \cdot \end{pmatrix} = -\overleftrightarrow{I} \tag{4.168}$$

In (4.168), $\overleftrightarrow{I}$, is the identity matrix and $\xi = 1 + \frac{(\omega + K)}{2W}$. The diffusion-reaction matrix on the left hand side of (4.168) is tridiagonal. The solution to (4.168) was introduced in Chaps. 1 and 2 and is given by:

$$\tilde{D}_{n,m} = \frac{-1}{W}\frac{\tau^{|n-m|+1}}{\tau^2 - 1} \tag{4.169}$$

with

$$\tau = \begin{cases} \xi - \left(\xi^2 - 1\right)^{1/2} & \text{if } \xi > 1 \\ \xi + \left(\xi^2 - 1\right)^{1/2} & \text{if } \xi < -1 \\ \xi + i\left(1 - \xi^2\right)^{1/2} & \text{if - } 1 < \xi < 1 \end{cases} \tag{4.170}$$

When K is positive, it is obvious that $\xi = 1 + \frac{(\omega + K)}{2W} > 1$ and τ is given by the first relation in (4.170). When $K < 0$, then we consider the inversion formula for the Laplace transform of some function, $F(\omega)$:

$$h(t) = \frac{1}{2\pi i}\int_{C-i\infty}^{C+i\infty} e^{\omega t}F(\omega)d\omega \tag{4.171}$$

In (4.171), the integration is along the imaginary axis with the real part of ω, C, taking any large positive value, thus enabling ξ to remain greater than

1 independently of the sign of K. We can also obtain the solution by considering $K = 0$, then (4.169) can we rewritten as:

$$\tilde{D}_{nm} = \frac{1}{\sqrt{\omega(\omega+4W)}}\left[\frac{2W}{(2W+\omega)+\sqrt{\omega(\omega+4W)}}\right]^{|n-m|} \tag{4.172}$$

The inverse Laplace transform of (4.172) is given by [96]:

$$D_{nm}(t) = e^{-2Wt} I_{|n-m|}(2Wt) \tag{4.173}$$

with $I_{|n-m|}$ being the modified Bessel function. Equation (4.173) is the Green's function for the diffusion problem along a discrete chain. If we now consider a non-zero reaction rate, K, we can make the change of variable $\omega' = \omega + K$ in (4.172). Here we used the fact that the sign of K does not impact the choice of the solution in (4.170) as discussed previously due to the nature of the inverse Laplace transform. We now use the relation $LT[e^{-at}h(t)] = F(\omega + a)$ with LT representing the Laplace transformation for all a's, positive or negative, to obtain:

$$D_{nm}(t) = e^{-Kt} e^{-2Wt} I_{|n-m|}(2Wt) \tag{4.174}$$

Let us now treat the case of a two states (two reaction rates) nonlinear reaction dynamics:

$$\frac{du_n}{dt} = W(u_{n+1} - 2u_n + u_{n-1}) - (K_1 + p(K_2 - K_1))u_n \tag{4.175}$$

Following Sect. 6.2, we can now address this problem in the limit of short times *i.e.*, time intervals during which we assume that the embedding parameter, p, is constant. We are now in a position to write the sought expression for the reaction-diffusion propagator with nonlinear reaction dynamics. The Green's function propagator in the short time interval is obtained in the form:

$$D_{nm}(t - t') = e^{-\left(K_1 + p(K_2 - K_1)(t-t'\right)} e^{-2W(t-t')} I_{|n-m|}(2W(t - t')) \tag{4.176}$$

Knowing an initial condition, the time evolution of the composition field in the one-dimensional chain of cells can therefore be obtained in an iterative manner according to:

$$u_n(t + \delta t) = \sum_m D_{nm}(\delta t) u_n(t) \tag{4.177}$$

where δt is small.

4.6.4 *Propagation and Propagation Failure in a One-Dimensional Chain of Cells*

Let us treat the problem of a chain of cells with variable calcium concentration due to intake and release by Endoplasmic Reticulae (ER). This case can be treated by considering the nonlinear reaction dynamics with $K_1 > 0$ and $K_2 < 0$. From the point of view of an isolated cell, when the calcium concentration is less than the critical value u_c, the ER deplete the endoplasm. When the calcium concentration exceeds the critical value, the ER releases calcium into the cell. We consider solutions to the nonlinear reaction dynamics-diffusion problem in the form of propagating waves, that is:

$$u_n(t) = \varphi(n - ct) \tag{4.178}$$

In (4.178), position is expressed in units of $a_0 = 1$ by the integer n. c is the wave velocity. Inserting this general form into (4.177) gives:

$$\varphi(n - c(t + \delta t)) = e^{-[K_1 + p(K_2 - K_1)]\delta t} e^{-2W\delta t} \sum_m I_{|n-m|}(2W\delta t)\varphi(m - ct) \tag{4.179}$$

It is convenient to define: $\zeta_n = n - ct$ and take $\delta t = 1/c$ as the time for a one step motion. With these definitions, (4.179) reduces to:

$$\varphi(\zeta_{n-1}) = e^{-[K_1 + p(K_2 - K_1)]\delta t} e^{-2W\delta t} \sum_m I_{|n-m|}(2W\delta t)\varphi(\zeta_m) \tag{4.180}$$

If we assume that the function φ is single valued then, $p = H(\varphi(\zeta_n - u_c)) = h(\zeta_n)$. We can choose without loss of generality, the point ζ_n where the composition is equal to the critical value for switching between rate K_1 and rate K_2:

$$\begin{aligned} \varphi(\zeta_n) &< u_c \text{ for } \zeta_n < 0 \\ \varphi(\zeta_n) &> u_c \text{ for } \zeta_n > 0 \end{aligned} \tag{4.181}$$

If follows that we can also write:

$$e^{-[K_1 + p(K_2 - K_1)]\delta t} = e^{-K_1\delta t} + h(\zeta_n)\left[e^{K_2\delta t} - e^{-K_1\delta t}\right] \tag{4.182}$$

Inserting all this into (4.179) gives:

$$\varphi(\zeta_{n-1}) = \left[e^{K_1/c} + h(\zeta_n)\left[e^{K_2 c} - e^{K_1/c}\right]\right] e^{-2W/c} \sum_m I_{|n-m|}(2W/c)\varphi(\zeta_m) \tag{4.183}$$

Equation (4.183) provides a means of calculating the velocity of the wave, c, in terms of the diffusion constant W and of the reaction rates, K_1 and K_2. We assume that u_c=0.5 and we now make a simple piece-wise linear approximation to the wave solution, φ:

$$\varphi(\zeta_n) = \begin{cases} 1 \text{ if } \zeta_n > 2b \\ 0 \text{ if } \zeta_n < -2b \\ \frac{1}{2} + \frac{\zeta_n}{4b} \text{ if } -2b < \zeta_n < 2b \end{cases} \tag{4.184}$$

The parameter b represents the width of the transition region between the two calcium concentrations 0 and 1. This function represents the propagation of a concentration front. With this solution, the summation in (4.183) becomes:

$$\sum_m I_{|n-m|}(2W/c)\varphi(\zeta_m) = \sum_{-2b}^{2b} I_{|n-m|}(2W/c)\left(\frac{1}{2}+\frac{\zeta_m}{4b}\right) + \sum_{m>2b} I_{|n-m|}(2W/c) \tag{4.185}$$

We rewrite (4.183) at two points along the concentration profile, namely $n = 1$ and $n = -1$. We obtain a set of two equations:

$$\begin{aligned} \frac{1}{2} &= e^{-K_2/c}e^{-2W/c}\left[\sum_{-2b}^{2b} I_{|1-m|}(2W/c)\left(\frac{1}{2}+\frac{\zeta_m}{4b}\right) + \sum_{m>2b} I_{|1-m|}(2W/c)\right] \text{ if } n=1 \\ \frac{1}{2}-\frac{1}{2b} &= e^{-K_1/c}e^{-2W/c}\left[\sum_{-2b}^{2b} I_{|-1-m|}(2W/c)\left(\frac{1}{2}+\frac{\zeta_m}{4b}\right) + \sum_{m>2b} I_{|-1-m|}(2W/c)\right] \text{ if } n=-1 \end{aligned} \tag{4.186}$$

After extensive algebraic manipulations, using the properties of the modified Bessel functions [97], the set of (4.186) can be simplified:

$$\begin{aligned} e^{\alpha}\left[e^{K_2/c}-1\right] &= I_{2b}(\alpha) + I_{2b+1}(\alpha) + \frac{1}{b\alpha}\sum_{p=1}^{2b} p^2 I_p(\alpha) \quad \text{if } n=1 \\ \left(1-\frac{1}{b}\right)e^{\alpha}e^{K_1/c} &= e^{\alpha} - (I_0(\alpha) + I_1(\alpha) + I_{2b}(\alpha) + I_{2b+1}(\alpha)) \\ &\quad - \frac{1}{b\alpha}\sum_{p=1}^{2b} p^2 I_p(\alpha) \quad \text{if } n=-1 \end{aligned} \tag{4.187}$$

where $\alpha = 2W/c$.

We can obtain a relationship between the wave velocity, width and the physical parameters of the problem, namely diffusion coefficient and reaction rates by taking the sum of both (4.187). This gives:

$$e^{\alpha}\left[2 - e^{K_2/c} - e^{K_1/c}\left(1 - \frac{1}{b}\right)\right] = I_0(\alpha) + I_1(\alpha) \tag{4.188}$$

To determine the wave velocity, it is easier to define $x = 1/c$ and recast (4.188) in the form:

$$f(x,b) = 2 - e^{K_2 x} - e^{K_1 x}\left(1 - \frac{1}{b}\right) = e^{-2Wx}(I_0(2Wx) + I_1(2Wx)) = g(x,W) \tag{4.189}$$

Using (4.189), we can study the condition for propagation of the compositional front wave as a function of the front width, diffusion and reaction parameters. For the sake of illustration, we plot in Fig. 4.22, the functions $f(x,b)$ and $g(x,W)$ representing the left-hand side and right-hand side of (4.189) for a variety of widths of the wave front, b, and diffusion constant W, as functions of x. For simplicity we also take the rate of calcium release and intake of the calcium stores to be equal in magnitude, i.e. $K_1=1$ and $K_2=-1$.

The intersection points between the functions $f(x,b)$ and $g(x,W)$ correspond to the solutions of (4.189). There is propagation failure of the compositional wave when the two curves do not intersect. For instance, a front with $b=2$ does not propagate unless the diffusion constant exceeds a value somewhere between 1 and 2. A physically unrealistic very narrow wave front ($b=1$) appears to have a single solution for $x=0$ independently of the diffusion coefficient. This solution corresponds to an infinite wave velocity, c. There appears to exist two solutions for the speed for fronts with $b>1$ and diffusion coefficients $W>2$. It is important to note

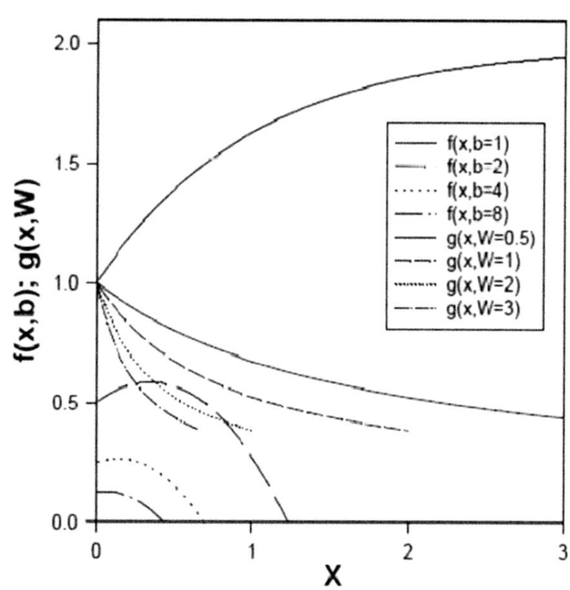

Fig. 4.22 Functions $f(x, b)$ and $g(x, b)$ for the case $K_1 = 1$, $K_2 = -1$ and $u_c = 0.5$

that the particular details of the functions f and g as well as the solutions of (4.189) may be artifacts due to the choice of a spatially-symmetric piece-wise linear function as an approximate solution of the concentration wave front. The main point that needs to be retained from this approximate analysis is that conditions in terms of diffusion and reaction dynamics may exist for which a wave front may experience propagation failure.

To investigate the failure of propagation further, we develop two numerical algorithms. The first one is a simple Finite Difference Time Domain (FDTD) method [1] applied to solving the nonlinear reaction-diffusion equation. All rates, W, K_1 and K_2 have dimension of inverse time. Since $W\delta t$ is a dimensionless quantity, transfer rates, reaction rates and time interval will be expressed as dimensionless quantities in the rest of this subsection. The second approach makes use of the propagator in (4.176) and (4.177). The second approach uses a short time approximation to the modified Bessel function [97], namely

$$I_{|n-m|}(2W\delta t) \cong \frac{(W\delta t)^{|n-m|}}{|n-m|!} \tag{4.190}$$

The nonlinear reaction dynamics is supplemented with a cap at a concentration of 1. The FDTD algorithm is significantly faster computationally than its Green's function counterpart since the evaluation of the discrete Laplacian involves only nearest neighbor cells. The Green's function algorithm, however, requires the computationally expensive summation of (4.177). We have found a computational compromise between accuracy and computing time by choosing a time step of δt=0.01. The convergence of the algorithm depends on the number of cell with non-zero contributions to the summation in (4.177). As the number of such contributions increases, convergence requires a decrease in the time step. Figure 4.23 illustrates the transition from propagation to propagation failure of the concentration wave.

The conditions of the calculations are given in the caption. These conditions differ from those of Fig. 4.22 as the later were obtained for an approximate analytical solution. More specifically, we use the critical concentration, u_c=0.3 to observe propagation and propagation failure with $K_1 = -K_2$=1 and W<1. The initial condition is a delta concentration located at cell 100 and exceeding the concentration threshold of the nonlinear reaction dynamics. Figure 4.23a shows the propagation of two symmetrical fronts from that initial condition. The deviations between the concentration profiles obtained with the FDTD and Green's function methods are the result of accumulation of errors due to the approximation made in (4.190). It is therefore clear that when W=0.7, K_1=1, and $K_2 = -1$, wave propagation takes place. The width of the wave front includes approximately 4–5 cells and is not described correctly by a linear function.

Because the transition from reaction rate K_1 to K_2 is not reversible, once cells reach a concentration of 1, they cannot revert to a lower concentration. In the context of bi-stability, the state associated with K_2 is more stable than the state K_1.

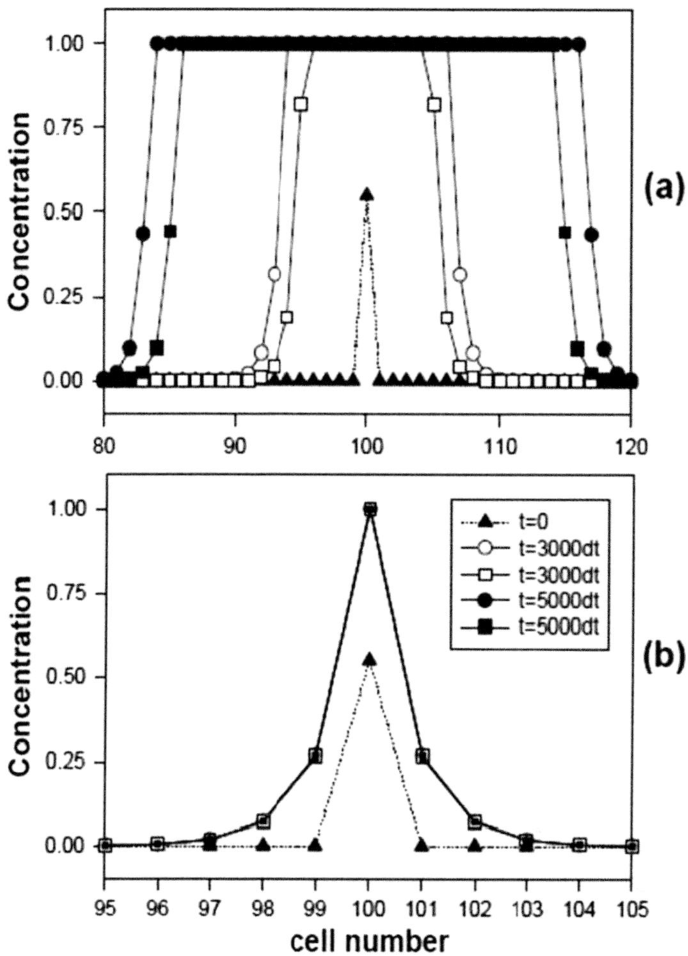

Fig. 4.23 Time evolution of calcium concentration profile for (**a**) $W = 0.7$, $K_1 = 1$, $K_2 = -1$ and (**b**) $W = 0.5$, $K_1 = 1$, $K_2 = -1$. The *open symbols* correspond to Finite Difference Time Domain (FDTD) solutions. The *closed symbols* correspond to solutions obtained with Green's function-based algorithm. The *closed triangles* are the initial conditions: $\delta t = 0.01$ and $u_c = 0.3$

A chain of cells with concentration of 1 throughout would not be able to revert to a concentration of zero via propagation of an inverted concentration front (low concentration propagating in a medium of high concentration). In that respect, the reaction model we have used is not symmetrical with respect to concentration. The propagation of concentration wave front can only occur in one direction (from low to high concentration). The propagation of concentration wave front is nonreciprocal.

Reducing the diffusion constant, W, from 0.7 to 0.5 leads to a transition to a non-propagating mode. Propagation failure leads to a concentration profile that does not evolve with time. Using the FDTD approach, the transition between front propagation and propagation failure occurs for a diffusion constant W=0.6109. We note that the two algorithms agree very well numerically since the number of cells contributing to the summation in (4.177) is small. The concentration profile at steady state has a FWHM of approximately 4–5 cells consistent with that of the propagating front.

In summary, from the numerical calculations, it is clear that the analytically derived (4.189) and corresponding Fig. 4.22 can only be used in a qualitative fashion due to the piece-wise symmetrical approximation made in defining the shape of the concentration front. However, this approach shows qualitatively that a transition from propagation to failure of propagation occurs as the diffusion constant decreases. From a numerical point of view, the Green's function approach is computationally more expensive than the FDTD method. The Green's function approach will give an advantage when considering more complex geometries.

Finally, both analytical and numerical approaches show propagation failure and non-reciprocity of a calcium wave front in an infinite chain of endothelial cells. This behavior is qualitatively comparable to other examples of propagation failure of front already reported in the literature [81, 82]. More specifically, our models show the known behavior of a transition from propagation to propagation failure when the diffusion rate is varied relative to the reaction rates. Although, we focused on calcium waves, the formalism and results presented in this section would apply to other type of signals (physical, mechanical, etc.) exhibiting rather general nonlinear bi-stable dynamics.

4.7 Nonlinear Interactions Between Elastic Waves and Multiple Spatio-Temporal Modulations of Stiffness in a One-Dimensional Waveguide

In Sect. 3.4, we introduced extrinsic topological phononic structures that were created by applying external stimuli that break symmetry. This was achieved by applying a time-varying periodic sinusoidal spatial modulation of the stiffness to a one-dimensional elastic medium. We have shown that the bulk elastic states of this time-dependent superlattice do not possess the conventional mirror symmetry in momentum space leading to non-reciprocity in the direction of propagation of the waves. A sinusoidal modulation is but one specific case of possible periodic modulations. Considering a general periodic modulation, as for example a square wave modulation, one can expand the spatial periodic functional form of the modulation into a Fourier series. This expansion becomes a superposition of a large number of sinusoidal modulations. These sinusoidal Fourier components however need to possess the same temporal evolution if one wants the overall periodic modulation to propagate without distortion. So here, we address a one-dimensional model of an elastic medium subjected to more than one sinusoidal spatio-temporal modulation. For the sake of tractability we consider only two such modulations. Again, the two modulations have different spatial characteristics but are subjected to identical temporal evolutions. We will show that the nonlinear interactions between elastic waves and the two spatio-temporal modulations can lead to a rich parameter space for the design and control of the conditions for breaking symmetry of elastic wave band structures.

In the long-wavelength limit, propagation of longitudinal elastic waves in a one-dimensional medium perturbed by a spatio-temporal modulation of its stiffness, $C(x,t)$, obeys the following equation of motion:

$$\rho\frac{\partial^2 u(x,t)}{\partial t^2} = \frac{\partial}{\partial x}\left(C(x,t)\frac{\partial u(x,t)}{\partial x}\right) \tag{4.191}$$

In (4.191), $u(x,t)$ is the displacement field and ρ is the mass density of the medium. Here, we choose a variation of the stiffness that is the superposition of two sinusoidal functions of position and time:

$$C(x,t) = C_0 + 2C_1 \sin(K_1 x + \Omega t) + 2C_2 \sin(K_2 x + \Omega t) \tag{4.192}$$

where C_0, C_1, and C_2 are positive constants. $K_i = \dfrac{2\pi}{L_i}$ where L_i is the period of the stiffness modulation $i = 1, 2$. Ω is a frequency associated with the velocity of the stiffness modulations, V. We reiterate that the two modulation possess the same velocity. The quantities K_i are independent of V. The sign of Ω determines the direction of propagation of the modulations. We now assume that the K_i's are not independent of each other as would be the case in a Fourier series. In particular, we choose the period $L_2 = L_1/2$ (*i.e.*, $K_2 = 2K_1$). The periodicity of the modulated one-dimensional medium $L = L_1 = 2L_2$ suggests that we should be seeking solutions of (4.191) in the form of Bloch waves:

$$u(x,t) = \sum_k \sum_g u(k,g,t) e^{i(k+g)x} \tag{4.193}$$

where $x \in [0, L]$. The wave number k is limited to the first Brillouin zone: $\left[\frac{-\pi}{L}, \frac{\pi}{L}\right]$ and $g = \frac{2\pi}{L} m$ with m being a positive or negative integer. With this choice of form for the solution and inserting (4.192) into (4.191), the equation of propagation takes the form:

$$\begin{aligned}\frac{\partial^2 u(k+g,t)}{\partial t^2} &+ v_a^2 (k+g)^2 u(k+g,t) = i\varepsilon\{f_1(k+g-K_1)u(k+g-K_1,t)e^{+i\Omega t}\\ &+\alpha f_2(k+g-K_2)u(k+g-K_2,t)e^{i\Omega t} + h_1(k+g+K_1)u(k+g+K_1,t)e^{-i\Omega t}\\ &+\alpha h_2(k+g+K_2)u(k+g+K_2,t)e^{-i\Omega t}\}\end{aligned} \tag{4.194}$$

where: $f_1(k)=K_1k+k^2$, $f_2(k)=K_2k+k^2$, $h_1(k)=K_1k-k^2$, $h_2(k)=K_2k-k^2$. In (4.194), we have defined: $v_a^2=\frac{C_0}{\rho}$ and $\varepsilon=\frac{C_1}{\rho}$. Furthermore, we have introduced the ratio: $\alpha=\frac{C_2}{C_1}$ with α being a finite number.

For the sake of analytical simplicity in the context of multiple time scale perturbation theory, we treat ε as a perturbation and write the displacement as a fourth order power series in the perturbation, namely:

$$\begin{aligned}u(k+g,\tau_0,\tau_1,\tau_2,\tau_3,\tau_4) = {} & u_0(k+g,\tau_0,\tau_1,\tau_2,\tau_3,\tau_4)\\ & +\varepsilon u_1(k+g,\tau_0,\tau_1,\tau_2,\tau_3,\tau_4)+\varepsilon^2u_2(k+g,\tau_0,\tau_1,\tau_2,\tau_3,\tau_4)\\ & +\varepsilon^3u_3(k+g,\tau_0,\tau_1,\tau_2,\tau_3,\tau_4)+\varepsilon^4u_4(k+g,\tau_0,\tau_1,\tau_2,\tau_3,\tau_4)\end{aligned} \tag{4.195}$$

In (4.195), u_i with $i=0,1,2,3,4$ are displacement functions expressed to zeroth-order, first-, second-, third- and fourth-order in the perturbation. We have also replaced the single time variable, t, by five variables representing different time scales: $\tau_n=\varepsilon^n t=\varepsilon^n\tau_0$ with $n=0,1,2,3,4$. We can subsequently decompose (4.194) into five equations: one equation to zeroth-order in ε, one equation to first-order in ε, a third equation to second-order in ε, a fourth equation to third-order in ε, and a fifth equation to fourth-order in ε. This decomposition is illustrated in Appendix 4, Chap. 3.

The zeroth-order equation:

$$\frac{\partial^2u_0(k+g,\tau_0,\tau_1,\tau_2,\tau_3,\tau_4)}{\partial\tau_0^2}+v_a^2(k+g)^2u_0(k+g,\tau_0,\tau_1,\tau_2,\tau_3,\tau_4)=0 \tag{4.196}$$

has solution

$$u_0(k+g,\tau_0,\tau_1,\tau_2,\tau_3,\tau_4)=a_0(k+g,\tau_1,\tau_2,\tau_3,\tau_4)e^{i\omega_0(k+g)\tau_0} \tag{4.197a}$$

with the eigen value:

$$\omega_0^2=v_a^2(k+g)^2 \tag{4.197b}$$

To first-order, the equation of motion is:

$$\begin{aligned}&\frac{\partial^2u_1(k+g,\tau_0,\tau_1,\tau_2,\tau_3,\tau_4)}{\partial\tau_0^2}+2\frac{\partial^2u_0(k+g,\tau_0,\tau_1,\tau_2,\tau_3,\tau_4)}{\partial\tau_1\partial\tau_0}\\ &+v_a^2(k+g)^2u_1(k+g,\tau_0,\tau_1,\tau_2,\tau_3,\tau_4)=i\{f_1(k+g-K_1)u_0(k+g-K_1,t)e^{+i\Omega\tau_0}\\ &+\alpha f_2(k+g-K_2)u_0(k+g-K_2,t)e^{i\Omega\tau_0}+h_1(k+g+K_1)u_0(k+g+K_1,t)e^{-i\Omega\tau_0}\\ &+\alpha h_2(k+g+K_2)u_0\left(k+g+K_2,t\right)e^{-i\Omega\tau_0}\}\end{aligned} \tag{4.198}$$

Inserting the zeroth-order solution (4.196) into (4.198) will make the second term on the right-hand side of the equation lead to a secular term. We eliminate this term by making u_0 independent of τ_1. Subsequently, we will make displacements at all order independent of τ_1. In anticipation of the appearance of similar secular terms in the fourth-order equation of motion, we will also make all displacements independent of τ_3. In fact we eliminate dependencies on the time scales with odd powers of ε.

With this, the solution of (4.198) is composed of a homogeneous solution and a particular solution:

$$
\begin{aligned}
u_1(k+g,\tau_0,\tau_2,\tau_4) &= u_{1,H}(k+g,\tau_0,\tau_2,\tau_4) + u_{1,P}(k+g,\tau_0,\tau_2,\tau_4)\\
&= a_1(k+g,\tau_2,\tau_4)e^{i\omega_0(k+g)\tau_0}\\
&\quad + i\frac{f_1(k+g-K_1)a_0(k+g-K_1,\tau_2,\tau_4)}{\omega_0^2(k+g)-(\omega_0(k+g-K_1)+\Omega)^2}e^{i(\omega_0(k+g-K_1)+\Omega)\tau_0}\\
&\quad + i\frac{\alpha f_2(k+g-K_2)a_0(k+g-K_2,\tau_2,\tau_4)}{\omega_0^2(k+g)-(\omega_0(k+g-K_2)+\Omega)^2}e^{i(\omega_0(k+g-K_2)+\Omega)\tau_0}\\
&\quad + i\frac{h_1(k+g+K_1)a_0(k+g+K_1,\tau_2,\tau_4)}{\omega_0^2(k+g)-(\omega_0(k+g+K_1)-\Omega)^2}e^{i(\omega_0(k+g+K_1)-\Omega)\tau_0}\\
&\quad + i\frac{\alpha h_2(k+g+K_2)a_0(k+g+K_2,\tau_2,\tau_4)}{\omega_0^2(k+g)-(\omega_0(k+g+K_2)-\Omega)^2}e^{i(\omega_0(k+g+K_2)-\Omega)\tau_0}
\end{aligned}
\tag{4.199}
$$

Note that the fractions in (4.199) may diverge. One may add a small imaginary term (damping) to the denominators to eliminate the divergence. The zero limit of the imaginary term would be eventually taken. For the sake of presenting the most compact equations, we have elected in the subsequent derivations to omit this damping term.

To second-order, the equation of motion takes the form:

$$
\begin{aligned}
&\frac{\partial^2 u_2(k+g,\tau_0,\tau_2,\tau_4)}{\partial\tau_0^2} + 2\frac{\partial^2 u_0(k+g,\tau_0,\tau_2,\tau_4)}{\partial\tau_2\partial\tau_0} + v_a^2(k+g)^2 u_2(k+g,\tau_0,\tau_2,\tau_4)\\
&= i\{f_1(k+g-K_1)u_1(k+g-K_1,\tau_0,\tau_2,\tau_4)e^{i\Omega\tau_0} + \alpha f_2(k+g-K_2)u_1(k+g\\
&\quad -K_2,\tau_0,\tau_2,\tau_4)e^{i\Omega\tau_0} + h_1(k+g+K_1)u_1(k+g+K_1,\tau_0,\tau_2,\tau_4)e^{-i\Omega\tau_0}\\
&\quad + \alpha h_2(k+g+K_2)u_1(k+g+K_2,\tau_0,\tau_2,\tau_4)e^{-i\Omega\tau_0}\}
\end{aligned}
\tag{4.200}
$$

We need to insert (4.199) into (4.200). Equation (4.200) is separated into two equations. The first equation is used to eliminate secular terms:

$$2\frac{\partial^2 u_0(k+g,\tau_0,\tau_2,\tau_4)}{\partial\tau_2\partial\tau_0} = -\frac{f_1(k+g-K_1)h_1(k+g)a_0(k+g,\tau_2,\tau_4)}{\omega_0^2(k+g-K_1)-(\omega_0(k+g)-\Omega)^2}e^{i\omega_0(k+g)\tau_0}$$
$$-\frac{\alpha f_2(k+g-K_2)h_2(k+g)a_0(k+g,\tau_2,\tau_4)}{\omega_0^2(k+g-K_2)-(\omega_0(k+g)-\Omega)^2}e^{i\omega_0(k+g)\tau_0}$$
$$-\frac{h_1(k+g+K_1)f_1(k+g)a_0(k+g,\tau_2,\tau_4)}{\omega_0^2(k+g+K_1)-(\omega_0(k+g)+\Omega)^2}e^{i\omega_0(k+g)\tau_0}$$
$$-\frac{\alpha h_2(k+g+K_2)f_2(k+g)a_0(k+g,\tau_2,\tau_4)}{\omega_0^2(k+g+K_2)-(\omega_0(k+g)+\Omega)^2}e^{i\omega_0(k+g)\tau_0} \tag{4.201}$$

This equation will lead to a second-order correction to the zeroth-order eigen value (4.197b). To do this we would express the zeroth-order amplitude in the complex form: $a_0(k+g,\tau_4)=\alpha_0(\tau_4)e^{i\varphi(k+g)\tau_2}$. With this, $u_0(k+g,\tau_0,\tau_2,\tau_4)=\alpha_0(\tau_4)e^{i\omega_0(k+g)\tau_0}e^{i\varphi(k+g)\tau_2}$.

The second equation takes the form:

$$\frac{\partial^2 u_2(k+g,\tau)}{\partial\tau_0^2}+v_a^2(k+g)^2u_2(k+g,\tau)=A_1+A_2+A_3+A_4 \tag{4.202}$$

where

$$A_1=\{if_1(k+g-K_1)a_1(k+g-K_1,\tau)e^{i(\omega_0(k+g-K_1)+\Omega)\tau_0}$$
$$+i\alpha f_2(k+g-K_2)a_1(k+g-K_2,\tau)e^{i(\omega_0(k+g-K_2)+\Omega)\tau_0}$$
$$+ih_1(k+g+K_1)a_1(k+g+K_1,\tau)e^{i(\omega_0(k+g+K_1)-\Omega)\tau_0}$$
$$+i\alpha h_2(k+g+K_2)a_1(k+g+K_2,\tau)e^{i(\omega_0(k+g+K_2)-\Omega)\tau_0}\}$$

and

$$A_2=\left\{-\frac{f_1(k+g-K_1)\alpha h_2(k+g-K_1+K_2)a_0(k+g-K_1+K_2,\tau)}{\omega_0^2(k+g-K_1)-(\omega_0(k+g-K_1+K_2)-\Omega)^2}e^{i(\omega_0(k+g-K_1+K_2))\tau_0}\right.$$
$$-\frac{\alpha h_2(k+g+K_2)f_1(k+g-K_1+K_2)a_0(k+g-K_1+K_2,\tau)}{\omega_0^2(k+g+K_2)-(\omega_0(k+g-K_1+K_2)+\Omega)^2}e^{i(\omega_0(k+g-K_1+K_2))\tau_0}$$
$$-\frac{\alpha f_2(k+g-K_2)h_1(k+g+K_1-K_2)a_0(k+g+K_1-K_2,\tau)}{\omega_0^2(k+g-K_2)-(\omega_0(k+g+K_1-K_2)-\Omega)^2}e^{i(\omega_0(k+g+K_1-K_2))\tau_0}$$
$$\left.-\frac{h_1(k+g+K_1)\alpha f_2(k+g+K_1-K_2)a_0(k+g+K_1-K_2,\tau)}{\omega_0^2(k+g+K_1)-(\omega_0(k+g+K_1-K_2)+\Omega)^2}e^{i(\omega_0(k+g+K_1-K_2))\tau_0}\right\}$$

and

$$A_3=\left\{-\frac{f_1(k+g-K_1)\alpha f_2(k+g-K_1-K_2)a_0(k+g-K_1-K_2,\tau)}{\omega_0^2(k+g-K_1)-(\omega_0(k+g-K_1+K_2)+\Omega)^2}e^{i(\omega_0(k+g-K_1-K_2)+2\Omega)\tau_0}\right.$$

$$-\frac{\alpha f_2(k+g-K_2)f_1(k+g-K_1-K_2)a_0(k+g-K_1-K_2,\tau)}{\omega_0^2(k+g-K_2)-(\omega_0(k+g-K_1-K_2)+\Omega)^2}e^{i(\omega_0(k+g-K_1-K_2)+2\Omega)\tau_0}$$

$$-\frac{f_1(k+g-K_1)f_1(k+g-2K_1)a_0(k+g-2K_1,\tau)}{\omega_0^2(k+g-K_1)-(\omega_0(k+g-2K_1)+\Omega)^2}e^{i(\omega_0(k+g-2K_1)+2\Omega)\tau_0}$$

$$\left.-\frac{\alpha f_2(k+g-K_2)\alpha f_2(k+g-2K_2)a_0(k+g-2K_2,\tau)}{\omega_0^2(k+g-K_2)-(\omega_0(k+g-2K_2)+\Omega)^2}e^{i(\omega_0(k+g-2K_2)+2\Omega)\tau_0}\right\}$$

and

$$A_4=\left\{-\frac{h_1(k+g+K_1)\alpha h_2(k+g+K_1+K_2)a_0(k+g+K_1+K_2,\tau)}{\omega_0^2(k+g+K_1)-(\omega_0(k+g+K_1+K_2)-\Omega)^2}e^{i(\omega_0(k+g+K_1+K_2)-2\Omega)\tau_0}\right.$$

$$-\frac{\alpha h_2(k+g+K_2)h_1(k+g+K_1+K_2)a_0(k+g+K_1+K_2,\tau)}{\omega_0^2(k+g+K_2)-(\omega_0(k+g+K_1+K_2)-\Omega)^2}e^{i(\omega_0(k+g+K_1+K_2)-2\Omega)\tau_0}$$

$$-\frac{h_1(k+g+K_1)h_1(k+g+2K_1)a_0(k+g+2K_1,\tau)}{\omega_0^2(k+g+K_1)-(\omega_0(k+g+2K_1)-\Omega)^2}e^{i(\omega_0(k+g+2K_1)-2\Omega)\tau_0}$$

$$\left.-\frac{\alpha h_2(k+g+K_2)\alpha h(k+g+2K_2)a_0(k+g+2K_2,\tau)}{\omega_0^2(k+g+K_2)-(\omega_0(k+g+2K_2)-\Omega)^2}e^{i(\omega_0(k+g+2K_2)-2\Omega)\tau_0}\right\}$$

For the sake of compactness, in the preceding relations we have replaced the dependency on τ_0, τ_2, τ_4 by a single symbol: τ.

We now seek the solution of (4.202) in the form of the sum of a homogenous part and a particular part:

$$u_2 = u_{2,H} + u_{2,P} \tag{4.203}$$

The homogenous solutions is:

$$u_{2,H}(k+g,\tau) = a_2(k+g,\tau)e^{i\omega_0(k+g)\tau_0}$$

The particular solution, $u_{2,p}$ includes several parts arising from the driving terms A_1, A_2, A_3, and A_4. The term A_1 leads to the following contributions to the particular solution:

$$u_{2,p}^{(1)} = \left\{ \frac{1}{\omega_0^2(k+g) - (\omega_0(k+g-K_1)+\Omega)^2} if_1(k+g-K_1) \right.$$
$$\times a_1(k+g-K_1,\tau)e^{i(\omega_0(k+g-K_1)+\Omega)\tau_0}$$
$$+\frac{1}{\omega_0^2(k+g) - (\omega_0(k+g-K_2)+\Omega)^2} i\alpha f_2(k+g-K_2)$$
$$\times a_1(k+g-K_2,\tau)e^{i(\omega_0(k+g-K_2)+\Omega)\tau_0}$$
$$+\frac{1}{\omega_0^2(k+g) - (\omega_0(k+g+K_1)-\Omega)^2} ih_1(k+g+K_1)$$
$$\times a_1(k+g+K_1,\tau)e^{i(\omega_0(k+g+K_1)-\Omega)\tau_0}$$
$$+\frac{1}{\omega_0^2(k+g) - (\omega_0(k+g+K_2)-\Omega)^2} i\alpha h_2(k+g+K_2)$$
$$\left. \times a_1(k+g+K_2,\tau)e^{i(\omega_0(k+g+K_2)-\Omega)\tau_0} \right\}$$

The term A_2 leads to the following contributions to the particular solution:

$$u_{2,p}^{(2)} = \frac{1}{\omega_0^2(k+g) - \omega_0^2(k+g-K_1+K_2)}$$
$$\times \left\{ -\frac{f_1(k+g-K_1)\alpha h_2(k+g-K_1+K_2)a_0(k+g-K_1+K_2,\tau)}{\omega_0^2(k+g-K_1) - (\omega_0(k+g-K_1+K_2)-\Omega)^2} \right.$$
$$\left. -\frac{\alpha h_2(k+g+K_2)f_1(k+g-K_1+K_2)a_0(k+g-K_1+K_2,\tau)}{\omega_0^2(k+g+K_2) - (\omega_0(k+g-K_1+K_2)+\Omega)^2} \right\} e^{i(\omega_0(k+g-K_1+K_2))\tau_0}$$
$$+\frac{1}{\omega_0^2(k+g) - \omega_0^2(k+g+K_1-K_2)}$$
$$\times \left\{ -\frac{\alpha f_2(k+g-K_2)h_1(k+g+K_1-K_2)a_0(k+g+K_1-K_2,\tau)}{\omega_0^2(k+g-K_2) - (\omega_0(k+g+K_1-K_2)-\Omega)^2} \right.$$
$$\left. -\frac{h_1(k+g+K_1)\alpha f_2(k+g+K_1-K_2)a_0(k+g+K_1-K_2,\tau)}{\omega_0^2(k+g+K_1) - (\omega_0(k+g+K_1-K_2)+\Omega)^2} \right\} e^{i(\omega_0(k+g+K_1-K_2))\tau_0}$$

The term A_3 gives:

$$u_{2,p}^{(3)}=\frac{1}{\omega_0^2(k+g)-(\omega_0(k+g-K_1-K_2)+2\Omega)^2}$$
$$\times\left\{-\frac{f_1(k+g-K_1)\alpha f_2(k+g-K_1-K_2)a_0(k+g-K_1-K_2,\tau)}{\omega_0^2(k+g-K_1)-(\omega_0(k+g-K_1-K_2)+\Omega)^2}\right.$$
$$\left.-\frac{\alpha f_2(k+g-K_2)f_1(k+g-K_1-K_2)a_0(k+g-K_1-K_2,\tau)}{\omega_0^2(k+g-K_2)-(\omega_0(k+g-K_1-K_2)+\Omega)^2}\right\}e^{i(\omega_0(k+g-K_1-K_2)+2\Omega)\tau_0}$$
$$-\frac{1}{\omega_0^2(k+g)-(\omega_0(k+g-2K_1)+2\Omega)^2}$$
$$\times\frac{f_1(k+g-K_1)f_1(k+g-2K_1)a_0(k+g-2K_1,\tau)}{\omega_0^2(k+g-K_1)-(\omega_0(k+g-2K_1)+\Omega)^2}e^{i(\omega_0(k+g-2K_1)+2\Omega)\tau_0}$$
$$-\frac{1}{\omega_0^2(k+g)-(\omega_0(k+g-2K_2)+2\Omega)^2}$$
$$\times\frac{\alpha f_2(k+g-K_2)\alpha f_2(k+g-2K_2)a_0(k+g-2K_2,\tau)}{\omega_0^2(k+g-K_2)-(\omega_0(k+g-2K_2)+\Omega)^2}e^{i(\omega_0(k+g-2K_2)+2\Omega)\tau_0}$$

The term A_4 leads to:

$$u_{2,p}^{(4)}=\frac{1}{\omega_0^2(k+g)-(\omega_0(k+g+K_1+K_2)-2\Omega)^2}$$
$$\times\left\{-\frac{h_1(k+g+K_1)\alpha h_2(k+g+K_1+K_2)a_0(k+g+K_1+K_2,\tau)}{\omega_0^2(k+g+K_1)-(\omega_0(k+g+K_1+K_2)-\Omega)^2}\right.$$
$$\left.-\frac{\alpha h_2(k+g+K_2)h_1(k+g+K_1+K_2)a_0(k+g+K_1+K_2,\tau)}{\omega_0^2(k+g+K_2)-(\omega_0(k+g+K_1+K_2)-\Omega)^2}\right\}e^{i(\omega_0(k+g+K_1+K_2)-2\Omega)\tau_0}$$
$$-\frac{1}{\omega_0^2(k+g)-(\omega_0(k+g+2K_1)-2\Omega)^2}$$
$$\times\frac{h_1(k+g+K_1)h_1(k+g+2K_1)a_0(k+g+2K_1,\tau)}{\omega_0^2(k+g+K_1)-(\omega_0(k+g+2K_1)-\Omega)^2}e^{i(\omega_0(k+g+2K_1)-2\Omega)\tau_0}$$
$$-\frac{1}{\omega_0^2(k+g)-(\omega_0(k+g+2K_2)-2\Omega)^2}$$
$$\times\frac{\alpha h_2(k+g+K_2)\alpha h(k+g+2K_2)a_0(k+g+2K_2,\tau)}{\omega_0^2(k+g+K_2)-(\omega_0(k+g+2K_2)-\Omega)^2}e^{i(\omega_0(k+g+2K_2)-2\Omega)\tau_0}$$

The wave equation to third-order is given by:

$$\frac{\partial^2 u_3(k+g,\tau)}{\partial\tau_0^2}+2\frac{\partial^2 u_1(k+g,\tau)}{\partial\tau_0\partial\tau_2}+v_a^2(k+g)^2u_3(k+g,\tau)$$
$$=i\{f_1(k+g-K_1)u_2(k+g-K_1,\tau)e^{i\Omega\tau_0}+\alpha f_2(k+g-K_2)u_2(k+g-K_2,\tau)e^{i\Omega\tau_0}$$
$$+h_1(k+g+K_1)u_2(k+g+K_1,\tau)e^{-i\Omega\tau_0}$$
$$+\alpha h_2(k+g+K_2)u_2(k+g+K_2,\tau)e^{-i\Omega\tau_0}\} \tag{4.204}$$

This equation is rewritten in the form:

$$\frac{\partial^2 u_3(k+g,\tau)}{\partial\tau_0^2}+2\frac{\partial^2 u_1(k+g,\tau)}{\partial\tau_0\partial\tau_2}+v_a^2(k+g)^2u_3(k+g,\tau)$$
$$=i\{B_1+B_2+B_3+B_4\}$$

We find:

$$\begin{aligned}\frac{\partial^2 u_1(k+g,\tau)}{\partial\tau_0\partial\tau_2}=&\frac{\partial a_1(k+g,\tau)}{\partial\tau_2}i\omega_0(k+g)e^{i\omega_0(k+g)\tau_0}\\
&+i\frac{f_1(k+g-K_1)}{\omega_0^2(k+g)-(\omega_0(k+g-K_1)+\Omega)^2}\alpha_0 i\varphi(k+g-K_1)\\
&\times e^{i\varphi(k+g-K_1)\tau_2}i(\omega_0(k+g-K_1)+\Omega)e^{i(\omega_0(k+g-K_1)+\Omega)\tau_0}\\
&+i\frac{\alpha f_2(k+g-K_2)}{\omega_0^2(k+g)-(\omega_0(k+g-K_2)+\Omega)^2}\alpha_0 i\varphi(k+g-K_2)\\
&\times e^{i\varphi(k+g-K_2)\tau_2}i(\omega_0(k+g-K_2)+\Omega)e^{i(\omega_0(k+g-K_2)+\Omega)\tau_0}\\
&+i\frac{h_1(k+g+K_1)}{\omega_0^2(k+g)-(\omega_0(k+g+K_1)-\Omega)^2}\alpha_0 i\varphi(k+g+K_1)\\
&\times e^{i\varphi(k+g+K_1)\tau_2}i(\omega_0(k+g+K_1)-\Omega)e^{i(\omega_0(k+g+K_1)-\Omega)\tau_0}\\
&+i\frac{\alpha h_2(k+g+K_2)}{\omega_0^2(k+g)-(\omega_0(k+g+K_2)-\Omega)^2}\alpha_0 i\varphi(k+g+K_2)\\
&\times e^{i\varphi(k+g+K_2)\tau_2}i(\omega_0(k+g+K_2)-\Omega)e^{i(\omega_0(k+g+K_2)-\Omega)\tau_0}\end{aligned} \tag{4.205}$$

We now use (4.199) and (4.203) to evaluate B_1, B_2, B_3, and B_4. We illustrate the process with the term B_1:

$$
\begin{aligned}
iB_1 &= if_1(k+g-K_1)u_2(k+g-K_1,\tau)e^{i\Omega\tau_0} \\
&= if_1(k+g-K_1)a_2(k+g-K_1,\tau)e^{i(\omega_0(k+g-K_1)+\Omega)\tau_0} \\
&\quad + if_1(k+g-K_1)\left\{\frac{1}{\omega_0^2(k+g-K_1)-(\omega_0(k+g-2K_1)+\Omega)^2}if_1(k+g-2K_1)\right. \\
&\quad \times a_1(k+g-2K_1,\tau)e^{i(\omega_0(k+g-2K_1)+2\Omega)\tau_0} \\
&\quad + \frac{1}{\omega_0^2(k+g-K_1)-(\omega_0(k+g-K_1-K_2)+\Omega)^2}i\alpha f_2(k+g-K_1-K_2) \\
&\quad \times a_1(k+g-K_1-K_2,\tau)e^{i(\omega_0(k+g-K_1-K_2)+2\Omega)\tau_0} \\
&\quad + \frac{1}{\omega_0^2(k+g-K_1)-(\omega_0(k+g)-\Omega)^2}ih_1(k+g)a_1(k+g,\tau)e^{i(\omega_0(k+g))\tau_0} \\
&\quad + \frac{1}{\omega_0^2(k+g-K_1)-(\omega_0(k+g-K_1+K_2)-\Omega)^2}i\alpha h_2(k+g-K_1+K_2) \\
&\quad \left.\times a_1(k+g-K_1+K_2,\tau)e^{i(\omega_0(k+g-K_1+K_2))\tau_0}\right\} \\
&\quad + if_1(k+g-K_1)\left\{\frac{1}{\omega_0^2(k+g-K_1)-\omega_0^2(k+g-2K_1+K_2)}\right. \\
&\quad \times\left\{-\frac{f_1(k+g-2K_1)\alpha h_2(k+g-2K_1+K_2)a_0(k+g-2K_1+K_2,\tau)}{\omega_0^2(k+g-2K_1)-(\omega_0(k+g-2K_1+K_2)-\Omega)^2}\right. \\
&\quad \left.-\frac{\alpha h_2(k+g-K_1+K_2)f_1(k+g-2K_1+K_2)a_0(k+g-2K_1+K_2,\tau)}{\omega_0^2(k+g-K_1+K_2)-(\omega_0(k+g-2K_1+K_2)+\Omega)^2}\right\} \\
&\quad \times e^{i(\omega_0(k+g-2K_1+K_2)+\Omega)\tau_0} \\
&\quad + \frac{1}{\omega_0^2(k+g-K_1)-\omega_0^2(k+g-K_2)} \\
&\quad \times\left\{-\frac{\alpha f_2(k+g-K_1-K_2)h_1(k+g-K_2)a_0(k+g-K_2,\tau)}{\omega_0^2(k+g-K_1-K_2)-(\omega_0(k+g-K_2)-\Omega)^2}\right. \\
&\quad \left.\left.-\frac{h_1(k+g)\alpha f_2(k+g-K_2)a_0(k+g-K_2,\tau)}{\omega_0^2(k+g)-(\omega_0(k+g-K_2)+\Omega)^2}\right\}e^{i(\omega_0(k+g-K_2)+\Omega)\tau_0}\right\} \\
&\quad + if_1(k+g-K_1)\left\{\frac{1}{\omega_0^2(k+g-K_1)-(\omega_0(k+g-2K_1-K_2)+2\Omega)^2}\right.
\end{aligned}
$$

$$\times \Bigg\{ -\frac{f_1(k+g-2K_1)\alpha f_2(k+g-2K_1-K_2)a_0(k+g-2K_1-K_2,\tau)}{\omega_0^2(k+g-2K_1)-(\omega_0(k+g-2K_1+K_2)+\Omega)^2}$$

$$-\frac{\alpha f_2(k+g-K_1-K_2)f_1(k+g-2K_1-K_2)a_0(k+g-2K_1-K_2,\tau)}{\omega_0^2(k+g-K_1-K_2)-(\omega_0(k+g-2K_1-K_2)+\Omega)^2}\Bigg\}$$

$$\times e^{i(\omega_0(k+g-2K_1-K_2)+3\Omega)\tau_0}-\frac{1}{\omega_0^2(k+g-K_1)-(\omega_0(k+g-3K_1)+2\Omega)^2}$$

$$\times \frac{f_1(k+g-2K_1)f_1(k+g-3K_1)a_0(k+g-3K_1,\tau)}{\omega_0^2(k+g-2K_1)-(\omega_0(k+g-3K_1)+\Omega)^2}e^{i(\omega_0(k+g-3K_1)+3\Omega)\tau_0}$$

$$-\frac{1}{\omega_0^2(k+g-K_1)-(\omega_0(k+g-K_1-2K_2)+2\Omega)^2}$$

$$\frac{\alpha f_2(k+g-K_1-K_2)\alpha f_2(k+g-K_1-2K_2)a_0(k+g-K_1-2K_2,\tau)}{\omega_0^2(k+g-K_1-K_2)-(\omega_0(k+g-K_1-2K_2)+\Omega)^2}$$

$$\times e^{i(\omega_0(k+g-K_1-2K_2)+3\Omega)\tau_0}\Bigg\}$$

$$+\,if_1(k+g-K_1)\Bigg\{\frac{1}{\omega_0^2(k+g-K_1)-(\omega_0(k+g+K_2)-2\Omega)^2}$$

$$\times \Bigg\{-\frac{h_1(k+g)\alpha h_2(k+g+K_2)a_0(k+g+K_2,\tau)}{\omega_0^2(k+g)-(\omega_0(k+g+K_2)-\Omega)^2}$$

$$-\frac{\alpha h_2(k+g-K_1+K_2)h_1(k+g+K_2)a_0(k+g+K_2,\tau)}{\omega_0^2(k+g-K_1+K_2)-(\omega_0(k+g+K_2)-\Omega)^2}\Bigg\}e^{i(\omega_0(k+g+K_2)-\Omega)\tau_0}$$

$$-\frac{1}{\omega_0^2(k+g-K_1)-(\omega_0(k+g+K_1)-2\Omega)^2}$$

$$\times \frac{h_1(k+g)h_1(k+g+K_1)a_0(k+g+K_1,\tau)}{\omega_0^2(k+g)-(\omega_0(k+g+K_1)-\Omega)^2}e^{i(\omega_0(k+g+K_1)-\Omega)\tau_0}$$

$$-\frac{1}{\omega_0^2(k+g-K_1)-(\omega_0(k+g-K_1+2K_2)-2\Omega)^2}$$

$$\times \frac{\alpha h_2(k+g-K_1+K_2)\alpha h(k+g-K_1+2K_2)a_0(k+g-K_1+2K_2,\tau)}{\omega_0^2(k+g-K_1+K_2)-(\omega_0(k+g-K_1+2K_2)-\Omega)^2}$$

$$\times e^{i(\omega_0(k+g-K_1+2K_2)-\Omega)\tau_0}\Bigg\}$$

Similarly complex expression will be found for the other driving terms. We note that all the secular terms of form $e^{i\omega_0(k+g)\tau_0}$ will be used to eliminate the first term on the right-hand side of (4.205). We are particularly interested in driving terms with the form $e^{i(\omega_0(k^*)\pm\Omega)\tau_0}$ and resonant amplitudes with denominators of the form: $\omega_0^2(k+g+K') - (\omega_0(k+g+K'') \pm 2\Omega)^2$. For instance, the term B_2 leads to terms of the form:

$$
\begin{aligned}
&i\alpha f_2(k+g-K_2)\left\{\frac{1}{\omega_0^2(k+g-K_2)-(\omega_0(k+g+K_1)-2\Omega)^2}\right.\\
&\quad\times\left\{-\frac{h_1(k+g+K_1-K_2)\alpha h_2(k+g+K_1K_2)a_0(k+g+K_1,\tau)}{\omega_0^2(k+g+K_1)-(\omega_0(k+g+K_1+K_2)-\Omega)^2}\right.\\
&\quad\left.-\frac{\alpha h_2(k+g)h_1(k+g+K_1)a_0(k+g+K_1,\tau)}{\omega_0^2(k+g)-(\omega_0(k+g+K_1)-\Omega)^2}\right\}e^{i(\omega_0(k+g+K_1)-\Omega)\tau_0}\\
&\quad-\frac{1}{\omega_0^2(k+g-K_2)-(\omega_0(k+g+2K_1-K_2)-2\Omega)^2}\\
&\quad\times\frac{h_1(k+g+K_1-K_2)h_1(k+g+2K_1-K_2)a_0(k+g+2K_1-K_2,\tau)}{\omega_0^2(k+g+K_1-K_2)-(\omega_0(k+g+2K_1-K_2)-\Omega)^2}\\
&\quad\times e^{i(\omega_0(k+g+2K_1-K_2)-\Omega)\tau_0}-\frac{1}{\omega_0^2(k+g-K_2)-(\omega_0(k+g+K_2)-2\Omega)^2}\\
&\quad\left.\times\frac{\alpha h_2(k+g)\alpha h(k+g+K_2)a_0(k+g+K_2,\tau)}{\omega_0^2(k+g)-(\omega_0(k+g+K_2)-\Omega)^2}e^{i(\omega_0(k+g+K_2)-\Omega)\tau_0}\right\}
\end{aligned}
$$

The term B_3 gives:

$$
\begin{aligned}
&ih_1(k+g+K_1)\left\{\frac{1}{\omega_0^2(k+g+K_1)-(\omega_0(k+g-K_2)+2\Omega)^2}\right.\\
&\quad\times\left\{-\frac{f_1(k+g)\alpha f_2(k+g-K_2)a_0(k+g-K_2,\tau)}{\omega_0^2(k+g)-(\omega_0(k+g-K_2)+\Omega)^2}\right.\\
&\quad\left.-\frac{\alpha f_2(k+g+K_1-K_2)f_1(k+g-K_2)a_0(k+g-K_2,\tau)}{\omega_0^2(k+g+K_1-K_2)-(\omega_0(k+g-K_2)+\Omega)^2}\right\}e^{i(\omega_0(k+g-K_2)+\Omega)\tau_0}\\
&\quad-\frac{1}{\omega_0^2(k+g+K_1)-(\omega_0(k+g-K_1)+2\Omega)^2}
\end{aligned}
$$

$$-\frac{f_1(k+g)f_1(k+g-K_1)a_0(k+g-K_1,\tau)}{\omega_0^2(k+g)-(\omega_0(k+g-K_1)+\Omega)^2}e^{i(\omega_0(k+g-K_1)+\Omega)\tau_0}$$

$$-\frac{1}{\omega_0^2(k+g+K_1)-(\omega_0(k+g+K_1-2K_2)+2\Omega)^2}$$

$$\times\frac{\alpha f_2(k+g+K_1-K_2)\alpha f_2(k+g+K_1-2K_2)a_0(k+g+K_1-2K_2,\tau)}{\omega_0^2(k+g+K_1-K_2)-(\omega_0(k+g+K_1-2K_2)+\Omega)^2}$$

$$\left.\times e^{i(\omega_0(k+g+K_1-2K_2)+\Omega)\tau_0}\right\}$$

Finally, B_4 gives:

$$i\alpha h_2(k+g+K_2)\left\{\frac{1}{\omega_0^2(k+g+K_2)-(\omega_0(k+g-K_1)+2\Omega)^2}\right.$$

$$\times\left\{-\frac{f_1(k+g-K_1+K_2)\alpha f_2(k+g-K_1)a_0(k+g-K_1,\tau)}{\omega_0^2(k+g-K_1+K_2)-(\omega_0(k+g-K_1)+\Omega)^2}\right.$$

$$\left.-\frac{\alpha f_2(k+g)f_1(k+g-K_1)a_0(k+g-K_1,\tau)}{\omega_0^2(k+g)-(\omega_0(k+g-K_1)+\Omega)^2}\right\}e^{i(\omega_0(k+g-K_1)+\Omega)\tau_0}$$

$$-\frac{1}{\omega_0^2(k+g+K_2)-(\omega_0(k+g-2K_1+K_2)+2\Omega)^2}$$

$$\times\frac{f_1(k+g-K_1+K_2)f_1(k+g-2K_1+K_2)a_0(k+g-2K_1+K_2,\tau)}{\omega_0^2(k+g-K_1+K_2)-(\omega_0(k+g-2K_1+K_2)+\Omega)^2}$$

$$\times e^{i(\omega_0(k+g-2K_1+K_2)+\Omega)\tau_0}-\frac{1}{\omega_0^2(k+g+K_2)-(\omega_0(k+g-K_2)+2\Omega)^2}$$

$$\left.\times\frac{\alpha f_2(k+g)\alpha f_2(k+g-K_2)a_0(k+g-K_2,\tau)}{\omega_0^2(k+g)-(\omega_0(k+g-K_2)+\Omega)^2}e^{i(\omega_0(k+g-K_2)+\Omega)\tau_0}\right\}$$

We are not solving for the complete third-order solution but we gain insight into some of the terms. For instance consider a driving term of the form: $\zeta e^{i(\omega_0(k+g+K_2)-\Omega)\tau_0}$ with $\zeta=\frac{1}{\omega_0^2(k+g-K_1)-(\omega_0(k+g+K_2)-2\Omega)^2}\xi$, then its contribution to the particular solution is: $u_{3,P}^*=\frac{1}{\omega_0^2(k+g)-(\omega_0(k+g+K_2)-\Omega)^2}\zeta e^{i(\omega_0(k+g+K_2)-\Omega)\tau_0}$. Now these solutions will drive the fourth-order equation:

$$\frac{\partial^2 u_4(k+g,\tau)}{\partial \tau_0^2} + 2\frac{\partial^2 u_2(k+g,\tau)}{\partial \tau_0 \partial \tau_2} + 2\frac{\partial^2 u_0(k+g,\tau)}{\partial \tau_4 \partial \tau_0}$$
$$+ \frac{\partial^2 u_0(k+g,\tau)}{\partial \tau_2^2} + v_a^2(k+g)^2 u_4(k+g,\tau)$$
$$= i\{f_1(k+g-K_1)u_3(k+g-K_1,\tau)e^{+i\Omega\tau_0} + \alpha f_2(k+g-K_2)$$
$$\times u_3(k+g-K_2,\tau)e^{i\Omega\tau_0} + h_1(k+g+K_1)u_3(k+g+K_1,\tau)e^{-i\Omega\tau_0}$$
$$+\alpha h_2(k+g+K_2)u_3\left(k+g+K_2,\tau\right)e^{-i\Omega\tau_0}\} \tag{4.206}$$

The driving term $i\alpha f_2(k+g-K_2)u_3(k+g-K_2,\tau)e^{i\Omega\tau_0}$ will lead to secular terms of the form: $i\alpha f_2(k+g-K_2)\frac{1}{\omega_0^2(k+g-K_2)-(\omega_0(k+g)-\Omega)^2}\frac{1}{\omega_0^2(k+g-K_1-K_2)-(\omega_0(k+g)-2\Omega)^2}$ $\xi(k+g-K_2)e^{i\omega_0(k+g)\tau_0}$.

In this expression: $\xi(k+g-K_2)=if_1(k+g-K_1-K_2)\left\{-\frac{h_1(k+g-K_2)\alpha h_2(k+g)a_0(k+g,\tau)}{\omega_0^2(k+g-K_2)-(\omega_0(k+g)-\Omega)^2}-\frac{\alpha h_2(k+g-K_1)h_1(k+g)a_0(k+g,\tau)}{\omega_0^2(k+g-K_1)-(\omega_0(k+g)-\Omega)^2}\right\}$. This secular term will need to cancel in part the secular terms arising from: $2\frac{\partial^2 u_0(k+g,\tau)}{\partial\tau_4\partial\tau_0}+\frac{\partial^2 u_0(k+g,\tau)}{\partial\tau_2^2}$ in (4.206). Note that for doing this, we will need to use $a_0(k+g,\tau_4)=\alpha_0(\tau_4)e^{i\varphi(k+g)\tau_2}=\alpha_0' e^{i\psi\tau_4}e^{i\varphi\tau_2}$ with the unknown quantity being ψ. Here we note that the quantity ξ is proportional to the quantity α. The secular terms proportional to α represent mixing of the two modulation. Indeed if $\alpha=0$, the problem reduces to that of a single modulation.

The complete set of secular driving terms proportional to α include resonances of the type:

$$\frac{1}{\omega_0^2(k+g-K_1-K_2)-(\omega_0(k+g)-2\Omega)^2}$$
$$\frac{1}{\omega_0^2(k+g-2K_1)-(\omega_0(k+g)-2\Omega)^2}$$
$$\frac{1}{\omega_0^2(k+g-2K_2)-(\omega_0(k+g)-2\Omega)^2}$$
$$\frac{1}{\omega_0^2(k+g+K_1+K_2)-(\omega_0(k+g)+2\Omega)^2}$$
$$\frac{1}{\omega_0^2(k+g+2K_1)-(\omega_0(k+g)+2\Omega)^2}$$
$$\frac{1}{\omega_0^2(k+g+2K_2)-(\omega_0(k+g)+2\Omega)^2}$$

Similarly to the case of the single modulation, these resonances correct the dispersion relation in an asymmetric way and may lead to asymmetric band gaps within the Brillouin zone resulting from hybridization between a zeroth-order mode and bands shifted in frequency by 2Ω (*i.e.*, $(\omega_0(k+g) \pm 2\Omega)$).

To verify the existence of such band gaps, we are considering the same extrinsic topological phononic structure as in Sect. 3.4 but replacing the single sinusoidal modulation by a modulation with the same period and same velocity but composed of a Gaussian function. The Fourier series of such a periodic function will include a number of sinusoidal terms with commensurate wave numbers. The band structure calculated with the SED method for this system is given in Fig. 4.24. This figure shows clearly the existence of an asymmetric hybridization gap between a zeroth-order mode and a band shifted in frequency by -2Ω. This model illustrates the notion of non-conventional band structure design through the interaction between Fourier components of general periodic spatial patterns of symmetry breaking spatio-temporal modulations.

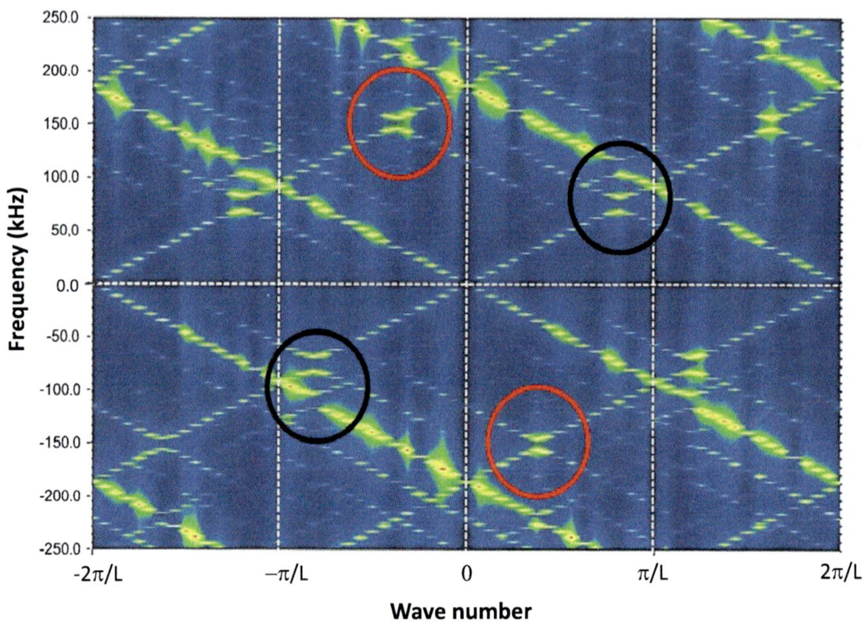

Fig. 4.24 Band structure calculated with the SED method for a system identical to that of Sect. 3.4 but subjected to a periodic modulation composed of a Gaussian function. The *black circles* highlight the hybridization gaps reported in Chap. 3. The *red circle*, indicates asymmetric band gaps due to the hybridization between a zeroth-order mode and a frequency shifted mode by -2Ω

References

1. P. A. Deymier (ed.), *Acoustic Metamaterials and Phononic Crystals, Chapter 7*, Springer Series in Solid-State Sciences, vol 173 (Springer, Berlin, 2013)
2. N. Boechler, G. Theocharis, C. Daraio, Bifurcation-based acoustic switching and rectification. Nat. Mater. **10**, 665 (2011)
3. N. Boechler, J.K. Eliason, A. Kumar, A.A. Maznev, K.A. Nelson, N. Fang, Interaction of a contact resonance of microspheres with surface acoustic waves. Phys. Rev. Lett. **111**, 036103 (2013)
4. N. Boechler, G. Theocharis, S. Job, P.G. Kevrekidis, M.A. Porter, C. Daraio, Discrete breathers in one-dimensional diatomic granular crystals. Phys. Rev. Lett. **104**, 244302 (2010)
5. G. Theocharis, N. Boechler, P.G. Kevrekidis, S. Job, M.A. Porter, C. Daraio, Intrinsic energy localization through discrete gap breathers in one-dimensional diatomic granular crystals. Phys. Rev. E **82**, 056604 (2010)
6. L. Ponson, N. Boechler, Y.M. Lai, M.A. Porter, P.G. Kevrekidis, C. Daraio, Nonlinear waves in disordered diatomic granular chains. Phys. Rev. E **82**, 021301 (2010)
7. T.K. Hsiao, H.K. Chang, S.C. Liou, M.W. Chu, S.C. Lee, C.W. Chang, Observation of room-temperature ballistic thermal conduction persisting over 8.3 μm in SiGe nanowires. Nat. Nanotechnol. **8**, 534 (2013)
8. I. Lisiecki, D. Polli, C. Yan, G. Soavi, E. Duval, G. Cerullo, M.P. Pileni, Coherent longitudinal acoustic phonons in three-dimensional supracrystals of cobalt nanocrystals. Nano Lett. **13**, 4914 (2013)
9. J. Ravichandran, A.K. Yadav, R. Cheaito, P.B. Rossen, A. Soukiassian, S.J. Suresha, J.C. Duda, B.M. Foley, C.H. Lee, Y. Zhu, A.W. Lichtenberger, J.E. Moore, D.A. Muller, D.G. Schlom, P.E. Hopkins, A. Majumdar, R. Ramesh, M.A. Zurbuchen, Crossover from incoherent to coherent phonon scattering in epitaxial oxide superlattices. Nat. Mater. **13**, 168 (2014)
10. A.A. Maznev, F. Hofmann, A. Jandl, K. Esfarjani, M.T. Bulsara, E.A. Fitzgerald, G. Chen, K.A. Nelson, Lifetime of sub-THz coherent acoustic phonons in a GaAs-AlAs superlattice. Appl. Phys. Lett. **102**, 041901 (2013)
11. C. Klieber, V.E. Gusev, T. Pezeril, K.A. Nelson, Nonlinear acoustics at GHz frequencies in a viscoelastic fragile glass former. Phys. Rev. Lett. **114**, 065701 (2015)
12. N. Swinteck, P.A. Deymier, K. Muralidharan, Phonon scattering in one-dimensional anharmonic crystals and superlattices: analytical and numerical study. J. Sound Vib. **135**, 041016 (2013)
13. B.L. Davis, M.I. Hussein, Nanophononic metamaterial: thermal conductivity reduction by local resonance. Phys. Rev. Lett. **112**, 055505 (2014)
14. M. Schaeffer, M. Ruzzene, Wave propagation in multistable magneto-elastic lattices. Int. J. Solids Struct. **56–57**, 78 (2015)
15. P.A. Deymier, K. Runge, N. Swinteck, K. Muralidharan, Rotational modes in a phononic crystal with fermion-like behavior. J. Appl. Phys. **115**, 163510 (2014)
16. A. Moussatov, V. Gusev, B. Castagnede, Self-induced hysteresis for nonlinear acoustic waves in cracked material. Phys. Rev. Lett. **90**, 124301 (2003)
17. P.-Y. Guerder, A.C. Deymier-Black, N.Z. Swinteck, J.O. Vasseur, O. Bou-Matar, K. Muralidharan, P.A. Deymier, Multi-phonon scattering processes in one-dimensional anharmonic biological superlattices: understanding the dissipation of mechanical waves in mineralized tissues. J. Mech. Behav. Biomed. Mater. **37**, 24 (2014)
18. V.Y. Zaitsev, A model of anomalous elastic nonlinearity of microinhomogeneous media. Acoust. Lett. **19**, 171 (1996)
19. N. Nadkarni, C. Daraio, D.M. Kochmann, Dynamics of periodic mechanical structures containing bistable elastic elements: from elastic to solitary wave propagation. Phys. Rev. E **90**, 023204 (2014)

20. N. Nadkarnia, A.F. Arrieta, C. Chong, D.M. Kochmann, C. Daraio, Unidirectional transition waves in bistable lattices. Phys. Rev. Lett. **116**, 244501 (2016)
21. D.C. Wallace, *Thermodynamics of Crystals* (Wiley, New York, 1972)
22. D.C. Wallace, Renormalization and statistical mechanics in many-particle systems. I. Hamiltonian perturbation method. Phys. Rev. **152**, 247 (1966)
23. A.A. Maradudin, A.E. Fein, Scattering of neutrons by an anharmonic crystal. Phys. Rev. **128**, 2589 (1962)
24. R.K. Narisetti, M.J. Leamy, M.J. Ruzzene, A perturbation approach for predicting wave propagation in one-dimensional nonlinear periodic structures. ASME J. Vib. Acoust. **132**, 031001 (2010)
25. K. Manktelow, M.J. Leamy, M. Ruzzene, Multiple scales analysis of wave-wave interactions in a cubically nonlinear monoatomic chain. Nonlinear Dyn. **63**, 193 (2011)
26. N.M. Krylov, N.N. Bogoliubov, in *Introduction to Nonlinear Mechanics*, trans. by S. Lefshetz (Princeton University Press, Princeton, NJ, 1947)
27. I.C. Khoo, Y.K. Wang, Multiple time scale analysis of an anharmonic crystal. J. Math. Phys. **17**, 222 (1976)
28. A.F. Vakakis, O.V. Gendelman, L.A. Bergman, D.M. McFarland, G. Kerschen, Y.S. Lee, *Nonlinear Targeted Energy Transfer in Mechanical and Structural Systems*, Solid Mechanics and Its Application Series (Springer, Netherland, 2009)
29. A.F. Vakakis, R.H. Rand, Nonlinear dynamics of a system of coupled oscillators with essential stiffness nonlinearity. Int. J. Nonlinear Mech. **39**, 1079 (2004)
30. O.V. Gendelman, T. Sapsis, A.F. Vakakis, L.A. Bergman, Enhanced passive targeted energy transfer in strongly nonlinear mechanical oscillators. J. Sound Vib. **330**, 1 (2011)
31. O.V. Gendelman, L.I. Manevitch, A.F. Vakakis, R. M'Closkey, Energy pumping in nonlinear mechanical oscillators: Part I—dynamics of the underlying Hamiltonian system. Trans. ASME **68**, 34 (2001)
32. G. Kerschen, A.F. Vakakis, Y.S. Lee, D.M. McFarland, J.J. Kowtko, L.A. Bergman, Energy transfer in a system of two coupled oscillators with essential nonlinearity: 1:1 resonance manifold and transient bridging orbits. Nonlinear Dyn. **42**, 283 (2005)
33. D. Laxalde, F. Thouverez, J.-J. Simou, Dynamics of a linear oscillator connected to a small strongly nonlinear hysteretic absorber. Int. J. Nonlinear Mech. **41**, 969 (2006)
34. P.N. Panagopoulos, A.F. Vakakis, S. Tsakirtzis, Transient resonant interactions of finite linear chains with essential nonlinear end attachments leading to passive energy pumping. Int. J. Solids Struct. **41**, 6505 (2004)
35. A.F. Vakakis, L.I. Manevitch, O. Gendelman, L. Bergman, Dynamics of linear discrete systems connected to local essential nonlinear attachments. J. Sound Vib. **264**, 559 (2003)
36. Y. Starosvetsky, M.A. Hasan, A.F. Vakakis, L.I. Manevitch, Strongly nonlinear beat phenomena and energy exchanges in weakly coupled granular chains on elastic foundations. SIAM J. Appl. Math. **72**, 337 (2012)
37. A. Seidel, H.H. Lin, D.H. Lee, Phonons in Hubbard ladders studied within the framework of the one-loop renormalization. Phys. Rev. B **71**, 22050 (2005)
38. M. Peyrard, A.R. Bishop, Statistical mechanics of a nonlinear model for DNA denaturation. Phys. Rev. Lett. **62**, 2755 (1989)
39. A. Dot, A. Borne, B. Boulanger, P. Segonds, C. Félix, K. Bencheikh, J.A. Levenson, Energetic and spectral properties of triple photon down conversion in a phase-matched KTiOPO4 crystal. Opt. Lett. **37**, 2334 (2012)
40. F. Boitier, A. Orieux, C. Autebert, A. Lemaître, E. Galopin, C. Manquest, C. Sirtori, I. Favero, G. Leo, S. Ducci, Electrically injected photon-pair source at room temperature. Phys. Rev. Lett. **112**, 183901 (2014)
41. A.R. Ten Cate, *Oral Histology: Development, Structure and Function* (Mosby, St Louis, MO, 1980)
42. J.D. Currey, *Bones: Structure and Mechanics* (Princeton University Press, Princeton, NJ, 2002)

43. A.C. Deymier-Black, *Study of the Elasto-plastic Properties of Mineralized Biomaterials via Synchrotron High-Energy X-Ray Diffraction*, PhD Dissertation, Department of Materials Science and Engineering, Northwestern University, Evaston, IL, 2011
44. I. Jager, P. Fratzl, Mineralized collagen fibrils: a mechanical model with a staggered arrangement of mineral particles. Biophys. J. **79**, 1737 (2000)
45. A. Boyde, K.S. Lester, An electron microscope study of fractured dentinal surfaces. Calcif. Tissue Res. **1**, 122 (1967)
46. R.M. Dillaman, R.D. Roer, D.M. Gay, Fluid movement in bone: theoretical and empirical. J. Biomech. **1**, 163 (1991)
47. L. Bertinetti et al., Surface structure, hydration, and cationic sites of nanohydroxyapatite: UHR-TEM, IR, and microgravimetric studies. J. Phys. Chem. C **111**, 4027 (2007)
48. A. De Simone, L. Vitagliano, R. Berisio, Role of hydration in collagen triple helix stabilization. Biochem. Biophys. Res. Commun. **372**, 121 (2008)
49. K.M. Ravikumar, W. Hwang, Region-specific role of water in collagen unwinding and assembly. Proteins **72**, 1320 (2008)
50. B. Kahler, M.V. Swain, A. Moule, Fracture-toughening mechanisms responsible for differences in work to fracture of hydrated and dehydrated dentine. J. Biomech. **36**, 229 (2003)
51. J.J. Kruzic et al., Crack blunting, crack bridging and resistance-curve fracture mechanics in dentin: effect of hydration. Biomaterials **24**, 5209 (2003)
52. J.S. Nyman et al., The influence of water removal on the strength and toughness of cortical bone. J. Biomech. **39**, 931 (2006)
53. S.G. Gevorkian et al., Stabilization and anomalous hydration of collagen fibril under heating. PLoS One **8**, 0078526 (2013)
54. C.A. Grant et al., Effects of hydration on the mechanical response of individual collagen fibrils. Appl. Phys. Lett. **92**, 233902 (2008)
55. F.C. Larche, J.W. Cahn, The interactions of composition and stress in crystalline solids. Acta Metall. **33**, 331 (1985)
56. F.C. Larche, Thermodynamics of stressed solids. Am. Ceram. Soc. Bull. **64**, 1344 (1985)
57. S.M. Bowman et al., The tensile behavior of demineralized bovine cortical bone. J. Biomech. **29**, 1497 (1996)
58. G.D. Fullerton et al., Micro-CT dilatometry measures of molecular collagen hydration using bovine extensor tendon. Med. Phys. **38**, 363 (2011)
59. J. Catanese III et al., Heterogeneity of the mechanical properties of demineralized bone. J. Biomech. **32**, 1365 (1999)
60. A.J. Heim, W.G. Matthews, Determination of the elastic modulus of native collagen fibrils via radial indentation. Appl. Phys. Lett. **89**, 181902 (2006)
61. J.H. Kinney, R.K. Nalla, J.A. Pople, T.M. Breunig, R.O. Ritchie, Age-related transparent root dentin: mineral concentrations, crystallite size and mechanical properties. Biomaterials **26**, 3363 (2005)
62. R.S. Weinstein et al., Endogenous glucocorticoids decrease skeletal angiogenesis, vascularity, hydration, and strength in aged mice. Aging Cell **9**, 147 (2010)
63. S.C. Manolagas, From estrogen-centric to aging and oxidative stress: a revised perspective of the pathogenesis of osteoporosis. Endocr. Rev. **31**, 266 (2010)
64. M.J. Bissell, A. Rizki, I.S. Mian, Tissue architecture: the ultimate regulator of breast epithelial function. Curr. Opin. Cell Biol. **15**, 753 (2003)
65. R.K. Hansen, M.J. Bissell, Tissue architecture and breast cancer: the role of extracellular matrix and steroid hormones. Endocr. Relat. Cancer **7**, 95 (2000)
66. C.M. Nelson, M.J. Bissell, Of extracellular matrix, scaffolds, and signaling: tissue architecture regulates development, homeostasis, and cancer. Annu. Rev. Cell. Dev. Biol. **22**, 287 (2006)
67. H.H.Q. Heng, Cancer genome sequencing: the challenges ahead. BioEssays **29**, 783 (2007)
68. M.H. Barcellos-Hoff, It takes a tissue to make a tumor: epigenetics, cancer and the microenvironment. J. Mammary Gland Biol. Neoplasia **6**, 213 (2001)

69. J. Kirshner, C.J. Chen, P. Liu, J. Huang, J.E. Shively, CEACAM1-4S, a cell-cell adhesion molecule, mediates apoptosis and reverts mammary carcinoma cells to a normal morphogenic phenotype in a 3D culture. Proc. Natl. Acad. Sci. U.S.A. **100**, 521 (2003)
70. L.A. Davidson, Integrating morphogenesis with underlying mechanics and cell biology. Curr. Top. Dev. Biol. **81**, 113 (2008)
71. A.G. Mikos, S.W. Herring, P. Ochareon, J. Elisseeff, H.H. Lu, R. Kandel, F.J. Schoen, M. Toner, D. Mooney, A. Atala, M.E. Van Dyke, D. Kaplan, G. Vunjak-Novakovic, Engineering complex tissues. Tissue Eng. **12**, 3307 (2006)
72. A. Ketchedjian, J.A. Linthurst, P. Krueger, E. Robinson, K. Croutch, L. Wolfinbarger, R. Hopkins, Recellularization of decellularized allograft scaffolds in ovine great vessel reconstructions. Ann. Thorac. Surg. **79**, 888 (2005)
73. Q.K. Tran, K. Ohashi, H. Watanabe, Calcium signaling in endothelial cells. Cardiovasc. Res. **48**, 13 (2000)
74. D.J. Watts, S. Strogatz, Collective dynamics of 'small-world' networks. Nature **393**, 440 (1998)
75. H.G. Othmer, L.E. Scriven, Instability and dynamic pattern in cellular networks. J. Theor. Biol. **32**, 507 (1971)
76. H.G. Othmer, L.E. Scriven, Interactions of reaction and diffusion in open systems. I & EC Fundam. **8**, 302 (1969)
77. A.E.R. Turing, The chemical basis of morphogenesis. Philos. Trans. R. Soc. Lond. Ser. B Biol. Sci. **237**, 37–72 (1952)
78. E. Plahte, Pattern formation in discrete cell lattices. J. Math. Biol. **43**, 411 (2001)
79. E. Plahte, L. Oyehaug, Pattern-generating travelling waves in a discrete multicellular system with lateral inhibition. Phys. D **226**, 117 (2007)
80. A. Carpio, L.L. Bonilla, Wave front depinning transition in discrete one-dimensional reaction-diffusion systems. Phys. Rev. Lett. **86**, 6034 (2001)
81. J.P. Keener, Propagation and its failure in coupled systems of discrete excitable cells. SIAM J. Appl. Math. **47**, 556 (1987)
82. K. Kladko, I. Mitkov, A.R. Bishop, Universal scaling of wave propagation failure in arrays of coupled nonlinear cells. Phys. Rev. Lett. **84**, 4505 (2000)
83. J.W. Cahn, J. Mallet-Paret, E.S. Van Vleck, Traveling wave solutions for systems of ODEs on a two-dimensional spatial lattice. SIAM J. Appl. Math. **59**, 455 (1998)
84. C. P. Fall, E. S. Marland, J. M. Wagner, J. J. Tyson (eds.), *Computational Cell Biology. Interdisciplinary Applied Mathematics*, vol 20 (Springer, New York, 2000)
85. J. Sneyd, K. Tsaneva-Atanasova, in *Understanding Calcium Dynamics: Experiments and Theory*, ed. by M. Falcke, D. Malchow (Springer, Berlin, 2003)
86. J. Keiser, G.D. Smith, S. Ponce-Dawson, J.E. Pearson, Saltatory propagation of Ca2+ waves by Ca2+ sparks. Biophys. J. **75**, 595 (1998)
87. S. Ponce-Dawson, J. Keizer, J.E. Pearson, Fire-diffuse-fire model of dynamics of intracellular calcium waves. Proc. Natl. Acad. Sci. U.S.A. **96**, 6060 (1999)
88. H. Oktem, A survey on piecewise-linear models of regulatory dynamical systems. Nonlinear Anal. **63**, 336 (2005)
89. H.M. Hardin, J.H. van Schuppen, System reduction of nonlinear positive systems by linearization and truncation, in *Positive Systems, Proceedings of the Second Multidisciplinary Symposium on Positive Systems: Theory and Applications (POSTA 06)*, ed. by C. Commault, N. Marchand. Lecture Notes in Control and Information Sciences, vol. 341 (2006), p. 431
90. J.G. Kirkwood, Statistical mechanics of fluid mixtures. J. Chem. Phys. **3**, 300 (1935)
91. R. Ravelo, J. Aguilar, M. Baskes, J.E. Angelo, B. Fultz, B.L. Holian, Free energy and vibrational entropy difference between ordered and disordered Ni_3Al. Phys. Rev. B. **57**, 862 (1998)
92. J.H. He, Some asymptotic methods for strongly nonlinear equations. Int. J. Mod. Phys. B **20**, 1141 (2006)

93. F. Shakeri, M. Dehghan, Solution of the delay differential equations via homotopy perturbation method. Phys. Scr. **75**, 551 (2007)
94. R.D. Mattuck, *A Guide to Feynman Diagrams in the Many-Body Problem* (McGraw-Hill, New York, 1992)
95. P. Ramond, *Field Theory: A Modern Primer* (The Benjamin/Cummings Publishing Company, Inc, Melo Park, CA, 1981)
96. L. Dobrzynski, Interface response theory of discrete composite systems. Surf. Sci. Rep. **6**, 119–157 (1986)
97. M. Abramowitz, I. A. Stegun (eds.), *Handbook of Mathematical Functions* (Dover Publications, New York, 1972)

Chapter 5
Wave Mixing

5.1 Introduction

Interactions between sound/elastic waves and other types of waves—physical (electromagnetic, electronic, etc.) or biological can lead to a wide range of very rich phenomena. Media supporting different types of waves and their sources can coherently convert energy between sound/elastic waves and other physical and biological waves. For example, unexpected biological morphogenesis responses of tissues to vibrations [1] have arisen by extending the science of sound to biological media, and may inspire revolutionary new therapeutic technologies that accelerate bone fracture or soft tissue lesion repair processes, or treatment of neurological disorders.

In this chapter, we consider three simple models that illustrate some of the unique phenomena that can arise when sound waves interact with biological waves (calcium waves) and two physical waves (electronic and magnetic waves). In particular, this chapter makes connection again with Chaps. 2 and 3, and the topology of waves as conventional sound waves may impart non-conventional topology to biological or physical waves.

In Sect. 5.2: "Effect of sound on gap junction-based intercellular signaling: topological calcium waves under acoustic irradiation", we present a previously unrecognized effect of sound waves on gap junction-based intercellular signaling such as in biological tissues composed of endothelial cells. The model is that of a one-dimensional chain of endothelial cells. We suggest that sound irradiation may, through temporal and spatial modulation of cell-to-cell conductance, create intercellular calcium waves with unidirectional signal propagation associated with nonconventional topologies. Non-reciprocity in calcium wave propagation induced by sound wave irradiation is demonstrated in the case of a linear and a nonlinear reaction diffusion model. This demonstration should be applicable to other types of gap junction-based intercellular signals and it is thought that it should be of help in interpreting a broad range of biological phenomena associated with the

P. Deymier, K. Runge, *Sound Topology, Duality, Coherence and Wave-Mixing*,
Springer Series in Solid-State Sciences 188, DOI 10.1007/978-3-319-62380-1_5

beneficial therapeutic effects of sound irradiation but also possibly the harmful effects of sound waves on health.

In Sect. 5.3: "Topological interpretation of phonon drag," we solve the time dependent Schroedinger equation for an electron in a one-dimensional spatio-temporally modulated potential. The modulation is interpreted as resulting from a directed coherent phonon. The electron wave function acquires a non-conventional torsional topology as the result of time-reversal symmetry breaking. This topology is associated with an asymmetric band structures which induces electron fluxes in bands that are completely filled.

In Sect. 5.4: "Phonon-magnon resonant processes with relevance to acoustic spin pumping," we illustrate the recently described phenomenon of resonant acoustic spin pumping due to resonant coupling between an incident elastic wave and spin waves in a ferromagnet medium. A classical one-dimensional discrete model of a ferromagnet with two forms of magneto-elastic coupling is treated to shed light on the conditions for resonance between phonons and magnons.

The interactions between sound/elastic waves and biological and physical waves are inherently nonlinear. The models in this chapter are therefore analyzed within the framework of the multiple time scale perturbation theory used extensively in other chapters. We are specifically focusing on the emergence of resonant conditions which can lead to topological effects on the biological and physical wave functions.

5.2 Effect of Sound on Gap-Junction Based Intercellular Signaling: Topological Calcium Waves Under Acoustic Irradiation

5.2.1 Background

The interactions between sound waves and living tissue were first observed and reported in the early twentieth century [2, 3]. Since then ultrasound-based therapies have been used to treat a very wide range of medical disorders [4]. Ultrasound stimulation accelerates the bone fracture repair process [5, 6]. Therapeutic ultrasound is also commonly employed to manage soft tissue lesions [7]. The ability of ultrasound to focus through the skull noninvasively has lately made it an attractive technique for neuromodulation to treat neurological disorders [8]. More recently, ultrasound has also been shown to remove amyloid plaque and restore memory in an Alzheimer's disease mouse model [9]. The therapeutic benefits of high intensity sound irradiation have been associated to date with thermal effects such as local heating due to absorption of sound energy by the tissue, but also with non-thermal effects [10]. Physical effects other than temperature in fluid-containing tissues include cavitation leading to concentration of the delocalized sound wave energy into high energy gaseous bubbles and acousto-hydrodynamics phenomena such as streaming or microstreaming. In the case of low-intensity ultrasound irradiation, mechanical effects resulting from the pressure field of the sound wave may induce mechano-transduction processes such as transient elevation of intracellular calcium in bone cells [11]. In addition, the acoustic pressure in solid tissue may lead to tension in cell membrane and alter membrane conductance [12]. Reversible suppression of transmission of cortical potentials along neural pathways was achieved by ultrasound irradiation [13]. A significant increase in gap junctional cell-to-cell communication in rat bone marrow stromal cells subjected to low intensity ultrasound irradiation was demonstrated [14]. There is also evidence that ultrasound modulates the ionic conductance of neurons and astrocytes and can stimulate electrical activity and calcium signaling in mice hippocampal slice cultures and *ex vivo* brains [15].

In this section, we present a simple model that suggests that sound waves can affect biological tissue and in particular can alter gap-junction based intercellular signaling. We discuss the mechanism by which a sound wave can modulate spatially and temporally the conductance between cells and impart the property of non-reciprocity in propagation to an intercellular wave as a consequence of its non-conventional topology. To date, these non-conventional topological properties have been mostly described and observed in physical waves such as electronic, electro-magnetic and acoustic waves in systems such as topological insulators [16–18]. The model presented here suggests that sound irradiation of some biological tissues may create intercellular waves with non-conventional topologies that can lead to unidirectional signal propagation. Following Chap. 4, we illustrate this mechanism in the case of an intercellular calcium wave. Calcium and calcium waves are biological regulators of differentiation and activity in numerous tissue types [19–23]. The mechanisms of initiation and propagation of intercellular calcium waves may involve cell-to-cell communication with intra-cellular messengers diffusing through gap junctions or extracellular messengers mediating paracrine signaling [24]. Intercellular calcium waves in tissues composed of endothelial cells originate from (a) feedback loops between Ca^{2+} and the messenger molecule Inositol 1,4,5-triphosphate IP_3 [25–27] and (b) cell-to-cell interactions via Ca^{2+} and IP_3 transport through gap junctions [28, 29]. In other tissues, cell-to-cell communication can circumvent the gap junction pathway involving other mechanisms such as paracrine signaling via extracellular messengers.

Turing noted the importance of studying the behavior of biological processes by considering the complementarity of both linear and nonlinear dynamical systems [30]: "Such systems (*with linear dynamics*) certainly have a special interest as giving the appearance of a pattern, but they are the exception rather than the rule. Most of an organism, most of the time, is developing from one pattern into another, rather than from homogeneity into a pattern. One would like to be able to follow this more general process (*nonlinear*) mathematically also. The difficulties are, however, such that one cannot hope to have any very embracing *theory* of such processes, beyond the statement of the equations. It might be possible, however, to treat a few particular cases in detail with the aid of a digital computer." (parenthetical comments added) Therefore, the models presented in this chapter address the interaction between acoustic waves and calcium waves from a theoretical point of view using a linear reaction diffusion model for the calcium dynamics in multicellular tissues as well as from a computational point of view in the case of a more complex and biologically realistic nonlinear reaction diffusion model.

5.2.2 Nonlinear Reaction Diffusion Model of Gap Junction Based Intercellular Calcium Waves

This subsection is devoted to the particular case of a nonlinear reaction diffusion model of calcium waves and on the effect of an acoustic wave on the propagation of gap junction based intercellular signals. The model of nonlinear calcium waves involves primarily the nonlinear intracellular reaction dynamics coupled with the intercellular diffusion of cytoplasmic calcium and IP_3. The nonlinear model used here employed an intracellular calcium release/intake pathway that is more complex than the one introduced in Chap. 4. This intracellular chemical reaction process is based on a model introduced by Politi et al. [31]. A schematic of this intracellular calcium pathway in the context of a multicellular structure with intercellular gap junction diffusion is shown in Fig. 5.1. The complete reaction-diffusion process involves primarily the intracellular reaction dynamics and the intercellular diffusion of cytoplasmic calcium and IP_3. For the sake of completeness, we describe this model in some detail. The intracellular calcium pathway starts with an extracellular agonist that combines with the G-protein-coupled receptors on the cell's membrane to activate phospholipase C (PLC). It is, in turn, able to catalyze the production of IP_3 [32]. IP_3 then can bind to the IP_3 receptor, IP_3R, to open calcium channels in the membrane of the Endoplasmic Reticulum (ER). This process releases stored Ca^{2+} into the cytosol. Meanwhile, the cytoplasmic Ca^{2+} can create both positive and negative feedback conditions in the production of IP_3. For the positive feedback condition, the cytoplasmic Ca^{2+} is capable of activating the PLC isoforms to release more IP_3 [33]. For the negative feedback condition, the increase in cytoplasmic Ca^{2+} can activate the IP_3 degradation via IP_3 3-kinase (IP_3K). Cells subsequently control the intracellular calcium level by buffering, sequestering in specialized compartments, and expelling to the extracellular space excess calcium thereby maintaining calcium homeostasis and resetting the signaling loop [34, 35].

The intracellular chemical reaction dynamics is formulated into a system of coupled differential equations involving four dynamical variables: the calcium concentration in the cytosol, c; the IP_3 concentration in the cytosol, p; the calcium concentration in the ER stores, s; and the fraction of IP_3R that has not been inactivated by Ca^{2+}, r. The rate equation for the IP_3 concentration takes the following form:

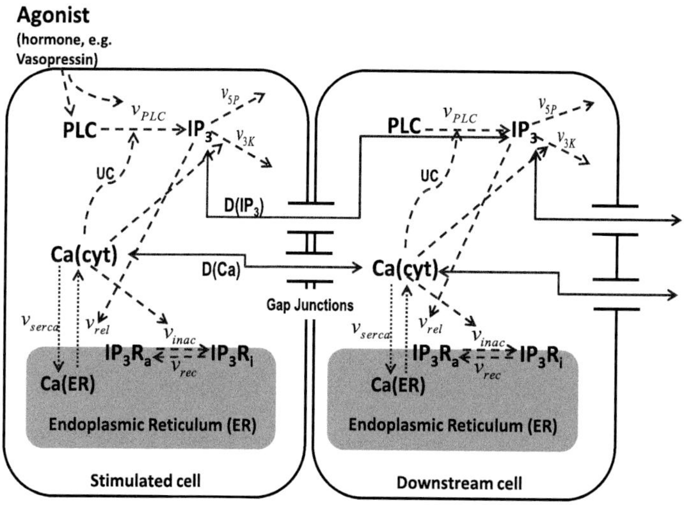

Fig. 5.1 Reaction/Diffusion process of Ca^{2+} and IP_3 metabolism included in the nonlinear model (after [31]). The *solid*, *dashed*, and *dotted arrows* indicate molecular diffusion, regulatory interactions, and reaction/transport steps respectively. The *bold* quantities indicate the following model variables: IP_3, the cytoplasmic IP_3: Ca(*cyt*), the free cytoplasmic Ca^{2+}: Ca(*ER*), the free Ca^{2+} in the ER; IP_3R_a, the active conformation of the IP_3R. The other abbreviations denote IP_3R_i, the inactive conformation of the IP_3R; v_{serca}, the active Ca^{2+} transport into the ER; v_{PLC}, the production rate of IP_3; v_{rel}, the rate of Ca^{2+} release through the IP_3R; v_{inac} and v_{rec}, the rates of Ca^{2+}-induced IP_3R inactivation and recovery, respectively; v_{5P} and v_{3K}, the rates of IP_3 dephosphorylation and phosphorylation, respectively; $D(IP_3)$ and $D(Ca)$, the gap-junction diffusion coefficient of IP_3 and Ca, respectively; and UC, the threshold of Ca needed to activate PLC

$$\frac{dp}{dt} = f(c,p) = v_{PLC} - v_{\deg} = v_{PLC} - (v_{5P} + v_{3K})$$

$$= \left(V_{PLC}\frac{c^2}{K_{PLC}^2 + c^2}\right) - \left(k_{5P} + k_{3K}\frac{c^2}{K_{3K}^2 + c^2}\right)p \tag{5.1}$$

where the v_{PLC} and $v_{\deg}$ represent the production and degradation rate of IP_3, respectively. V_{PLC} is the maximum production rate of PLC that depends on the agonist concentration. K_{PLC} characterizes the sensitivity of PLC to Ca^{2+}. Phosphorylation and dephosphorylation lead to degradation of IP_3 with the rates v_{3K} and v_{5P}, respectively. The phosphorylation rate is described by a Hill function with the rate constant k_{3k} and the half-saturation constant K_{3K} [36]. The rate equation for the cytoplasmic Ca^{2+} takes the following form:

$$\frac{dc}{dt} = g(c,p) = v_{rel} - v_{serca} = \left[k_1\left(r\frac{c}{K_a + c}\frac{p}{K_p + p}\right)^3 + k_2\right](s - c) - V_{serca}\frac{c^2}{K_{serca}^2 + c^2}. \tag{5.2}$$

v_{serca} represents the rate for the active Ca^{2+} transport into the ER and v_{rel}, the rate of Ca^{2+} release through the IP_3R. V_{serca} and K_{serca} are the maximum SERCA pump rate and half-activation constant, respectively.

For the sake of simplicity, the total calcium concentration in the cell, c_{tot}, is assumed to be conserved and is represented as $c_{tot} = c + \beta s$, where β is the ratio of effective cytoplasmic volume to effective ER volume (both accounting for Ca^{2+} buffering). Therefore, the calcium concentration in the ER store can be expressed as

$$s = \frac{c_{tot} - c}{\beta}. \tag{5.3}$$

The dynamics of IP_3R inactivation by Ca^{2+} is modeled as follow:

$$\frac{dr}{dt} = v_{rec} - v_{inac} = \frac{1}{\tau_r}\left(1 - r\frac{K_i + c}{K_i}\right). \tag{5.4}$$

v_{inac} and v_{rec} are the rates of Ca^{2+}-induced IP_3R inactivation and recovery, respectively.

The simulation of intracellular oscillations reported herein uses the model parameters reported by Politi [31]. Reference [31] discusses extensively the physiological role of both positive and negative Ca^{2+}/IP_3 feedbacks in relation to the enhancement of the range of frequency encoding of the agonist stimulus. However, since the frequency encoding supported by the positive feedback appears to be more robust against variations in the parameters describing the calcium kinetics, we have limited ourselves to the model supporting calcium positive feedback. Positive feedback is achieved in the model by setting the phosphorylation rate constant k_{3k} to zero. We anticipate that the effect of an acoustic wave on Ca^{2+} signal propagation in a multicellular structure does not depend qualitatively on this choice since, as will be seen below, the acoustic wave impacts the cell-to-cell conductance. All the parameters of the model are summarized in Table 5.1. The numerical solutions of (5.1), (5.2) and (5.4) are obtained by using a 4th-order Runge-Kutta algorithm with time step size, $\Delta t = 0.01$ s.

Politi's model of intracellular calcium pathway is integrated into a multicellular model by considering the phenomenon of diffusion via gap junctions of both Ca^{2+} and IP_3 driven by the concentration gradients between neighboring cells. The multicellular structure considered here is composed of a single linear chain of N cells with periodic boundary conditions (PBC). In developing this model, we have in mind blood vessels where the endothelial monolayer is, in reality, a two-dimensional structure (wrapped in a tube) which, in the case of capillaries, may be approximated by a one-dimensional chain of endothelial cells since the cross section approaches the cell dimension. In such a chain, in which every cell is connected to two other cells (diffusion between nearest neighbor cells), one can write the one-dimensional time-dependent reaction/diffusion equation for Ca^{2+} and IP_3:

Table 5.1 Values of parameters used in the nonlinear reaction diffusion model

Parameters	Description	Value
IP_3 dynamics parameters		
K_{3K}	Half-activation constant of IP_3K	0.4 μM
k_{3K}	IP_3 phosphorylation rate constant	0
k_{5P}	IP_3 dephosphorylation rate constant	0.66 s^{-1}
K_{PLC}	Half-activation constant of PLC	0.2 μM
V_{PLC}	Maximum production rate of IP_3	1.5 μM s^{-1}
Ca^{2+} transport and structural parameters		
β	Ratio of effective volumes ER/cytosol	0.185
V_{serca}	Maximal SERCA pump rate	0.9 μM s^{-1}
K_{serca}	Half-activation constant	0.1 μM
c_{tot}	Total Ca^{2+} concentration	2 μM
IP_3R parameters		
k_1	Maximal rate of Ca^{2+} release	1.11 s^{-1}
k_2	Ca^{2+} leak	0.0203 s^{-1}
K_a	Ca^{2+} binding to activating site	0.08 μM
K_i	Ca^{2+} binding to inhibiting site	0.4 μM
K_p	IP_3 binding	0.13 μM
τ_r	Characteristic time IP_3R inactivation	12.5 s
Reference diffusion parameters		
D_c^*	Diffusion coefficient rate of Ca^{2+}	0.005 s^{-1}
D_p^*	Diffusion coefficient rate of IP_3	$10D_c^*$
UC	Threshold of Ca^{2+} to activate PLC	0.057 μM

$$\frac{\partial c}{\partial t} = D_c \frac{\partial^2 c}{\partial x^2} + g(c,p) = D_c^* [c(x_{i+1}, t_n) - 2c(x_i, t_n) + c(x_{i-1}, t_n)] + g(c,p), \quad (5.5)$$

and

$$\frac{\partial p}{\partial t} = D_p \frac{\partial^2 p}{\partial x^2} + f(c,p) = D_p^* [p(x_{i+1}, t_n) - 2p(x_i, t_n) + p(x_{i-1}, t_n)] + f(c,p), \quad (5.6)$$

where D_c and D_p are again the diffusion coefficients of Ca^{2+} and IP_3. "x" and "t" are the position and time variables. Equations (5.5) and (5.6) have been discretized in space and time using finite differences. "$t_n = n\Delta t$" refers to the discretized time line with the time step Δt. $D_c/\Delta x^2$ and $D_p/\Delta x^2$ are defined as the diffusion coefficient rate of Ca^{2+} and IP_3 with unit "per second", "s^{-1}", which we denote D_c^* and D_p^*, respectively. Δx is the nearest neighbor cell center to cell center distance. Since diffusion occurs via gap junctions, we can assume that it is limited to nearest neighboring cells. We further assume that the distribution of gap junctions in the plasma membrane is spatially uniform and that the diffusion coefficients are constants independent of cell number. Note that the mobility of Ca^{2+} through gap junction is restricted in comparison to that of IP_3 because of the higher buffering capacity of cytoplasm for Ca^{2+} than for IP_3 [37]. Thus, IP_3 diffuses much faster than Ca^{2+} [38]. For the sake of simplicity, we set $D_p^* = 10D_c^*$ in our model. We have verified that our combined reaction-diffusion algorithm has fully converged for the time step $\Delta t = 0.01$ s.

To induce trains of calcium waves in the multicellular one-dimensional chain, we initially stimulate a single cell in the center of the chain with the agonist. PLC of the stimulated cell is activated initially by the extracellular agonist to induce intracellular Ca^{2+}/IP_3 oscillations. All other cells in the chain are not initially stimulated ($V_{PLC} = 0$). The reaction dynamics of the stimulated cell increases its calcium concentration. Diffusion of Ca^{2+} between the stimulated cell and its neighboring cells elevates their Ca^{2+} concentration. We enable the propagation of a train of calcium waves by introducing a threshold based on the calcium concentration for inducing Ca^{2+}/IP_3 positive feedback in the neighboring cells (Fig. 5.1). When the cytoplasmic Ca^{2+} concentration reaches a value exceeding a threshold, UC, the positive feedback of cytoplasmic Ca^{2+} is activated to increase the production rate of IP_3. If the cytoplasmic Ca^{2+} concentration is below the threshold, PLC enzymes are not activated. The introduction of this threshold enables the synchronized development of collective spatio-temporal response of the multicellular architecture. This extension is based not only on diffusion but also on an additional amplification mechanism through the generation of IP_3 and the Ca^{2+}-dependent activation of PLC [39]. The general behavior of this nonlinear reaction diffusion model of calcium wave propagation and the numerical methods for solving the nonlinear coupled differential equations have been detailed in reference [40]. The Fortran 90 computer program used for these calculations is given in Appendix 1.

In the example case, the model one-dimensional tissue is composed of a chain of 401 cells. Initially, an intracellular calcium oscillation is induced by an extracellular agonist in cell 201. In Fig. 5.2(top), we present snapshots (at the same time step = 30,000 Δt) of the nonlinear calcium waves propagating in absence of acoustic irradiation. In this case, the nonlinear calcium wave takes the form of two trains of pulses propagating symmetrically from the stimulated cell (201). The important feature, here, is that the calcium waves in the absence of an acoustic wave are symmetric in space and time. This feature is characteristic of reciprocal propagation of the intercellular signals. We now consider the case of the model tissue subjected to an acoustic wave propagating from the left to the right of the chain of cells. Let us consider a sound wave with a frequency Ω and a wave number K, the spatially and temporally varying pressure field is given by $P(x,t) = P_0 \cos(Kx + \Omega t)$. Here the wave number is $K = \frac{2\pi}{L}$ with L being the wavelength of the sound wave. This pressure field may lead to a modification [13] and modulation of the gap junction cell-to-cell conductance, however, we need to identify its functional form. It has been shown that gap junctional conductance of Hensen cells is correlated with membrane strain [41]. In that study, experimental data demonstrated a nearly linear relation between the transjunctional conductance and membrane strain up to values of 1.3. Considering a linear Hooke's law type relation between pressure and membrane strain and a linear strain-conductance relationship,

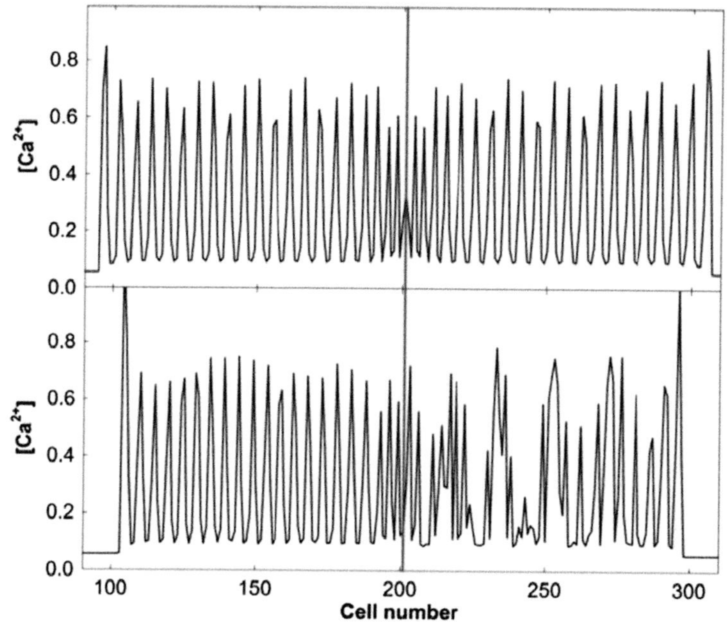

Fig. 5.2 Sound irradiation breaks the reciprocity of nonlinear calcium waves. *Top* Snapshot of the concentration of cytosolic calcium [Ca^{2+}] at time t = 30,000Δt ($\Delta t = 0.01$ s) resulting from the stimulation of cell 201 (*red line*) in absence of acoustic irradiation. *Bottom* Snapshot of the concentration of cytosolic calcium at the same time in presence of acoustic irradiation. PBC are used but the simulation is interrupted before the calcium wave can reach the cells 1 and 401. The concentrations are in units of μM

one expects a linear relation between pressure and gap junctional conductance. Therefore, when subjected to sound irradiation, the spatially and temporally varying diffusion coefficients of the Ca^{2+} and IP_3 are chosen to take a functional form given by the following equations:

$$D_c(x,t) = D_{0c} + 2D_{1c}\cos(Kx + \Omega t) \tag{5.7a}$$

$$D_p(x,t) = D_{0p} + 2D_{1p}\cos(Kx + \Omega t) \tag{5.7b}$$

D_{0c} and D_{0p} are the diffusion constants in absence of the sound wave that were used in the absence of acoustic irradiation. D_{1c} and D_{1p} reflect the variation in membrane conduction due to a given level of the maximum acoustic pressure. The cosine function in (5.7a, 5.7b) parallels the cosine functional form of the pressure wave. When considering spatially and temporally dependent diffusion coefficients, (5.5) and (5.6) take the form:

$$\frac{\partial c}{\partial t} = D_c^*(x_i, t)[c(x_{i+1}, t_n) - 2c(x_i, t_n) + c(x_{i-1}, t_n)] + \frac{1}{2}(D_c^*(x_{i+1}, t) - D_c^*(x_{i-1}, t))\frac{1}{2}(c(x_{i+1}, t) - c(x_{i-1}, t)) + g(c,p),$$

$$\frac{\partial p}{\partial t} = D_p^*(x_i, t)[p(x_{i+1}, t_n) - 2p(x_i, t_n) + p(x_{i-1}, t_n)] + \frac{1}{2}\left(D_p^*(x_{i+1}, t) - D_p^*(x_{i-1}, t)\right)\frac{1}{2}(p(x_{i+1}, t) - p(x_{i-1}, t))f(c,p)$$

The second terms in the right-hand side of the preceding equations are advective terms involving products of the gradient in composition with the gradient in diffusion coefficient. We first neglect these advective terms and simply insert (5.7a, 5.7b) into (5.5) and (5.6). We therefore show below that such advective terms are not necessary for breaking symmetry in the propagation of calcium waves. We will also show in Sect. 5.2.3 that advective terms are not necessary to achieve symmetry breaking in our simplified linear reaction/diffusion model. We have also conducted numerical calculations with the advective terms and verified that symmetry breaking also occurs. In our numerical calculations, we use: $\frac{D_{1c}}{D_{0c}} = \frac{D_{1p}}{D_{0p}} = 0.6$. Under these conditions, we have identified the existence of resonant conditions that lead to disruption of the reciprocity in propagation of the calcium waves. For instance, in the case of irradiation by an acoustic wave with $L = 36\Delta x$ and $\Omega = 7 \times 10^{-3} rad/sec$. Figure 5.2(bottom) shows clearly that the symmetry of the trains of calcium pulses is broken. The train of pulses propagating toward the left retains its integrity although with a reduced wavelength (inter-peak distance). In contrast the calcium wave propagation to the right has completely lost its integrity. Pulses are still propagating but without a well-defined wavelength. In the presence of directed acoustic irradiation, non-reciprocal propagation of nonlinear intercellular calcium waves is therefore observed. For a typical intercellular spacing: $\Delta x \sim 30$ μm, the conditions at which symmetry breaking is observed correspond to a wavelength $L = 1.08$ mm (a wavenumber $K = 5.817$ rad/mm^{-1}). This condition cannot be realized with a monochromatic sound wave for which the relation $\Omega = cK$ with $c \sim 1500$ m/s would hold. In the previous relation, we take the speed of sound in the biological medium to be that in water. The condition for symmetry breaking can be achieved as the low frequency beat that arises from the interference resulting from the superposition of two waves that are travelling in opposite directions, namely $\sin(k_1 x + f_1 t) + \sin(k_2 x - f_2 t) = 2\sin\left[\left(\frac{k_1+k_2}{2}\right)x + \left(\frac{f_1-f_2}{2}\right)t\right] \cos\left[\left(\frac{k_1-k_2}{2}\right)x + \left(\frac{f_1+f_2}{2}\right)t\right]$. The two waves have slightly different frequencies, f_1 and f_2, and wavenumbers, k_1 and k_2, but they both travel with the same wave speed, c. The envelope of the superposition of waves given by the sine function in the right hand-side of the preceding equation will have a wave number comparable to those of the superposed waves. Even if the frequencies f_1 and f_2, are reasonably high, the envelope will impact the gap junction cell-to-cell conductance with the low beat frequency $\Omega = ½(f_1 - f_2)$. With wavenumbers $k_1 \sim k_2 \sim 5.817$ rad/mm^{-1}, and the speed of sound of $c \sim 1500$ *m/sec*, both slightly differing frequencies f_1 and f_2 are on the order of 1.5 MHz which is typical of sonographic instruments which operate in the range of 1–18 MHz. A plausible scenario that would lead to the condition for symmetry breaking may consist of the interference due to multiple reflections of an incident ultrasonic beam with central frequency of 1.5 MHz with even narrow bandwidth. Our model shows numerically the phenomenon of non-reciprocity in calcium wave propagation induced by sound wave irradiation. To gain further insight into this phenomenon from a theoretical point of view, we consider in the next section a simplified analytical model to describe the interaction between an acoustic wave and a linearized version of the reaction diffusion model.

5.2.3 Linear Reaction Diffusion Model of Calcium Waves

Following Turing's seminal work on linear diffusion-reaction models for modeling morphogenesis [30], we employ a linear reaction diffusion model to describe the spatio-temporal behavior of the Ca^{2+}/IP_3 coupling in tissues composed of endothelial cells. Similarly to Turing, we assume that the system is not very far from a homogeneous equilibrium and that the linear model describes small variations in composition about that equilibrium. The matter of the multicellular chain is assumed to be continuously distributed. Under these conditions, the dynamics of the small excursions in the concentrations of Ca^{2+} and IP_3 about the equilibrium state, C and P, is described by the linearly coupled set of differential equations:

$$\frac{\partial C(x,t)}{\partial t} = \frac{\partial}{\partial x}\left[D_c(x,t)\frac{\partial C(x,t)}{\partial x}\right] - rP(x,t) \tag{5.8a}$$

$$\frac{\partial P(x,t)}{\partial t} = \frac{\partial}{\partial x}\left[D_p(x,t)\frac{\partial P(x,t)}{\partial x}\right] + rC(x,t) \tag{5.8b}$$

This linear model represents a composite regulatory process between Ca^{2+} and IP_3 that leads to effective compositional oscillations. This simplified linear reaction diffusion model lumps many of the fine detail control systems of cellular calcium [24] into a single reaction rate constant r. The linear model replaces the functions f and g, by $g(C,P) = aC - rP$ and $f(C,P) = rC + bP$. We neglect terms proportional to C in g [i.e., (5.8a)] and P in f [i.e., (5.8b)] because they would correspond to degradation leading to damping of the oscillations. The remaining terms on the right-hand side of (5.8a, 5.8b) model the feedback loop between Ca^{2+} and IP_3 that produced oscillatory behavior. The primary goal of this analytical study is to identify the resonant conditions that result in symmetry breaking. Oscillatory behavior in composition and not damping is the primary physical phenomenon that is sensitive to resonant interaction with an external oscillatory acoustic stimulus.

In these equations, x marks the position along the one-dimensional tissue, t denotes the time. The first terms in the right hand-side of the (5.8a, 5.8b) model the diffusion of Ca^{2+} and IP_3 across the cellular membranes via gap junctions. D_c and D_p are the diffusion coefficients associated with that process. When subjected to irradiation by a sound wave with a frequency Ω and a wave number K, the spatially and temporally varying pressure field will lead to a modulation of the gap junction cell-to-cell conductance. This modulation is again represented by the spatio-temporal variations of the diffusion coefficients given in (5.7a, 5.7b). For the sake of analytical simplicity, we take $D_{1c}=D_{1p}=\varepsilon$, that is the modulation of cell-to-cell conduction is the same for Ca^{2+} and IP_3 but not necessarily the unmodulated diffusivities, D_{0c} and D_{0p}. The periodicity of the modulated diffusion coefficients suggests that we should be seeking solutions of (5.8a, 5.8b) in the form of Bloch waves:

$$C(x,t) = \sum_k \sum_g C(k,g,t) e^{i(k+g)x} \tag{5.9a}$$

$$P(x,t) = \sum_k \sum_g P(k,g,t) e^{i(k+g)x} \tag{5.9b}$$

where $x \in [0, L]$. The wave number k is limited to the first Brillouin zone: $\left[\frac{-\pi}{L}, \frac{\pi}{L}\right]$ and $g = \frac{2\pi}{L} m$ with m being a positive or negative integer. With these, (5.8a and 5.8b) take the form:

$$\begin{aligned}\frac{\partial C(k+g,t)}{\partial t} &= -(k+g)^2 D_{0c} C(k+g,t) - rP(k+g,t) \\ &\quad + \varepsilon\left[f(k+g-K)e^{i\Omega t} C(k+g-K,t)\right. \\ &\quad \left.+h(k+g+K)e^{-i\Omega t} C(k+g+K,t)\right]\end{aligned} \tag{5.10a}$$

$$\begin{aligned}\frac{\partial P(k+g,t)}{\partial t} &= -(k+g)^2 D_{0p} P(k+g,t) + rC(k+g,t) \\ &\quad + \varepsilon\left[f(k+g-K)e^{i\Omega t} P(k+g-K,t)\right. \\ &\quad \left.+h(k+g+K)e^{-i\Omega t} P(k+g+K,t)\right]\end{aligned} \tag{5.10b}$$

The functions $f(k)$ and $h(k)$ are defined by $f(k+g) = -(k+g)^2 - K(k+g)$ and $h(k+g) = -(k+g)^2 + K(k+g)$. For the sake of analytical tractability, we treat ε as a perturbation and use multiple time scale perturbation theory [42] (also see Appendix 4 in Chap. 3) to shed light on the effect of the sound wave on the biological waves. In absence of sound irradiation, $\varepsilon = 0$, the wave behavior of the Ca^{2+} and IP_3 concentration excursions take the form:

$$C_0(k+g,t) = a_0\xi(k+g)e^{i\omega_0(k+g)t} \tag{5.11a}$$

$$P_0(k+g,t) = (-1)a_0\zeta(k+g)e^{i\omega_0(k+g)t} \tag{5.11b}$$

with

$$\xi(k+g) = \sqrt{\alpha \mp i\sqrt{r^2-\alpha^2}} \tag{5.12a}$$

$$\zeta(k+g) = \sqrt{\alpha \pm i\sqrt{r^2-\alpha^2}} \tag{5.12b}$$

and a dispersion relation

$$\omega_0(k+g) = \pm\omega_r + i\omega_i \tag{5.13}$$

In (5.12a, 5.12b) and (5.13), we have defined the following quantities: $\alpha = (k+g)^2\frac{(D_{0c}-D_{0p})}{2}$, the real frequency $\omega_r = \sqrt{r^2-\alpha^2}$ and the imaginary part of the frequency $\omega_i = (k+g)^2\frac{(D_{0c}+D_{0p})}{2}$. a_0 is a constant amplitude. The functions ξ and ζ lead to wave number dependent phase differences between the oscillations of Ca^{2+} and the oscillations of IP_3. For instance, for a very long wavelength mode $k+g\sim 0$, the Ca^{2+} and IP_3 composition incursions are $\frac{\pi}{2}$ out of phase. The $\pm$ in front of the real part of the frequency of the calcium and IP_3 waves represents two types of solutions that propagate in opposite directions. The imaginary part of the frequency results in the damping of the compositional waves. The multiple time scale perturbation approach is carried out to first and second order with details reported in Appendix 2. The sound wave is shown to lead to an alteration of the dispersion relation of the Ca^{2+} and IP_3 waves given by (5.11a, 5.11b). These alterations occur at wave numbers k' and k'' which satisfy the equations:

$$\omega_0(k'+g+K) = \omega_0(k'+g)+\Omega \tag{5.14a}$$

$$\omega_0(k''+g-K) = \omega_0(k''+g)-\Omega \tag{5.14b}$$

The most remarkable result is that while k' and k'' are solutions of (5.14a) and (5.14b), $-k'$ and $-k''$ are not. This means that the actual dispersion relation of the Ca^{2+} and IP_3 waves perturbed by the acoustic wave is not symmetrical about the origin $k = 0$. This asymmetry tells us that there exist intervals of frequency and wave numbers whereby the propagation of Ca^{2+} and IP_3 waves in one direction differs from the propagation in the opposite direction. Irradiation by an acoustic wave has broken the directional symmetry of the biological waves leading to non-reciprocity in the propagation of the intercellular signal. It is worth noting that since the dispersion relation ω_0 could be a reasonably shallow function of k, the resonant conditions (5.14a, 5.14b) may correspond to large values of $k = k'(k'')$ and small values of Ω that do not satisfy the relation $\Omega = ck$. The small value of Ω may be attained by the superposition of counter propagative acoustic waves with slightly

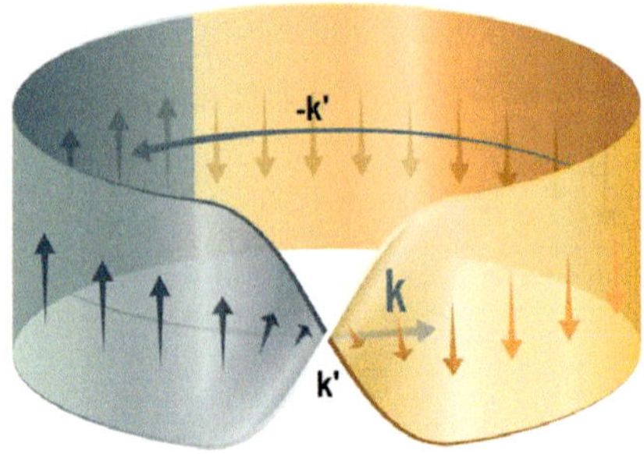

Fig. 5.3 Sound-induced twist in the topology of linear calcium waves: Schematic illustration of the k-space manifold supporting calcium waves in the one-dimensional tissue subjected to acoustic irradiation. The manifold takes the general form of a strip with a single twist centered on the wavenumber k'. The geometric phase change of π of the amplitude of the wave is represented by the change in orientation of a vector field (*arrows*) tangent to the manifold as the vector is transported across k'. The amplitude of the calcium wave function does not accumulate a phase across $-k'$. A similar topology is expected at k''

differing frequencies and wave numbers as was discussed in Sect. 5.2.2. Equations (5.14a, 5.14b) provide also a physical interpretation of the origin of symmetry breaking in terms of scattering phenomena. Considering (5.14a), the acoustic modulation with wave number, K, scatters a calcium wave with wave number $k' + g$ to produce a calcium wave with wave number $k' + g + K$. This scattering process conserves the wave number. In the absence of symmetry breaking, the scattering cross section for scattering of left-moving and right-moving calcium waves would be equal. However, with the introduction of the acoustic wave, which breaks the time-reversal symmetry, the scattering cross sections are no longer the same. The resulting asymmetry can also be explained within the context of topology as we discuss next.

The analytical model analyzed herein enables us to shed light on the topology of the calcium wave function in wave number space. The amplitude of the calcium waves changes sign (i.e., accumulate a π geometric phase [43]) as one is moving along wave number space through the resonant conditions (5.14a) and (5.14b) (i.e., through k' and k''). On the other hand, the amplitude is not accumulating any geometric phase along a path in wave number space that goes through $-k'$ and $-k''$. This observation indicates that the manifold in wave number space that supports the calcium wave function exhibits a non-conventional topology. This manifold possesses a twist at k' and k'' but not at $-k'$ and $-k''$ as illustrated schematically in Fig. 5.3. Parallel transport of a vector field that may represent the wave amplitude along a closed path in k space (here a circle) can be used to describe this topology. The geometric phase change of the amplitude of the calcium wave is represented by the change in orientation of the vector tangent to the manifold as it is parallel transported along the path. This behavior is phenomenologically equivalent to topological insulators more commonly known in the physical sciences [16–18].

The linear model so far has included advective terms. A simplified version of that model which neglects advection takes the form given by (5.15a, 5.15b):

$$\frac{\partial C(x,t)}{\partial t} = D_c(x,t)\frac{\partial^2 C(x,t)}{\partial x^2} - rP(x,t) \tag{5.15a}$$

$$\frac{\partial P(x,t)}{\partial t} = D_p(x,t)\frac{\partial^2 P(x,t)}{\partial x^2} + rC(x,t) \tag{5.15b}$$

The treatment of this simplified problem with the multiple time scale perturbation is similar to that presented for the more general case in Appendix 2. The major difference only lies in a simplified form of the functions $f(k)$ and $h(k)$ which are now defined by $f(k+g) = -(k+g)^2$ and $h(k+g) = -(k+g)^2$. Neglecting advection is then determined not to affect the observation of symmetry breaking in the direction of propagation. These results support the observation of symmetry breaking in the direction of propagation of calcium waves observed in the numerical simulation of the nonlinear reaction diffusion model of Sect. 5.2.2. The nonconventional topology imparted to calcium waves by the sound irradiation is isomorphic to the nonconventional topology of sound waves subjected to a spatio-temporal modulation of the elastic coefficients discussed in Chap. 3.

In an attempt to generalize the analysis, we have also considered (but not reported here) a linear model system whereby the reaction rate, r and not the diffusion coefficients of Ca^{2+} and IP_3 is modulated spatio-temporally. This calculation also showed non-reciprocity in calcium wave propagation. Observation of asymmetric calcium wave propagation in linear model systems with either modulated conductance or intracellular reaction rate suggests that symmetry breaking due to sound irradiation is likely to be insensitive to the details of the biological mechanism that is responsive to the acoustic waves. Since all linear models (i.e., with spatio-temporal modulation of the diffusion coefficient with or without advection, or with spatio-temporal modulation of reaction rate) and the nonlinear models (with advection and without advection) considered in the present study show non-reciprocity in the propagation of calcium waves when subjected to an acoustic wave, it is likely that the phenomenon of symmetry breaking due to sound irradiation exhibits some robustness with respect to the biological complexity of the intercellular and intracellular dynamics.

It is thought that (a) the observation of non-reciprocity in calcium signaling in a realistic nonlinear reaction diffusion model of calcium waves and (b) the existence of a similar behavior in the simplified linear model of calcium waves, should be of some help in interpreting real biological phenomena associated with the beneficial therapeutic effects of sound irradiation but also possibly the harmful effects of sound waves on health. The evaluation of the risks of routine prenatal ultrasound imaging on a fetus during pregnancy has not considered the possible effect on intercellular signaling identified here [44]. It is also known that exposure to traffic noise in addition to that of air pollution leads to an increase in risk of cardio-vascular diseases [45, 46]. The mechanisms driving the effect of noise exposure on the function of the cardio-vascular system are essentially unknown. Finally, vibroacoustic disease (VAD) is a pathology that develops in individuals excessively exposed to low frequency sound [47]. VAD is associated with the abnormal growth of extra-cellular matrices (collagen and elastin), in the absence of an inflammatory process, seen in blood vessels, cardiac structures, trachea, lung, and kidney. VAD appears to be a mechano-transduction disease resulting from impairment of mechanically mediated signaling in exposed tissues [47]. The effects of the physical parameters of the sound on signaling, such as frequency or intensity (amplitude), are essentially unknown.

We also note that the observed mechanism of non-reciprocity in the propagation of intercellular calcium signals should be applicable to other types of intercellular signals. Given that our analysis does not depend on the specific chemical species involved in signaling, it is likely that our results apply to other signaling molecules such as oxidizing chemical agents that are thought to play roles similar to calcium in plant cells [48, 49].

5.3 Topological Interpretation of Phonon Drag

At low temperature, the enhancement in thermoelectric power of semiconductors [50] (see Fig. 5.4) has been interpreted as the result of an interaction between coherent phonons and electrons, the so-called phonon drag effect [51].

This effect has been observe in numerous other materials. Phonon drag was also shown recently to enhance the Seebeck effect of other forms of excitations, such as magnons [52, 53]. Phonon drag contributes to the thermoelectric effect via scattering and exchange of momentum between phonons and other excitations such as electrons or magnons. Phonon-electron scattering contribution dominates in the temperature region where phonon-phonon scattering is reduced and where phonon-electron scattering is dominates, hence the low temperature.

In this section, we reinterpret the phenomenon of phonon-electron scattering picture in the context of topological waves. The interaction between phonons and electrons is traditionally interpreted in terms of inter-band transitions in the space of the time-independent wave functions of the electronic structure. These time-independent wave functions exhibit the conventional topology imparted by the underlying material structure and the transitions are constrained by the usual scattering selection rules. Here, we demonstrate unidirectional propagation of electrons in a one-dimensional time-dependent super-lattice by breaking the symmetry of their dispersion behavior. The model of an electron in a modulation of the potential that varies with time and space is reminiscent of the sliding lattice potential of the Thouless pump in quantum mechanics [54]. In our model, the time-dependent super-lattice forms as a result of a propagating coherent phonon. In contrast to the scattering transition interpretation, the time-dependence of the interaction between the electron and the spatio-temporal modulation of the electrostatic potential it sees, is included in the electronic wave function itself leading to a more comprehensive non-conventional topological interpretation of the states and

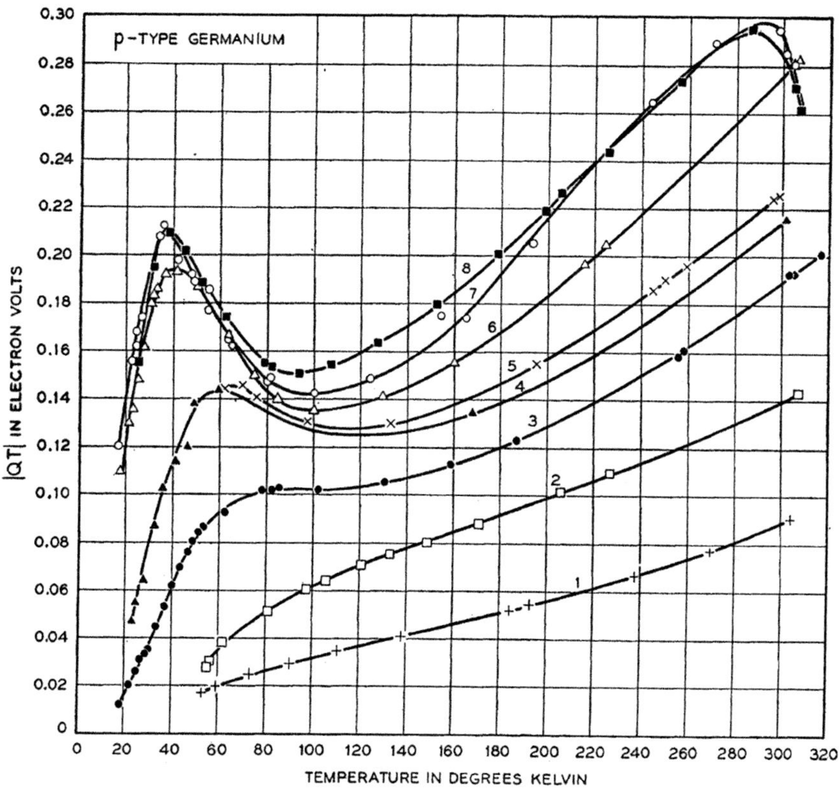

Fig. 5.4 Measured values of the product QT, where Q is the thermoelectric power and T is the temperature, as functions of temperature for p-type Germanium samples of decreasing dopant concentration (1–8). From [50]

of their topological constraints. The non-conventional topology is associated with an asymmetry in the electronic band structure that leads to electron fluxes.

We consider the one-dimensional case of an electron in a crystal through which a coherent longitudinal phonon propagates in some determined direction, e.g., the direction of a temperature gradient. The displacement field of the lattice associated with the passing phonon would result in a spatio-temporal modulation of the crystal electrostatic potential. The electronic states are therefore those in a time-dependent super-lattice with a period of the phonon wavelength and with a frequency that is associated with the phonon speed. To simplify that picture, we limit ourselves to the continuum limit of the material subjected to a sinusoidal modulation of the electrostatic potential. Therefore, we want to solve the one-dimensional Schroedinger equation:

$$i\hbar\frac{\partial}{\partial t}\psi(x,t) = \frac{-\hbar^2}{2m}\frac{\partial^2}{\partial x^2}\psi(x,t) + V(x,t)\psi(x,t) \tag{5.16}$$

with a spatio-temporal modulation of the potential:

$$V(x,t) = V_0 2\cos(Kx + \Omega t). \tag{5.17}$$

Here, $K = \frac{2\pi}{L}$ where L is the period of the modulation, Ω is the frequency modulation and its sign determines the direction of propagation of the modulation. In the long-wavelength limit and for a potential modulation that results from a directed phonon propagating in the underlying medium, we have $\Omega = VK$ where V is the velocity of sound.

The periodicity of the modulation suggests that we should be seeking solutions of (5.15a) in the form of Bloch waves: $\psi(x,t) = \sum_k \sum_g \varphi(k,g,t)e^{i(k+g)x}$ where $x \in [0,L]$. The wave number k is limited to the first Brillouin zone: $\left[\frac{-\pi}{L}, \frac{\pi}{L}\right]$ and $g = \frac{2\pi}{L}l$ with l being an integer. With Bloch wave solutions and inserting (5.17) into (5.16) leads to the Fourier domain equation:

$$i\hbar\frac{\partial}{\partial t}\varphi(k^*,t) = \frac{\hbar^2}{2m}(k^*)^2\varphi(k^*,t) + V_0\left[e^{i\Omega t}\varphi(k^* - K,t) + e^{-i\Omega t},\varphi(k^* + K,t)\right] \tag{5.18}$$

with $k^* = k + g$. In the case of a phonon with small amplitude, we take $V_0 = \varepsilon$ as a small perturbation. We solve (5.18) using multiple time scale perturbation theory [42] up to second order (also see Appendix 4 in Chap. 3). The wave function is written as a second order power series in ε, namely:

$$\varphi(k^*,\tau_0,\tau_1,\tau_2) = \varphi^{(0)}(k^*,\tau_0,\tau_1,\tau_2) + \varepsilon\varphi^{(1)}(k^*,\tau_0,\tau_1,\tau_2) + \varepsilon^2\varphi^{(2)}(k^*,\tau_0,\tau_1,\tau_2) \tag{5.19}$$

Here $\varphi^{(i)}$ with $i = 0, 1, 2$ are wave functions expressed to zeroth, first and second order. We have also replaced the single time variable, t, by three variables representing different time scales: $\tau_0 = t$, $\tau_1 = \varepsilon t$, and $\tau_2 = \varepsilon^2 t = \varepsilon^2\tau_0$. With this: $\frac{\partial\varphi}{\partial t} = \frac{\partial\varphi}{\partial\tau_0} + \varepsilon\frac{\partial\varphi}{\partial\tau_1} + \varepsilon^2\frac{\partial\varphi}{\partial\tau_2}$.

We can subsequently decompose (5.18) into equations to zeroth, first and second order in ε. The zeroth order equation is Schroedinger equation in absence of a potential:

$$i\hbar\frac{\partial}{\partial\tau_0}\varphi^{(0)}(k^*,\tau_0,\tau_1,\tau_2) = \frac{\hbar^2}{2m}(k^*)^2\varphi^{(0)}(k^*,\tau_0,\tau_1,\tau_2) \tag{5.20}$$

Its solutions take the form:

$$\varphi^{(0)}(k^*, \tau_0, \tau_1, \tau_2) = a_0(k^*, \tau_1, \tau_2)e^{-i\frac{E_0(k^*)}{\hbar}\tau_0} \tag{5.21}$$

where $E_0(k^*) = \frac{\hbar^2}{2m}(k^*)^2$. To first order, (5.18) reads:

$$\begin{aligned} i\hbar\frac{\partial}{\partial\tau_0}\varphi^{(1)}(k^*, \tau_0, \tau_1, \tau_2) - \frac{\hbar^2}{2m}(k^*)^2\varphi^{(1)}(k^*, \tau_0, \tau_1, \tau_2) \\ = -i\hbar\frac{\partial}{\partial\tau_1}\varphi^{(0)}(k^*, \tau_0, \tau_1, \tau_2) + e^{i\Omega\tau_0}\varphi^{(0)}(k^* - K, \tau_0, \tau_1, \tau_2) \\ + e^{-i\Omega\tau_0}\varphi^{(0)}(k^* + K, \tau_0, \tau_1, \tau_2) \end{aligned} \tag{5.22}$$

The term $\frac{\partial}{\partial\tau_1}\varphi^{(0)}$ leads to secular terms that can be eliminated by ensuring that φ is independent of τ_1 at all orders in ε. Under this condition, the first order solutions are the sum of solutions to the homogeneous equation and particular solutions: $\varphi^{(1)}(k^*, \tau_0, \tau_2) = \varphi_H^{(1)}(k^*, \tau_0, \tau_2) + \varphi_P^{(1)}(k^*, \tau_0, \tau_2)$. The homogeneous solutions take a form similar to the zeroth order solutions:

$\varphi_H^{(1)}(k^*, \tau_0, \tau_2) = a_1(k^*, \tau_2)e^{-i\frac{E_0(k^*)}{\hbar}\tau_0}$.

The particular solutions are given by:

$$\varphi_P^{(1)}(k^*, \tau_0, \tau_2) = b_1(k^*, \tau_2)e^{i\left(\Omega - \frac{E_0(k^*-K)}{\hbar}\right)\tau_0} + b'_1(k^*, \tau_2)e^{-i\left(\Omega + \frac{E_0(k^*+K)}{\hbar}\right)\tau_0} \tag{5.23}$$

with the resonant coefficients:

$$b_1 = \frac{a_0(k^* - K, \tau_2)}{E_0(k^* - K) - \hbar\Omega - E_0(k^*)}$$

$$b'_1 = \frac{a_0(k^* + K, \tau_2)}{E_0(k^* + K) + \hbar\Omega - E_0(k^*)}$$

To second order, one has to solve:

$$\begin{aligned} i\hbar\frac{\partial}{\partial\tau_0}\varphi^{(2)}(k^*, \tau_0, \tau_2) - \frac{\hbar^2}{2m}(k^*)^2\varphi^{(2)}(k^*, \tau_0, \tau_2) = -i\hbar\frac{\partial}{\partial\tau_2}\varphi^{(0)} \\ (k^*, \tau_0, \tau_2) + e^{i\Omega\tau_0}\varphi^{(1)}(k^* - K, \tau_0, \tau_2) + e^{-i\Omega\tau_0}\varphi^{(1)}(k^* + K, \tau_0, \tau_2) \end{aligned} \tag{5.24}$$

Inserting (5.23) into (5.24) leads to secular terms. The term in $\frac{\partial}{\partial\tau_2}\varphi^{(0)}$ also leads to a secular term. We combine all secular terms and set them equal to zero. Replacing $\varphi^{(0)}$ by (5.21) results in the condition for eliminating secular terms:

$$i\hbar\frac{\partial a_0(k^*,\tau_2)}{\partial\tau_2}=a_0(k^*,\tau_2)\left[\frac{1}{E_0(k^*)+\hbar\Omega-E_0(k^*-K)}+\frac{1}{E_0(k^*)-\hbar\Omega-E_0(k^*+K)}\right]. \tag{5.25}$$

Equation (5.25) leads to a second order correction of the zeroth order solution. To progress further, we seek amplitudes in the form: $a_0(k^*,\tau_2)=\alpha(k^*)e^{i\delta\tau_2}$. Inserting this form into (5.25) gives:

$$\delta=-\frac{1}{\hbar}\left[\frac{1}{E_0(k^*)+\hbar\Omega-E_0(k^*-K)}+\frac{1}{E_0(k^*)-\hbar\Omega-E_0(k^*+K)}\right].$$

With this result, the zeroth order energy is corrected to the form:

$$E(k^*)=E_0(k^*)+\varepsilon^2\left[\frac{1}{E_0(k^*)+\hbar\Omega-E_0(k^*-K)}+\frac{1}{E_0(k^*)-\hbar\Omega-E_0(k^*+K)}\right]. \tag{5.26}$$

We note that only the second term in the bracket of (5.26) can diverge within the Brillouin zone. This asymmetry reflects a breaking of symmetry in wave number space due to the directionality of the propagation of the phonon that modulates the electron potential. A divergence occurs at k_c when the condition: $E_0(k_c+g)-\hbar\Omega-E_0(k_c+g+K)$ is satisfied. This divergence is indicative of the formation of a gap in the band structure. It is the band folding due to the spatial modulation of the potential which enables overlap and hybridization between the energy shifted Bloch modes and the original Bloch modes of the electron without the time dependency of the spatial modulation. The hybridization opens gaps in a band structure that has lost its mirror symmetry about the origin of the Brillouin zone. Starting at $k=0$ at the bottom of the band, the wave function transitions from a state corresponding to a zeroth-order type wave, with orbital part, $e^{-i\frac{E_0(k^*)}{\hbar}\tau_0}$, to a wave having the characteristics of the energy-shifted first-order wave with orbital part $e^{-i\left(\pm\Omega+\frac{E_0(k^*+K)}{\hbar}\right)\tau_0}$ (see Fig. 5.5).

The modulation breaks the symmetry with respect to wave number of the electronic band structure. The topology of electronic waves subjected to a spatio-temporal modulation of the potential can again be represented by a manifold in wave number space composed of a closed loop strip due to the periodicity of the Brillouin zone with a twist located at the wave number corresponding to the hybridization gap. The phase of the electronic waves is associated with the change in angle of a vector field parallel transported along the manifold. Following a path in the Brillouin zone, leads to the accumulation of a π phase. One therefore needs to follow a closed path in wave number space which length is $2\times\frac{2\pi}{L}$ to accumulate a phase of 2π and recover the original wave function.

The breaking of symmetry of the band structure impacts electron transport. Indeed, considering a conventional solid within the semi-classical model of

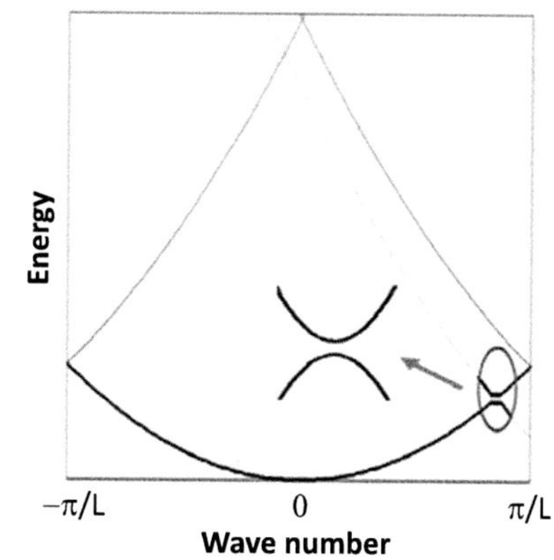

Fig. 5.5 Schematic representation of the electronic band structure with a spatio-temporal modulation of the potential. The *dashed line* corresponds to energy-shifted first-order electronic waves

electron dynamics, the total contribution to the electric current density from a three dimensional filled band is [55]:

$$\vec{j} = -e \int \frac{d\vec{k}}{4\pi^3} \frac{1}{\hbar} \frac{\partial E\left(\vec{k}\right)}{\partial \vec{k}} \tag{5.27}$$

The integral in (5.27) is over the Brillouin zone. It is well known that the integral over the Brillouin zone of the gradient of a periodic function must vanish. As a consequence, filled bands do not contribute to the transport properties of a solid [55]. If we consider that the Fermi energy is much larger than the energy at which the hybridization gap occurs, the bands in Fig. 5.5 are assumed to be filled. The gradient $\frac{\partial E(k)}{\partial k}$ is not symmetrical and its integral over the Brillouin zone does not vanish. Such a filled band contributes to a charge current and these currents have been observed in response to phonon or strain pulses [56–60].

5.4 Phonon-Magnon Resonant Processes with Relevance to Acoustic Spin Pumping

5.4.1 Magnon Background

The production of spin polarized currents is of tremendous importance for the transport of information in the form of the spin instead of electron charges. The process of generation of spin currents is an essential component in the development of spintronics. To date, several approaches have been investigated to generate spin currents. The first one utilizes non-local spin injection [61, 62]: spin current is created by electrons flowing in a ferromagnet and diffusing into an adjacent normal metal. A second set of methods generates pure spin currents from spin pumping. A processing magnetization (e.g., driven into ferromagnetic resonance by an external magnetic field) injects a non-equilibrium spin current into an adjacent normal metal via spin angular momentum transfer [63–65]. Another popular method that currently fosters intense efforts is the spin Hall effect (SHE), where unpolarized electrons flowing in a normal metal are asymmetrically scattered via spin-orbit coupled impurities, resulting in the generation of a transverse pure spin current [66, 67]. All the above methods seek to generate pure spin currents from electrical voltages and electron flows. In fact, recent developments in the field of spintronics have demonstrated that heat (i.e., acoustic phonons) could be used to pump spin currents from (insulating or metallic) ferromagnets. The exploration of the spin Seebeck effect (SSE) [68]—i.e., the generation of spin currents using heat—has revealed that chargeless spin currents can be efficiently carried by magnons in magnetic insulators such as Yttrium Iron Garnet (YIG) [69]. Combining elastic waves with magnons in magnetic insulators offers exciting opportunities to the development of chargeless information control since no electrical voltages or charge current flows are needed [70]. Elastic waves have also been used recently to coherently excite the magnetization of ferromagnets, induce ultrafast demagnetization, or reorient the magnetization direction [71, 72].

In this context, the role of phonon-magnon coupling is particularly inspiring. Indeed, it has recently been shown that the resonant absorption of elastic waves by a ferromagnet can also drive a spin current [73]. This approach has been termed acoustic spin pumping [ASP] [74]. ASP is a resonant equivalent of the SSE. The fundamental mechanism underlying ASP results from the coupling between an elastic wave (phonons) and spin waves (magnons) through the magneto-elastic effect [75]. Of other significant relevance are the role played by phonon-magnon interactions and the magneto-elastic effect in the phenomenon of phonon drag enhancement of SSE [76] or the development of the field of spincaloritronics [77, 78].

Several theories of the magneto-elastic effect have relied on (a) approximating the ferromagnet as a continuum and (b) applying the quasistatic approximation whereby magnons are at equilibrium with respect to the slower phonons [79–81]. More recently methods based on molecular and spin-dynamics have been employed to investigate numerically phonon-magnon interactions. These numerical methods solve the equations of motion for atoms and spins simultaneously [82–84]. Some of these studies have shown softening and damping of magnon modes due to lattice vibrations and the existence of coupled phonon-magnon modes with identical frequencies [82]. In light of the importance of phonon-magnon interactions for a wide range of emerging scientific and technical fields, we have developed an analytical approach to shed light on resonant phonon-magnon processes in models of ferromagnet media. In this section, we focus on a one-dimensional model composed of atomic sites that support lattice vibrations and spin precession. We solve the coupled magneto-elastic equations of motion for the spin degrees of freedom and the atomic displacement using analytical mathematical methods. We have considered two cases, namely a case with a nonlinear magneto-elastic coupling and another one with linear coupling. While analyzing the linear system is straightforward, the nonlinear system requires the use of the multiple time scale perturbation theory [42, 85]. This approach enables us to obtain analytical solutions for the wave representation of the nonlinearly interacting phonons and magnons.

5.4.2 *Model*

5.4.2.1 Hamiltonian

We consider a one dimensional discrete model of a medium that can support coupled spin waves and elastic waves (see Fig. 5.6). This is used as a model of ferromagnetic materials with magneto-elastic coupling. This model is constituted of a chain of atoms where each atomic site "n" is characterized by a spin, $\vec{S}_n$ and a displacement, u_n. We assume that the model is limited to first nearest neighbor interactions. The Hamiltonian for this system takes the form:

$$H = -2\sum\nolimits_n \sum\nolimits_{i,j=x,y,z} J_{n,n+1}^{i,j}(u_{n+1}, u_n) S_n^i S_{n+1}^j + \frac{1}{2}\sum\nolimits_n \beta(u_{n+1} - u_n)^2. \quad (5.28)$$

In this equation, the atoms are interacting elastically via a harmonic potential with a uniform, constant stiffness, β. The remaining part of the Hamiltonian is that of a Heisenberg model where the nearest neighbor spin exchange coupling constant $J_{n,n+1}^{i,j}(u_{n+1}, u_n)$ depends on the displacement of the atoms. The superscripts, i and j run over all directions x, y, and z.

We consider two cases for the magneto-elastic coupling, namely case I (5.29a) and case II (5.29b):

$$J_{n,n+1}^{i,j}(u_{n+1}, u_n) = J_{n,n+1}(u_{n+1}, u_n) = J\delta_{ij} - K(u_{n+1} - u_n)\delta_{ij} \quad (5.29a)$$

$$J_{n,n+1}^{i,j}(u_{n+1}, u_n) = J\delta_{ij} - K_T(u_{n+1} - u_n)(1 - \delta_{ij}) \quad (5.29b)$$

In the first case the coupling between the displacement and the spins is limited to terms in the Hamiltonian of the form: $S_n^i S_{n+1}^i$. In case I, the magneto-elastic coupling is restricted to diagonal terms in the components of the spin degrees of freedom to the same coordinates. Case II corresponds to interactions limited to

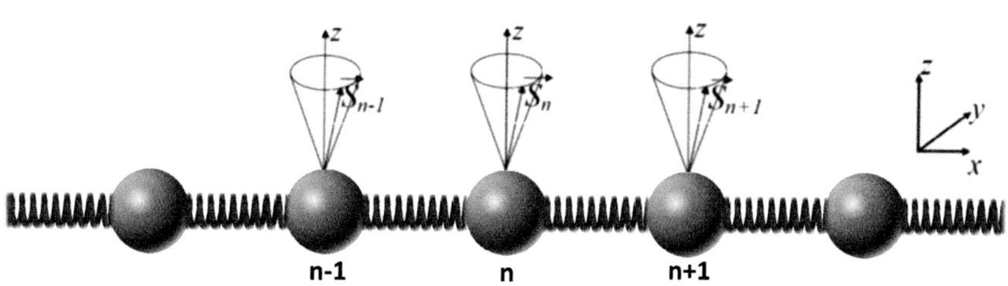

Fig. 5.6 Schematic illustration of the one dimensional discrete atomic model supporting coupled spin waves and elastic waves. The X axis is defined along the chain of atoms

cross terms with respect to components of the spin degrees of freedom. The magneto-elastic coupling term in the Hamiltonian involves only terms of the form: $S_n^i S_{n+1}^{j \neq i}$. J, K and K_T are positive constants. The minus sign in (5.29a, 5.29b) indicates that the coupling between adjacent spin weakens as the separation distance increases.

The Hamiltonian for case I takes the specific form:

$$H = -2\sum_n J_{n,n+1}(u_n, u_{n+1})\vec{S}_n.\vec{S}_{n+1} + \frac{1}{2}\sum_n \beta(u_{n+1} - u_n)^2 \tag{5.30}$$

This Hamiltonian, then, simplifies to:

$$H = -2J\sum_n \vec{S}_n.\vec{S}_{n+1} + 2K\sum_n (u_{n+1} - u_n)\vec{S}_n.\vec{S}_{n+1} + \frac{1}{2}\sum_n \beta(u_{n+1} - u_n)^2. \tag{5.31}$$

The second term in (5.31) represents the magneto-elastic interactions between spin and displacement. As discussed previously, magneto-elastic coupling is composed only of terms diagonal in the components of the spin degrees of freedom, that is, these interactions involve only pairs of components: xx, yy, and zz by virtue of the dot product.

In the case II considered here, the Hamiltonian is written as

$$\begin{aligned} H = & -2J\sum_n \vec{S}_n.\vec{S}_{n+1} + 2K_T\sum_n (u_{n+1} - u_n)\left(S_n^x S_{n+1}^y + S_n^y S_{n+1}^x + S_n^z S_{n+1}^x \right. \\ & \left. + S_n^x S_{n+1}^z + S_n^z S_{n+1}^y + S_n^y S_{n+1}^z\right) + \frac{1}{2}\sum_n \beta(u_{n+1} - u_n)^2. \end{aligned} \tag{5.32}$$

The second term in (5.32) involves products of different components of the spin degrees of freedom, that is, these interactions involve only pairs of components of the type: xy, yz, and xz.

5.4.2.2 Equations of Motion for the Spin Degrees of Freedom

The dynamics of a spin is given by Landau-Lifshitz equation:

$$\frac{\partial \vec{S}_n}{\partial t} = -\gamma \vec{S}_n \times \vec{h}_n \tag{5.33a}$$

where

$$\vec{h}_n = \frac{1}{\gamma\hbar} \cdot \frac{\partial H}{\partial \vec{S}_n} \tag{5.33b}$$

is the effective magnetic field at site "n". In (5.33), γ (>0) and $\hbar$ stand for the atomic gyromagnetic ratio and the reduced Planck's constant, respectively. For the sake of simplicity, we neglect magnetic damping.

Using the Hamiltonian of (5.31) that corresponds to case I, the spin equation of motion becomes:

$$\frac{\partial \vec{S}_n}{\partial t} = 2\frac{J}{\hbar}\vec{S}_n \times \left(\vec{S}_{n-1} + \vec{S}_{n+1}\right) - 2\frac{K}{\hbar}\left[(u_{n+1} - u_n)\vec{S}_n \times \vec{S}_{n+1} + (u_n - u_{n-1})\vec{S}_n \times \vec{S}_{n-1}\right] \tag{5.34}$$

We seek a solution in the form:

$$\vec{S}_n = \vec{S}_o + \vec{\varepsilon}_n = \begin{pmatrix} 0 \\ 0 \\ S_z^0 \end{pmatrix} + \begin{pmatrix} \varepsilon_{n,x} \\ \varepsilon_{n,y} \\ \varepsilon_{n,z} \end{pmatrix} \text{ with } \varepsilon_{n,x}, \varepsilon_{n,y}, \varepsilon_{n,z} << S_z^0 \tag{5.35}$$

$\vec{S}_o$ is the magnetic moment of the spin and is oriented along the z axis in the positive direction. $\varepsilon_{n,z}$ must be a second-order term to satisfy the conservation of the norm of $\vec{S}_n$.

We expand terms of the form $\vec{S}_n \times \vec{S}_{n-1}$ up to second-order in ε:

$$\vec{S}_n \times \vec{S}_{n-1} \sim \begin{pmatrix} -S_z^0(\varepsilon_{n-1,y} - \varepsilon_{n,y}) \\ +S_z^0(\varepsilon_{n-1,x} - \varepsilon_{n,x}) \\ 0 \end{pmatrix} + \begin{pmatrix} 0 \\ 0 \\ \varepsilon_{n,x}\varepsilon_{n-1,y} - \varepsilon_{n,y}\varepsilon_{n-1,x} \end{pmatrix} \tag{5.36}$$

With this, (5.33a) expressed in component form becomes:

$$\frac{\partial \varepsilon_{n,x}}{\partial t} = -2\frac{J}{\hbar}S_z^0\left(\varepsilon_{n+1,y} - 2\varepsilon_{n,y} + \varepsilon_{n-1,y}\right) + 2\frac{K}{\hbar}S_z^0\left[(u_{n+1} - u_n)\left(\varepsilon_{n+1,y} - \varepsilon_{n,y}\right) + (u_n - u_{n-1})\left(\varepsilon_{n-1,y} - \varepsilon_{n,y}\right)\right] \tag{5.37a}$$

$$\frac{\partial \varepsilon_{n,y}}{\partial t} = +2\frac{J}{\hbar}S_z^0(\varepsilon_{n+1,x} - 2\varepsilon_{n,x} + \varepsilon_{n-1,x}) - 2\frac{K}{\hbar}S_z^0[(u_{n+1} - u_n)(\varepsilon_{n+1,x} - \varepsilon_{n,x}) + (u_n - u_{n-1})(\varepsilon_{n-1,x} - \varepsilon_{n,x})] \tag{5.37b}$$

$$\frac{\partial \varepsilon_{n,z}}{\partial t} = 2\frac{J}{\hbar}\left[\varepsilon_{n,x}\left(\varepsilon_{n-1,y} + \varepsilon_{n+1,y}\right) - \varepsilon_{n,y}\left(\varepsilon_{n-1,x} + \varepsilon_{n+1,x}\right)\right] \tag{5.37c}$$

The first term in the right-hand side of (5.37a) and (5.37b) are the usual first-order terms that appear in the equations of motion of spin waves. All other terms that involve the constant K are second-order terms resulting from the coupling between displacement and spin. Equation (5.37c) is second-order and couples the directions x and y to direction z.

For case II, application of (5.33b) to the Hamiltonian given by (5.32) results in

$$\vec{h}_n = -2\frac{J}{\hbar}\left(\vec{S}_{n-1} + \vec{S}_{n+1}\right) + 2\frac{K_T}{\hbar}\left\{(u_{n+1} - u_n)\vec{V}_{n+1} + (u_n - u_{n-1})\vec{V}_{n-1}\right\} \tag{5.38}$$

where the vectors $\vec{V}_n$ are defined as $\begin{pmatrix} S_n^y + S_n^z \\ S_n^x + S_n^z \\ S_n^x + S_n^y \end{pmatrix}$. Using (5.35), we approximate to zeroth-order this vector by the constant vector: $\begin{pmatrix} S_z^0 \\ S_z^0 \\ 0 \end{pmatrix}$. This vector is independent of location along the chain of atoms. Using this, we obtain the equations of motion for the x and y components to first-order in displacement:

$$\frac{\partial \varepsilon_{n,x}}{\partial t} = -2\frac{J}{\hbar} S_z^0 \left(\varepsilon_{n+1,y} - 2\varepsilon_{n,y} + \varepsilon_{n-1,y}\right) + 2\frac{K_T}{\hbar}\left(S_z^0\right)^2 \left(u_{n+1} - u_{n-1}\right) \quad (5.39a)$$

$$\frac{\partial \varepsilon_{n,y}}{\partial t} = +2\frac{J}{\hbar} S_z^0 \left(\varepsilon_{n+1,x} - 2\varepsilon_{n,x} + \varepsilon_{n-1,x}\right) - 2\frac{K_T}{\hbar}\left(S_z^0\right)^2 \left(u_{n+1} - u_{n-1}\right) \quad (5.39b)$$

We do not have to consider an equation of motion for the component $\varepsilon_{n,z}$ since in case II, all equations can be expressed to first-order only.

5.4.2.3 Equations of Motion for the Displacement

In case I, the force acting on an atom "n" is obtained as:

$$F_n = -\frac{\partial H}{\partial u_n} = +\beta(u_{n+1} - u_n) - \beta(u_n - u_{n-1}) + 2K\left(\vec{S}_n.\vec{S}_{n+1} - \vec{S}_{n-1}.\vec{S}_n\right) \quad (5.40)$$

Assuming that each atom has the same mass m, their motion is described by the equation:

$$m\frac{\partial^2 u_n}{\partial t^2} = \beta(u_{n+1} - 2u_n + u_{n-1}) + 2K\left(\vec{S}_n.\vec{S}_{n+1} - \vec{S}_{n-1}.\vec{S}_n\right) \quad (5.41)$$

Utilizing (5.35), (5.41) takes its final form:

$$m\frac{\partial^2 u_n}{\partial t^2} = \beta(u_{n+1} - 2u_n + u_{n-1}) + 2K\left(S_z^0(\varepsilon_{n+1,z} - \varepsilon_{n-1,z}) + \varepsilon_{n,x}(\varepsilon_{n+1,x} - \varepsilon_{n-1,x}) + \varepsilon_{n,y}\left(\varepsilon_{n+1,y} - \varepsilon_{n-1,y}\right)\right) \quad (5.42)$$

When $K = 0$, one recovers the usual equation of motion for a monoatomic one-dimensional harmonic chain. The second term in the right-hand side of (5.42) models the magneto-elastic coupling to second-order.

Equations (5.37a–5.37c) and (5.42) form the complete set of equations describing the coupled motion of spins and atoms in case I. In this case, the magneto-elastic coupling is nonlinear and we will solve these equations within the context of multiple time scale perturbation theory in Sects. 5.4.3 and 5.4.4.

In case II, limiting the interaction terms to first-order in the spin degrees of freedom, one gets the equation of motion for the displacement:

$$m\frac{\partial^2 u_n}{\partial t^2} = \beta(u_{n+1} - 2u_n + u_{n-1}) + 2K_T S_z^0\left((\varepsilon_{n+1,x} - \varepsilon_{n-1,x}) + \left(\varepsilon_{n+1,y} - \varepsilon_{n-1,y}\right)\right) \tag{5.43}$$

Equations (5.39a, 5.39b) and (5.43) are used to model the dynamics of the displacement and the spin degrees of freedom in case II. Because these equations are linear, solutions are straightforward and will be derived in Sect. 5.4.6.

5.4.3 *Multiple Time Scale Perturbation Method Applied to Case I*

We apply the multiple time scale perturbation theory to the nonlinear equations of motion of case I. We assume that the constant K that couples the spins and the displacement is small. Subsequently, we expand the spin degrees of freedom and the displacement in power series of K up to second-order:

$$\varepsilon_{n,i} = \varepsilon_{n,i}^{(0)} + K\varepsilon_{n,i}^{(1)} + K^2\varepsilon_{n,i}^{(2)}, i = x, y, z \tag{5.44a}$$

$$u_n = u_n^{(0)} + Ku_n^{(1)} + K^2u_n^{(2)} \tag{5.44b}$$

The quantities $\varepsilon_{n,i}^{(0)}, \varepsilon_{n,i}^{(1)}, \varepsilon_{n,i}^{(2)}, u_n^{(0)}, u_n^{(1)}, u_n^{(2)}$ are functions of three time related variables defined as: $\tau_0 = t$, $\tau_1 = Kt$, and $\tau_2 = K^2t$.

The first-order and second-order time derivatives in the left-hand side of (5.37a–5.37c) and (5.42) are then rewritten as:

$$\frac{\partial^2 u_n}{\partial t^2} = \frac{\partial^2 u_n^{(0)}}{\partial {\tau_0}^2} + K\left(\frac{\partial^2 u_n^{(1)}}{\partial {\tau_0}^2} + 2\frac{\partial^2 u_n^{(0)}}{\partial \tau_0 \partial \tau_1}\right) + K^2\left(\frac{\partial^2 u_n^{(1)}}{\partial {\tau_0}^2} + 2\frac{\partial^2 u_n^{(1)}}{\partial \tau_0 \partial \tau_1} + 2\frac{\partial^2 u_n^{(0)}}{\partial \tau_0 \partial \tau_2} + \frac{\partial^2 u_n^{(0)}}{\partial {\tau_1}^2}\right) \tag{5.45a}$$

$$\frac{\partial \varepsilon}{\partial t} = \frac{\partial \varepsilon^{(0)}}{\partial \tau_0} + K\left(\frac{\partial \varepsilon^{(1)}}{\partial \tau_0} + \frac{\partial \varepsilon^{(0)}}{\partial \tau_1}\right) + K^2\left(\frac{\partial \varepsilon^{(2)}}{\partial \tau_0} + \frac{\partial \varepsilon^{(1)}}{\partial \tau_1} + \frac{\partial \varepsilon^{(0)}}{\partial \tau_2}\right) \tag{5.45b}$$

We insert (5.45a, 5.45b) and (5.44a, 5.44b) into (5.37a) and (5.42) and separate the terms that are independent of K (zeroth-order terms in K), from the first-order terms in K, from the terms in K^2. This operation leads to three sets of four equations.

The equations of motion to zeroth-order in K are:

$$\frac{\partial \varepsilon_{n,x}^{(0)}}{\partial \tau_0} = -2\frac{J}{\hbar}S_z^0\left(\varepsilon_{n+1,y}^{(0)} - 2\varepsilon_{n,y}^{(0)} + \varepsilon_{n-1,y}^{(0)}\right) \tag{5.46a}$$

$$\frac{\partial \varepsilon_{n,y}^{(0)}}{\partial \tau_0} = +2\frac{J}{\hbar}S_z^0\left(\varepsilon_{n+1,x}^{(0)} - 2\varepsilon_{n,x}^{(0)} + \varepsilon_{n-1,x}^{(0)}\right) \tag{5.46b}$$

$$\frac{\partial \varepsilon_{n,z}^{(0)}}{\partial \tau_0} = +2\frac{J}{\hbar}\left(\varepsilon_{n,x}^{(0)}\left(\varepsilon_{n+1,y}^{(0)} + \varepsilon_{n-1,y}^{(0)}\right) - \varepsilon_{n,y}^{(0)}\left(\varepsilon_{n+1,x}^{(0)} + \varepsilon_{n-1,x}^{(0)}\right)\right) \tag{5.46c}$$

$$\frac{\partial^2 u_n^{(0)}}{\partial {\tau_0}^2} = \omega_M^2\left(u_{n+1}^{(0)} - 2u_n^{(0)} + u_{n-1}^{(0)}\right) \tag{5.46d}$$

We have introduced in (5.46d) the characteristic frequency: $\omega_M = \sqrt{\frac{\beta}{m}}$. To first-order in K, we obtain:

$$\frac{\partial \varepsilon_{n,x}^{(1)}}{\partial \tau_0} + \frac{\partial \varepsilon_{n,x}^{(0)}}{\partial \tau_1} = -2\frac{J}{\hbar}S_z^0\left(\varepsilon_{n+1,y}^{(1)} - 2\varepsilon_{n,y}^{(1)} + \varepsilon_{n-1,y}^{(1)}\right) + \frac{2}{\hbar}S_z^0\Big[\left(u_{n+1}^{(0)} - u_n^{(0)}\right)\left(\varepsilon_{n+1,y}^{(0)} - \varepsilon_{n,y}^{(0)}\right)$$
$$-\left(u_n^{(0)} - u_{n-1}^{(0)}\right)\left(\varepsilon_{n,y}^{(0)} - \varepsilon_{n-1,y}^{(0)}\right)\Big] \tag{5.47a}$$

$$\frac{\partial \varepsilon_{n,y}^{(1)}}{\partial \tau_0} + \frac{\partial \varepsilon_{n,y}^{(0)}}{\partial \tau_1} = +2\frac{J}{\hbar}S_z^0\left(\varepsilon_{n+1,x}^{(1)} - 2\varepsilon_{n,x}^{(1)} + \varepsilon_{n-1,x}^{(1)}\right) - \frac{2}{\hbar}S_z^0\Big[\left(u_{n+1}^{(0)} - u_n^{(0)}\right)\left(\varepsilon_{n+1,x}^{(0)} - \varepsilon_{n,x}^{(0)}\right)$$
$$-\left(u_n^{(0)} - u_{n-1}^{(0)}\right)\left(\varepsilon_{n,x}^{(0)} - \varepsilon_{n-1,x}^{(0)}\right)\Big] \tag{5.47b}$$

$$\frac{\partial \varepsilon_{n,z}^{(1)}}{\partial \tau_0} + \frac{\partial \varepsilon_{n,z}^{(0)}}{\partial \tau_1} = 2\frac{J}{\hbar}\Big[\varepsilon_{n,x}^{(1)}\left(\varepsilon_{n+1,y}^{(0)} + \varepsilon_{n-1,y}^{(0)}\right) + \varepsilon_{n,x}^{(0)}\left(\varepsilon_{n+1,y}^{(1)} + \varepsilon_{n-1,y}^{(1)}\right) - \varepsilon_{n,y}^{(1)}\left(\varepsilon_{n+1,x}^{(0)} + \varepsilon_{n-1,x}^{(0)}\right)$$
$$-\varepsilon_{n,y}^{(0)}\left(\varepsilon_{n+1,x}^{(1)} + \varepsilon_{n-1,x}^{(1)}\right)\Big] \tag{5.47c}$$

$$\frac{\partial^2 u_n^{(1)}}{\partial {\tau_0}^2} + 2\frac{\partial^2 u_n^{(0)}}{\partial \tau_0 \partial \tau_1} = \omega_M^2\left(u_{n+1}^{(1)} - 2u_n^{(1)} + u_{n-1}^{(1)}\right) + \frac{2}{m}S_z^0\left(\varepsilon_{n+1,z}^{(0)} - \varepsilon_{n-1,z}^{(0)}\right)$$
$$+\frac{2}{m}\left[\varepsilon_{n,x}^{(0)}\left(\varepsilon_{n+1,x}^{(0)} - \varepsilon_{n-1,x}^{(0)}\right) + \varepsilon_{n,y}^{(0)}\left(\varepsilon_{n+1,y}^{(0)} - \varepsilon_{n-1,y}^{(0)}\right)\right] \tag{5.47d}$$

The equations to the order of K^2 are:

$$\frac{\partial \varepsilon_{n,x}^{(2)}}{\partial \tau_0} + \frac{\partial \varepsilon_{n,x}^{(1)}}{\partial \tau_1} + \frac{\partial \varepsilon_{n,x}^{(0)}}{\partial \tau_2} = -2\frac{J}{\hbar}S_z^0\left(\varepsilon_{n+1,y}^{(2)} - 2\varepsilon_{n,y}^{(2)} + \varepsilon_{n-1,y}^{(2)}\right)$$
$$+\frac{2}{\hbar}S_z^0\Big[\left(u_{n+1}^{(0)} - u_n^{(0)}\right)\left(\varepsilon_{n+1,y}^{(1)} - \varepsilon_{n,y}^{(1)}\right) + \left(u_{n+1}^{(1)} - u_n^{(1)}\right)\left(\varepsilon_{n+1,y}^{(0)} - \varepsilon_{n,y}^{(0)}\right)$$
$$-\left(u_n^{(0)} - u_{n-1}^{(0)}\right)\left(\varepsilon_{n,y}^{(1)} - \varepsilon_{n-1,y}^{(1)}\right) - \left(u_n^{(1)} - u_{n-1}^{(1)}\right)\left(\varepsilon_{n,y}^{(0)} - \varepsilon_{n-1,y}^{(0)}\right)\Big], \tag{5.48a}$$

$$\frac{\partial \varepsilon_{n,y}^{(2)}}{\partial \tau_0}+\frac{\partial \varepsilon_{n,y}^{(1)}}{\partial \tau_1}+\frac{\partial \varepsilon_{n,y}^{(0)}}{\partial \tau_2}=+2\frac{J}{\hbar}S_z^0\left(\varepsilon_{n+1,x}^{(2)}-2\varepsilon_{n,x}^{(2)}+\varepsilon_{n-1,x}^{(2)}\right)$$
$$-\frac{2}{\hbar}S_z^0\left[\left(u_{n+1}^{(0)}-u_n^{(0)}\right)\left(\varepsilon_{n+1,x}^{(1)}-\varepsilon_{n,x}^{(1)}\right)+\left(u_{n+1}^{(1)}-u_n^{(1)}\right)\left(\varepsilon_{n+1,x}^{(0)}-\varepsilon_{n,x}^{(0)}\right)\right.$$
$$\left.-\left(u_n^{(0)}-u_{n-1}^{(0)}\right)\left(\varepsilon_{n,x}^{(1)}-\varepsilon_{n-1,x}^{(1)}\right)-\left(u_n^{(1)}-u_{n-1}^{(1)}\right)\left(\varepsilon_{n,x}^{(0)}-\varepsilon_{n-1,x}^{(0)}\right)\right] \tag{5.48b}$$

$$\frac{\partial \varepsilon_{n,z}^{(2)}}{\partial \tau_0}+\frac{\partial \varepsilon_{n,z}^{(1)}}{\partial \tau_1}+\frac{\partial \varepsilon_{n,z}^{(0)}}{\partial \tau_2}=2\frac{J}{\hbar}\left[\varepsilon_{n,x}^{(2)}\left(\varepsilon_{n+1,y}^{(0)}+\varepsilon_{n-1,y}^{(0)}\right)+\varepsilon_{n,x}^{(0)}\left(\varepsilon_{n+1,y}^{(2)}+\varepsilon_{n-1,y}^{(2)}\right)\right.$$
$$+\varepsilon_{n,x}^{(1)}\left(\varepsilon_{n+1,y}^{(1)}+\varepsilon_{n-1,y}^{(1)}\right)-\varepsilon_{n,y}^{(2)}\left(\varepsilon_{n+1,x}^{(0)}+\varepsilon_{n-1,x}^{(0)}\right)$$
$$\left.-\varepsilon_{n,y}^{(0)}\left(\varepsilon_{n+1,x}^{(2)}+\varepsilon_{n-1,x}^{(2)}\right)-\varepsilon_{n,y}^{(1)}\left(\varepsilon_{n+1,x}^{(1)}+\varepsilon_{n-1,x}^{(1)}\right)\right] \tag{5.48c}$$

$$\frac{\partial^2 u_n^{(2)}}{\partial \tau_0{}^2}+2\frac{\partial^2 u_n^{(1)}}{\partial \tau_0 \partial \tau_1}+2\frac{\partial^2 u_n^{(0)}}{\partial \tau_0 \partial \tau_2}+\frac{\partial^2 u_n^{(0)}}{\partial \tau_1{}^2}=\omega_M^2\left(u_{n+1}^{(2)}-2u_n^{(2)}+u_{n-1}^{(2)}\right)$$
$$+\frac{2}{m}S_z^0\left(\varepsilon_{n+1,z}^{(1)}-\varepsilon_{n-1,z}^{(1)}\right)+\frac{2}{m}\left[\varepsilon_{n,x}^{(0)}\left(\varepsilon_{n+1,x}^{(1)}-\varepsilon_{n-1,x}^{(1)}\right)+\varepsilon_{n,x}^{(1)}\left(\varepsilon_{n+1,x}^{(0)}-\varepsilon_{n-1,x}^{(0)}\right)\right.$$
$$\left.+\varepsilon_{n,y}^{(0)}\left(\varepsilon_{n+1,y}^{(1)}-\varepsilon_{n-1,y}^{(1)}\right)+\varepsilon_{n,y}^{(1)}\left(\varepsilon_{n+1,y}^{(0)}-\varepsilon_{n-1,y}^{(0)}\right)\right] \tag{5.48d}$$

5.4.4 *Solutions of the Equations of Motion in Case I*

5.4.4.1 Zeroth-Order in *K*

Here we solve the set of (5.46a–5.46d). Equations (5.46a–5.46c) and (5.46d) are not coupled so we can solve them separately. For (5.46a, 5.46b) we assume that the x and y components of the spin degree of freedom are expressed as:

$$\varepsilon_{n,x}^{(0)}(\tau_0,\tau_1,\tau_2)=X(\tau_1,\tau_2)e^{ikna}e^{-i\omega_0'\tau_0}+\bar{X}(\tau_1,\tau_2)e^{-ikna}e^{i\omega_0'\tau_0} \tag{5.49a}$$
$$\varepsilon_{n,y}^{(0)}(\tau_0,\tau_1,\tau_2)=Y(\tau_1,\tau_2)e^{ikna}e^{-i\omega_0'\tau_0}+\bar{Y}(\tau_1,\tau_2)e^{-ikna}e^{i\omega_0'\tau_0} \tag{5.49b}$$

In (5.49a, 5.49b), a stands for the spacing between two atoms in the chain. $\bar{X}$ and $\bar{Y}$ are the complex conjugates of X and Y. Inserting (5.49a, 5.49b) into (5.46a, 5.46b) yields two sets of linear equations in X and Y and $\bar{X}$ and $\bar{Y}$. The Eigen values associated with these sets of equations are given by:

$$\omega_0'=8\frac{J}{\hbar}S_z^0 sin^2\left(\frac{ka}{2}\right) \tag{5.50}$$

In the limit of small wave number k, one recovers the usual quadratic dispersion relation for spin waves or magnons of long wavelength. The Eigen vectors are found to be $Y = -iX$ and $\bar{Y} = i\bar{X}$. Then, the solutions (5.49a, 5.49b) take the specific form:

$$\varepsilon_{n,x}^{(0)} = \varepsilon_o(\tau_1, \tau_2)e^{ikna}e^{-i\omega_0'\tau_0} + \bar{\varepsilon}_0(\tau_1, \tau_2)e^{-ikna}e^{i\omega_0'\tau_0} \tag{5.51a}$$

$$\varepsilon_{n,y}^{(0)} = -i\varepsilon_o(\tau_1, \tau_2)e^{ikna}e^{-i\omega_0'\tau_0} + i\bar{\varepsilon}_0(\tau_1, \tau_2)e^{-ikna}e^{i\omega_0'\tau_0} \tag{5.51b}$$

where ε_o is some yet unknown function and $\bar{\varepsilon}_0$ its complex conjugate.

If one inserts (5.51a, 5.51b) into (5.46c), the right-hand side term is analytically zero and one finds that $\varepsilon_{n,z}^{(0)}$ is constant. It can be taken to be equal to zero.

Finally, we also chose a general solution for the zeroth-order displacement given by:

$$u_n^{(0)}(\tau_0, \tau_1, \tau_2) = A_0(\tau_1, \tau_2)e^{ikna}e^{-i\omega_0\tau_0} + \bar{A}_0(\tau_1, \tau_2)e^{-ikna}e^{i\omega_0\tau_0} \tag{5.52}$$

Inserting this expression into the zeroth-order equation of motion (5.46d) yields the well-known Eigen values for elastic waves or phonons in a monatomic harmonic crystal:

$$\omega_0 = 2\omega_M \sin\left(\frac{ka}{2}\right) \tag{5.53}$$

5.4.4.2 First-Order in *K*

We can now insert the zeroth-order solutions obtained in the preceding subsection into the first-order equations (5.47a–5.47d). By doing this, we restrict our calculation to solutions whereby the magnons and phonons that interact with each other have the same wave number, k. For instance, (5.47d) can be rewritten as:

$$\frac{\partial^2 u_n^{(1)}}{\partial \tau_0{}^2} - \left(u_{n+1}^{(1)} - 2u_n^{(1)} + u_{n-1}^{(1)}\right) = 2i\omega_0\left(\frac{\partial A_0}{\partial \tau_1}e^{ikna}e^{-i\omega_0\tau_0} + \frac{\partial \bar{A}_0}{\partial \tau_1}e^{-ikna}e^{i\omega_0\tau_0}\right) \tag{5.54}$$

The solution of the homogeneous equation (left-hand side of the equation equal to zero) is of the form:

$$u_{n,h}^{(1)}(\tau_0, \tau_1, \tau_2) = B_0(\tau_1, \tau_2)e^{ikna}e^{-i\omega_0\tau_0} + \bar{B}_0(\tau_1, \tau_2)e^{-ikna}e^{i\omega_0\tau_0} \tag{5.55}$$

The term in parenthesis on the right-hand side of (5.54) leads to secular terms unless one sets it to zero by imposing $\frac{\partial A_0}{\partial \tau_1} = 0$ and $\frac{\partial \bar{A}_0}{\partial \tau_1} = 0$. This implies that the quantities A_0 and $\bar{A}_0$ are functions of τ_2 only. This constrains the zeroth-order solution (5.52) to have the form:

$$u_n^{(0)}(\tau_0, \tau_2) = A_0(\tau_2)e^{ikna}e^{-i\omega_0\tau_0} + \bar{A}_0(\tau_2)e^{-ikna}e^{i\omega_0\tau_0} \tag{5.56}$$

Similarly, the same constraint applies to the solution of the homogeneous equation (5.55):

$$u_{n,h}^{(1)}(\tau_0, \tau_2) = B_0(\tau_2)e^{ikna}e^{-i\omega_0\tau_0} + \bar{B}_0(\tau_2)e^{-ikna}e^{i\omega_0\tau_0} \tag{5.57}$$

We now solve the first-order equations of motion for the spin degrees of freedom. We begin with (5.47a). We rewrite it in the form:

$$\begin{aligned}\frac{\partial \varepsilon_{n,x}^{(1)}}{\partial \tau_0} + 2\frac{J}{\hbar}S_z^0\left(\varepsilon_{n+1,y}^{(1)} - 2\varepsilon_{n,y}^{(1)} + \varepsilon_{n-1,y}^{(1)}\right) = -\frac{\partial \varepsilon_{n,x}^{(0)}}{\partial \tau_1} + \frac{2}{\hbar}S_z^0\Big[\left(u_{n+1}^{(0)} - u_n^{(0)}\right)\left(\varepsilon_{n+1,y}^{(0)} - \varepsilon_{n,y}^{(0)}\right) \\ - \left(u_n^{(0)} - u_{n-1}^{(0)}\right)\left(\varepsilon_{n,y}^{(0)} - \varepsilon_{n-1,y}^{(0)}\right)\Big]\end{aligned} \tag{5.58}$$

The solution of the homogeneous equation is given by the expression:

$$\varepsilon_{n,x,h}^{(1)} = \eta_0(\tau_1, \tau_2)e^{ikna}e^{-i\omega_0'\tau_0} + \overline{\eta}_0(\tau_1, \tau_2)e^{-ikna}e^{i\omega_0'\tau_0}$$

Inserting (5.51a) into the first term of the right-hand side of (5.58) will lead to secular terms unless one imposes the constraints: $\varepsilon_o(\tau_1, \tau_2) = \varepsilon_o(\tau_2)$ and $\overline{\varepsilon}_0(\tau_1, \tau_2) = \overline{\varepsilon}_0(\tau_2)$. This results in the reformulation of the zeroth-order solution:

$$\varepsilon_{n,x}^{(0)} = \varepsilon_o(\tau_2)e^{ikna}e^{-i\omega_0'\tau_0} + \overline{\varepsilon}_0(\tau_2)e^{-ikna}e^{i\omega_0'\tau_0} \tag{5.59}$$

as well as the reformulation of the homogeneous solution:

$$\varepsilon_{n,x,h}^{(1)} = \eta_0(\tau_2)e^{ikna}e^{-i\omega_0'\tau_0} + \overline{\eta}_0(\tau_2)e^{-ikna}e^{i\omega_0'\tau_0} \tag{5.60}$$

We finally rewrite (5.58) by calculating the term between the square brackets. After lengthy algebraic manipulations, (5.58) becomes:

$$\begin{aligned}\frac{\partial \varepsilon_{n,x}^{(1)}}{\partial \tau_0} + 2\frac{J}{\hbar}S_z^0\left(\varepsilon_{n+1,y}^{(1)} - 2\varepsilon_{n,y}^{(1)} + \varepsilon_{n-1,y}^{(1)}\right) = 2\frac{J}{\hbar}S_z^0 2(\sin(2ka) \\ -2\sin(ka))\left(\varepsilon_0 A_0 e^{2ikna}e^{-i(\omega_0+\omega_0')\tau_0}\right. \\ \left. + \overline{\varepsilon}_0\bar{A}_0 e^{-2ikna}e^{i(\omega_0+\omega_0')\tau_0}\right)\end{aligned} \tag{5.61}$$

We proceed in the same way for the equation of motion (5.47b). We find that we need to reformulate the zeroth-order solution:

$$\varepsilon_{n,y}^{(0)} = -i\varepsilon_o(\tau_2)e^{ikna}e^{-i\omega_0'\tau_0} + i\bar{\varepsilon}_0(\tau_2)e^{-ikna}e^{i\omega_0'\tau_0} \tag{5.62}$$

as well as the homogeneous solution:

$$\varepsilon_{n,y,h}^{(1)} = -i\eta_0(\tau_2)e^{ikna}e^{-i\omega_0'\tau_0} + i\bar{\eta}_0(\tau_2)e^{-ikna}e^{i\omega_0'\tau_0} \tag{5.63}$$

The equation of motion (5.47b) is rewritten in the form:

$$\begin{aligned}\frac{\partial\varepsilon_{n,y}^{(1)}}{\partial\tau_0} - 2\frac{J}{\hbar}S_z^0\left(\varepsilon_{n+1,x}^{(1)} - 2\varepsilon_{n,x}^{(1)} + \varepsilon_{n-1,x}^{(1)}\right) = -2\frac{J}{\hbar}S_z^0 2i(\sin(2ka)\\ -2\sin(ka))\left(\varepsilon_0A_0e^{2ikna}e^{-i(\omega_0+\omega_0')\tau_0} - \bar{\varepsilon}_0\bar{A}_0e^{-2ikna}e^{i(\omega_0+\omega_0')\tau_0}\right)\end{aligned} \tag{5.64}$$

Particular solutions of (5.61) and (5.64) can be introduced in the form:

$$\varepsilon_{n,x,p}^{(1)} = X'(\tau_2)e^{2ikna}e^{-i(\omega_0+\omega_0')\tau_0} + \bar{X}'(\tau_2)e^{-2ikna}e^{i(\omega_0+\omega_0')\tau_0} \tag{5.65a}$$

$$\varepsilon_{n,y,p}^{(1)} = Y'(\tau_2)e^{2ikna}e^{-i(\omega_0+\omega_0')\tau_0} + \overline{Y}'(\tau_2)e^{-2ikna}e^{i(\omega_0+\omega_0')\tau_0} \tag{5.65b}$$

Inserting these particular solutions into (5.61) and (5.64) and solving for X', Y', $\overline{X}', \overline{Y}'$ yields:

$$X' = -i\varepsilon_0A_0f(k) \tag{5.66a}$$

$$Y' = -iX' = -\varepsilon_0A_0f(k) \tag{5.66b}$$

$$\overline{X}' = i\bar{\varepsilon}_0\bar{A}_0f(k) \tag{5.66c}$$

$$\overline{Y}' = i\overline{X}' = -\bar{\varepsilon}_0\bar{A}_0f(k) \tag{5.66d}$$

where we define $f(k) = \frac{\frac{4}{\hbar}S_z^0(\sin(2ka)-2\sin(ka))}{\frac{8J}{\hbar}S_z^0\sin^2(ka)-(\omega_0+\omega_0')+i\psi}$. We have introduced a damping term in the definition of $f(k)$ since its denominator may become zero. We will take the limit $\psi \to 0$ at the end of our derivation.

Adding the homogeneous and particular solutions gives the first order solutions:

$$\begin{aligned}\varepsilon_{n,x}^{(1)} &= \eta_0(\tau_2)e^{ikna}e^{-i\omega_0'\tau_0} + \bar{\eta}_0(\tau_2)e^{-ikna}e^{i\omega_0'\tau_0}\\ &\quad - f(k)\left(i\varepsilon_0A_0e^{2ikna}e^{-i(\omega_0+\omega_0')\tau_0} - i\bar{\varepsilon}_0\bar{A}_0e^{-2ikna}e^{i(\omega_0+\omega_0')\tau_0}\right)\end{aligned} \tag{5.67a}$$

$$\begin{aligned}\varepsilon_{n,y}^{(1)} &= -i\eta_0(\tau_2)e^{ikna}e^{-i\omega_0'\tau_0} + i\bar{\eta}_0(\tau_2)e^{-ikna}e^{i\omega_0'\tau_0}\\ &\quad - f(k)\left(\varepsilon_0A_0e^{2ikna}e^{-i(\omega_0+\omega_0')\tau_0} + \bar{\varepsilon}_0\bar{A}_0e^{-2ikna}e^{i(\omega_0+\omega_0')\tau_0}\right)\end{aligned} \tag{5.67b}$$

We can now address the solutions of the first-order equation (5.47c). We set $\frac{\partial\varepsilon_{n,z}^{(0)}}{\partial\tau_1} = 0$ (i.e., make $\varepsilon_{n,z}^{(0)}$ independent of τ_1) to eliminate secular solutions. Inserting the zeroth-order and the first-order solutions into (5.47c) results in:

$$\frac{\partial \varepsilon_{n,z}^{(1)}}{\partial \tau_0} = \frac{8J}{\hbar} f(k)\varepsilon_0\bar{\varepsilon}_0(\cos(ka) - 2\cos(2ka))\left(A_0 e^{ikna} e^{-i\omega_0\tau_0} + \bar{A}_0 e^{-ikna} e^{i\omega_0\tau_0}\right) \tag{5.68}$$

Solutions of (5.68) are then easily obtained to within a constant of integration which may be taken equal to zero:

$$\varepsilon_{n,z}^{(1)} = \frac{8J}{\hbar} f(k)\varepsilon_0\bar{\varepsilon}_0(\cos(ka) - 2\cos(2ka))\left(\frac{1}{i\omega_0}\right)\left(-A_0 e^{ikna} e^{-i\omega_0\tau_0} + \bar{A}_0 e^{-ikna} e^{i\omega_0\tau_0}\right) \tag{5.69}$$

5.4.4.3 Second-Order in *K*

We first deal with the (5.48d) for phonons. The terms with derivatives with respect to τ_1 are equal to zero. Inserting the zeroth-order and the first-order solutions determined previously and after extensive algebraic manipulations we obtain:

$$\begin{aligned}\frac{\partial^2 u_n^{(2)}}{\partial \tau_0^2} - \omega_M^2\left(u_{n+1}^{(2)} - 2u_n^{(2)} + u_{n-1}^{(2)}\right) &= \\ -2\frac{\partial^2 u_n^{(0)}}{\partial \tau_0 \partial \tau_2} - \frac{8}{m} f(k)g(k)\varepsilon_0\bar{\varepsilon}_0&\left(A_0 e^{ikna} e^{-i\omega_0\tau_0} + \bar{A}_0 e^{-ikna} e^{i\omega_0\tau_0}\right)\end{aligned} \tag{5.70}$$

where the function $g(k)$ is given by the expression:

$$g(k) = \frac{4J}{\hbar} S_z^0 \frac{1}{\omega_0}(\cos(ka) - \cos(2ka))\sin(ka) + \sin(2ka) - \sin(ka) \tag{5.71}$$

We subsequently define:

$$A_0(\tau_2) = \alpha(\tau_2)e^{-i\varphi(\tau_2)}$$

and

$$\bar{A}_0(\tau_2) = \alpha(\tau_2)e^{+i\varphi(\tau_2)}$$

With this definition the term:

$$-2\frac{\partial^2 u_n^{(0)}}{\partial\tau_0\partial\tau_2} = -2\left(-i\omega_0\frac{\partial A_0(\tau_2)}{\partial\tau_2}e^{ikna}e^{-i\omega_0\tau_0} + i\omega_0\frac{\partial\bar{A}_0(\tau_2)}{\partial\tau_2}e^{-ikna}e^{i\omega_0\tau_0}\right)$$
$$= e^{ikna}e^{-i\omega_0\tau_0}2\omega_0 e^{-i\varphi}\left(\frac{\partial\varphi}{\partial\tau_2}\alpha + i\frac{\partial\alpha}{\partial\tau_2}\right) + e^{-ikna}e^{i\omega_0\tau_0}2\omega_0 e^{i\varphi}\left(\frac{\partial\varphi}{\partial\tau_2}\alpha - i\frac{\partial\alpha}{\partial\tau_2}\right) \quad (5.72)$$

The homogeneous equation [i.e., left-hand term of (5.70) set equal to zero] has solutions of the form $e^{ikna}e^{-i\omega_0\tau_0}$ and $e^{-ikna}e^{i\omega_0\tau_0}$, so to avoid secular terms the right-hand side of (5.70) must be equal to zero. This gives the two relations:

$$2\omega_0 e^{-i\varphi}\left(\frac{\partial\varphi}{\partial\tau_2}\alpha + i\frac{\partial\alpha}{\partial\tau_2}\right) = \frac{8}{m}f(k)g(k)\varepsilon_0\bar{\varepsilon}_0\alpha e^{-i\varphi}$$

and

$$2\omega_0 e^{i\varphi}\left(\frac{\partial\varphi}{\partial\tau_2}\alpha - i\frac{\partial\alpha}{\partial\tau_2}\right) = \frac{8}{m}f(k)g(k)\varepsilon_0\bar{\varepsilon}_0\alpha e^{i\varphi}$$

Since $\varepsilon_0\bar{\varepsilon}_0$ is real, one must have: $\frac{\partial\alpha}{\partial\tau_2} = 0$ and $\alpha = \alpha_0$ is a constant. Both preceding equations give the same relation:

$$\frac{\partial\varphi}{\partial\tau_2} = \frac{4}{m\omega_0}f(k)g(k)\varepsilon_0\bar{\varepsilon}_0.$$

This equation leads to a linear dependency of φ on the time τ_2 to within an arbitrary constant (which is taken to be zero):

$$\varphi = \frac{4}{m\omega_0}f(k)g(k)\varepsilon_0\bar{\varepsilon}_0\tau_2 = \frac{4}{m\omega_0}f(k)g(k)\varepsilon_0\bar{\varepsilon}_0K^2\tau_0 = K^2h(k)\tau_0 \quad (5.73)$$

This implies that the zeroth-order solution given by (5.56) for the displacement includes a frequency shift. The complete zeroth-order solution for the displacement is now:

$$u_n^{(0)}(\tau_0,\tau_2) = \alpha_0 e^{ikna}e^{-i\left(\omega_0+K^2h(k)\right)\tau_0} + \bar{\alpha}_0 e^{-ikna}e^{i\left(\omega_0+K^2h(k)\right)\tau_0} \quad (5.74)$$

The frequency of the phonon is shifted from ω_0 to $\omega_0 + K^2h(k)$.

We now address the second-order equations of motion for $\varepsilon_{n,x}^{(2)}$ and for $\varepsilon_{n,y}^{(2)}$, that is, (5.48a, 5.48b). For this we need to insert the expressions for the complete zeroth-order displacement (5.74) and the first-order displacement given by (5.57). We also need the zeroth-order solutions for ε [(5.59) and (5.62)] and their first-order expressions given by (5.67a, 5.67b). Both equations of motion become:

$$\frac{\partial \varepsilon_{n,x}^{(2)}}{\partial \tau_0}+2\frac{J}{\hbar}S_z^0\left(\varepsilon_{n+1,y}^{(2)}-2\varepsilon_{n,y}^{(2)}+\varepsilon_{n-1,y}^{(2)}\right)=-\frac{\partial \varepsilon_{n,x}^{(0)}}{\partial \tau_2}-\frac{4}{\hbar}S_z^0(\sin(2ka)$$
$$-2\sin(ka))\left[(\eta_0 A_0+\varepsilon_0 B_0)e^{i2kna}e^{-i\left(\omega_0+\omega_0'\right)\tau_0}-(\bar{\eta}_0\bar{A}_0+\bar{\varepsilon}_0\bar{B}_0)e^{-i2kna}e^{i\left(\omega_0+\omega_0'\right)\tau_0}\right]$$
$$-i\frac{4}{\hbar}S_z^0 f(k)(\sin(3ka)-\sin(2ka)-\sin(ka))\left(\varepsilon_0 A_0^2 e^{i3kna}e^{-i\left(2\omega_0+\omega_0'\right)\tau_0}\right.$$
$$\left.-\bar{\varepsilon}_0\bar{A}_0^2 e^{-i3kna}e^{i\left(2\omega_0+\omega_0'\right)\tau_0}\right)-i\frac{4}{\hbar}S_z^0 f(k)A_0\bar{A}_0(\sin(2ka)$$
$$-2\sin(ka))\left(-\varepsilon_o e^{ikna}e^{-i\omega_0'\tau_0}+\bar{\varepsilon}_0 e^{-ikna}e^{i\omega_0'\tau_0}\right) \tag{5.75a}$$

$$\frac{\partial \varepsilon_{n,y}^{(2)}}{\partial \tau_0}-2\frac{J}{\hbar}S_z^0\left(\varepsilon_{n+1,x}^{(2)}-2\varepsilon_{n,x}^{(2)}+\varepsilon_{n-1,x}^{(2)}\right)=-\frac{\partial \varepsilon_{n,y}^{(0)}}{\partial \tau_2}-i\frac{4}{\hbar}S_z^0(\sin(2ka)$$
$$-2\sin(ka))\left[(\eta_0 A_0+\varepsilon_0 B_0)e^{i2kna}e^{-i\left(\omega_0+\omega_0'\right)\tau_0}-(\bar{\eta}_0\bar{A}_0+\bar{\varepsilon}_0\bar{B}_0)e^{-i2kna}e^{i\left(\omega_0+\omega_0'\right)\tau_0}\right]$$
$$-\frac{4}{\hbar}S_z^0 f(k)(\sin(3ka)-\sin(2ka)-\sin(ka))\left(\varepsilon_0 A_0^2 e^{i3kna}e^{-i\left(2\omega_0+\omega_0'\right)\tau_0}\right.$$
$$\left.+\bar{\varepsilon}_0\bar{A}_0^2 e^{-i3kna}e^{i\left(2\omega_0+\omega_0'\right)\tau_0}\right)+\frac{4}{\hbar}S_z^0 f(k)A_0\bar{A}_0(\sin(2ka)$$
$$-2\sin(ka))\left(\varepsilon_o e^{ikna}e^{-i\omega_0'\tau_0}+\bar{\varepsilon}_0 e^{-ikna}e^{i\omega_0'\tau_0}\right) \tag{5.75b}$$

Only the terms in $e^{ikna}e^{-i\omega_0'\tau_0}$ and $e^{-ikna}e^{i\omega_0'\tau_0}$ will give secular solutions. To eliminate these solutions, we set:

$$0=-\frac{\partial \varepsilon_{n,x}^{(0)}}{\partial \tau_2}-i\frac{4}{\hbar}S_z^0 f(k)A_0\bar{A}_0(\sin(2ka)-2\sin(ka))\left(-\varepsilon_o e^{ikna}e^{-i\omega_0'\tau_0}+\bar{\varepsilon}_0 e^{-ikna}e^{i\omega_0'\tau_0}\right)$$
$$0=-\frac{\partial \varepsilon_{n,y}^{(0)}}{\partial \tau_2}+\frac{4}{\hbar}S_z^0 f(k)A_0\bar{A}_0(\sin(2ka)-2\sin(ka))\left(\varepsilon_o e^{ikna}e^{-i\omega_0'\tau_0}+\bar{\varepsilon}_0 e^{-ikna}e^{i\omega_0'\tau_0}\right)$$

Utilizing the expressions previously derived for $\varepsilon_{n,x}^{(0)}$ and $\varepsilon_{n,y}^{(0)}$, these two conditions reduce to:

$$\frac{\partial \varepsilon_0}{\partial \tau_2}=i\frac{4}{\hbar}S_z^0 f(k)A_0\bar{A}_0(\sin(2ka)-2\sin(ka))\varepsilon_0(\tau_2)$$
$$\frac{\partial \bar{\varepsilon}_0}{\partial \tau_2}=-i\frac{4}{\hbar}S_z^0 f(k)A_0\bar{A}_0(\sin(2ka)-2\sin(ka))\bar{\varepsilon}_0(\tau_2)$$

Similarly to finding the second-order solution for the displacement, we define: $\varepsilon_0(\tau_2)=\lambda(\tau_2)e^{-i\varphi'(\tau_2)}$ and $\bar{\varepsilon}_0(\tau_2)=\lambda(\tau_2)e^{i\varphi'(\tau_2)}$. Inserting these expressions into the preceding conditions, leads to taking $\lambda(\tau_2)=\lambda_0$ (i.e., a constant) and

$$\varphi'(\tau_2) = -\frac{4}{\hbar}S_z^0 f(k)\alpha_0^2(\sin(2ka) - 2\sin(ka))\tau_2 = K^2 h'(k)\tau_0 \tag{5.76}$$

To arrive at that equation, we have assumed that the constant of integration is zero.

Equation (5.76) introduces a correction to the zeroth-order solutions for $\varepsilon_{n,x}^{(0)}$ and $\varepsilon_{n,y}^{(0)}$ which become:

$$\varepsilon_{n,x}^{(0)} = \lambda_o e^{ikna} e^{-i\left(\omega_0'+K^2h'(k)\right)\tau_0} + \overline{\lambda}_0 e^{-ikna} e^{i\left(\omega_0'+K^2h'(k)\right)\tau_0} \tag{5.77a}$$

$$\varepsilon_{n,y}^{(0)} = -i\lambda_o e^{ikna} e^{-i\left(\omega_0'+K^2h'(k)\right)\tau_0} + i\overline{\lambda}_0(\tau_2) e^{-ikna} e^{i\left(\omega_0'+K^2h'(k)\right)\tau_0} \tag{5.77b}$$

Similarly to the phonons, the magnons are frequency shifted.

The denominator of the function $f(k)$ was written as:

$$d(k) = 8\frac{J}{\hbar}S_z^0 \sin^2(ka) - \left(\omega_0 + \omega_0'\right) + i\psi \tag{5.78}$$

The introduction of the small damping, $i\psi$, enables us to overcome the divergence at the resonance. Indeed, we can take the following limit using Sokhotski's formula [86]:

$$\lim_{\psi\to 0}\frac{1}{d(k)} = \lim_{\psi\to 0}\frac{1}{8\frac{J}{\hbar}S_z^0\sin^2(ka) - \left(\omega_0(k) + \omega_0'(k)\right) + i\psi}$$
$$= \left[\frac{1}{8\frac{J}{\hbar}S_z^0\sin^2(ka) - \left(\omega_0(k) + \omega_0'(k)\right)}\right]_{pv} - i\pi\delta\left(8\frac{J}{\hbar}S_z^0\sin^2(ka) - \left(\omega_0(k) + \omega_0'(k)\right)\right).$$

In the previous equation, the bracket $[\ ]_{pv}$ means Cauchy's principal value.

The shift in frequency of the magnons and phonons becomes finite with a damping at the resonance. The imaginary term with the delta function leads to a finite lifetime for the phonon and magnon excitations at the resonance.

5.4.5 *Discussion of Results for Case I*

5.4.5.1 Dispersion with Phonon-Magnon Interactions

First of all, we summarize the findings of Sect. 5.4.3. We have seen that the dispersion relations of the magnons and of the phonons are shifted due to their mutual interaction. Both dispersion relations are re-expressed in the form:

$$\omega_0'^{*}(k) = \omega_0'(k) - K^2\frac{2}{\hbar}S_z^0\alpha_0^2[\xi(k)]_{pv}, \tag{5.79}$$

$$\omega_0^{*}(k) = \omega_0(k) + K^2\frac{\frac{4}{\hbar}S_z^0}{m\omega_M^2}\lambda_0^2[\xi(k)\zeta(k)]_{pv}. \tag{5.80}$$

The frequency shift of the phonons depends on the square of the amplitude of the magnons and vice versa. In (5.79) and (5.80), the corrected dispersion relations involve two functions of the wave number:

$$\xi(k) = \frac{\sin(2ka) - 2\sin(ka)}{\sin^2(ka) - \sin^2\left(\frac{ka}{2}\right) - \rho\sin\left(\frac{ka}{2}\right)}$$

and $\zeta(k) = \frac{\sin(2ka)-2\sin(ka)}{\sin^2(ka)-\sin^2\left(\frac{ka}{2}\right)-\rho\sin\left(\frac{ka}{2}\right)} \times \frac{\frac{1}{2}(\cos(ka)-\cos(2ka))\sin(ka)+\rho\sin\left(\frac{ka}{2}\right)(\sin(2ka)-\sin(ka))}{\sin^2\left(\frac{ka}{2}\right)}$

with the dimensionless quantity ρ defined as: $\rho = \frac{2\sqrt{\beta/m}}{\frac{8J}{\hbar}S_z^0}$. The sign of these functions determines whether the frequencies of the nonlinear coupled system are softened or hardened compared to the dispersion of the non-interacting phonons and magnons. In Fig. 5.7, we plot these functions along with the denominator of these very same functions. We vary ka in the positive region of the Brillouin zone: $[0, \pi]$. The parameter $\rho = \frac{2\sqrt{\beta/m}}{\frac{8J}{\hbar}S_z^0}$ measures the relative value of the cutoff frequencies of the magnons and the phonons in the chain of atoms. We investigate the cases: $\rho = 4, 2, 1, 0.4, 0.04$.

Figure 5.7a illustrates the condition for phonon-magnon resonance as a function of the reduced wave number ka and ratio of cutoff frequencies ρ. When $\rho > 1$, at a given ka, the frequency of a phonon is always larger than that of a magnon. The denominator of the functions ξ and ζ is always non-zero and negative. ξ is positive over the entire Brillouin zone leading to a softening of the magnon frequency [see (5.79)]. ζ is positive for small ka and changes sign beyond a threshold wave number. Phonon-magnon coupling hardens the phonons at low ka and softens them beyond the threshold. At $\rho = 1$, a transition occurs. This corresponds to phonon and magnon bands with the same cutoff frequency. The denominator takes on a value approaching 0 at a non-vanishing wave number ($ka = 1.047$). A near resonance condition is reached when a phonon with frequency $\omega_0(ka = 1.047)$ and a magnon with frequency $\omega'_0\,(ka = 1.047)$ interact to create a magnon of frequency $\omega'_0\,(ka = 2.094) = \omega_0\,(ka = 1.047) + \omega'_0\,(ka = 1.047)$. This resonance condition is illustrated in the inset of Fig. 5.7a. The function ξ remains positive but exhibits a large maximum near $ka = 1.047$. The function ζ also shows a positive maximum near the resonance condition but changes sign with increasing wave number. The frequencies of the interacting magnons are softened over the complete Brillouin zone (5.79). The frequency of the phonons increases at small wave number and decreases at larger ones in comparison to the frequency of the non-interacting phonons (5.80). As ρ decreases below the value of 1, the denominator possesses

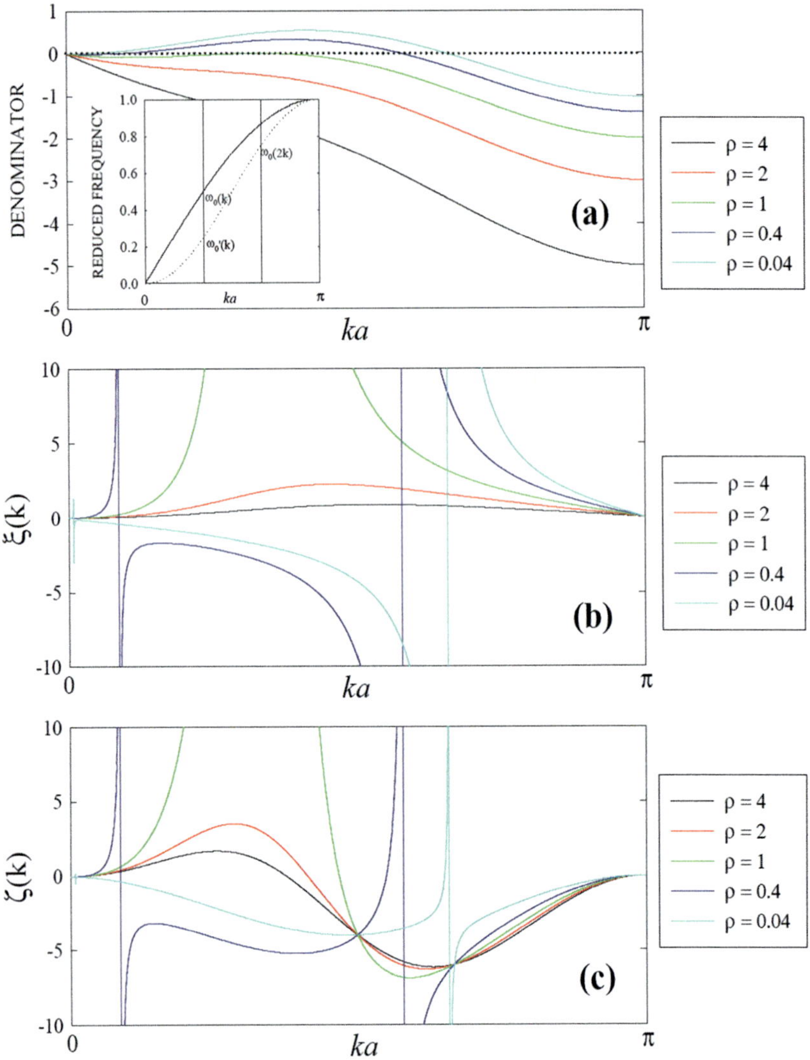

Fig. 5.7 (**a**) The denominator of functions ξ and ζ versus reduced wave number ka for different values of the parameter ρ. The inset represents the magnon (*dotted line*) and the phonon (*solid line*) dispersion curves at $\rho = 1$. The condition for resonance $\omega_0'(2k) = \left(\omega_0(k) + \omega_0'(k)\right)$ is highlighted with *vertical lines*. (**b**) ξ and (c) ζ as functions of ka and ρ

two zeros, one at low ka and one at large ka. These correspond to two resonance conditions. The low ka resonance evolves towards longer wavelength as ρ decreases. The high ka resonance moves towards the upper edge of the Brillouin zone. The functions ξ and ζ change sign several times over the complete Brillouin zone. These are the conditions that may be representative of most common ferromagnets, as will be discussed now.

Table 5.2 Values of the parameter $\rho = \frac{2\sqrt{\beta/m}}{\frac{8J}{\hbar}S_z^0} = \frac{1}{4}\frac{T_D}{T_C}$ for some representative ferromagnet materials

Material	T_D(K)	T_C(K)	ρ
YIG	600	550	0.2727
Ni	450	627	0.1794
Fe	470	1043	0.1126
Co	385	1388	0.0693

The parameter ρ can be related phenomenologically to the Curie temperature, T_C and the Debye temperature, T_D. Ferromagnetic order breaks down at non-zero temperature for a one-dimensional Ising model with nearest neighbor interactions. However, a one-dimensional Ising model in the presence of a small external field exhibits ferromagnetism at finite temperature. At high temperature, the magnetic susceptibility follows a Curie-Weiss form with a Curie temperature defined as $T_C = \frac{2J}{k_B}$ where k_B is Boltzmann's constant [87]. In the preceding expression, the factor 2 relates the number of nearest neighbors of a given spin site. This is the form of the Curie temperature that would be obtained erroneously when the one-dimensional system is treated with a mean field approach. In spite of the inability of the mean field model to capture the physics of the one-dimensional system, we employ this temperature as a phenomenological measure of the thermal equivalent of the energy of our magnetic excitations. Defining the Debye frequency: $\upsilon_D = \frac{1}{2\pi}2\sqrt{\frac{\beta}{m}}$ enables us to introduce the Debye temperature: $T_D = \frac{h\upsilon_D}{k_B}$. With these, and taking $S_z^0 = 1$, the parameter ρ can be expressed in terms of Curie and Debye temperatures as $\rho = \frac{1}{4}\frac{T_D}{T_C}$. In table 5.2, we report some values of the parameter ρ for some representative ferromagnet materials.

We note that these values are less than 1 and fall within the lower range of the cases we discuss in Fig. 5.7. Again, we have used a one-dimensional-system mean field formula for the Curie temperature. Should we have considered three-dimensional structures, $T_C = \frac{zJ}{k_B}$ where z is the number of nearest neighbors. This number would range from 6 to 8 to 12 for simple cubic, body-centered cubic and face centered cubic structures, effectively increasing the value of ρ by factors of 3, 4, or 6, respectively. Even with such multiplicative factors, the parameter ρ would remain less than 1 for common ferromagnet media.

5.4.5.2 Effect of an External Magnetic Field

We can also consider the nonlinear phonon-magnon interactions in the presence of an external magnetic field, $\vec{b}$ oriented along the z direction. In this case, (5.33b) becomes: $\vec{h}_n = \vec{b} + \frac{1}{\gamma\hbar}\cdot\frac{\partial H}{\partial \vec{S}_n}$. This adds the linear terms: $\gamma|b_z|\varepsilon_{n,y}$ and $-\gamma|b_z|\varepsilon_{n,x}$ to the right hand sides of (5.37a) and (5.37b), respectively. $|b_z|$ is the intensity of the external field. These linear terms do not affect the form of the results concerning the frequency shifts in the dispersion relations obtained with the multiple time scale perturbation approach used previously. To account for these linear terms, one simply needs to replace the magnon dispersion relation given by (5.50) by: $\omega'_0 = \frac{8J}{\hbar} S_z^0 sin^2\left(\frac{ka}{2}\right) + \gamma \mid b_z \mid$. The external field shifts the non-interacting magnon dispersion curve by a constant proportional to the zth component of the field.

By employing this zeroth-order magnon dispersion relation, we can rewrite the function, $d(k)$, defined by (5.78) in the form:

$$d(k) = \omega'_0(2k) - \gamma|b_z| - \left(\omega_0(k) + \omega'_0(k)\right) + i\psi. \tag{5.81}$$

In absence of damping and magnetic field, this expression gives a condition for the divergence (resonance) of the function, $f(k)$. This divergence corresponds to a process involving a phonon with wave number, k and frequency $\omega_0(k)$ interacting with a magnon with the same wave number but a frequency $\omega'_0(k)$ to produce a magnon with wave number $2k$ and frequency, $\omega'_0(2k) = \left(\omega_0(k) + \omega'_0(k)\right)$. This corresponds to a single-phonon-two-magnon resonant process [88]. Keeping this picture of phonon-magnon interactions as a basis for our discussion, the application of an external magnetic field effectively modifies the phonon resonant frequency to a value given by $\omega_0(k)+\gamma|b_z|$. Therefore using a fixed frequency phonon to excite magnons, one can apply a variable external field to reach resonance. On the other hand, one can apply a fixed magnetic field and change the phonon frequency to attain the resonance condition. This is illustrated by rewriting (5.54) in the form of the following resonance condition:

$$\Delta\omega'_0 = \omega'_0(2k) - \omega'_0(k) = \omega_0(k) + \gamma \mid b_z \mid. \tag{5.82}$$

Note that in this relation the left-hand term is independent of an applied magnetic field. Consider now that the phonon frequency, $\omega_0(k)$ is fixed. That is, we excite the system with a device such as an interdigitated transducer that emits elastic waves over a narrow band (i.e., produces monochromatic elastic waves with a nearly fixed frequency and wave number). In that case, the quantity $\Delta\omega'_0$ is also a constant. Let us assume that the system is subjected to an applied field, b_z^0 such that the resonance condition is not met, that is: $\Delta\omega'_0(k) \neq \omega_0(k) + \gamma \mid b_z^0 \mid$. It is therefore possible to detune positively or negatively the magnitude of the external field, $\mid b_z^0 \mid \pm\Delta b$ to approach the condition for resonance given by (5.82). In [73], Weiler et al. have studied the attenuation of a surface acoustic wave (SAW) launched in a ferromagnetic cobalt film as a function of the magnitude of an external magnetic field applied in some specific direction. The frequency of the SAW was constant and determined by the interdigitated transducer that emitted it. These authors observed a resonant absorption of the SAW at a distinct value of the magnetic field strength in accordance with our one-phonon-two-magnon resonant process prediction. In [75], a similar study was conducted for a nickel film. The strength of the magnetic field necessary to achieve resonance was shown to increase with increasing SAW frequency. This could be explained in the context of our one-phonon-two-magnon resonant process. Indeed, in the long wavelength limit, $\Delta\omega'_0(k)$ increases rapidly with increasing wave number owing to the quadratic form of the magnon dispersion relation. In that same limit, the frequency of the phonons increases linearly with the wave number. One therefore needs to increase the value of γb_z (i.e., the strength of the external field) as one increases the wave number (i.e., phonon frequency) to maintain the condition for resonance.

5.4.6 *Analysis of Case II*

In Sects. 5.4.3–5.4.5, we have addressed the coupling between phonons and magnons in case I, whereby the interactions between these excitations are nonlinear. Here, we derive the solutions of the equations of motion (5.39a, 5.39b) and (5.43) which correspond to a linear coupling between phonons and magnons. The linear interactions arose from the fact that we limited the magneto-elastic coupling to cross terms in the components of the spin in the system's Hamiltonian. In this case, we assume that the spin degrees of freedom and the displacement take a plane wave form:

$$\varepsilon_{n,x} = \varepsilon_{o,x} e^{ikna} e^{i\omega t} \tag{5.83a}$$

$$\varepsilon_{n,y} = \varepsilon_{o,y} e^{ikna} e^{i\omega t} \tag{5.83b}$$

$$u_n = A_o e^{ikna} e^{i\omega t} \tag{5.83c}$$

Inserting these solutions into the equations of motion leads to a simple Eigen value problem. The frequency of the linearly coupled phonon and magnon is subsequently obtained as:

$$\omega^2 = \frac{\omega_0'^2 + \omega_0^2}{2} \pm \frac{1}{2} \sqrt{\left(\omega_0'^2 - \omega_0^2\right)^2 + \frac{\frac{8}{\hbar} S_z^0 B^2}{m} \omega_0'} \tag{5.84}$$

where $B = 4K_T S_z^0 \sin(ka)$.

In (5.84), we have used the expressions for the Eigen values of the phonons and magnons in absence of interactions given by (5.50) and (5.53). If an external magnetic field is applied, the frequency ω_0' is simply augmented by the constant term $\gamma \mid b_z^0 \mid$. Equation (5.84) indicates that a resonance between a single magnon and a single phonon will occur when $\omega_0'(k) = \omega_0(k)$, that is when the elastic wave and the magnetic wave have identical phase velocities. This resonance is more easily seen when performing a Taylor expansion of (5.84) to second-order in K_T:

$$\omega^2 = \frac{\omega_0'^2 + \omega_0^2}{2} \pm \frac{1}{2} \mathrm{sgn}\left\{\left(\omega_0'^2 - \omega_0^2\right)\right\}\left(\omega_0'^2 - \omega_0^2\right) \pm \mathrm{sgn}\left\{\left(\omega_0'^2 - \omega_0^2\right)\right\} \frac{\frac{2}{\hbar} S_z^0 B^2}{m} \frac{\omega_0'}{\left(\omega_0'^2 - \omega_0^2\right)} \tag{5.85}$$

Depending on the sign of $\omega_0'^2 - \omega_0^2$ (i.e., $\mathrm{sgn}\{(\omega_0'^2 - \omega_0^2)\}$), the first two terms give $\omega_0'^2$ and ω_0^2. The third terms represents the effect of the resonance. The effect of magneto-elastic coupling on the dispersion relations of the phonons and magnons in case II is independent of the amplitude of the waves.

We can also consider the application of an external magnetic field. An external magnetic field can be applied to the ferromagnet medium to tune the dispersion curve of the magnons with the aim of achieving resonance under the condition of a stimulating acoustic wave with a fixed frequency. The observations concerning the dependence of strength of the applied magnetic field on frequency of the acoustic wave made in Sect. 5.4.5.2 for the one-phonon-two-magnon resonance also transpose to the case of a one-phonon-single magnon resonance.

5.4.7 Phonon-Magnon Interaction Summary

We considered the effect of coupling between elastic waves and magnetic waves in a one-dimensional discrete model system of a ferromagnet medium. Two cases are addressed. The first case corresponds to a system where the magnons and phonons interact nonlinearly. The nonlinear interaction is due to a displacement-dependent spin exchange coupling constant that involves only diagonal terms in the components of the spin. The second case leads to linear interactions resulting from a displacement-dependent spin exchange coupling constant that involves off-diagonal terms in the components of the spin. We have developed an approach based on the multiple time scale perturbation method to calculate analytically the effect of the nonlinear magneto-elastic coupling on the dispersion relation of magnons and phonons. For the sake of mathematical tractability, we have limited our study to the interaction between magnons and phonons with the same wave number. We have determined the conditions that lead to phonon-magnon resonance. In absence of an external magnetic field, the resonance for the first case is associated with a one-phonon-two-magnon resonant process. Such two magnon-one phonon processes are characteristic of the interaction between longitudinal phonons with magnons in nearest neighbor Heisenberg models where the exchange coupling constant is modulated with respect to the lattice displacements [89]. Our one-phonon-two-magnon resonance leads to a frequency shift of the magnon and phonon bands that depends on the magnitude of the elastic and magnetic waves. The frequency shift also depends on the relative magnitude of the cutoff frequencies of the two excitations. For the second case (i.e., linear coupling), the resonance includes only one phonon and one magnon. The effect of magneto-elastic coupling results in modifications to the dispersion relations that do not depend on the magnitude of the waves. We have also investigated the effect of an external field on that linear resonant condition. The behavior of the linear system with respect to the strength of the magnetic field is similar to that of the nonlinear system. We have demonstrated that similarly to observations reported in the literature concerning the phenomena of resonant spin pumping with coherent elastic waves [73, 75], variation of the strength of an externally applied magnetic field at a fixed frequency of the elastic wave can be used to tune the system toward resonance. In addition, the strength of the external field at resonance increases with increasing acoustic wave frequency.

By providing analytical solutions for the spin degrees of freedom and the lattice dynamics in model ferromagnetic atomic systems, our study enables a deeper insight into the complex interplay between magnons and phonons interacting via the magneto-elastic effect as mechanism for spin pumping. This insight offers perspective in the design of ferromagnet media that may optimize the energy conversion between phonons and magnons. In particular, one may be able to engineer the phonon and magnon band structures of the ferromagnet medium to control the conditions of resonance. We have in mind composite structures constituted of ferromagnetic and normal materials that can be tailored to exhibit band gaps in their magnon band structure [90]. Similarly these types of composites may also act as phononic crystals that can exhibit simultaneously band gaps in their phonon band structure [91]. Wave localization resulting from these anomalous band structures may subsequently be employed to enhance magneto-elastic coupling and therefore the efficiency of energy conversion between phonon and magnons in acoustic spin pumping devices.

Appendix 1: Fortran 90 Program Used to Model the Nonlinear Reaction-Diffusion Dynamics of a Chain of Cells Subjected to a Spatio-Temporal Modulation of Intercellular Diffusion Coefficient Induced by a Sound Wave

```
!-----------------------------------------------------------
!Politi model (positive feedback)used for simulating calcium wave
propagation in a single chain of cells  induced by a single
stimulation.  The chain is subjected to a spatio-temporal modula-
tion of the intercellular diffusion coefficient induced by a sound
wave.
!-----------------------------------------------------------
!-----------------------------------------------------------
!Definition of model parameters (see Table 5.1)
!-----------------------------------------------------------
! k3k is IP3 phosphorylation rate constant
! k5p is IP3 dephosphorylation rate constant
! kplc is half-activation constant of PLC
! vs is the maximal SERCA pump rate
```

```
! k1 is maximal rate of calcium release
! k2 is maximal rate of calcium leak
! ka is calcium binding to activating site
! kp is IP3 binding
! ks half-activation constant
! vplc is the maximal production rate of IP3
! svplc is rescaled maximal PLC activity
! tp is the characteristic time of IP3 turnover
! eta is the ratio of the maxiaml IP3K rate to the total maximal
degradation rate of IP3
! v0 is the constant influx
! fi is the stimulation-dependent influx
! kpm is the half-activation constant
! vpm is the maximal PMCA pump rate
! epsilon is the strength of plasma membrane fluxes
! beta is the ratio of effective volumes ER/cytosol
! tr is the characteristic time IP3R inactivation
! ki is Calcium binding to inhibiting site

         IMPLICIT NONE
         double precision::fr,uc1,h,c,p,r,tp,k3k,k5p,kplc,vplc,&
     k1,k2,ka,kp,vs,ks,eta,v0,k,ki,ctot,&
     fi,kpm,vpm,svplc,epsilon,beta,tr,&
     temp1IP3,temp1Ca,temp1ER,temp1In,&
     temp2IP3,temp2Ca,temp2ER,temp2In,&
     kIP31,tempIP31,kCa1,tempCa1,kER1,tempER1,kIn1,tempIn1,&
     kIP32,tempIP32,kCa2,tempCa2,kER2,tempER2,kIn2,tempIn2,&
     kIP33,tempIP33,kCa3,tempCa3,kER3,tempER3,kIn3,tempIn3,&
     kIP34,kCa4,kER4,kIn4,a,b,d,yu,derivIP3, derivCa, &
     derivER, derivIn,cn,pn,d2Ca,d2IP3,coeffCa,coeffIP3,dt
!--------------------------------------------------
! Diffusion coefficients and modulation parameters
!--------------------------------------------------
         double precision::Dc,Dp
         double precision::M1,M2,M3,M4
         INTEGER::n,i,j,nc,nls,jm,jp
         dimension::c(1000,2),p(1000,2),r(1000),vplc(1000),uc1(1000)
         dimension::Dc(1000),Dp(1000)
!------------------------------------
!Output files, calcium, IP3 and R concentrations
!------------------------------------
         open(unit=1, file='ca.dat',status='old')
         open(unit=2, file='ip3.dat',status='old')
         open(unit=3, file='r.dat',status='old')
```

```
!------------------------------------
!numerical values of reaction parameter
!------------------------------------
        kp=0.13
        ks=0.1
        k=0.4
        beta=0.185
        epsilon=0.0
        kpm=0.12
        vpm=0.01
        v0=0.0004
        ctot=2.0
        k3k=0.0
        k5p=0.66
        kplc=0.2
        vs=0.9
        fi=0.0047
        k1=1.11
        k2=0.0203
        ka=0.08
        ki=0.4
        tr=12.5

!Diff Coef at time t=0
        coeffCa=0.005
        coeffIP3=0.05
! Parameters for the sound induced modulation of the diffusion
coefficients
! Amplitude of the modulation
        M1=0.0030
        M2=0.03
! Period of the modulation (in units of cell-to-cell distance)
        M3=36.0
! Velocity of the modulation
        M4=0.004

!------------------------------------
!    number of cells in backbone
!------------------------------------
        nc=401
        open(unit=4,file='diff.dat',status='old')
        do i=1,nc
         Dc(i)=(M1*cos((6.283185307/M3)*(i-201)))+coeffCa
         Dp(i)=(M2*cos((6.283185307/M3)*(i-201)))+coeffIP3
        write(4,*)i,Dc(i),Dp(i)
```

```
        enddo
        close(unit=4)

!-------------------------------
!    location of stimulation
!-------------------------------
        nls=201
!---------------------
!    inital pulse
!---------------------
        do j=1,nc
        c(j,1)=0.05
        p(j,1)=0.05
        r(j)=0.85
        enddo

!-------------------------
!    Threshold
!-------------------------
        do j=1,nc
        uc1(j)=0.057
        enddo
!-------------------------
!    following vplc
!-------------------------
        fr=1.5
!------------------------
!     time step size (sec)
!------------------------
        dt=0.01
!------------------------
!    Reaction Kinetics step size
!------------------------
        h=dt
!-------------------
!    Time Loop
!-------------------
! number of time iterations
        n=30000
!Modulation of the diffusion coefficients
        do i=1,n
         do j=1,nc
          Dc(j)=(M1*cos((6.283185307/M3)*(1.*(j-201)-M4*i)))&
    +coeffCa
```

```
          Dp(j)=(M2*cos((6.283185307/M3)*(1.*(j-201)-M4*i)))&
     +coeffIP3
          enddo
!--------------------------------
!    Cells supporting reactions
!--------------------------------
         do j=1,nc
         if(vplc(j)==fr) go to 100
         if (j==nls) vplc(j)=fr
         if(c(j,1)<uc1(j) .and. j/=nls) vplc(j)=0.0
         if(c(j,1)>=uc1(j) .and. j/=nls) vplc(j)=fr
100      continue

!-------------------
!Call Reaction Kinetics Subroutines
!-------------------
          call derivfunction1(c(j,1),p(j,1),vplc(j),derivIP3)
          kIP31=h*derivIP3
          tempIP31=p(j,1)+0.5*kIP31
          call derivfunction2(c(j,1),p(j,1),r(j),vplc(j),derivCa)
          kCa1=h*derivCa
          tempCa1=c(j,1)+0.5*kCa1
          call derivfunction3(c(j,1),r(j),derivIn)
          kIn1=h*derivIn
          tempIn1=r(j)+0.5*kIn1
          call derivfunction1(tempCa1,tempIP31,vplc(j),derivIP3)
          kIP32=h*derivIP3
          tempIP32=p(j,1)+0.5*kIP32
          call derivfunction2(tempCa1,tempIP31,&
          tempIn1,vplc(j),derivCa)
          kCa2=h*derivCa
          tempCa2=c(j,1)+0.5*kCa2
          call derivfunction3(tempCa1,tempIn1,derivIn)
          kIn2=h*derivIn
          tempIn2=r(j)+0.5*kIn2
          call derivfunction1(tempCa2,tempIP32,vplc(j),derivIP3)
          kIP33=h*derivIP3
          tempIP33=p(j,1)+kIP33
          call derivfunction2(tempCa2,tempIP32,&
          tempIn2,vplc(j),derivCa)
          kCa3=h*derivCa
          tempCa3=c(j,1)+kCa3
          call derivfunction3(tempCa2,tempIn2,derivIn)
          kIn3=h*derivIn
          tempIn3=r(j)+kIn3
```

```
        call derivfunction1(tempCa3,tempIP33,vplc(j),derivIP3)
        kIP34=h*derivIP3
        call derivfunction2(tempCa3,tempIP33,&
        tempIn3,vplc(j),derivCa)
        kCa4=h*derivCa
        call derivfunction3(tempCa3,tempIn3,derivIn)
        kIn4=h*derivIn

        p(j,2)=p(j,1)+(kIP31+2.0*(kIP32+kIP33)+kIP34)/6.0
        c(j,2)=c(j,1)+(kCa1+2.0*(kCa2+kCa3)+kCa4)/6.0
        r(j)=r(j)+(kIn1+2.0*(kIn2+kIn3)+kIn4)/6.0
        enddo

!-----------------------------------------------------
!Diffusion---loop over the cells in chain
!-----------------------------------------------------
        do j=1,nc
        jp=j+1
        jm=j-1
        if(jp>nc)jp=1
        if(jm<1)jm=nc
        d2Ca=Dc(j)*(c(jp,1)-2*c(j,1)+c(jm,1))
        d2Ca=d2Ca+0.5*(Dc(jp)-Dc(jm))*0.5*(c(jp,1)-c(jm,1))
        d2IP3=Dp(j)*(p(jp,1)-2*p(j,1)+p(jm,1))
        d2IP3=d2IP3+0.5*(Dp(jp)-Dp(jm))*0.5*(p(jp,1)-p(jm,1))

        c(j,2)=c(j,2)+d2Ca*dt
        p(j,2)=p(j,2)+d2IP3*dt

        enddo
!------------------------
!    Time shift
!------------------------
        do j=1,nc
        c(j,1)=c(j,2)
        p(j,1)=p(j,2)

!----------------------------
!    end of time loop
!----------------------------
        enddo
! Write the concentration profiles at end of time loop
        do j=1,nc
        write(1,*)j,c(j,1)
        write(2,*)j,p(j,1)
```

```
          write(3,*)j,r(j)
          enddo

          end
!-----------------------------------------
! Reaction kinetics subroutines
!-----------------------------------------
          subroutine derivfunction1(c,p,vplc,derivIP3)
          double precision, intent(out)::derivIP3
          double precision, intent(in):: c,p,vplc
          double precision:: tp,k3k,k5p,kplc,&
     eta,k,ki,svplc,a,b,d
!positive feedback intracellular pathway
          k=0.4
          k3k=0.0
          k5p=0.66
          kplc=0.2
          tp=1.0/(k5p+k3k)
          eta=k3k/(k3k+k5p)
          svplc=vplc*tp
          derivIP3=(((svplc*c**2)/(kplc**2+c**2))&
     -((eta*c**2)/(k**2+c**2)+(1-eta))*p)/tp
           return
           end subroutine derivfunction1

           subroutine derivfunction2(c,p,r,vplc,derivCa)
           double precision, intent(out)::derivCa
           double precision, intent(in):: c,p,r,vplc
           real:: tp,k3k,k5p,kplc,&
     k1,k2,ka,kp,vs,ks,eta,v0,k,ki,fi,kpm,&
     vpm,svplc,epsilon,beta,tr,e,f,g,h,ctot
          kp=0.13
          ks-0.1
          k=0.4
          epsilon=0.0
          kpm=0.12
          vpm=0.01
          v0=0.0004
          ctot=2.0
          beta=0.185
!positive feedback
          k3k=0.0
          k5p=0.66
          kplc=0.2
          vs=0.9
```

```
            fi=0.0047
            k1=1.11
            k2=0.0203
            ka=0.08
            tp=1.0/(k5p+k3k)
            svplc=vplc*tp
           derivCa=(k1*((r*c*p)/((ka+c)*(kp+p)))**3+k2)*(((ctot-c)/
beta)-c)&
       -((vs*c**2)/(ks**2+c**2))&
       +epsilon*(v0+(fi*svplc)-((vpm*c**2)/(kpm**2+c**2)))
            return
            end subroutine derivfunction2

            subroutine derivfunction3(c,r,derivIn)
            double precision, intent(out)::derivIn
            double precision, intent(in):: c,r
            real:: tp,k3k,k5p,kplc,vplc,&
       k1,k2,ka,kp,vs,ks,eta,v0,k,ki,&
       fi,kpm,vpm,svplc,epsilon,beta,tr,j
!      positive feedback intracellular pathway
            ki=0.4
            tr=12.5
            derivIn=(1-r*(ki+c)/ki)/tr
            return
            end subroutine derivfunction3
```

Appendix 2: Multiple Time Scale Perturbation Theory Analysis of Calcium Wave in an Acoustic Field

For the sake of analytical tractability, we treat ε in (5.10a, 5.10b) of the main text as a perturbation and write the concentration excursions C and P as a second order power series in the perturbation, namely:

$$C(k+g,\tau_0,\tau_1,\tau_2) = C_0(k+g,\tau_0,\tau_1,\tau_2) + \varepsilon C_1(k+g,\tau_0,\tau_1,\tau_2) + \varepsilon^2 C_2(k+g,\tau_0,\tau_1,\tau_2) \quad (5.86a)$$

$$P(k+g,\tau_0,\tau_1,\tau_2) = P_0(k+g,\tau_0,\tau_1,\tau_2) + \varepsilon P_1(k+g,\tau_0,\tau_1,\tau_2) + \varepsilon^2 P_2(k+g,\tau_0,\tau_1,\tau_2) \quad (5.86b)$$

In (5.86a, 5.86b), X_i with $i=0,1,2$ and $X = C, P$ are the concentrations excursions expressed to zeroth-order, first-order and second-order in the perturbative time, respectively. Because the sought solution might have several time characteristics, multiple time scale perturbation theory expands the problem onto a series of time scales expressed in powers of the perturbation parameter times time. Considering the problem to second-order in perturbation, we have replaced the single time variable, t, by three variables representing different time scales: $\tau_0 = t$, $\tau_1 = \varepsilon t$, and $\tau_2 = \varepsilon^2 t = \varepsilon^2 \tau_0$. We can subsequently decompose equations (5.10a, 5.10b) into three equations: one equation to zeroth-order in ε, one equation to first-order in ε, and a third equation to second-order in ε. The zeroth-order equation represents the dynamics of Ca^{2+} and IP_3 concentration excursions in absence of sound irradiation. To zeroth-order, the solutions take the form similar to that reported in the main text:

$$C_0(k+g,\tau_0,\tau_1,\tau_2) = a_0^c(\tau_1,\tau_2)\xi(k+g)e^{i\omega_0(k+g)\tau_0} \tag{5.87a}$$

$$P_0(k+g,\tau_0,\tau_1,\tau_2) = (-1)a_0^p(\tau_1,\tau_2)\zeta(k+g)e^{i\omega_0(k+g)\tau_0} \tag{5.87b}$$

with

$$\xi(k+g) = \sqrt{\alpha \mp i\sqrt{r^2-\alpha^2}} \tag{5.88a}$$

$$\zeta(k+g) = \sqrt{\alpha \pm i\sqrt{r^2-\alpha^2}} \tag{5.88b}$$

and a dispersion relation

$$\omega_0(k+g) = \pm\omega_r + i\omega_i \tag{5.89}$$

In (5.88a, 5.88b) and (5.89), we have defined the following quantities: $\alpha = (k+g)^2\frac{(D_{0c}-D_{0p})}{2}$, the real frequency $\omega_r = \sqrt{r^2-\alpha^2}$ and the imaginary part of the frequency $\omega_i = (k+g)^2\frac{(D_{0c}+D_{0p})}{2}$. To zeroth-order, the amplitude functions obey: $a_0^c(\tau_1,\tau_2) = a_0^p(\tau_1,\tau_2) = a_0(\tau_1,\tau_2)$, however, we will keep the upper scripts to differentiate between calcium and IP_3 amplitudes since to higher order in perturbation theory, these quantities may differ.

The first-order equations are used to solve for C_1 and P_1:

$$\begin{aligned}&\frac{\partial C_1(k+g,\tau_0,\tau_1,\tau_2)}{\partial\tau_0} + (k+g)^2 D_{0c}C_1(k+g,\tau_0,\tau_1,\tau_2) + rP_1(k+g,\tau_0,\tau_1,\tau_2)\\ &= -\frac{\partial C_0(k+g,\tau_0,\tau_1,\tau_2)}{\partial\tau_1} + f(k+g-K)e^{i(\Omega+\omega_0(k+g-K))\tau_0}a_0^c(\tau_1,\tau_2)\xi(k+g-K,\tau_1,\tau_2)\\ &+h(k+g+K)e^{i\left(-\Omega+\omega_0(k+g+K)\tau_0\right.}a_0^c(\tau_1,\tau_2)\xi(k+g+K,\tau_1,\tau_2)\end{aligned} \tag{5.90a}$$

$$\frac{\partial P_1(k+g,\tau_0,\tau_1,\tau_2)}{\partial \tau_0}+(k+g)^2 D_{0p}P_1(k+g,\tau_0,\tau_1,\tau_2)-rC_1(k+g,\tau_0,\tau_1,\tau_2)$$
$$=-\frac{\partial P_0(k+g,\tau_0,\tau_1,\tau_2)}{\partial \tau_1}-f(k+g-K)e^{i(\Omega+\omega_0(k+g-K))\tau_0}a_0^p(\tau_1,\tau_2)\zeta(k+g-K,\tau_1,\tau_2)$$
$$-h(k+g+K)e^{i\left(-\Omega+\omega_0(k+g+K)\tau_0\right.}a_0^p(\tau_1,\tau_2)\zeta(k+g+K,\tau_1,\tau_2) \tag{5.90b}$$

The first term in the right-hand side of these two equations lead to secular solutions and need to be eliminated. This is achieved by imposing on the zeroth-order solutions, as well as subsequent order solutions, independence of the time scale τ_1. The other right-hand side terms serve as driving forces for the first-order solutions. The solutions of (5.90a, 5.90b) are therefore the sum of the solutions to the homogeneous solutions (i.e., without the driving force) and of particular solutions. The solutions to the homogeneous equations have forms similar to the solutions given by (5.87a, 5.87b).

The first-order particular solutions take the form:

$$C_1^{(p)}(k+g,\tau_0,\tau_2)=b_1(k+g)e^{i(\Omega+\omega_0(k+g-K))\tau_0}+b_1'(k+g)e^{i\left(-\Omega+\omega_0(k+g+K)\tau_0\right.} \tag{5.91a}$$

$$P_1^{(p)}(k+g,\tau_0,\tau_2)=d_1(k+g)e^{i(\Omega+\omega_0(k+g-K))\tau_0}+d_1'(k+g)e^{i\left(-\Omega+\omega_0(k+g+K)\tau_0\right.} \tag{5.91b}$$

with

$$b_1(k+g)=\frac{1}{\Delta_1(k+g)}\left[f(k+g-K)a_0^c(\tau_2)\xi(k+g-K)\{i(\Omega+\omega_0(k+g-K))\right.$$
$$\left.+(k+g)^2D_{0p}\}+rf(k+g-K)a_0^p(\tau_2)\zeta(k+g-K)\right] \tag{5.92a}$$

$$b_1'(k+g)=\frac{1}{\Delta_2(k+g)}\left[h(k+g+K)a_0^c(\tau_2)\xi(k+g+K)\{i(-\Omega+\omega_0(k+g+K))\right.$$
$$\left.+(k+g)^2D_{0p}\}+rh(k+g-K)a_0^p(\tau_2)\zeta(k+g+K)\right] \tag{5.92b}$$

$$d_1(k+g)=\frac{1}{\Delta_1(k+g)}\left[-f(k+g-K)a_0^p(\tau_2)\zeta(k+g-K)\{i(\Omega+\omega_0(k+g-K))\right.$$
$$\left.+(k+g)^2D_{0c}\}+rf(k+g-K)a_0^c(\tau_2)\xi(k+g-K)\right] \tag{5.92c}$$

$$b_1'(k+g)=\frac{1}{\Delta_2(k+g)}\left[-h(k+g+K)a_0^p(\tau_2)\zeta(k+g+K)\{i(-\Omega+\omega_0(k+g+K))\right.$$
$$\left.+(k+g)^2D_{0c}\}+rh(k+g+K)a_0^c(\tau_2)\xi(k+g+K)\right] \tag{5.92d}$$

and

$$\Delta_1(k+g)=r^2+\left\{i(\Omega+\omega_0(k+g-K))+(k+g)^2D_{0c}\right\}\left\{i(\Omega+\omega_0(k+g-K))+(k+g)^2D_{0p}\right\} \tag{5.93a}$$

$$\Delta_2(k+g)=r^2+\left\{i(-\Omega+\omega_0(k+g+K))+(k+g)^2D_{0c}\right\}\left\{i(-\Omega+\omega_0(k+g+K))+(k+g)^2D_{0p}\right\} \tag{5.93b}$$

The conditions $\Delta_1=0$ and $\Delta_2=0$ lead to resonances.

The second-order equations take the form:

$$\begin{aligned}&\frac{\partial C_2(k+g,\tau_0,\tau_2)}{\partial\tau_0}+(k+g)^2D_{0c}C_2(k+g,\tau_0,\tau_2)+rP_2(k+g,\tau_0,\tau_2)\\&=-\frac{\partial C_0(k+g,\tau_0,\tau_2)}{\partial\tau_2}+f(k+g-K)e^{i\Omega\tau_0}C_1(k+g-K,\tau_2)\\&\quad+h(k+g+K)e^{-i\Omega\tau_0}C_1(k+g+K\tau_2)\end{aligned} \tag{5.94a}$$

$$\begin{aligned}&\frac{\partial P_2(k+g,\tau_0,\tau_2)}{\partial\tau_0}+(k+g)^2D_{0c}P_2(k+g,\tau_0,\tau_2)+rC_2(k+g,\tau_0,\tau_2)\\&=-\frac{\partial P_0(k+g,\tau_0,\tau_2)}{\partial\tau_2}+f(k+g-K)e^{i\Omega\tau_0}P_1(k+g-K,\tau_2)\\&\quad+h(k+g+K)e^{-i\Omega\tau_0}P_1(k+g+K\tau_2)\end{aligned} \tag{5.94b}$$

The first term in the left-hand side of both equations is proportional to $e^{i\omega_0(k+g)\tau_0}$ and therefore leads to secular behavior. Inserting (5.91a, 5.91b) into (5.94a, 5.94b) leads to additional secular terms. Regrouping these terms and setting them to zero results in the following set of two equations:

$$\frac{\partial a_0^c(\tau_2)}{\partial\tau_2}=a_0^c(\tau_2)F_1(k+g)+a_0^p(\tau_2)F_2(k+g) \tag{5.95a}$$

$$\frac{\partial a_0^p(\tau_2)}{\partial\tau_2}=a_0^p(\tau_2)G_2(k+g)+a_0^c(\tau_2)G_1(k+g) \tag{5.95b}$$

with

$$F_1(k+g)=\frac{h^2(k+g)}{\Delta_2(k+g-K)}F_p+\frac{f^2(k+g)}{\Delta_1(k+g+K)}G_p$$

$$F_2(k+g)=r\gamma\left[\frac{h^2(k+g)}{\Delta_2(k+g-K)}+\frac{f^2(k+g)}{\Delta_1(k+g+K)}\right]$$

$$G_1(k+g)=-r\gamma^{-1}\left[\frac{h^2(k+g)}{\Delta_2(k+g-K)}+\frac{f^2(k+g)}{\Delta_1(k+g+K)}\right]$$

$$G_2(k+g)=\frac{h^2(k+g)}{\Delta_2(k+g-K)}F_c+\frac{f^2(k+g)}{\Delta_1(k+g+K)}G_c$$

and

$$F_{p,c} = i(-\Omega + \omega_0(k+g)) + (k+g-K)^2 D_{0p.c}$$
$$G_{p,c} = i(\Omega + \omega_0(k+g)) + (k+g+K)^2 D_{0p,c}$$
$$\gamma(k+g) = \frac{\zeta(k+g)}{\xi(k+g)} = e^{\pm i\varphi_0}$$
$$\gamma^{-1} = \gamma^* = e^{\mp i\varphi_0}$$

We now chose the following forms for the calcium and IP_3 amplitudes $a_0^c(\tau_2) = Ae^{i\delta\tau_2}e^{+i\varphi_2}$ and $a_0^p(\tau_2) = Ae^{i\delta\tau_2}e^{-i\varphi_2}$ such that both amplitudes have the same second-order frequency shift, δ, but differ in their phase. We saw earlier that to zeroth-order in perturbation these amplitudes were identical. However, we kept the upper script "*c*" and "*p*" throughout our derivation to reflect the possibility of a phase difference, φ_2, that arises from carrying the perturbation expansion up to second-order. Inserting these forms for the amplitudes into (5.95a, 5.95b) results in the following corrections to the zeroth-order frequency and phases of the Ca^{2+} and IP_3 waves:

$$\omega_0^{cor}(k+g) = \omega_0(k+g) - \varepsilon^2\left[\frac{h^2(k+g)}{\Delta_2(k+g-K)}\left\{r\sin(2\varphi_2 \mp \varphi_0) + \frac{i}{2}(F_p + F_c)\right\} + \frac{f^2(k+g)}{\Delta_1(k+g+K)}\left\{r\sin(2\varphi_2 \mp \varphi_0) + \frac{i}{2}(G_p + G_c)\right\}\right] \tag{5.96}$$

$$\cos(2\varphi_2 \mp \varphi_0) = -\frac{1}{r}\frac{\left[\frac{h^2(k+g)}{\Delta_2(k+g-K)}(F_p - F_c) + \frac{f^2(k+g)}{\Delta_1(k+g+K)}(G_p - G_c)\right]}{\frac{h^2(k+g)}{\Delta_2(k+g-K)} + \frac{f^2(k+g)}{\Delta_1(k+g+K)}} \tag{5.97}$$

We note that when approaching the resonance conditions $\Delta_1(k+g+K) \to 0$ and $\Delta_2\,(k+g-K) \to 0$, the phase given by (5.97) does not diverge. The divergence of (5.96) at the two possible resonances is indicative of two significant alterations of the unperturbed dispersion relation of the Ca^{2+} and IP_3 waves that result from the spatio-temporal modulation of the cell-to-cell conductance due to the applied acoustic wave. These alterations occur at wave numbers k' and k'' which satisfy the equations:

$$\omega_0(k'+g+K) = \omega_0(k'+g) + \Omega \tag{5.98a}$$
$$\omega_0(k''+g-K) = \omega_0(k''+g) - \Omega \tag{5.98b}$$

References

1. M.R. Bailey, V.A. Khokhlova, O.A. Sapozhnikov, S.G. Kargl, L.A. Crum, Physical mechanisms of the therapeutic effect of ultrasound (a review). Acoust. Phys. **49**, 369 (2003)
2. R.W. Wood, A.L. Loomis, The physical and biological effects of high frequency sound waves of great intensity. Lond. Edinb. Dublin Philos. Mag. J. Sci. **4**, 417 (1927)
3. E.N. Harvey, A.L. Loomis, High frequency sound waves of small intensity and their biological effects. Nature **121**, 622 (1928)
4. S. Mitragotri, Healing sound: the use of ultrasound in drug delivery and other therapeutic applications. Nat. Rev. **4**, 255 (2005)
5. J.D. Heckman, J.P. Ryaby, J. McCabe, J.J. Frey, R.F. Kilcoyne, Acceleration of tibial fracture-healing by non-invasive, low-intensity pulsed ultrasound. J. Bone Joint Surg. **76-A**, 26 (1994)
6. F. Padilla, R. Puts, L. Vico, K. Raum, Stimulation of bone repair with ultrasound: a review of the possible mechanic effects. Ultrasonics **54**, 1125 (2014)
7. C.A. Speed, Therapeutic ultrasound in soft tissue lesions. Rheumatology **40**, 1331 (2001)
8. Y. Tufail, A. Yoshihiro, S. Pati, W.J. Tyler, Ultrasonic neuromodulation by brain stimulation with transcranial ultrasound. Nat. Protoc. **6**, 1453 (2011)
9. G. Leinenga, J. Götz, Scanning ultrasound removes amyloid-β and restores memory in an Alzheimer's disease mouse model. Sci. Transl. Med. **7**, 1 (2015)
10. W.J. Fry, V.J. Wulff, D. Tucker, F.J. Fry, Physical factors involved in ultrasonically induced changes in living systems: I. Identification of non-temperature effects. J. Acoust. Soc. Am. **22**, 867 (1950)
11. J. Parvizi, V. Parpura, J.F. Greenleaf, M.E. Bolander, Calcium signaling is required for ultrasound-stimulated aggrecan synthesis by rat chondrocytes. J. Orthop. Res. **20**, 51 (2002)
12. M.A. Dinno, M. Dyson, S.R. Young, A.J. Mortimer, J. Hart, L.A. Crum, The significance of membrane changes in the safe and effective therapeutic and diagnostic ultrasound. Phys. Med. Biol. **34**, 1543 (1989)
13. F.J. Fry, H.W. Ades, W.J. Fry, Production of reversible changes in the central nervous system by ultrasound. Science **127**, 83 (1958)
14. K. Sena, S.R. Angle, A. Kanaji, C. Aher, D.G. Karwo, D.R. Sumner, A.S. Virdi, Low-intensity pulsed ultrasound and cell-to-cell communication in bone marrow stromal cells. Ultrasonics **51**, 639 (2011)
15. W.J. Tyler, Y. Tufail, M. Finsterwald, M.L. Tauchmann, E.J. Olson, C. Majestic, Remote excitation of neuronal circuits using low-intensity, low-frequency ultrasound. PLoS One **3**, 1 (2008)
16. M.Z. Hasan, C.L. Kane, Colloquium: Topological insulators. Rev. Mod. Phys. **82**, 3045 (2010)
17. A.B. Khanikaev, S.H. Mousavi, W.-K. Tse, M. Kargarian, A.H. MacDonald, G. Shvets, Photonic topological insulators. Nat. Mater. **12**, 233 (2013)
18. H.C. Manoharan, Topological insulators: a romance with many dimensions. Nat. Nanotech. **5**, 477 (2010)
19. A. Trewavas, Le Calcium, c'est la vie: calcium makes waves. Plant Physiol. **120**, 1–6 (1999)
20. H.C. Blair, L.J. Robinson, C.L.-H. Huang, L. Sun, P.A. Friedman, P.H. Schlesinger, M. Zaid, Calcium and bone disease. Biofactors **37**, 159 (2011)
21. E. Scemes, C. Giaume, Astrocyte calcium waves: what they are and what they do. Glia **54**, 716 (2006)
22. A.H. Carnell-Bell, S.M. Finkbeiner, M.S. Cooper, S.J. Smith, Glutamate induces calcium waves in cultured astrocytes: long-range glial signaling. Science **247**, 470 (1990)
23. S.E. Webb, A.L. Miller, Calcium signaling during embryonic development. Nat. Rev. Mol. Cell Biol. **4**, 539 (2003)
24. L. Leybaert, M.J. Sanderson, Intercellular Ca2+ waves: mechanisms and functions. Physiol. Rev. **92**, 1359 (2012)
25. I. Bezprozvanny, J. Watras, B.E. Ehrlich, Bell-shaped calcium-response curves of IP3 and calcium-gated channels from endoplasmic reticulum of cerebellum. Nature **351**, 751 (1991)

26. G.L. Combettes, L.L.C. dynamics, Spatio-temporal organization from the subcellular to the organ level. Int. Rev. Cytol. **261**, 193 (2007)
27. E.A. Finch, T.J. Turner, S.M. Goldin, Calcium as a coagonist of inositol 1,4,5-trisphosphate-induced calcium release. Science **252**, 443 (1991)
28. W.H. Evans, P.E.M. Martin, Gap junctions: structure and function (review). Mol. Membr. Biol. **19**, 121 (2002)
29. E. Decrock, D.V. Krysko, M. Vinken, A. Kaczmarek, G. Crispino, M. Bol, N. Wang, M. De Bock, E. De Vuyst, C.C. Naus, V. Rogiers, P. Vandenabeele, C. Erneux, F. Mammano, G. Bultynck, L. Leybaert, Transfer of IP3 through gap junctions is critical, but not sufficient, for the spread of apoptosis. Cell Death Differ. **19**, 947 (2012)
30. M.A. Turing, The chemical basis of morphogenesis. Phil. Trans. R. Soc. B **237**, 37 (1952)
31. A. Politi, L.D. Gaspers, A.P. Thomas, T. Hofer, Models of IP3 and Ca2+ oscillations: frequency encoding and identification of underlying feedbacks. Biophys. J. **90**, 3120 (2006)
32. S.G. Rhee, Regulation of phosphoinositide-specific phospholipase C. Annu. Rev. Biochem. **70**, 281–312 (2001)
33. A.T. Harootunian, J.P. Kao, S. Parajape, R.Y. Tsien, Generation of calcium oscillations in fibroblasts by positive feedback between calcium and IP3. Science **251**, 75 (1991)
34. G.L. Combettes, L. Leybaert, Calcium dynamics: spatio-temporal organization from the subcellular to the organ level. Int. Rev. Cytol. **261**, 193 (2007)
35. A. Verkhratsky, R.K. Orkand, H. Kettenmann, Glial calcium: homeostasis and signaling function. Physiol. Rev. **78**, 99 (1998)
36. D. Communi, V. Vanweyenberg, C. Erneux, D-myo-inositol-1,4,5-trisphosphate 3-kinase A is activated by receptor activation trough a calcium: calmodulin-dependent protein kinase II phosphorylation mechanism. EMBO J. **16**, 1943 (1997)
37. M.J. Sanderson, A.C. Charles, S. Boitano, E.R. Dirksen, Mechanisms and function of intercellular calcium signaling. Mol. Cell. Endocrinol. **98**, 173 (1994)
38. N.L. Allbritton, T. Meyer, L. Stryer, Range of messenger action of calcium ion and inositol 1, 4, 5,-trisphosphate. Science **258**, 1812 (1992)
39. M.J. Berridge, Inositol trisphosphate and calcium signaling. Nature **361**, 315 (1993)
40. J. Long, P.A. Deymier, K. Runge, J.B. Hoying, Regulation of the frequency and wavelength of calcium waves propagating in networks on interconnected cells: a simulation study. Trends Cell Mol. Biol. **9**, 19 (2014)
41. H.-B. Zhao, J. Santos-Sacchi, Effect of membrane tension in gap junctional conductance of supporting cells in Corti's organ. J. Gen. Physiol. **112**, 447 (1998)
42. R.H.G. Helleman, E.W. Montroll, On a nonlinear perturbation theory without secular terms. Physica **74**, 22 (1974)
43. M.V. Berry, Quantal phase factors accompanying adiabatic changes. Proc. R. Soc. A **392**, 45 (1984)
44. J.P. Newnham, S.F. Evans, C.A. Michael, F.J. Stanley, L.I. Landau, Effects of frequent ultrasound during pregnancy: a randomized controlled trial. Lancet **342**, 887 (1993)
45. H. Kälsch, F. Hennig, S. Moebus, S. Möhlenkamp, N. Dragano, H. Jakobs, M. Memmesheimer, R. Erbel, K.H. Jöckel, B. Hoffmann, Are air pollution and traffic noise independently associated with atherosclerosis: the Heinz Nixdorf Recall Study. Eur. Heart J. **35**, 853 (2014)
46. M. Sørensen, Z.J. Andersen, R.B. Nordsborg, S.S. Jensen, K.G. Lillelund, R. Beelen, E.B. Schmidt, A. Tjønneland, K. Overvad, O. Raaschou-Nielsen, Road traffic noise and incident myocardial infarction: a prospective cohort study. PLoS One **7**, e39283 (2012)
47. M. Alves-Pereira, N.A.A.C. Branco, Vibroacoustic disease: biological effects of infrasound and low-frequency noise explained by mechanotransduction cellular signaling. Prog. Biophys. Mol. Biol. **93**, 256 (2006)
48. C.L. Vestergaard, H. Flyvbjerg, I.M. Møller, Intracellular signaling by diffusion: can waves of hydrogen peroxide transmit intracellular information in plant cells? Front. Plant. Sci. **3**, 1 (2012)

49. S. Neill, R. Desikan, J. Hancock, Hydrogen peroxide signaling. Curr. Opin. Plant Biol. **5**, 388 (2002)
50. T.H. Geballe, G.W. Hull, Seebeck effect in Germanium. Phys. Rev. **94**, 1134 (1954)
51. C. Herring, Theory of the thermoelectric power of semiconductors. Phys. Rev. **96**, 1163 (1954)
52. H. Adachi, K. Uchida, E. Saitoh, J. Ohe, S. Takahashi, S. Maekawa, Gigantic enhancement of spin Seebeck effect by phonon drag. Appl. Phys. Lett. **97**, 252506 (2010)
53. C.M. Jaworski, J. Yang, S. Mack, D.D. Awschalom, R.C. Myers, J.P. Heremans, Spin-Seebeck effect: a phonon driven spin distribution. Phys. Rev. Lett. **106**, 186601 (2011)
54. D.J. Thouless, Quantization of particle transport. Phys. Rev. B **27**, 6083–6087 (1983)
55. N.W. Ashcroft, N.D. Mermin, *Solid State Physics* (Holt, Rinehart and Winston, London, 1976)
56. M.T. Greenaway, A.G. Balanov, D. Fowler, A.J. Kent, T.M. Fromhold, Using acoustic waves to induce high-frequency current oscillations in superlattices. Phys. Rev. B **81**, 235313 (2010)
57. D.M. Moss, A.V. Akimov, B.A. Glavin, M. Henini, A.J. Kent, Ultrafast strain-induced current in a GaAs Schottky diode. Phys. Rev. Lett. **106**, 066602 (2011)
58. G.D. Mahan, L. Lindsay, D.A. Broido, The Seebeck coefficient and phonon drag in silicon. J. Appl. Phys. **116**, 245102 (2014)
59. D.R. Fowler, A.V. Akimov, A.G. Balanov, M.T. Greenaway, M. Henini, T.M. Fromhold, A.J. Kent, Semiconductor charge transport driven by a picosecond strain pulse. Appl. Phys. Lett. **92**, 232104 (2008)
60. C.L. Poyser, A.V. Akimov, A.G. Balanov, R.P. Campion, A.J. Kent, A weakly coupled semiconductor superlattice as a harmonic hypersonic-electrical transducer. New J. Phys. **17**, 083064 (2015)
61. M. Johnson, R.H. Silsbee, Interfacial charge-spin coupling: injection and detection of spin magnetization in metals. Phys. Rev. Lett. **55**, 1790 (1985)
62. F.J. Jedema, A.T. Filip, B.J. van Wees, Electrical spin injection and accumulation at room temperature in an all-metal mesoscopic spin valve. Nature **410**, 345 (2001)
63. Y. Tserkovnyak, A. Brataas, G. Bauer, Enhanced Gilbert damping in thin ferromagnetic films. Phys. Rev. Lett. **88**, 117601 (2002)
64. A. Brataas, Y. Tserkovnyak, G. Bauer, B. Halperin, Spin battery operated by ferromagnetic resonance. Phys. Rev. B **66**, 60404 (2002)
65. S. Mizukami, Y. Ando, T. Miyazaki, Effect of spin diffusion on Gilbert damping for a very thin permalloy layer in Cu/permalloy/Cu/Pt films. Phys. Rev. B **66**, 104413 (2002)
66. T. Jungwirth, J. Wunderlich, K. Olejnik, Spin Hall effect devices. Nat. Mater. **11**, 382 (2012)
67. A. Hoffman, Spin hall effects in metals. IEEE Trans. Magn. **49**, 5172 (2013)
68. K. Uchida, S. Takahashi, K. Harii, J. Ieda, W. Koshibae, K. Ando, S. Maekawa, E. Saitoh, Observation of the spin Seebeck effect. Nature **455**, 778 (2008)
69. K. Uchida et al., Spin Seebeck insulator. Nat. Mater. **9**, 894 (2010)
70. Y. Kajiwara et al., Transmission of electrical signals by spin-wave interconversion in a magnetic insulator. Nature (London) **464**, 262 (2010)
71. A.V. Scherbakov et al., Coherent magnetization precession in ferromagnetic (Ga,Mn)As induced by picosecond acoustic pulses. Phys. Rev. Lett. **105**, 117204 (2010)
72. J.-W. Kim, M. Vomir, J.-Y. Bigot, Ultrafast magnetoacoustics in nickel films. Phys. Rev. Lett. **109**, 166601 (2012)
73. M. Weiler, H. Huebl, F.S. Goerg, F.D. Czeschka, R. Gross, S.T.B. Goennenwein, Spin pumping with coherent elastic waves. Phys. Rev. Lett **108**, 176601 (2012)
74. K. Uchida, H. Adachi, T. An, T. Ota, M. Toda, B. Hillebrands, S. Maekawa, E. Saitoh, Long-range spin Seebeck effect and acoustic spin pumping. Nat. Mater. **10**, 737 (2011)
75. M. Weiler, L. Dreher, C. Heeg, H. Huebl, R. Gross, M.S. Brandt, S.T.B. Goennenwein, Elastically driven ferromagnetic resonance in nickel thin films. Phys. Rev. Lett **106**, 117601 (2011)
76. H. Adachi, K. Uchida, E. Saitoh, J. Ohe, S. Takahashi, S. Maekawa, Gigantic enhancement of spin Seebeck effect by phonon drag. Appl. Phys. Lett **97**, 252506 (2010)

77. T. An, V.I. Vasyuchka, K. Uchida, A.V. Chumak, K. Yamaguchi, K. Harii, J. Ohe, M.B. Jungfleisch, Y. Kajiwara, H. Adachi, B. Hillebrands, S. Maekawa, E. Saitoh, Unidirectional spin-wave heat conveyer. Nat. Mater. **12**, 549 (2013)
78. G.W. Bauer, E. Saitoh, B.J. van Wees, Spin caloritronics. Nat. Mater. **11**, 391 (2012)
79. P.A. Fedders, Theory of acoustic resonance and dispersion in bulk ferromagnets. Phys. Rev. B **9**, 3835 (1974)
80. T. Kobayashi, R.C. Barker, J.L. Bleustein, A. Yelon, Ferromagnetoelastic resonance in thin films. I. Formal treatment. Phys. Rev. B **7**, 3273 (1973)
81. H.F. Tiersten, Coupled magnetomechanical equations for magnetically saturated insulators. J. Math. Phys. **5**, 1298 (1964)
82. D. Perera, D.P. Landau, D.M. Nicholson, G.M. Stocks, M. Eisenbach, J. Yin, G. Brown, Phonon-magnon interactions in body centered cubic iron: a combined molecular and spin dynamics study. J. Appl. Phys. **115**, 17D124 (2014)
83. P.-W. Ma, S.L. Dudarev, C.H. Woo, Spin-lattice dynamics model for magnon-phonon-electron heat transfer on a million atom scale. J. Appl. Phys. **111**, 07D114 (2012)
84. P.-W. Ma, C.H. Woo, S.L. Dudarev, Large-scale simulation of the spin-lattice dynamics in ferromagnetic iron. Phys. Rev. B **78**, 024434 (2008)
85. I.C. Khoo, Y.K. Wang, Multiple time scale analysis of an anharmonic crystal. J. Math. Phys. **17**, 222 (1976)
86. V.S. Vladimirov, *Equations of Mathematical Physics* (Marcel Dekker, New York, 1971)
87. R. Skomski, *Simple Models of Magnetism* (Oxford University Press, Oxford, 2008)
88. R. Silberglitt, Effect of spin waves on the phonon energy spectrum of a Heisenberg ferromagnet. Phys. Rev. **188**, 786 (1969)
89. L.M. Woods, Magnon-phonon effects in ferromagnetic manganites. Phys. Rev. B **65**, 014409 (2001)
90. J.O. Vasseur, L. Dobrzynski, B. Djafari-Rouhani, H. Puszkarski, Magnon band structure of periodic composites. Phys. Rev. B **54**, 1043 (1996)
91. P. A. Deymier (ed.), *Acoustic Metamaterials and Phononic Crystals*, Springer series in solid-state sciences, vol 173 (Springer, Berlin, 2013)

Chapter 6
Acoustic Analogues

6.1 Introduction

The previous chapters illustrate with simple models the emergence of a new science of sound that accounts not only for spectral and refractive characteristics of waves but also addresses the amplitude and phase of these waves. In this chapter, these concepts are taken further and related to the development of acoustic analogues of other physical phenomena ranging from quantum phenomena to general relativity. These analogues may offer perspectives for applications and technological developments of this new science of sound. In this chapter, we first introduce the concept of phase bit (φ-bit) based on the fermion-like behavior of phonons in some elastic structures as an analogue of a quantum bit (qubit). The analogy with the notion of spin suggests the possibility of developing quantum information technologies based on superposition of elastic waves as well the capability of achieving exponentially parallel algorithms utilizing the non-separability of elastic waves. We then consider the analogy between fermion-like elastic waves (waves with spinor characteristics) supported by a structure subjected to a spatio-temporal modulation of its properties. We show the analogy between the equation that drives the dynamics of this elastic system and the Dirac equation for a charged particle including an electromagnetic field. The modulation can be used to tune the spinor part of the elastic waves suggesting that it can work as a gauge field analogue. Staying within the realm of analogies with quantum phenomena, we explore the acousto-hydrodynamics of bubbles in a fluid irradiated with an acoustic standing wave. We argue that the secondary sound field emitted by bubbles may lead to self-interaction that can modify the translational motion of bubbles. This phenomenon is reminiscent of the pilot wave model of quantum mechanics and suggests the possibility of acoustic bubbles to exhibit particle-wave duality. Finally, we review within the context of simple models the analogy between the propagation of acoustic waves in moving fluids and general relativity. We relate these concepts to the model of a one-dimensional harmonic crystal subjected to a directed spatio-temporal

P. Deymier, K. Runge, *Sound Topology, Duality, Coherence and Wave-Mixing*,
Springer Series in Solid-State Sciences 188, DOI 10.1007/978-3-319-62380-1_6

modulation of its stiffness. The dynamics of this system is described within the context of interpretations based on differential geometry.

6.2 Analogue of Quantum Bit (qubit): Phase Bit (φ-bit)

Quantum information science exploits quantum-mechanical phenomena to perform computing operations. Quantum bit (qubit)-based computing platforms capitalize on the phenomenon of "superposition." Thanks to superposition, a quantum computer can process a number of calculations simultaneously beyond what can be achieved with transistor-based processors. Superposition is essentially the ability of a system to be in multiple states—that is, instead of encoding information in the form of a "0" or a "1" in conventional computers, a qubit can store and process a "0" and a "1" at the same time. When qubits are entangled, the state of an array of N qubits cannot be separated into a tensor product of states of individual qubits. With entangled states one can achieve the very desirable characteristic of quantum exponential complexity. Entangled states of an N qubit system can simultaneously store and process 2^N bits of information, a colossal number as N becomes large. However, two quantum-computing barriers must be broken for quantum computers to achieve their promise by making: (1) qubits stable on macro timescales without necessitating cryogenic temperature to prevent decoherence of the states; and (2) numerous at the scale of massively parallel processing without decoherence as well. Current qubit platforms use a variety of quantum systems ranging from nuclear/electron spin, trapped ions, superconducting states, to photon polarization. Phonon-based qubits can be created by entangling a phonon with the spin states of a single atom embedded into a resonant mechanical cavity [1–4]. These phonon-based devices manipulate quantum information as it transfers from phonon to matter, and vice versa; however, their conventional qubits are still encumbered with the major challenge of maintaining qubit coherence.

In Chap. 3, we noted that one can create elastic waves, in simple one-dimensional harmonic crystal with masses attached to a rigid substrate, with spin-like characteristics related to the direction of propagation. These elastic waves are called pseudospins. A pseudospin describes how waves with broken symmetry and therefore non-conventional topology are distributed between different "internal" degrees of freedom or microscopic states, such as polarization, direction of propagation, sublattices, or layers. These additional degrees of freedom result in a half-integer spin analogue reflecting the fermionic nature of low-energy quasiparticles. Unlike electron spin, pseudospin is not associated with any intrinsic property of particles; rather it arises from the substructure in space (e.g., sublattices) and time in which the wave lives. The state of a pseudospin is observable and measurable (without the wave function collapse of quantum mechanics). We have seen in Chap. 3 that transmission coefficient measurement can provide the necessary information to determine the superposition of states in terms of forward propagating and backward propagating components of the wave. Pseudospins can be prepared in superposition of states without suffering from quantum decoherence; they are nearly coherent in the elastic regime. An elastic pseudospin can serve as a perfect qubit analogue and we call it a phase bit or φ-bit.

We first use a simple model algorithm, the Deutsch-Jozsa algorithm [5] that illustrates the parallelism in quantum computing associated with a superposition of states. This algorithm is then mapped onto an elastic pseudospin to demonstrate the capability of φ-bits. The objective of this algorithm is to determine if a function $f(x)$ defined on the domain $x \epsilon \{0, 1\}$ which can take the values range $\{0, 1\}$ is balanced or constant. The function is constant if $f(0) = f(1) = 0$ or 1. The function is balanced if $f(0) = 0$ and $f(1) = 1$ or $f(0) = 1$ and $f(1) = 0$. Answering that question using classic computation approach requires evaluation of the function serially for the two possible values of x. On the other hand, a quantum computation would evaluate the function f for a superposition of two states $a|0\rangle + b|1\rangle$ where $|0\rangle$ and $|1\rangle$ represent the values $x = 0$ and $x = 1$. A single evaluation would determine the nature of the function. To illustrate this, a first qubit is prepared from a pure state $|1\rangle$ is transformed into a superposition of states $\frac{1}{\sqrt{2}}(|0\rangle - |1\rangle)$ by applying a Hadamard gate transformation. An external oracle operator performs the operation, U_f, dependent on the function f, is applied to the qubit. The operator U_f maps: $|x\rangle \longrightarrow (-1)^{f(x)}|x\rangle$. We then obtain the following cases:

(a) For a constant function

$$f(0) = f(1) = 0, \quad |0\rangle - |1\rangle \longrightarrow |0\rangle - |1\rangle$$
$$f(0) = f(1) = 1, \quad |0\rangle - |1\rangle \longrightarrow -(|0\rangle - |1\rangle)$$

(b) For a balanced function

$$f(0) = 0, f(1) = 1, \quad |0\rangle - |1\rangle \longrightarrow |0\rangle + |1\rangle$$
$$f(0) = 1, f(1) = 0, \quad |0\rangle - |1\rangle \longrightarrow -(|0\rangle + |1\rangle)$$

To within a general phase of π, the oracle operator produces different superpositions of states depending on whether the function is constant or balanced. Since a superposition of quantum states is not directly measurable, one needs to apply the Hadamard gate to these states. This procedure results into the measureable pure state $|1\rangle$ if the function is constant or $|0\rangle$ if the function is balanced. This algorithm is illustrated in Fig. 6.1 (Top).

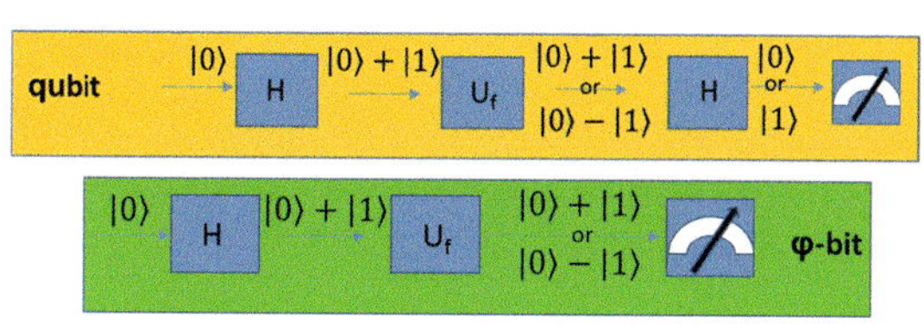

Fig. 6.1 *Top* Schematic representation of Deutsch-Jozsa algorithm for a single qubit; *Bottom* same as above but for a single φ-bit

Table 6.1 Spinor states of a φ-bit in the two limits $k \to 0$ and $k \to \infty$

	$e^{+ikx}e^{+i\omega_k t}$	$e^{-ikx}e^{+i\omega_k t}$	$e^{+ikx}e^{-i\omega_k t}$	$e^{-ikx}e^{-i\omega_k t}$
$\xi_{k\to 0}$	$\begin{pmatrix}1\\1\end{pmatrix}$	$\begin{pmatrix}1\\1\end{pmatrix}$	$\begin{pmatrix}-1\\1\end{pmatrix}$	$\begin{pmatrix}-1\\1\end{pmatrix}$
$\xi_{k\to\infty}$	$\begin{pmatrix}1\\0\end{pmatrix}$	$\begin{pmatrix}0\\1\end{pmatrix}$	$\begin{pmatrix}0\\1\end{pmatrix}$	$\begin{pmatrix}-1\\0\end{pmatrix}$

Considering now a single φ-bit, all its possible spinor states were reported in Table 6.1. The table below expresses these states in the two limits $k \to 0$ and $k \to \infty$.

This table suggests that a single φ-bit can be employed to perform a Deutsch-Jozsa algorithm by manipulating wave number and frequency. The state $\xi_{k\to\infty} = \begin{pmatrix}0\\1\end{pmatrix}$ for an incident plane wave $e^{+ikx}e^{-i\omega_k t}$ represents the initial state $|1\rangle$. This is a wave propagating in the direction k with a negative velocity $(-\omega_k)$. By tuning down the frequency, $\omega_k \to \alpha$ one equivalently reduces the wave number $k \to 0$. The state approaches a standing wave: $\xi_{k\to 0} = \begin{pmatrix}-1\\1\end{pmatrix}$. This transformation is equivalent to the application of a Hadamard gate operation on the pure state $|1\rangle$ resulting in the superposition $|0\rangle - |1\rangle$ (to within a global phase). The oracle operator would leave this superposition of state intact if the function, f, is constant. The oracle operator would change the direction of the velocity of the incident plane wave to $e^{+ikx}e^{+i\omega_k t}$ and simultaneously the spinor state to $\xi_{k\to 0} = \begin{pmatrix}1\\1\end{pmatrix}$ if the function is balanced. The states $\xi_{k\to 0} = \begin{pmatrix}-1\\1\end{pmatrix}$ and $\xi_{k\to 0} = \begin{pmatrix}1\\1\end{pmatrix}$ can be differentiated by measuring transmission or reflection coefficients. Therefore, in contrast with quantum systems, there is no need for further application of the Hadamard gate to project the superposition of states onto a measurable pure state. This algorithm is illustrated in Fig. 6.1 (bottom).

While algorithms exploiting the property of superposition of states of individual qubit or φ-bit are useful the full power of quantum computing lies in the ability to operate on exponentially complex entangled states of multiple qubits. We have seen in Chap. 3 that interaction between elastically coupled φ-bits can be used to create the superposition of multiple pseudospins. This is of great significance because similarly to quantum systems the elastic coupling of N pseudospins can be used to form an effective Hilbert space (2^N dimensional), which is the outer product of the Hilbert spaces of the constitutive pseudospins. It is then possible to form superpositions of pseudospin states that are non-separable in that space. Such a non-separable superposition of states cannot be factored into the tensor product of individual pseudospin states thus achieving the very desirable characteristic of quantum-like exponential complexity associated with the notion of entanglement.

We review below a two-qubit algorithm, in the spirit of the Deutsch-Jozsa algorithm, that exploits quantum entanglement [6]. We then argue that non-separability of φ-bit pseudospins can be used to support this type of algorithm. Let us define the problem at hand. One considers a function of two variables, $f(x_1, x_2)$ with both $x_1, x_2 \in \{0, 1\}$. The function can take on two values, namely 0 and 1. There are 16 possibilities for realizing such a function. For example, one can have $f(0,0) = f(0,1) = f(1,0) = f(1,1) = 0$ or $f(0,0) = 1$ and $f(0,1) = f(1,0) = f(1,1) = 0$ (as we all permutations of 0s and 1 for the value of f) or $f(0,0) = f(0,1) = 1$ and $f(1,0) = f(1,1) = 0$ (with all permutations of 0s and 1s) or $f(0,0) = f(0,1) = f(1,0) = 1$ and $f(1,1) = 0$ (with all permutations of 0s and 1) or finally $f(0,0) = f(0,1) = f(1,0) = f(1,1) = 1$.

These functions can be organized into classes labelled $[0,4], [1,3], [2,2], [3,1], [4,0]$ where the first number in the square bracket stands for the number of values of f equal to 1 and the second number representing the number of values of f equal to 0. The functions in classes $[0,4], [2,2], [4,0]$ are said to be even. The functions in classes $[1,3]$ and $[3,1]$ are said to be odd. Among the 16 possible functions, 8 are even and 8 are odd. The question that is posed is: for a given function, how does one find if the function is even or odd? Again, answering that question using a classic computation approach requires evaluation of the function serially for the two possible values of x_1 and x_2. The objective is to evaluate the nature of the function with a single operation using quantum superposition and entanglement. We note that the four possible inputs of the function $(0,0), (0,1), (1,0), (1,1)$ can be mapped on the basis of the Hilbert space of two qubits: $|0\rangle_1 \otimes |0\rangle_2, |0\rangle_1 \otimes |1\rangle_2, |1\rangle_1 \otimes |0\rangle_2, |1\rangle_1 \otimes |1\rangle_2$. The subscript refers to qubits 1 and 2, respectively. For the sake of simplicity, in the rest of this subsections we will drop the tensor product symbol $\otimes$. The state of the two-qubit system is now represented as a 4×1 vector:

$$\begin{pmatrix} a \\ b \\ c \\ d \end{pmatrix} = a|0\rangle_1|0\rangle_2 + b\,|0\rangle_1|1\rangle_2 + c\,|1\rangle_1|0\rangle_2 + d\,|1\rangle_1|1\rangle_2$$

We prepare the system in the tensor product state:

$$(|0\rangle + |1\rangle)_1 (|0\rangle - |1\rangle)_2 = \begin{pmatrix} 1 \\ 1 \end{pmatrix} \otimes \begin{pmatrix} 1 \\ -1 \end{pmatrix} = \begin{pmatrix} 1 \\ -1 \\ 1 \\ -1 \end{pmatrix} \tag{6.1}$$

by applying the Hadamard gate operation on each qubit originally in state $|0\rangle_1$ and $|1\rangle_2$. The oracle call is similar to the one described above for the Deutsch-Jozsa algorithm. Each function f is encoded by a unitary transformation, U_f, that acts on two-qubit states. The unitary transformation is given by:

$$U_f = \begin{pmatrix} (-1)^{f(0,0)} & 0 & 0 & 0 \\ 0 & (-1)^{f(0,1)} & 0 & 0 \\ 0 & 0 & (-1)^{f(1,0)} & 0 \\ 0 & 0 & 0 & (-1)^{f(1,1)} \end{pmatrix} \quad (6.2)$$

There are 16 matrices of the type given by (6.2). Even functions in the classes $[0,4]$, $[2,2]$, $[4,0]$ when acting on the product state given by (6.1) give two-qubit states that remain separable (i.e., product states). For instance in the case of the function, $f(0,0)=f(0,1)=0$ and $f(1,0)=f(1,1)=1$ in the class $[2,2]$, the oracle operation takes the form:

$$\begin{pmatrix} 1 & 0 & 0 & 0 \\ 0 & 1 & 0 & 0 \\ 0 & 0 & -1 & 0 \\ 0 & 0 & 0 & -1 \end{pmatrix} \begin{pmatrix} 1 \\ -1 \\ 1 \\ -1 \end{pmatrix} = \begin{pmatrix} 1 \\ -1 \\ -1 \\ 1 \end{pmatrix} = \begin{pmatrix} 1 \\ 1 \end{pmatrix} \otimes \begin{pmatrix} 1 \\ -1 \end{pmatrix} \quad (6.3)$$

Odd functions in the classes $[1,3]$, $[3,1]$ lead to transformations that create non-separable two-qubit states. For instance, the function $f(0,0)=0$, $f(0,1)=0$, $f(1,0)=0$, $f(1,1)=1$ yield the oracle operation:

$$\begin{pmatrix} 1 & 0 & 0 & 0 \\ 0 & 1 & 0 & 0 \\ 0 & 0 & 1 & 0 \\ 0 & 0 & 0 & -1 \end{pmatrix} \begin{pmatrix} 1 \\ -1 \\ 1 \\ -1 \end{pmatrix} = \begin{pmatrix} 1 \\ -1 \\ 1 \\ 1 \end{pmatrix} \quad (6.4)$$

To identify whether a function f is even or odd, one just needs to identify is the final state is separable or entangled. The challenge for quantum systems lies now in the measurability of the final states. There is no unambiguous single measurement of entangled states of quantum systems. To distinguish between separable and non-separable superpositions of states, one needs to make multiple measurements and obtain a statistical representation of the superpositions.

This drawback is overcome by using the spinor states (pseudospins) of two coupled one-dimensional harmonic chains a and b, introduced in Sect. 3.6.4. When the orbital state of the elastic plane wave is defined by $k=0$ and $\omega=2\delta$, the spinor state of the system is taking the form of a tensor product of the states of two φ-bits [see (3.96)]:

$$a_0 2\delta \begin{pmatrix} 1 \\ -1 \\ -1 \\ 1 \end{pmatrix} = a_0 2\delta \begin{pmatrix} 1 \\ -1 \end{pmatrix} \otimes \begin{pmatrix} 1 \\ -1 \end{pmatrix}$$

This state is isomorphic to the separable two-qubit state of (6.3).

When the state of the elastic plane wave is defined by a general wave number and frequency, the state of the coupled chain takes the non-separable form:

$$\begin{pmatrix} s_1^a s_1^b \sqrt{\pm}\sqrt{\pm} \\ s_1^a s_2^b \sqrt{\pm}\sqrt{\mp} \\ s_2^a s_1^b \sqrt{\mp}\sqrt{\pm} \\ s_2^a s_2^b \sqrt{\mp}\sqrt{\mp} \end{pmatrix}$$

Where $\sqrt{+} = \sqrt{\omega + \beta k}$ and $\sqrt{-} = \sqrt{\omega - \beta k}$ and $s_1^{a,b}$ and $s_2^{a,b}$ take on the values +1 or −1. The upper scripts refer to the chains, a and b. This non-separable state cannot be expressed in the form of the tensor product of two φ-bit states. Separable and non-separable states can be distinguished by measuring directly transmission/reflection coefficients of elastic waves in the two coupled chains. This elastic system can therefore be used to implement an analogue of a two-qubit algorithm with entanglement. The oracle operator is realized by operating on the frequency of the plane wave state of the elastic system to achieve either a measurable separable spinor state (even function f) or a measurable non-separable spinor state (odd function).

This simple model vindicates the possibility of utilizing exponential complexity in the states of elastic systems to perform parallel quantum-like algorithms.

6.3 Gauge Fields Analogue for Phonons

The interaction between charged particles and the electromagnetic field can be described by an extension of the Dirac equation. In the minimum coupling representation of this interaction, coupling is achieved by subtracting the charge times the electromagnetic four-potential (vector potential and scalar potential of the field) from the four-momentum (related to space-time derivatives) in the Lagrangian or Hamiltonian of the system [7]. It has been shown that dynamical modulation of the dielectric properties of optical materials achieves gauge fields analogues. These field analogues enable the control of neutral particles like photons [8–11] in ways analogous to charged quantum particles such as electrons. We show below that spatio-temporal modulation of the elastic properties of a medium composed of a one-dimensional harmonic chain which masses are attached via elastic side springs to a rigid substrate leads also to a gauge field analogue for phonons. For this, we use the simple model derived in Sect. 3.5.1. We recall several equations derived in that subsection.

To zeroth-order in perturbation of the modulation of the side springs, the Dirac equation reads:

$$\left[\boldsymbol{\sigma}_x \frac{\partial}{\partial \tau_0} + i\beta\boldsymbol{\sigma}_y(ik^*) - i\alpha_0 \boldsymbol{I}\right]\psi^{(0)}(k^*, \tau_0, \tau_1, \tau_2) = 0 \tag{6.5}$$

with solutions $\psi^{(0)}(k^*, \tau_0, \tau_1, \tau_2) = a^{(0)}(k^*, \tau_1, \tau_2)e^{i\omega_0\tau_0}$ and the spinorial amplitude $a^{(0)}(k^*, \tau_1, \tau_2) = \begin{pmatrix} a_1^{(0)} \\ a_2^{(0)} \end{pmatrix}$. To second-order, the secular term $-\boldsymbol{\sigma}_x \frac{\partial\psi^{(0)}(k^*, \tau_0, \tau_2)}{\partial \tau_2}$ needs to be canceled by the terms proportional to $e^{i\omega_0\tau_0}$ arising from the first-order driving force: $\boldsymbol{I}\left[\psi^{(1)}(k^* + K, \tau_0, \tau_2)e^{i\Omega\tau_0} - \psi^{(1)}(k^* - K, \tau_0, \tau_2)e^{-i\Omega\tau_0}\right]$. This leads to the system of (3.70a, b).

$$\frac{\partial a_1^{(0)}(k^*)}{\partial \tau_2} = ia_1^{(0)}(k^*)G' + ia_2^{(0)}(k^*)F$$

$$\frac{\partial a_2^{(0)}(k^*)}{\partial \tau_2} = ia_1^{(0)}(k^*)F + ia_2^{(0)}(k^*)G$$

where F, G, G' are functions of k^*.

These equations impose second-order corrections onto the amplitude of the zeroth-order solution. We multiply these relations by $e^{i\omega_0\tau_0}$ to obtain them in terms of $\psi^{(0)}$:

$$\frac{\partial \psi_1^{(0)}(k^*)}{\partial \tau_2} = i\psi_1^{(0)}(k^*)G' + i\psi_2^{(0)}(k^*)F \tag{6.6a}$$

$$\frac{\partial \psi_2^{(0)}(k^*)}{\partial \tau_2} = i\psi_1^{(0)}(k^*)F + i\psi_2^{(0)}(k^*)G \tag{6.6b}$$

Equation (6.6a, 6.6b) are reformulated as follows:

$$\frac{\partial \psi_1^{(0)}(k^*)}{\partial \tau_2} = i\psi_1^{(0)}(k^*)\left(\frac{G + G'}{2}\right) - i\psi_1^{(0)}(k^*)\left(\frac{G - G'}{2}\right) + i\psi_2^{(0)}(k^*)F \tag{6.7a}$$

$$\frac{\partial \psi_2^{(0)}(k^*)}{\partial \tau_2} = i\psi_1^{(0)}(k^*)F + i\psi_2^{(0)}(k^*)\left(\frac{G + G'}{2}\right) + i\psi_2^{(0)}(k^*)\left(\frac{G - G'}{2}\right) \tag{6.7b}$$

We now multiply (6.7a, 6.7b) by ε^2 and recombine them with the zeroth-order equation (6.5).

$$\frac{\partial \psi_1^{(0)}(k^*)}{\partial \tau_0} - i\beta k^* \psi_1^{(0)}(k^*) - i\alpha_0 \psi_2^{(0)}(k^*) + \varepsilon^2 \left\{ \frac{\partial \psi_1^{(0)}(k^*)}{\partial \tau_2} - i\psi_1^{(0)}(k^*)\left(\frac{G+G'}{2}\right) + i\psi_1^{(0)}(k^*)\left(\frac{G-G'}{2}\right) - i\psi_2^{(0)}(k^*)F \right\} = 0 \tag{6.8a}$$

$$\frac{\partial \psi_2^{(0)}(k^*)}{\partial \tau_0} + i\beta k^* \psi_2^{(0)}(k^*) - i\alpha_0 \psi_1^{(0)}(k^*) + \varepsilon^2 \left\{ \frac{\partial \psi_2^{(0)}(k^*)}{\partial \tau_2} - \psi_1^{(0)}(k^*)F - i\psi_2^{(0)}(k^*)\left(\frac{G+G'}{2}\right) - i\psi_2^{(0)}(k^*)\left(\frac{G-G'}{2}\right) \right\} = 0 \tag{6.8b}$$

This procedure reconstructs the perturbative series of the Dirac equation of the modulated system in terms of the solution corrected to second-order. We therefore rename $\psi^{(0)}$ by $\psi'^{(0)}$ to reflect the corrected nature of the solution. Equation (6.8a, 6.8b) can be rearranged in the form:

$$\left[\boldsymbol{\sigma}_x\left(\frac{\partial}{\partial t} - i\varnothing_{k^*}\right) + i\beta\boldsymbol{\sigma}_y(ik^* - iA_{k^*}) - i(\alpha_0 + m_{k^*})\boldsymbol{I}\right]\psi'^{(0)}(k^*, t) = 0 \tag{6.9}$$

with $\begin{pmatrix} \varnothing_{k^*} \\ A_{k^*} \end{pmatrix} = \begin{pmatrix} \frac{\varepsilon^2}{2}(G+G') \\ \frac{\varepsilon^2}{2\beta}(G-G') \end{pmatrix}$ and $m_{k^*} = \varepsilon^2 F$. To obtain (6.9), we have also used: $\frac{\partial}{\partial t} = \frac{\partial}{\partial \tau_0} + \varepsilon^2 \frac{\partial}{\partial \tau_2}$. Equation (6.9) shows that the dynamics of elastic waves in a harmonic chain grounded to a substrate via side springs, whose stiffness is modulated in space and time, is, due to second-order corrections, isomorphic to Dirac equation in Fourier domain for a charged quasiparticle including an electromagnetic field. The potential $\begin{pmatrix} \varnothing_{k^*} \\ A_{k^*} \end{pmatrix}$ acts as an artificial gauge field with the quantity $\varnothing_{k^*}$ playing the role of the electrostatic potential and A_{k^*} the role of a scalar form of the vector potential. The parentheses $\left(\frac{\partial}{\partial t} - i\varnothing_{k^*}\right)$ and $(ik^* - iA_{k^*})$ are the Fourier transforms of the usual minimal coupling rule. $\alpha_0 + m_{k^*}$ is the dressed mass of the quasiparticle. The mechanical system provides a mechanism for exchange of energy between the main chain modes and the side springs. The side springs lead to the formation of a fermion-like quasiparticle while their modulation provides a field through which quasiparticles interact. The strength and nature of the interaction is controllable through the independent modulation parameters, α_1, Ω and K.

The mechanical system allows for the exploration of a large parameter space of scalar QFT as the functions $\varnothing_{k^*}$ and A_{k^*} can be varied by manipulating the spatio-temporal modulation of the side spring stiffness. We imagine, therefore, that this classical phononic system can be employed to examine the behavior of scalar QFT from weak to strong coupling regimes, as well as, at all intermediate couplings.

Further, the capacity to separate the ratio of the effective potentials opens avenues for the experimental realization of scalar fields whose behavior could previously only have been theorized.

6.4 Particle-Wave Duality with Acoustic Bubbles

In this section, we address a potential analogy with the quantum mechanical feature of particle-wave duality. Critical developments in wave–particle duality in macroscopic systems, a phenomenon thought to be restricted to a peculiar microscale behavior, have been developed for 2-D hydrodynamic systems [12–17]. Bouncing drops of oil "walk" on an excited liquid oil surface near a Faraday instability. Each droplet is accompanied by a surface wave that pilots its motion. Quantization emerges from the dynamic constraint imposed on the oil drop by its monochromatic pilot wave field. The droplet and the pilot wave interaction is but one example of classical phenomena displaying the fundamentally probabilistic nature of quantum mechanical phenomena: single-particle diffraction, tunneling, singly/doubly quantized orbits, orbital level splitting, spin and multimodal statistics.

We anticipate that the self-interaction of a sound source with its associated wave or the interaction between sound sources through their associated waves can lead to effects reminiscent of the quantum mechanical concept of particle–wave duality in macroscopic systems. The highly nonlinear case of the behavior of bubbles in a fluid irradiated by an acoustic wave (i.e., the so-called acoustic bubbles) may extend the concept of particle–wave duality from 2-D to 3-D in the context of an acousto-hydrodynamic quantum analogue. In 3-D, gas bubbles in liquids excited with primary sound waves behave as 3-D nonlinear dynamical macroscopic objects accompanied by their own secondary sound field [18]. It is probable that bubble trajectories display the probabilistic nature of quantum mechanical phenomena due to the nonlinear self-interaction between the bubble and its own pilot wave, or the bubble–pilot wave with its environment. We make an attempt here to illustrate the behavior of acoustic bubbles in a standing wave field by developing simple models as a small step toward a more comprehensive understanding of the acoustic bubbles' interactions. We first consider the behavior of a single bubble in an acoustic standing wave. The second simple model includes bubble-to-bubble interactions in a chain of bubbles that forms in a standing wave.

6.4.1 Dynamical Model of a Single Bubble in an Acoustic Standing Wave Field

Bubbles in an acoustic field are subjected to an effective acoustic radiation force

$$F(\mathbf{r}, t) = -V(t)\nabla P(\mathbf{r}, t) \tag{6.10}$$

where $V(t)$ is the time dependent volume of the bubble and $\nabla P(r, t)$ is the local instantaneous pressure gradient. Under the influence of this effective force, a bubble may undergo oscillatory translations in response to an oscillatory pressure field associated with the sound wave as well as its own volume pulsation. This instantaneous oscillatory motion is expected to occur at the frequency of the acoustic field about a position that drifts slowly. The net translational drift of the bubble results from the fact that the time average of the acoustic radiation force, the so-called Bjerknes force [18], is nonzero. In absence of any other forces such as drag, the dynamics of the bubble can be simply described by the equation of motion:

$$m_A\ddot{\mathbf{r}} = F(\mathbf{r}, t) \tag{6.11}$$

where $m_A = \rho V_A$ is the apparent mass of the bubble with ρ being the density of the fluid (a constant for an incompressible fluid) and $V_A = \frac{1}{2}V(t)$ being the apparent volume. Inserting (6.10) into (6.11), one notes that the translational motion of the bubble becomes independent of the volume of the bubble, namely

$$\ddot{\mathbf{r}} = -\frac{2}{\rho}\nabla P(\mathbf{r}, t) \tag{6.12}$$

We consider a one-dimensional acoustic standing wave field with frequency ω and wave number $k = \frac{2\pi}{\lambda}$ (λ being the wavelength):

$$P(x, t) = 2P_a \sin\omega t \sin kd \tag{6.13}$$

In this equation, P_a is the pressure amplitude of the acoustic wave. Assuming that the bubble is located initially near an antinode $d = \frac{2n+1}{4}\lambda + x$ of the standing wave, (6.12) can be simplified to describe the translational motion, $x(t)$ of the bubble in one dimension along the direction of the wave field:

$$\ddot{x} = -\frac{2}{\rho}\frac{\partial P(x, t)}{\partial x} = \frac{4P_a}{\rho}k \sin\omega t \sin kx \tag{6.14}$$

In the limit of small kx (i.e., long acoustic wavelength and small displacements), (6.14) can be linearized:

$$\ddot{x} - \frac{4P_a}{\rho}k^2 \sin\omega t x = 0 \tag{6.15}$$

Defining the variable z by $2z = \omega t - \frac{\pi}{2}$, i.e., redefining the origin of time or introducing a phase which does not impact the search for the Eigen values, (6.15) can be rewritten into the conventional Mathieu's equation [19]:

$$\frac{d^2x}{dz^2} - 2q\cos 2zx = 0 \tag{6.16}$$

with $q = \frac{8P_a k^2}{\rho\omega^2} = \frac{8P_a}{\rho v^2}$. In defining the parameter q we have used the dispersion relation for the acoustic wave in the long wavelength limit: $\omega = vk$. The parameter q depends only on the physical characteristics of the fluid and the pressure amplitude of the acoustic wave. Equation (6.16) is a simplified form of Mathieu's equation. The coefficient $2q\cos 2z$ is a periodic function of time and Floquet's theorem can be used to seek solutions in the form:

$$x(z) = e^{i\gamma z} f(z) \tag{6.17}$$

where $f(z)$ is a periodic function of period π and γ is the characteristic exponent. Expanding the periodic function f in a Fourier series: $f(z) = \sum_{K=-\infty}^{K=+\infty} c_K e^{i2Kz}$ and inserting the sought solution (6.17) into (6.16) leads to an Eigen value problem with the recurrence relation [20]:

$$c_K(2K+\gamma)^2 + q(c_{K+1} + c_{K-1}) = 0 \tag{6.18}$$

This Eigen value problem can be written in matrix form:

$$Hc = 0 \tag{6.19}$$

by defining the tridiagonal operator H:

$$\begin{pmatrix} 0 & q & (-4+\gamma)^2 & q & 0 & . & . & . \\ . & 0 & q & (-2+\gamma)^2 & q & 0 & . & . \\ . & . & 0 & q & \gamma^2 & q & 0 & . \\ . & . & . & 0 & q & (2+\gamma)^2 & q & 0 \\ . & . & . & . & 0 & q & (4+\gamma)^2 & q \end{pmatrix} \tag{6.20}$$

and the vector

$$c = \begin{pmatrix} . \\ . \\ c_{-2} \\ c_{-1} \\ c_0 \\ c_1 \\ c_2 \\ . \\ . \end{pmatrix} \tag{6.21}$$

The Eigen value problem produces nontrivial solutions when the determinant of the matrix H is equal to zero: $\det H = 0$. The H matrix is infinite so one can approximate the problem by employing a truncated version of H to some order n: $\det H_n = 0$. For instance, to second-order ($n = 2$), we can write $\det H_2 = \begin{vmatrix} \gamma^2 & q \\ q & (2+\gamma)^2 \end{vmatrix} = 0$. This leads to the following approximate solutions for the characteristic exponent: $\gamma_1 = -1 \pm \sqrt{1+q}$ and $\gamma_2 = -1 \pm \sqrt{1-q}$. We consider the case of small q. Indeed for water using $\rho = 1000$ kg/m^3 and $v \sim 1500$ m/s, $q \sim 3.5510^{-9}\ P_a$ and a typical acoustic pressure amplitude of 2.3 atm leads to $q \sim 8.16 \times 10^{-4}$. Therefore, when q is small, the possible values of the characteristic exponents are real:

$$\gamma = \begin{cases} \pm\frac{q}{2} \\ -2\pm\frac{q}{2} \end{cases} \tag{6.22}$$

A real characteristic exponents imparts an additional periodic character to the motion of the bubble $x(t)$ through the prefactor $e^{i\gamma z}$ in (6.17). Of particular interest here, is the case when $\gamma = \pm\frac{q}{2}$. Let us take $\gamma = +\frac{q}{2}$, the bubble displacement truncated to second order now reads:

$$x(z) \sim c_0 e^{i\frac{q}{2}z} + c_1 e^{i\left(2+\frac{q}{2}\right)z} \tag{6.23}$$

Since q is small, the second term of (6.23) corresponds to high frequency oscillations with a frequency that differs only slightly from that of the acoustic standing wave. Solving for the truncated Eigen vector (c_0, c_1) for small q, through (6.19) leads to $c_1 = -\frac{q}{4}c_0$ with c_0 indeterminate. This indicates that the low frequency oscillations are representative of an unstable dynamical equilibrium. Indeed, consider (6.16), $x = 0$, is a stable solution. The onset of the low frequency oscillations and the amplitude of these oscillations depend on the initial conditions of the velocity of the bubble.

6.4.2 *Bubble in a Chain of Interacting Bubbles*

We now consider a chain of bubbles organizing along the direction of the acoustic standing wave. The bubbles are assumed to be large compared to their resonant size. Under this condition, chains of bubbles form due to attractive secondary Bjerknes forces which bind the bubbles. Due to bubble-to-bubble interactions one expects the slow motion of a bubble in the chain to be now coupled leading to the formation of a frequency band. The pressure radiated by a radially oscillating bubble, "b", at a radial distance, r, is given by [18]:

$$P_b = \rho \frac{R}{r} \left(R\ddot{R} + 2\dot{R}^2 \right) \tag{6.24}$$

We consider the limit of small radial oscillation amplitudes, $R(t) = R_0(1 + \varepsilon(t))$ where R_0 is the radius at equilibrium. Considering the dynamics of the bubble radius in the regime of the linearized form of Rayleigh-Plesset equation, $\varepsilon(t)$ is solution of the equation:

$$\ddot{\varepsilon} + \omega_0^2 \varepsilon = -\frac{1}{\rho R_0^2} P(x, t) \tag{6.25}$$

In that equation, the pressure is given by (6.13) and the characteristic frequency of the bubble is defined through the relation: $\omega_0^2 = \frac{1}{\rho R_0^2}\left[3\gamma_0\left(P_0 + \frac{2\sigma}{R_0}\right) - \frac{2\sigma}{R_0}\right]$, with γ_0, σ, and P_0 standing for the ratio of the specific heat of the gas at constant pressure to that at constant volume, the gas-water surface tension, and the static pressure [18]. For a bubble at the antinode of a standing wave and in the limit of very small displacements, we take $\sin kd \sim 1$ and find: $\varepsilon(t) = \frac{2P_A}{\rho R_0^2} \frac{1}{\omega^2 - \omega_0^2} \sin \omega t$. Using the linearized $R(t)$ and its derivatives, discarding the second term in (6.24), we approximate the radiated pressure by:

$$P_b \sim -2P_a \frac{\omega^2}{\omega^2 - \omega_0^2} \frac{R_0}{r} \sin \omega t \tag{6.26}$$

In a chain of bubbles, we assume that a bubble "n" is subjected only to the radiation pressure of its nearest neighbors, "$n - 1$" and "$n + 1$". The equation of motion of the bubble "n" takes the form:

$$\ddot{x}_n = -\frac{2}{\rho} \left(\frac{\partial P(x_n, t)}{\partial x_n} + \frac{\partial P_{b,n+1}(r(x_{n+1}, x_n), t)}{\partial x_n} + \frac{\partial P_{b,n-1}(r(x_n, x_{n-1}), t)}{\partial x_n} \right) \tag{6.27}$$

In (6.27), we have defined the radial distances $r(x_{n+1}, x_n) = \frac{\lambda}{2} + (x_{n+1} - x_n)$ and $r(x_n, x_{n-1}) = \frac{\lambda}{2} + (x_n - x_{n-1})$ where $\lambda/2$ is the equilibrium distance between bubbles in the chain (antinode to antinode distance of the standing wave). We have also assumed that the coupling between bubbles in the chain does not affect their Eigen frequency and that all bubbles oscillate in phase. Equation (6.27) becomes:

$$\ddot{x}_n + \frac{4P_a}{\rho} \frac{2R_0}{(\lambda/2)^3} \frac{\omega^2}{\omega^2 - \omega_0^2} \sin \omega t (x_{n+1} - 2x_n + x_{n-1}) - \frac{4P_a}{\rho} k^2 \sin \omega t x_n = 0 \tag{6.28}$$

In deriving (6.28), we have used the approximation: $\frac{1}{\left[1+\frac{x_{n+1}-x_n}{\lambda}\right]^2} \sim 1 - 2\frac{x_{n+1}-x_n}{\lambda}$. The second term in (6.28) introduces spectral dispersion. This is best seen if we seek solutions in the form:

$$x_n(t) = x(\kappa, t)e^{i\beta n\lambda} \tag{6.29}$$

with $\beta \epsilon \left[-\frac{\pi}{\lambda}, \frac{\pi}{\lambda}\right]$. With this, (6.28) reduces to:

$$\ddot{x}(\beta, t) - \frac{4P_a}{\rho}k^2\left(1 + \frac{2\omega^2}{\left(\omega^2 - \omega_0^2\right)}\frac{8R_0}{\pi^2\lambda}\sin^2\frac{\beta\lambda}{2}\right)\sin\omega t x(\beta, t) = 0 \tag{6.30}$$

Again, introducing the variable z, the equation of motion (6.30) takes the same form as (6.16):

$$\frac{d^2x(\beta, z)}{dz^2} - 2q'\cos 2zx(\beta z) = 0 \tag{6.31}$$

with $q'(\beta) = q\left(1 + \frac{2\omega^2}{\left(\omega^2 - \omega_0^2\right)}\frac{8R_0}{\pi^2\lambda}\sin^2\frac{\beta\lambda}{2}\right)$. The slow oscillatory translational motion of bubbles in the chain is dispersive with frequencies given by: $\gamma(\beta)\omega = \frac{q'(\beta)}{2}\omega$. The translational oscillations form a band in the interval: $\frac{q}{2}\omega\left[1, 1 + \frac{2\omega^2}{\left(\omega^2 - \omega_0^2\right)}\frac{8R_0}{\pi^2\lambda}\right]$.

Up to this point, we have restricted the bubble-bubble interaction to first nearest neighbors. However, since the bubble-to-bubble separation distance in a chain is on the order of the wavelength of the secondary sound waves emitted by the radially oscillating bubbles, other secondary Bjerknes forces beyond the first-nearest neighbor interaction should also be accounted for. In water, one does not expect damping of the secondary Bjerknes forces beyond distance amounting to many wavelengths i.e., many inter-bubble distances. If we account for interactions to second-nearest, third-nearest, and higher-order nearest neighbors, then (6.28) reads:

$$\ddot{x}_n + \frac{4P_a}{\rho}\frac{2R_0}{(\lambda/2)^3}\frac{\omega^2}{\omega^2 - \omega_0^2}\sin\omega t\sum_{p=1}^{\infty}\left(x_{n+p} - 2x_n + x_{n-p}\right) - \frac{4P_a}{\rho}k^2\sin\omega t x_n = 0 \tag{6.32}$$

The dispersion relation includes now higher-order harmonics through the long range interactions:

$$\frac{q'(\kappa)}{2}\omega = \frac{q}{2}\omega\left(1 + \frac{2\omega^2}{\left(\omega^2 - \omega_0^2\right)}\frac{8R_0}{\pi^2\lambda}\sum_{p=1}^{\infty}\sin^2 p\frac{\kappa\lambda}{2}\right) \tag{6.33}$$

Long range bubble-to-bubble interactions through their emitted sound broadens the band through the summation in (6.33).

6.4.3 Acousto-hydrodynamics of a Bubble

In Sects. 6.4.1 and 6.4.2, we have considered simple models giving rise to slow motion of bubbles in a fluid subjected to acoustic irradiation. We now present a more general formalism for the acousto-hydrodynamics of bubbles introduced in reference [21]. This model is used to derive the dynamical equations which couple the radial oscillations of a bubble with its translational motion in the fluid. Although there are more complex models of this system [22–24], the model shown below may serve as a simpler foundation for understanding the self-interaction of a bubble with the secondary sound field it generates.

One considers an incompressible fluid containing a spherical bubble. Let $x(t)$ determine the translational motion of the center of mass of the bubble along a one-dimensional axis, X. Let $R(t)$ be the time dependent radius of the bubble.

The radial velocity of the bubble boundary at point A on Fig. 6.2 is given by:

$$v_r = \dot{R} + \dot{x}\cos\theta \tag{6.34}$$

The overdot in (6.34) denotes the time derivative. For an irrotational fluid flow, the radial velocity can be expressed in terms of the radial gradient of a velocity potential, ϕ, such that

$$\frac{\partial\phi}{\partial r} = \dot{R} + \dot{x}\cos\theta \tag{6.35}$$

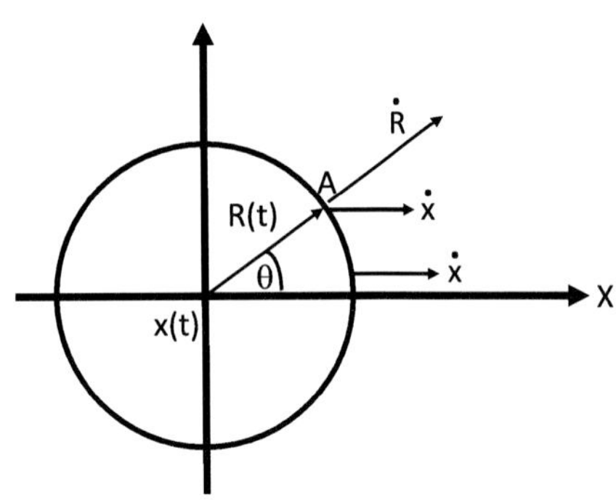

Fig. 6.2 Schematic representation of the spherical bubble undergoing radial and translational motion

The radial derivative is taken at $r = R(t)$. Because the fluid is incompressible, the velocity has zero divergence and the velocity potential must satisfy the Laplace equation:

$$\Delta\phi = 0 \tag{6.36}$$

The general form of such as potential is

$$\phi = \frac{a(t)}{r} + \frac{b(t)\cos\theta}{r^2} \tag{6.37}$$

Inserting (6.37) into (6.35) leads to:

$$a(t) = -\dot{R}R^2 \text{ and } b(t) = -\frac{\dot{x}R^3}{2} \tag{6.38}$$

One now seeks the Lagrangian of the fluid displaced by the bubble:

$$L = T - U \tag{6.39}$$

Where the kinetic energy of the fluid is given by:

$$T = \frac{1}{2}\int_{V\,fluid} \rho v^2 dV = \frac{1}{2}\rho\int_{V\,fluid} v^2 dV = \frac{1}{2}\rho\int_{V\,fluid} (\nabla\phi)^2 dV \tag{6.40}$$

ρ is the constant density of the fluid. In (6.40), $(\nabla\phi)^2 = \nabla\,.\,(\phi\nabla\phi) - \phi\Delta\phi = \nabla\,.\,(\phi\nabla\phi)$ because of (6.36). The kinetic energy is rewritten as:

$$T = \frac{1}{2}\rho\int_{V\,fluid} \nabla.(\phi\nabla\phi)dV = \frac{1}{2}\rho\int_{V\,fluid} \overrightarrow{\text{div}}\,(\phi\nabla\phi)dV \tag{6.41}$$

Using Green's theorem, the integral over the volume of the fluid is reformulated in terms of an integral over the surface of the fluid (i.e., the surface of the bubble):

$$T = \frac{1}{2}\rho\int_{S\,bubble} \phi\vec{n}.\nabla\phi dS \tag{6.42}$$

Inserting (6.35), (6.37) and (6.38) into (6.42) yields:

$$T = 2\pi\rho R^3\left(\dot{R}^2 + \frac{\dot{x}^2}{6}\right) \tag{6.43}$$

For the sake of simplicity, let us consider a potential energy of the bubble arising from the effect of the pressure felt by the bubble, P_b:

$$U = -P_s(x)V_b \tag{6.44}$$

The time dependent volume of the bubble is given by: $V_b = \frac{4\pi R^3}{3}$. The equations describing the radial and translational motion of the bubble are given by Lagrange equations,

$$\frac{\partial}{\partial t}\left(\frac{\partial L}{\partial \dot{R}}\right) - \frac{\partial L}{\partial R} = 0$$

$$\frac{\partial}{\partial t}\left(\frac{\partial L}{\partial \dot{x}}\right) - \frac{\partial L}{\partial x} = 0$$

resulting in the system of coupled second-order differential equations:

$$R\ddot{R} + \frac{3}{2}\dot{R}^2 - \frac{P_b}{\rho} = \frac{\dot{x}^2}{4} \tag{6.45a}$$

$$\ddot{x} + 3\frac{\dot{R}\dot{x}}{R} = \frac{2}{\rho}\frac{dP_b}{dx} \tag{6.45b}$$

In an acoustic field, in addition to the pressure of the primary field, the translational motion of the center of mass of the bubble may also be driven by external forces which may result from the self-interaction of the bubble with the pressure associated with the secondary sound field it generates. In turn, the translational motion impacts the radial oscillation of the bubbles which subsequently affects also the translational dynamics. This simple model illustrates the possibility of the translational motion of the bubble to be piloted by its secondary sound field. The bubble translational motion may therefore be driven by the interference of its secondary sound field with the environment. Translation motion of a massive object driven by interference phenomena involving its own accompanying sound field is reminiscent of phenomena associated with particle-wave duality in quantum mechanics.

6.5 General Relativity Analogues

There are a large number of analogue models of general relativity [25]. Here we focus on some simple models of acoustic waves which dynamics can be equivalently described within a geometrical interpretation.

6.5.1 *Acoustic Waves in a Moving Fluid*

Let us first consider the propagation of acoustic waves in a fluid in motion [26]. A fluid in motion is characterized by the mass and momentum conservation equations:

$$\frac{\partial \rho}{\partial t} + \nabla \cdot (\rho \vec{v}) = 0 \tag{6.46a}$$

$$\frac{\partial \vec{v}}{\partial t} + (\vec{v} \cdot \nabla)\vec{v} = -\frac{1}{\rho} \nabla P \tag{6.46b}$$

In (6.46a, 6.46b), ρ, $\vec{v}$, and P are the density, velocity of the fluid and the pressure in the fluid. For the sake of simplicity, we are neglecting external forces and viscous effects. We are also assuming that the fluid flow is irrotational, $\nabla \times \vec{v} = 0$ and:

$$\vec{v} = \nabla \psi \tag{6.47}$$

where ψ is the velocity potential. We consider a low amplitude acoustic disturbance around a fluid flow background $(\rho_0, P_0, \psi_0, \vec{v}_0)$, namely:

$$\rho = \rho_0 + \varepsilon \rho_1 + \ldots \tag{6.48a}$$

$$P = P_0 + \varepsilon P_1 + \ldots \tag{6.48b}$$

$$\psi = \psi_0 + \varepsilon \psi_1 + \ldots \tag{6.48c}$$

$$\vec{v} = \vec{v}_0 + \varepsilon \vec{v}_1 + \ldots \tag{6.48d}$$

Inserting (6.48a–6.48d) into (6.46a) and separating that equation into a zeroth-order and first-order equation in ε, yields:

$$\frac{\partial \rho_0}{\partial t} + \nabla \cdot (\rho_0 \vec{v}_0) = 0 \tag{6.49a}$$

$$\frac{\partial \rho_1}{\partial t} + \nabla \cdot (\rho_1 \vec{v}_0 + \rho_0 \vec{v}_1) = \frac{\partial \rho_1}{\partial t} + \nabla \cdot (\rho_1 \nabla \psi_0 + \rho_0 \nabla \psi_1) = 0 \tag{6.49b}$$

We now modify the momentum conservation equation (6.46b) by using the vectorial identity:

$$\nabla(\vec{U}.\vec{V}) = (\vec{V}.\nabla)\vec{U} + (\vec{U}.\nabla)\vec{V} + \vec{V} \times (\nabla \times \vec{U}) + \vec{U} \times (\nabla \times \vec{V})$$

If we take $\vec{U} = \vec{V} = \vec{v}$, the last two terms vanish for an irrotational flow and the identity reduces to:

$$(\vec{v} \cdot \nabla)\vec{v} = \frac{1}{2}\nabla.v^2 \tag{6.50}$$

Equation (6.46b) then takes the form:

$$\frac{\partial \vec{v}}{\partial t} + \frac{1}{2}\nabla.v^2 = -\frac{1}{\rho}\nabla P \tag{6.51}$$

The left-hand side of (6.51) is expanded up to first-order in ε and becomes:

$$\frac{\partial \vec{v}}{\partial t} + \frac{1}{2}\nabla.v^2 \sim \frac{\partial \vec{v}_0}{\partial t} + \frac{1}{2}\nabla.v_0^2 + \varepsilon\left(\frac{\partial \vec{v}_1}{\partial t} + \frac{1}{2}\nabla.(\vec{v}_0.\vec{v}_1)\right)$$

Similarly, we expand the right-hand side of (6.51):

$$-\frac{1}{\rho}\nabla P \sim -\frac{1}{\rho_0\left(1+\varepsilon\frac{\rho_1}{\rho_0}\right)}(\nabla P_0 + \varepsilon \nabla P_1) \sim -\frac{1}{\rho_0}\nabla P_0 - \varepsilon\left(\frac{1}{\rho_0}\nabla P_1 - \frac{\rho_1}{\rho_0^2}\nabla P_1\right) \tag{6.52}$$

The term to first-order in ε in (6.52) can be rewritten by using the constitutive relations:

$$P_0 = c_s^2 \rho_0 \tag{6.53a}$$

$$P_1 = c_s^2 \rho_1 \tag{6.53b}$$

Indeed the speed of sound c_s can be expressed in terms of the density and the bulk modulus, B, as: $c_s^2 = \frac{B}{\rho}$. The bulk modulus is the inverse of the compressibility $K = \frac{1}{B} = -\frac{1}{V}\frac{\partial V}{\partial P} = \frac{1}{\rho}\frac{\partial \rho}{\partial P}$. In the preceding relation and V is the volume of the fluid. With these $c_s^2 = \frac{\partial P}{\partial \rho}$.

Equation (6.52) reduces to:

$$-\frac{1}{\rho}\nabla P \sim -\frac{1}{\rho_0}\nabla P_0 - \varepsilon \nabla\left(\frac{P_1}{\rho_0}\right) \tag{6.54}$$

To first-order, the momentum conservation equation takes the form:

$$\frac{\partial \vec{v}_1}{\partial t} + \frac{1}{2}\nabla.(\vec{v}_0.\vec{v}_1) = -\nabla\left(\frac{P_1}{\rho_0}\right) \tag{6.55}$$

Equation (6.55) is now expressed in term of velocity potential by using (6.47):

$$\nabla\left(\frac{\partial\psi_1}{\partial t}+\nabla\psi_0\nabla\psi_1\right)=-\nabla\left(\frac{P_1}{\rho_0}\right)$$

which after integration gives:

$$\frac{\partial\psi_1}{\partial t}+\nabla\psi_0\nabla\psi_1=-\frac{P_1}{\rho_0}$$

or

$$\rho_0\left(\frac{\partial\psi_1}{\partial t}+\nabla\psi_0\nabla\psi_1\right)=-P_1 \tag{6.56}$$

We can now combine the first-order mass conservation equation (6.49b), first-order momentum conservation equation (6.56) and the first-order constitutive relation (6.53a) to obtain the second-order partial differential equation:

$$\frac{\partial}{\partial t}\left[\frac{\rho_0}{c_s^2}\left(\frac{\partial\psi_1}{\partial t}+\nabla\psi_0\nabla\psi_1\right)\right]=\nabla\left[\rho_0\nabla\psi_1-\frac{\rho_0}{c_s^2}\nabla\psi_0\left(\frac{\partial\psi_1}{\partial t}+\nabla\psi_0\nabla\psi_1\right)\right] \tag{6.57}$$

All coefficients in (6.57) but c_s dependent on time and spatial coordinate. Equation (6.57) can be used to solve for the acoustic perturbation, $(\rho_1, P_1, \psi_1, \vec{v}_1)$, about the flow background: $(\rho_0, P_0, \psi_0, \vec{v}_0)$. We note that (6.57) is not invariant upon applying the transformation: $t\rightarrow -t$. The fluid flow breaks time-reversal symmetry. In absence of fluid flow, $\nabla\psi_0=0$ and ρ_0 is uniform and independent of time. Equation (6.57) reduces to the usual equation:

$$\frac{\partial^2\psi_1}{\partial t^2}=c_s^2\nabla^2\psi_1$$

for acoustic perturbations propagating at the speed c_s.

One can show algebraically that (6.57) that describes the propagation of acoustic waves in a moving fluid can be rewritten in a form analogous to the equation that represents a massless scalar field propagating in a curved manifold in space-time in absence of external forces (see Appendix 1):

$$\frac{1}{\sqrt{|g|}}\frac{\partial}{\partial x_\mu}\left(\sqrt{|g|}g^{\mu\nu}\frac{\partial\psi_1}{\partial x_\nu}\right)=0 \tag{6.58}$$

or in a more compact form:

$$\frac{1}{\sqrt{|g|}}\partial_\mu\left(\sqrt{|g|}g^{\mu\nu}\partial_\nu\psi_1\right)=0 \tag{6.59}$$

Equation (6.59) employs Einstein notation for summation over repeating indices. We have also defined the contravariant 4-vector $x^\mu = t, x, y, z$ for $\mu = 0, 1, 2, 3$. g is defined as the determinant of the covariant metric tensor $[g_{\mu\nu}]$. One obtains the components of the covariant metric tensor $[g_{\mu\nu}]$ as the inverse of $[g^{\mu\nu}]$, that is, using $g_{\mu\gamma}g^{\gamma\nu} = \delta^\nu_\mu$.

$g^{\mu\nu}$ are the components of the contravariant metric tensor:

$$[g^{\mu\nu}] = \frac{1}{\rho_0 c_s}\begin{pmatrix} -1 & -\partial_x\psi_0 & -\partial_y\psi_0 & -\partial_z\psi_0 \\ -\partial_x\psi_0 & c_s^2-(\partial_x\psi_0)^2 & -\partial_x\psi_0\partial_y\psi_0 & -\partial_x\psi_0\partial_z\psi_0 \\ -\partial_y\psi_0 & -\partial_y\psi_0\partial_x\psi_0 & c_s^2-(\partial_y\psi_0)^2 & -\partial_y\psi_0\partial_z\psi_0 \\ -\partial_z\psi_0 & -\partial_z\psi_0\partial_x\psi_0 & -\partial_z\psi_0\partial_y\psi_0 & c_s^2-(\partial_z\psi_0)^2 \end{pmatrix} \tag{6.60}$$

g' is defined as the determinant of $[g^{\mu\nu}]$. Using (6.60), one finds $\sqrt{|g'|} = \frac{\rho_0^2}{c_s}$. Equation (6.59) establishes an analogy that can draw from general relativity and the properties of curved spaces.

The covariant metric tensor is therefore

$$\left[g_{\mu\nu}\right] = \frac{\rho_0}{c_s}\begin{pmatrix} -\left[c_s^2-\left((\partial_x\psi_0)^2+(\partial_y\psi_0)^2+(\partial_z\psi_0)^2\right)\right] & -\partial_x\psi_0 & -\partial_y\psi_0 & -\partial_z\psi_0 \\ -\partial_x\psi_0 & 1 & 0 & 0 \\ -\partial_y\psi_0 & 0 & 1 & 0 \\ -\partial_z\psi_0 & 0 & 0 & 1 \end{pmatrix} \tag{6.61}$$

The covariant metric tensor enables us to write an expression for a length element in the curved space:

$$ds^2 = g_{\mu\nu}dx^\mu dx^\nu \tag{6.62}$$

Using (6.61) one finds:

$$ds^2 = \frac{\rho_0}{c_s}\left[-c_s^2dt^2+(\partial_x\psi_0 dt-dx)^2+(\partial_y\psi_0 dt-dy)^2+(\partial_z\psi_0 dt-dz)^2\right] \tag{6.63}$$

One could have arrived at this result by considering the propagation of an acoustic waves in a moving medium. For this, we first consider the acoustic wave equation in a medium at rest:

$$\frac{\partial^2 \psi}{\partial t^2} = c_s^2 \left(\frac{\partial^2 \psi}{\partial x^2} + \frac{\partial^2 \psi}{\partial x^2} + \frac{\partial^2 \psi}{\partial x^2} \right)$$

This equation is invariant under a local Lorentz transformation:

$$\begin{aligned} x' &= (x - vt)\gamma \\ y' &= y \\ z' &= z \\ t' &= \left(t - \frac{v}{c_s^2} x \right) \gamma \end{aligned}$$

with $\gamma = \frac{1}{\sqrt{1 - \left(\frac{v}{c_s}\right)^2}}$. One therefore expects a measure of distance to also be invariant under such a transformation. It is easy to show that the measure of such a line element is given by:

$$ds^2 = dx^2 + dy^2 + dz^2 - c_s^2 dt^2$$

We define by v_{0x}, v_{0y}, and v_{0z} the components of the velocity of the medium at some moving point x', y', z'. The coordinates of the moving point can be related to coordinates at rest by the Galilean transformation:

$$\begin{aligned} dx' &= dx + v_{0x} dt \\ dy' &= dy + v_{0y} dt \\ dz' &= dz + v_{0z} dt \\ dt' &= dt \end{aligned}$$

Inserting these relations in the measure of the line element above gives (6.18).

The model above is but one example of a geometrical interpretation in terms of an effective metric of a dynamical physical system supporting wave-like excitations. It can be shown [27] that linearization of a scalar field around some background representing a dynamical system described by a Lagrangian or a hyperbolic second-order partial differential equation (PDE) can also be interpreted in terms of an effective Lorentzian geometry. This geometry is the consequence of the perturbative normal mode analysis of the field. This concept is illustrated below.

6.5.2 Lagrangian Approach

We consider a scalar field $\phi(x_\mu)$ with $\mu = 1, 2, 3, \ldots$ and a dynamical system described by the Lagrangian $L(\phi, \partial_\mu\phi)$. We expand the Lagrangian about some background state $(\phi_0, \partial_\mu\phi_0)$ up to second-order:

$$\delta L = L(\phi, \partial_\mu\phi) - L(\phi_0, \partial_\mu\phi_0) = \frac{\partial L}{\partial\phi}\bigg|_0 (\phi - \phi_0) + \frac{\partial L}{\partial(\partial_\mu\phi)}\bigg|_0 (\partial_\mu\phi - \partial_\mu\phi_0)$$

$$+\frac{1}{2}\left(\frac{\partial^2 L}{\partial\phi^2}\bigg|_0 (\phi - \phi_0)^2 + 2\frac{\partial^2 L}{\partial\phi\partial(\partial_\mu\phi)}\bigg|_0 (\phi - \phi_0)(\partial_\mu\phi - \partial_\mu\phi_0) + \frac{\partial^2 L}{\partial(\partial_\mu\phi)\partial(\partial_\nu\phi)}\bigg|_0\right.$$

$$\left.(\partial_\mu\phi - \partial_\mu\phi_0)(\partial_\nu\phi - \partial_\nu\phi_0)\right) + \ldots \tag{6.64}$$

The field and its derivatives are also expanded to second-order:

$$\phi - \phi_0 = \varepsilon\phi_1 + \frac{\varepsilon^2}{2}\phi_2 + \ldots \tag{6.65a}$$

$$\partial_\mu\phi - \partial_\mu\phi_0 = \varepsilon\partial_\mu\phi_1 + \frac{\varepsilon^2}{2}\partial_\mu\phi_2 + \ldots \tag{6.65b}$$

ε is a small perturbative parameter. Inserting expressions (6.65a, 6.65b) into (6.64) and regrouping the first- and second-order terms in ε yields:

$$\begin{aligned}\delta L(\phi_1, \partial_\mu\phi_1) = \; & \varepsilon\left(\frac{\partial L}{\partial\phi}\bigg|_0 \phi_1 + \frac{\partial L}{\partial(\partial_\mu\phi)}\bigg|_0 \partial_\mu\phi_1\right) \\ & + \frac{\varepsilon^2}{2}\left(\frac{\partial L}{\partial\phi}\bigg|_0 \phi_2 + \frac{\partial L}{\partial(\partial_\mu\phi)}\bigg|_0 \partial_\mu\phi_2\right) \\ + \frac{\varepsilon^2}{2}\left(\frac{\partial^2 L}{\partial\phi^2}\bigg|_0 \phi_1\phi_1 + 2\frac{\partial^2 L}{\partial\phi\partial(\partial_\mu\phi)}\bigg|_0 \phi_1\partial_\mu\phi_1 \right. & \left. + \frac{\partial^2 L}{\partial(\partial_\mu\phi)\partial(\partial_\nu\phi)}\bigg|_0 \partial_\mu\phi_1\partial_\nu\phi_1\right) + \ldots\end{aligned} \tag{6.66}$$

Equation (6.66) employs Einstein notation for summation of repeating indices. Applying the principle of least action (S) to this system leads to the variational problem:

$$\delta S = \int_A^B \prod_\eta dx_\eta \delta L(\phi_1, \partial_\mu\phi_1) = 0 \tag{6.67}$$

and the condition $\phi_1(A) = \phi_1(B) = 0$, where A and B are fixed points.

After inserting (6.66) into (6.67), we obtain a number of integral terms to evaluate. First, we calculate via integration by parts the term, I_1, arising from the second terms in the first parenthesis of (6.66):

$$I_1 = \int_A^B \prod_\eta dx_\eta \frac{\partial L}{\partial(\partial_\mu \phi)}\bigg|_0 \partial_\mu \phi_1 = \frac{\partial L}{\partial(\partial_\mu \phi)}\bigg|_0 \phi_1|_A^B - \int_A^B \prod_\eta dx_\eta \partial_\mu \left(\frac{\partial L}{\partial(\partial_\mu \phi)}\bigg|_0 \right) \phi_1$$

The scalar field is zero at the fixed points and the integral I_1 simplifies to

$$I_1 = -\int_A^B \prod_\eta dx_\eta \partial_\mu \left(\frac{\partial L}{\partial(\partial_\mu \phi)}\bigg|_0 \right) \phi_1 \tag{6.68}$$

Similarly, we can show for the integral of the second terms in the second parenthesis of (6.66) that

$$I_2 = \int_A^B \prod_\eta dx_\eta \frac{\partial L}{\partial(\partial_\mu \phi)}\bigg|_0 \partial_\mu \phi_2 = -\int_A^B \prod_\eta dx_\eta \partial_\mu \left(\frac{\partial L}{\partial(\partial_\mu \phi)}\bigg|_0 \right) \phi_2 \tag{6.69}$$

Using (6.68), the integral of the first parenthesis in (6.66) now reduces to:

$$\int_A^B \prod_\eta dx_\eta \left(\frac{\partial L}{\partial \phi}\bigg|_0 \phi_1 + \frac{\partial L}{\partial(\partial_\mu \phi)}\bigg|_0 \partial_\mu \phi_1 \right) = \int_A^B \prod_\eta dx_\eta \left(\frac{\partial L}{\partial \phi}\bigg|_0 - \partial_\mu \left(\frac{\partial L}{\partial(\partial_\mu \phi)}\bigg|_0 \right) \right) \phi_1$$

This integral is analytically zero because $\frac{\partial L}{\partial \phi}\Big|_0 - \partial_\mu \left(\frac{\partial L}{\partial(\partial_\mu \phi)}\Big|_0 \right) = 0$ is the Euler-Lagrange equation for the background field. By virtue of (6.69), the integral of the second parenthesis in (6.66) will also vanish.

We also need to evaluate the integral I_3 arising from the second term in the third parenthesis in (6.66). Again, we use integration by parts and the fact that the field is zero at the fixed points to obtain:

$$\begin{aligned}
I_3 &= 2\int_A^B \prod_\eta dx_\eta \frac{\partial^2 L}{\partial \phi \partial(\partial_\mu \phi)}\bigg|_0 \phi_1 \partial_\mu \phi_1 = -2\int_A^B \prod_\eta dx_\eta \partial_\mu \left(\frac{\partial^2 L}{\partial \phi \partial(\partial_\mu \phi)}\bigg|_0 \phi_1 \right) \phi_1 \\
&= -2\int_A^B \prod_\eta dx_\eta \partial_\mu \left(\frac{\partial^2 L}{\partial \phi \partial(\partial_\mu \phi)}\bigg|_0 \right) \phi_1 \phi_1 - 2\int_A^B \prod_\eta dx_\eta \frac{\partial^2 L}{\partial \phi \partial(\partial_\mu \phi)}\bigg|_0 \partial_\mu \phi_1 \phi_1 \\
&= -2\int_A^B \prod_\eta dx_\eta \partial_\mu \left(\frac{\partial^2 L}{\partial \phi \partial(\partial_\mu \phi)}\bigg|_0 \right) \phi_1 \phi_1 - I_3
\end{aligned} \tag{6.70}$$

So

$$I_3 = -\int_A^B \prod_\eta dx_\eta \partial_\mu \left(\frac{\partial^2 L}{\partial \phi \partial(\partial_\mu \phi)}\bigg|_0 \right) \phi_1 \phi_1 \tag{6.71}$$

Finally, we also integrate by parts the integral,

$$\begin{aligned} I_4 &= \int_A^B \prod_\eta dx_\eta \frac{\partial^2 L}{\partial(\partial_\mu \phi)\partial(\partial_\nu \phi)}\bigg|_0 \partial_\mu \phi_1 \partial_\nu \phi_1 \\ &= -\int_A^B \prod_\eta dx_\eta \partial_\mu \left(\frac{\partial^2 L}{\partial(\partial_\mu \phi)\partial(\partial_\nu \phi)}\bigg|_0 \partial_\nu \phi_1 \right) \phi_1 \end{aligned}$$

With all this, (6.67) becomes:

$$\delta S = \frac{\varepsilon^2}{2} \int_A^B \prod_\eta dx_\eta \left\{ -\partial_\mu \left(\frac{\partial^2 L}{\partial(\partial_\mu \phi)\partial(\partial_\nu \phi)}\bigg|_0 \partial_\nu \phi_1 \right) + \left[\frac{\partial^2 L}{\partial \phi^2}\bigg|_0 - \partial_\mu \left(\frac{\partial^2 L}{\partial \phi \partial(\partial_\mu \phi)}\bigg|_0 \right) \right] \phi_1 \right\} \phi_1 = 0 \tag{6.72}$$

Equation (6.72) leads to the equation of motion:

$$\partial_\mu \left(\frac{\partial^2 L}{\partial(\partial_\mu \phi)\partial(\partial_\nu \phi)}\bigg|_0 \partial_\nu \phi_1 \right) - \left[\frac{\partial^2 L}{\partial \phi^2}\bigg|_0 - \partial_\mu \left(\frac{\partial^2 L}{\partial \phi \partial(\partial_\mu \phi)}\bigg|_0 \right) \right] \phi_1 = 0 \tag{6.73}$$

We now introduce an effective contravariant metric tensor and establish the identity:

$$\sqrt{|g|} g^{\mu\nu} = f^{\mu\nu} = \frac{\partial^2 L}{\partial(\partial_\mu \phi)\partial(\partial_\nu \phi)}\bigg|_0 \tag{6.74}$$

and multiply (6.73) by $\frac{1}{\sqrt{|g|}}$ to obtain:

$$\frac{1}{\sqrt{|g|}} \partial_\mu \left(\sqrt{|g|} g^{\mu\nu} \partial_\nu \phi_1 \right) - \frac{1}{\sqrt{|g|}} \left[\frac{\partial^2 L}{\partial \phi^2}\bigg|_0 - \partial_\mu \left(\frac{\partial^2 L}{\partial \phi \partial(\partial_\mu \phi)}\bigg|_0 \right) \right] \phi_1 = 0 \tag{6.75}$$

The first term in (6.75) is the d'Alembertian and defining the background field dependent potential $V(\phi_0) = \frac{1}{\sqrt{|g|}}\left[\frac{\partial^2 L}{\partial \phi^2}\bigg|_0 - \partial_\mu\left(\frac{\partial^2 L}{\partial\phi\partial(\partial_\mu\phi)}\bigg|_0\right)\right]$, the equation of motion takes the geometrical form:

$$(\square - V(\phi_0))\phi_1 = 0 \tag{6.76}$$

Equation (6.74) can be used to define the covariant matric tensor as follows. In d spatial dimensions and one (1) time dimension, we have:

$$(|g|)^{\frac{d-1}{2}} = -\det\left(\frac{\partial^2 L}{\partial\phi\partial(\partial_\mu\phi)}\bigg|_0\right) \tag{6.77}$$

Therefore the contravariant metric tensor is

$$g^{\mu\nu}(\phi_0) = \left\{-\det\left(\frac{\partial^2 L}{\partial\phi\partial(\partial_\mu\phi)}\bigg|_0\right)^{\frac{-1}{d-1}}\left(\frac{\partial^2 L}{\partial\phi\partial(\partial_\mu\phi)}\bigg|_0\right)\right\} \tag{6.78}$$

Inverting (6.78), leads to the covariant metric tensor:

$$g_{\mu\nu}(\phi_0) = \left\{-\det\left(\frac{\partial^2 L}{\partial\phi\partial(\partial_\mu\phi)}\bigg|_0\right)^{\frac{1}{d-1}}\left(\frac{\partial^2 L}{\partial\phi\partial(\partial_\mu\phi)}\bigg|_0\right)^{-1}\right\} \tag{6.79}$$

Note that the case of one space dimension $d = 1$ (i.e., $\frac{1}{d-1} \to \infty$), is ill-defined unless, $\det f^{\mu\nu} = \det\left(\frac{\partial^2 L}{\partial\phi\partial(\partial_\mu\phi)}\bigg|_0\right) = 1$. In Sect. 6.5.4, we will address the case of the one dimensional harmonic crystal subjected of to a spatio-temporal modulation of its stiffness i.e., the time-dependent superlattice. This system will be tractable for two reasons, namely $\det f^{\mu\nu} = 1$ but also in the context of multiple time scale perturbation theory, the number of temporal dimensions exceeds 1.

6.5.3 *Second-Order Partial Differential Equation*

In the case of a dynamical system that does not have a Lagrangian but obeys a general second-order hyperbolic PDE that supports wave excitations:

$$F(x_\mu, \phi, \partial_\mu\phi, \partial_\nu\partial_\mu\phi) = 0 \tag{6.80}$$

it is still possible to obtain a geometrical description by defining an effective metric [27].

Linearizing (6.80) around some solution:

$$\phi(x_\mu, t) = \phi_0(x_\mu, t) + \varepsilon\phi_1(x_\mu, t) + \ldots \tag{6.81}$$

yields

$$\left.\frac{\partial F}{\partial(\partial_\mu\partial_\nu\phi)}\right|_0 \partial_\mu\partial_\nu\phi_1 + \left.\frac{\partial F}{\partial(\partial_\mu\phi)}\right|_0 \partial_\mu\phi_1 + \left.\frac{\partial F}{\partial(\phi)}\right|_0 \phi_1 = 0 \tag{6.82}$$

Equation (6.82) can be rewritten as:

$$\partial_\mu\left\{\left.\frac{\partial F}{\partial(\partial_\mu\partial_\nu\phi)}\right|_0 \partial_\nu\phi_1\right\} + \left\{\left.\frac{\partial F}{\partial(\partial_\mu\phi)}\right|_0 - \partial_\mu\left(\left.\frac{\partial F}{\partial(\partial_\mu\partial_\nu\phi)}\right|_0\right)\right\}\partial_\mu\phi_1$$
$$+ \left.\frac{\partial F}{\partial(\phi)}\right|_0 \phi_1 = 0 \tag{6.83}$$

We can now define

$$f^{\mu\nu} = \sqrt{|g|}g^{\mu\nu} = \left.\frac{\partial F}{\partial(\partial_\mu\partial_\nu\phi)}\right|_0 \tag{6.84}$$

The second and third terms in (6.83) may be viewed as forming the analogue of a vector potential and a scalar potential, respectively.

6.5.4 *Time-Dependent Superlattice*

In Chap. 3, we studied the time-dependent superlattice within the context of multiple time scale perturbation theory. In this subsection, we interpret this approach in terms of an effective Lorentzian geometry.

6.5.4.1 Temporal Geometrical Description

The multiple time scale perturbation theory of the time-dependent super-lattice is developed in Sect. 3.4.3.2. Here we recall some of the results of that subsection. The displacement is expanded as a second-order power series in the perturbation (i.e., modulation):

$$u(k+g,\tau_0,\tau_1,\tau_2) = u_0(k+g,\tau_0,\tau_1,\tau_2) + \varepsilon u_1(k+g,\tau_0,\tau_1,\tau_2) + \varepsilon^2 u_2(k+g,\tau_0,\tau_1,\tau_2) + \ldots \tag{6.85}$$

u_i with $i=0,1,2$ are displacement functions expressed to zeroth-order, first-order and second-order in perturbation. We have also replaced the single time variable, t, by three variables representing different time scales: $\tau_0 = t$, $\tau_1 = \varepsilon t$, and $\tau_2 = \varepsilon^2 t = \varepsilon^2 \tau_0$. The problem at hand is now a $d=1$ spatial dimension plus 3 temporal dimensions problem. The introduction of multiple time variables may be understood as follows. In a conventional perturbation series with a single time variable, t:

$u(k+g,t) = u_0(k+g,t) + \varepsilon u_1(k+g,t) + \varepsilon^2 u_2(k+g,t) + \ldots$ one can capture non-linear amplitude–frequency interaction in systems with time-invariant amplitude and phase. In contrast, a perturbation expansion with multiple time scales such as (6.85) is able to capture time-varying amplitude behavior.

To zeroth-order, the wave equation is that of the unmodulated one dimensional system:

$$\frac{\partial^2 u_0(k+g,\tau_0,\tau_1,\tau_2)}{\partial \tau_0^2} + v_a^2 (k+g)^2 u_0(k+g,\tau_0,\tau_1,\tau_2) = 0 \tag{6.86}$$

To eliminate secular terms in the first-order form of the wave equations, we assume that the displacement,

$$u_0(k+g,\tau_0,\tau_2) = a_0(k+g,\tau_2) e^{i\omega_0(k+g)\tau_0} \tag{6.87}$$

is not a function of τ_1 anymore. The problem now reduces to $d=1$ spatial dimension plus 2 temporal dimensions (τ_0, τ_2).

To second-order the dynamical equation is:

$$\frac{\partial^2 u_2(k+g,\tau_0,\tau_2)}{\partial {\tau_0}^2} + \omega_0^2(k+g) u_2(k+g,\tau_0,\tau_2) + 2\frac{\partial^2 u_0(k+g,\tau_0,\tau_2)}{\partial \tau_2 \partial \tau_0} = i\left\{f(k')u_1(k',\tau_0,\tau_2)e^{i\Omega\tau_0} + h(k'')u_1(k'',\tau_0,\tau_2)e^{-i\Omega\tau_0}\right\} \tag{6.88}$$

where $f(k) = Kk + k^2$, $h(k) = Kk - k^2$, $k' = k+g-K$ and $k'' = k+g+K$. When inserting the appropriate first-order solutions into (6.88), one obtains terms of the form $e^{i\omega_0(k+g)\tau_0}$ in the right-hand side of the equation. These terms lead to secular behavior that can be cancelled by equating them to the third term in the left-hand side of the equation:

$$2\frac{\partial^2 u_0(k+g,\tau_0,\tau_2)}{\partial\tau_2\partial\tau_0} = -\left\{f(k')h(k+g)\left(\frac{1}{\omega_0^2(k')-(\omega_0(k+g)-\Omega)^2}\right)\right.$$
$$\left.+h(k'')f(k+g)\left(\frac{1}{\omega_0^2(k'')-(\omega_0(k+g)+\Omega)^2}\right)\right\}$$
$$\times\, a_0(k+g,\tau_2)e^{i\omega_0(k+g)\tau_0} \qquad (6.89)$$

Equation (6.89) can be reformulated using (6.87) as

$$2\frac{\partial^2 u_0(k+g,\tau_0,\tau_2)}{\partial\tau_2\partial\tau_0} = M(k+g,\Omega,K)u_0(k+g,\tau_0,\tau_2) \qquad (6.90)$$

This equation leads to a correction of the first-order solution u_0. We can gain insight in the meaning of the time scale τ_2 by inserting (6.87) into (6.90) resulting in:

$$i\frac{\partial a_0(k+g,\tau_2)}{\partial\tau_2} = H(k+g,\Omega,K)a_0(k+g,\tau_2) \qquad (6.91)$$

with $H(k+g,\Omega,K) = \frac{M(k+g,\Omega,K)}{2\omega_0(k+g)}$. Equation (6.91) is isomorphic to the time dependent Schroedinger equation with an Hamiltonian H. τ_2 is therefore the time variable associated with the scattering of phonons by an effective Hamiltonian representing the effect of the spatio-temporal modulation to second-order.

We can now reconstruct the perturbative series of the wave equation of the modulated system in terms of solution corrected to second order. For this, we multiply (6.91) by ε^2 and combine it with (6.86). We also rename u_0 by u_0' to reflect the corrected nature of the solution. The wave equation including second-order corrections becomes:

$$\frac{\partial^2 u_0'(k+g,\tau_0,\tau_2)}{\partial\tau_0^2} + 2\varepsilon^2\frac{\partial^2 u_0'(k+g,\tau_0,\tau_2)}{\partial\tau_2\partial\tau_0}$$
$$+\left\{v_a^2(k+g)^2+\varepsilon^2 M(k+g,\Omega,K)\right\}u_0'(k+g,\tau_0,\tau_2)=0 \qquad (6.92)$$

For reasons that will be apparent in the upcoming derivations, we divide throughout (6.92) by $\frac{1}{\varepsilon^2}$ and obtain:

$$\frac{1}{\varepsilon^2}\frac{\partial^2 u_0'(k+g,\tau_0,\tau_2)}{\partial\tau_0^2} + \frac{\partial^2 u_0'(k+g,\tau_0,\tau_2)}{\partial\tau_2\partial\tau_0}$$
$$+\frac{\partial^2 u_0'(k+g,\tau_0,\tau_2)}{\partial\tau_0\partial\tau_2} + \left\{\frac{1}{\varepsilon^2}v_a^2(k+g)^2 + M(k+g,\Omega,K)\right\}u_0'(k+g,\tau_0,\tau_2)=0$$
$$(6.93)$$

The second and third terms of (6.93) are obtained by recalling the origin of the second term of (6.92) (see Appendix 3 in Chap. 3).

This equation is the equivalent of (6.80) for our time dependent elastic superlattice. Expanding (6.92) around the phonon vacuum i.e., no elastic displacement, then u_0' plays the role of ϕ_1 in Sect. 6.5.3. We can then define the contravariant metric tensor using relation (6.84).

We find:

$$[f^{\mu\nu}] = \begin{pmatrix} \frac{1}{\varepsilon^2} & 1 \\ 1 & 0 \end{pmatrix} \tag{6.94}$$

The determinant of that matrix is -1, so we can define the contravariant metric tensor:

$$[g^{\mu\nu}] = \begin{pmatrix} \frac{1}{\varepsilon^2} & 1 \\ 1 & 0 \end{pmatrix} \tag{6.95}$$

We take the inverse of $[g^{\mu\nu}]$ to obtain the covariant metric tensor:

$$\left[g_{\mu\nu}\right] = \begin{pmatrix} 0 & -1 \\ -1 & \frac{1}{\varepsilon^2} \end{pmatrix} \tag{6.96}$$

The temporal line element in the τ_0, τ_2 space is then calculated from (6.62):

$$ds^2 = (d\tau_2, d\tau_0)\begin{pmatrix} 0 & -1 \\ -1 & \frac{1}{\varepsilon^2} \end{pmatrix}\begin{pmatrix} d\tau_2 \\ d\tau_0 \end{pmatrix} = \frac{1}{\varepsilon^2}d\tau_0^2 - 2d\tau_0 d\tau_2 \tag{6.97}$$

The off-diagonal terms in the metric tensor lead to a line element that is characteristic of a non-Euclidian τ_0, τ_2 space. Recalling (6.91), the sign of $d\tau_2$ depends on the sign of the Hamiltonian, H. H contains resonant terms (6.89) which change sign across resonances. We have seen that these resonances are associated with band gaps in the band structure of the time dependent superlattice. Therefore, as one varies the wave vector across the gap, the scattering time element $d\tau_2$ must change sign as well. The length of the temporal line element ds as a function of the wave vector parameter is shorter than the un-modulated time element, $d\tau_0$ on one side of the gap and longer on the other side. The curvature of the τ_0, τ_2 space changes across the gap. This curvature change mirrors the manifold with a twist in wave number space that characterizes the change in phase of the elastic waves across the gap.

6.5.4.2 Space-Multiple Times (1+2) Geometrical Model

In Appendix 2, we will show that one can represent the trajectory of a system in a Euclidian space subjected to forces resulting from some potential as a free moving object propagating in a curved space. In this subsection, we consider the purely geometrical interpretation of the propagation of elastic waves in the time-dependent superlattice. That is, we introduce the potential into the curvature of space-time. We start with (6.92) which is reformulated as:

$$\frac{\partial^2 \psi(k+g,\tau_0,\tau_2)}{\partial \tau_0^2} + 2\varepsilon^2 \frac{\partial^2 \psi(k+g,\tau_0,\tau_2)}{\partial \tau_2 \partial \tau_0} + N(k+g,\varepsilon,\Omega,K)\psi(k+g,\tau_0,\tau_2) = 0 \tag{6.98}$$

Here, we have replace the symbol u_0' by ψ for ease of notation. We want to derive (6.98) from purely geometrical arguments. In the three-dimensional space-time (ψ, τ_0, τ_2), we define the line element:

$$dl^2 = d\psi^2 - N\psi^2 \frac{1}{1-\varepsilon^2}\left(d\tau_0^2 - \frac{1}{\varepsilon^2} d\tau_2^2\right) \tag{6.99}$$

We consider a curve between two fixed point A and B. The length along the curve is:

$$L = \int_A^B dl = \int_A^B \sqrt{\left(\frac{\partial \psi}{\partial \lambda}\right)^2 - N\psi^2 \frac{1}{1-\varepsilon^2}\left(\left(\frac{\partial \tau_0}{\partial \lambda}\right)^2 - \frac{1}{\varepsilon^2}\left(\frac{\partial \tau_2}{\partial \lambda}\right)^2\right)}\, d\lambda = \int_A^B F d\lambda \tag{6.100}$$

In (6.100) the curve is described in terms of the parameter, λ. To find a minimum of the length, $\delta L = 0$, we utilize the Euler-Lagrange equation:

$$\frac{\partial}{\partial \lambda}\left(\frac{\partial F}{\partial\left(\frac{\partial \psi}{\partial \lambda}\right)}\right) + \frac{\partial}{\partial \lambda}\left(\frac{\partial F}{\partial\left(\frac{\partial \tau_0}{\partial \lambda}\right)}\right) + \frac{\partial}{\partial \lambda}\left(\frac{\partial F}{\partial\left(\frac{\partial \tau_2}{\partial \lambda}\right)}\right) - \frac{\partial F}{\partial \psi} - \frac{\partial F}{\partial \tau_0} - \frac{\partial F}{\partial \tau_2} = 0 \tag{6.101}$$

$$\frac{\partial}{\partial \lambda}\left\{2\frac{\partial \psi}{\partial \lambda} - 2N\psi^2 \frac{1}{1-\varepsilon^2}\left(\frac{\partial \tau_0}{\partial \lambda} - \frac{1}{\varepsilon^2}\frac{\partial \tau_2}{\partial \lambda}\right)\right\} + 2N\psi \frac{1}{1-\varepsilon^2}\left(\left(\frac{\partial \tau_0}{\partial \lambda}\right)^2 - \frac{1}{\varepsilon^2}\left(\frac{\partial \tau_2}{\partial \lambda}\right)^2\right)$$
$$-\frac{1}{2}\frac{1}{G}\frac{\partial G}{\partial \lambda}\left\{2\frac{\partial \psi}{\partial \lambda} - 2N\psi^2 \frac{1}{1-\varepsilon^2}\left(\frac{\partial \tau_0}{\partial \lambda} - \frac{1}{\varepsilon^2}\frac{\partial \tau_2}{\partial \lambda}\right)\right\} = 0 \tag{6.102}$$

We have defined $G = F^2$. We now chose $d\lambda = dl$ such that $F = G = 1$ and $\frac{\partial G}{\partial \lambda} = 0$. Equation (6.102) reduces to the first two terms. Expanding the derivative, $\frac{\partial}{\partial \lambda}$, one gets:

$$2\frac{\partial^2\psi}{\partial l^2} - 2N2\psi\frac{\partial\psi}{\partial l}\frac{1}{1-\varepsilon^2}\left(\frac{\partial\tau_0}{\partial l} - \frac{1}{\varepsilon^2}\frac{\partial\tau_2}{\partial l}\right) - 2N\psi^2\frac{1}{1-\varepsilon^2}\left(\frac{\partial^2\tau_0}{\partial l^2} - \frac{1}{\varepsilon^2}\frac{\partial^2\tau_2}{\partial l^2}\right)$$
$$+2N\psi\frac{1}{1-\varepsilon^2}\left(\left(\frac{\partial\tau_0}{\partial l}\right)^2 - \frac{1}{\varepsilon^2}\left(\frac{\partial\tau_2}{\partial l}\right)^2\right) = 0 \tag{6.103}$$

This is the equation of a geodesic. Along the geodesic $dl = vdt$ with a constant v. Furthermore, we recall that $\tau_0 = t$ and $\tau_2 = \varepsilon^2 t$. Inserting these into (6.103) results in:

$$\frac{\partial^2\psi}{\partial t^2} + N\psi = \frac{\partial^2\psi}{\partial\tau_0^2} + 2\varepsilon^2\frac{\partial^2\psi}{\partial\tau_0\partial\tau_2} + N\psi = 0 \tag{6.104}$$

Equation (6.104) is indeed (6.98). To obtain (6.94), we used the following:

$$\frac{\partial\tau_0}{\partial l} - \frac{1}{\varepsilon^2}\frac{\partial\tau_2}{\partial l} = \frac{1}{v}\left(\frac{\partial\tau_0}{\partial t} - \frac{1}{\varepsilon^2}\frac{\partial\tau_2}{\partial t}\right) = \frac{1}{v}\left(1 - \frac{1}{\varepsilon^2}\varepsilon^2\right) = 0$$
$$\frac{\partial^2\tau_0}{\partial t^2} = \frac{\partial^2\tau_2}{\partial t^2} = 0$$

and

$$\frac{1}{1-\varepsilon^2}\left(\left(\frac{\partial\tau_0}{\partial t}\right)^2 - \frac{1}{\varepsilon^2}\left(\frac{\partial\tau_2}{\partial t}\right)^2\right) = \frac{1}{1-\varepsilon^2}\left(1 - \frac{1}{\varepsilon^2}\varepsilon^4\right) = 1.$$

The latter approximation maintains (6.104) to second-order in ε. We note that it is the introduction of two time scales τ_0, τ_2 which enables us to obtain (6.104) from $\delta L = 0$. In Chap. 3, we have shown that [see (3.65)] the frequency of elastic waves in the time-dependent superlattice corrected to second-order is given by:

$$\omega_0^*(k+g) = \omega_0(k+g) + \frac{\varepsilon^2}{2\omega_0(k+g)}\left\{\begin{array}{l} f(k')h(k+g)\dfrac{1}{\omega_0^2(k') - (\omega_0(k+g) - \Omega)^2} \\ +h(k'')f(k+g)\dfrac{1}{\omega_0^2(k'') - (\omega_0(k+g) + \Omega)^2} \end{array}\right\}$$

with

$$\omega_0(k+g) = v_a|k+g|$$

With this the quantity, $N(k+g,\varepsilon,\Omega,K)$, can be reformulated as:

$$N(k+g,\varepsilon,\Omega,K) = \omega_0^2(k+g) + \varepsilon^2 \left\{ \begin{array}{l} f(k')h(k+g)\dfrac{1}{\omega_0^2(k') - (\omega_0(k+g) - \Omega)^2} \\ +h(k'')f(k+g)\dfrac{1}{\omega_0^2(k'') - (\omega_0(k+g) + \Omega)^2} \end{array} \right\}$$

or to second-order

$$N(k+g,\varepsilon,\Omega,K) \sim \left(\omega_0^*(k+g)\right)^2 \tag{6.105}$$

The line element given by (6.99) can then be rewritten as:

$$dl^2 = d\psi^2 - \psi^2\left(\omega_0^*(k+g)\right)^2 \frac{1}{1-\varepsilon^2}\left(d\tau_0^2 - \frac{1}{\varepsilon^2}d\tau_2^2\right) \tag{6.106}$$

This line element defines the metric tensor in the three dimensional space (ψ, τ_0, τ_2),

$$[g] = \begin{pmatrix} 1 & 0 & 0 \\ 0 & -\psi^2\left(\omega_0^*(k+g)\right)^2 \dfrac{1}{1-\varepsilon^2} & 0 \\ 0 & 0 & \dfrac{1}{\varepsilon^2}\psi^2\left(\omega_0^*(k+g)\right)^2 \dfrac{1}{1-\varepsilon^2} \end{pmatrix} \tag{6.107}$$

Equation (6.106) is reminiscent of the usual line element in polar coordinates, (ρ, θ):

$$ds^2 = d\rho^2 + \rho^2 d\theta^2$$

with the role of the radial variable, ρ, played by ψ (i.e., the amplitude of the elastic wave). The angular variable is related to the two temporal variables scales τ_0, τ_2 via:

$$d\theta^2 = -\left(\omega_0^*(k+g)\right)^2 \frac{1}{1-\varepsilon^2}\left(d\tau_0^2 - \frac{1}{\varepsilon^2}d\tau_2^2\right) \tag{6.108}$$

This result is different from (6.97) as the line element of (6.99) includes the effect of the potential: $V(\psi) = \frac{1}{2}N(k+g,\varepsilon,\Omega,K)\psi^2$ in the geometrical description.

We also note that with $d\tau_0 = dt$ and $d\tau_2 = \varepsilon^2 dt$, the line element $d\theta^2 = -\left(\omega_0^*(k+g)\right)^2 dt^2$ (i.e., $d\theta = i\omega_0^*(k+g)dt$). The line element (6.106) reduces to the line element in the complex plane in polar coordinates.

We can see this from another point of view. We can rewrite (6.108) as a non-Euclidian Minkowski-like metric with two time-related variables:

$$d\theta^2 = dX_0^2 - \frac{1}{\varepsilon^2} dX_2^2 \tag{6.109}$$

with $X_0 = i\omega_0^*(k+g)\frac{1}{\sqrt{1-\varepsilon^2}}\tau_0$ and $X_2 = i\omega_0^*(k+g)\frac{1}{\sqrt{1-\varepsilon^2}}\tau_2$. The space spanned by the variables X_0 and X_2 is conical:

$$X_0^2 - \frac{1}{\varepsilon^2} X_2^2 = 0 \tag{6.110}$$

which implies $\tau_2 = \pm\varepsilon^2\tau_0$. The positive solution being the one that defined τ_2. Equation (6.106) becomes

$$dl^2 = d\psi^2 - \psi^2 d\theta^2 \tag{6.111}$$

Again, this is the polar coordinate representation in the complex plane of: $(\psi, i\theta)$. The temporal variation of ψ was found to be: $\psi = \alpha(k+g)e^{i\omega_0^*(k+g)t}$. This is a normal mode of the time dependent superlattice with second-order correction to the frequency which follows a nonlinear trajectory in the (ψ, t) plane. One can describe this mode geometrically in ways similar to those of Appendix 2.

For instance, we can obtain the wave equation (6.104) from a purely geometrical argument in the space of ψ by introducing the line element (6.130):

$$ds^2 = 2(E - V(\psi))d\psi^2 \tag{6.112}$$

This wave equation is the equation of a geodesic in a one dimensional curved space, ψ, with a metric $g = \sqrt{2(E - V(\psi))}$. This point is discussed in detail in Appendix 2.

Finally, we recall from Chap. 3, that $\omega_0^*(k+g)$ is not a symmetric function of the wave number. The band structure exhibits band gaps that are not symmetric with respect to wave number origin. Furthermore, the amplitude $\alpha(k+g)$ changes sign as the wave number varies across the asymmetric band gap leading to accumulation of phase. There are, therefore, two topologies to consider in this model. The topology of the wave amplitude with respect to wave number, which we have seen to be that of a closed twisted strip, and the topology of the spatio-temporal space supporting the dynamics of the model, which we have addressed partially in this section (and Appendix 2). The former is independent of time as the phase considered excludes the dynamical phase. The latter will depend of the wave number as seen in the k-dependency of the metric tensor given for instance by (6.107).

Appendix 1: Laplacian and d'Alembertian in a General Coordinate System

Let us consider a 3-vector $x^{\mu} = x\,,y\,,z$ for $\mu = 1\,,2\,,3$ in a general coordinate system. The Laplacian is given by [28]:

$$\Delta u = \frac{1}{\sqrt{|g|}} \partial_{\mu} \left(\sqrt{|g|} g^{\mu\nu} \partial_{\nu} u \right) \tag{6.113}$$

Let us apply (6.113) to the well know spherical coordinate system, (r, θ, φ). The covariant metric tensor is given by:

$$\left[g_{\mu\nu} \right] = \begin{pmatrix} g_{rr} & 0 & 0 \\ 0 & g_{\theta\theta} & 0 \\ 0 & 0 & g_{\varphi\varphi} \end{pmatrix} = \begin{pmatrix} 1 & 0 & 0 \\ 0 & r^2 & 0 \\ 0 & 0 & r^2 sin^2\theta \end{pmatrix} \tag{6.114}$$

The contravariant metric tensor is the inverse of $[g_{\mu\nu}]$:

$$[g^{\mu\nu}] = \frac{1}{r^4 sin^2\theta} \begin{pmatrix} r^4 sin^2\theta & 0 & 0 \\ 0 & r^2 sin^2\theta & 0 \\ 0 & 0 & r^2 \end{pmatrix} \tag{6.115}$$

The determinant of the covariant metric tensor is $g = r^4 \sin^2\theta$. Inserting this determinant and (6.115) into (6.113), leads to the expression for the Laplacian in spherical coordinates:

$$\Delta u = \frac{\partial^2 u}{\partial r^2} + \frac{2}{r}\frac{\partial u}{\partial r} + 2\frac{\cos\theta}{r^2 \sin\theta}\frac{\partial u}{\partial \theta} + \frac{\partial^2 u}{\partial \theta^2} \tag{6.116}$$

If we now consider the 4-vector $x^{\mu} = t\,,x\,,y\,,z$ for $\mu = 0\,,1\,,2\,,3$ in a general coordinate system. The d'Alembertian is given by [28]:

$$\square u = \frac{1}{\sqrt{|g|}} \partial_{\mu} \left(\sqrt{|g|} g^{\mu\nu} \partial_{\nu} u \right) \tag{6.117}$$

The condition $\square u = 0$ leads to a wave equation in any general system of coordinate.

Appendix 2: Geometrical Representation of the Dynamics of One-Dimensional Harmonic Systems

The Lagrangian density, $\mathcal{L}\left(x, t, u, \frac{\partial u}{\partial t}, \frac{\partial u}{\partial x}\right)$, for the continuous one dimensional elastic string (i.e., long wavelength limit of the one dimensional harmonic crystal) is given by:

$$\mathcal{L} = \frac{1}{2}\rho\left(\frac{\partial u}{\partial t}\right)^2 - \frac{1}{2}\beta^2\left(\frac{\partial u}{\partial x}\right)^2 \tag{6.118}$$

$u(x, t)$ is the displacement field. ρ and β^2 are the mass density and stiffness, respectively.

The equations of motion are derived from Euler-Lagrange equation:

$$\frac{\partial}{\partial t}\left(\frac{\partial \mathcal{L}}{\partial\left(\frac{\partial u}{\partial t}\right)}\right) + \frac{\partial}{\partial x}\left(\frac{\partial \mathcal{L}}{\partial\left(\frac{\partial u}{\partial x}\right)}\right) - \frac{\partial \mathcal{L}}{\partial u} = 0 \tag{6.119}$$

$\mathcal{L}$ is independent of the displacement u so the last term in (6.119) is zero. The elastic wave equation of motion is obtained in the usual form:

$$\rho\frac{\partial^2 u}{\partial t^2} - \beta^2\frac{\partial^2 u}{\partial x^2} = 0 \tag{6.120}$$

If we define $u(x, t) = \int dk\ u(k, t)e^{ikx}$, then the wave equation reduces to:

$$\rho\frac{\partial^2 u(k, t)}{\partial t^2} + (\beta k)^2 u(k, t) = 0 \tag{6.121}$$

This equation can be obtained from a Lagrangian density of the form:

$$\mathcal{L} = \frac{1}{2}\rho\left(\frac{\partial u(k, t)}{\partial t}\right)^2 - V(u) \tag{6.122}$$

with the k-dependent quadratic potential $V(u) = \frac{1}{2}(\beta k)^2 u^2$. We now note that with the form (6.122) the Lagrangian density is now only an explicit function of u and $\frac{\partial u}{\partial t}$.

Inserting (6.122) into (6.119) yields the wave equation:

$$\rho\frac{\partial^2 u(k, t)}{\partial t^2} + \frac{\partial V(u)}{\partial u} = 0 \tag{6.123}$$

The one dimensional harmonic system is conservative and the constant energy is given by the Hamiltonian:

$$E = T + V = \frac{1}{2}\rho\left(\frac{\partial u(k, t)}{\partial t}\right)^2 + V(u) \tag{6.124}$$

In (6.124), T is the kinetic energy. During harmonic motion energy passes from kinetic form into potential form and vice versa maintaining the overall energy constant. The crux of this appendix lies in the possibility of deriving the wave equation (6.123) from another Lagrangian density form, namely:

$$\mathcal{L}\left(u,\frac{\partial u}{\partial t}\right)=2(E-V(u))\left(\frac{\partial u}{\partial t}\right)^2 \tag{6.125}$$

For pedagogical reason we apply Euler-Lagrange equation step-by-step. That equation reads: $\frac{\partial}{\partial t}\left(\frac{\partial \mathcal{L}}{\partial\left(\frac{\partial u}{\partial t}\right)}\right)+-\frac{\partial \mathcal{L}}{\partial u}=0$. The first terms is obtained in the form:

$$\frac{\partial}{\partial t}\left(\frac{\partial \mathcal{L}}{\partial\left(\frac{\partial u}{\partial t}\right)}\right)=\frac{\partial}{\partial t}\left(2(E-V(u))2\frac{\partial u}{\partial t}\right)=2(E-V(u))2\frac{\partial^2 u}{\partial t^2}$$

The second term is simply: $\frac{\partial \mathcal{L}}{\partial u}=-2\frac{\partial V(u)}{\partial u}\left(\frac{\partial u}{\partial t}\right)^2$. The equation of motion then becomes:

$$2(E-V(u))2\frac{\partial^2 u}{\partial t^2}+2\frac{\partial V(u)}{\partial u}\left(\frac{\partial u}{\partial t}\right)^2=0 \tag{6.126}$$

Using (6.124) (energy conservation), we can replace $2(E-V(u))=\rho\left(\frac{\partial u}{\partial t}\right)^2$ and simplify (6.126) by $\left(\frac{\partial u}{\partial t}\right)^2$. This yields again the wave equation (6.123).

It is worth noting that the Lagrangian density (6.122) in absence of a potential reduces to:

$$\mathcal{L}=\frac{1}{2}\rho\left(\frac{\partial u}{\partial t}\right)^2 \tag{6.127}$$

Equation (6.127) implies uniform motion at constant velocity. This is free motion in a Euclidian space. The importance of (6.125) resides in its similarity to (6.127). Indeed, the right hand side of (6.125) can be defined as the kinetic energy:

$$T=\frac{1}{2}\left(\frac{\partial u'}{\partial t}\right)^2=2(E-V(u))\left(\frac{\partial u}{\partial t}\right)^2 \tag{6.128}$$

In the previous equation, we have defined the square of a new segment length along the axis of displacements by:

$$du'^2=2(E-V(u))du^2 \tag{6.129}$$

The quantity $g=2(E-V(u))$ is the metric that relates the square of the length element du^2 to du'^2. The length du, varies as a function of u (or t) for fixed time intervals dt. This is illustrated in Fig. 6.3. For a solution $u=\sin t$ of (6.121) using $\rho=\beta k=1$, the displacement increases toward its maximum in a non-uniform way.

The intervals du shortens to zero for fixed time intervals dt as one approaches the maximum displacement.

Finally, we consider the optimization problem of minimizing the length along the u' axis between two points A and B:

$$L = \int_A^B du' = \int_A^B \sqrt{2(E - V(u))\left(\frac{\partial u}{\partial \lambda}\right)^2} d\lambda = \int_A^B F d\lambda \tag{6.130}$$

In (6.130), we have introduced a parameter λ which enables us to move along the axis u'. This optimization problem can also be solved using Euler-Lagrange equation: $\frac{\partial}{\partial \lambda}\left(\frac{\partial F}{\partial\left(\frac{\partial u}{\partial \lambda}\right)}\right) + -\frac{\partial F}{\partial u} = 0$ which yields the equation:

$$\frac{\partial^2 u}{\partial \lambda^2} - \frac{1}{2}\frac{1}{g}\frac{\partial g}{\partial u}\left(\frac{\partial u}{\partial \lambda}\right)^2 = \frac{1}{F}\frac{\partial F}{\partial \lambda}\frac{\partial u}{\partial \lambda} \tag{6.131}$$

This is the equation of a geodesic in a curved space which metric is g.

Let us choose, $d\lambda = du'$, then by (6.130), $F = 1$ and $\frac{\partial F}{\partial \lambda} = 0$. Equation (6.131) reduces to

$$\frac{\partial^2 u}{\partial u'^2} - \frac{1}{2}\frac{1}{g}\frac{\partial g}{\partial u}\left(\frac{\partial u}{\partial u'}\right)^2 = 0 \tag{6.132}$$

or

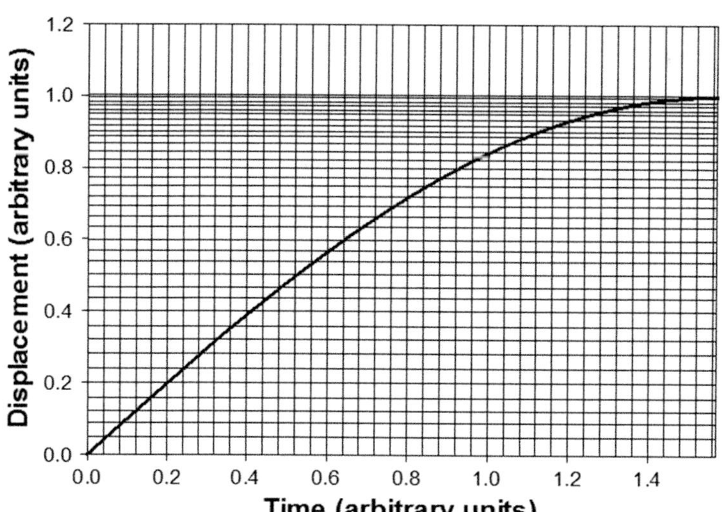

Fig. 6.3 Displacement $u = \sin t$ showing that equally spaced intervals of time Δt lead to a non-uniform set of intervals: Δu

$$\frac{\partial^2 u}{\partial u'^2} + \frac{1}{2(E - V(u))} \frac{\partial V}{\partial u} \left(\frac{\partial u}{\partial u'} \right)^2 = 0 \tag{6.133}$$

Along the geodesic the velocity is a constant (free motion), v such that $du' = vdt$. Inserting that relation in (6.133) and also using energy conservation (6.124) yields again the wave equation (6.123).

It is possible to construct curved spaces in some dimension by embedding them in higher dimensional spaces. The additional dimension add a curvature term to the usual metric of flat space. To illuminate this concept and the meaning of (6.130) we consider the two dimensional space $(u, v = \frac{dv}{dt})$ in which the one dimensional space describing the displacement of an oscillator is embedded, namely u. For the sake of simplicity, we take all physical constants equal to 1.

The conserved energy of the oscillator gives:

$$T + V = v^2 + u^2 = 2E \tag{6.134}$$

We note the well-known result that the oscillator follows a circular trajectory in the (u, v) plane. A line element on the trajectory circle is defined as (Fig. 6.4):

$$ds^2 = dv^2 + du^2 \tag{6.135}$$

Differentiating (6.134) results in:

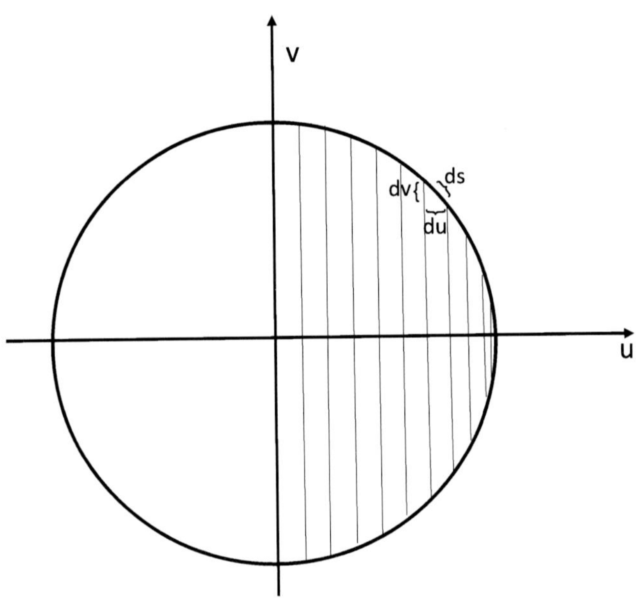

Fig. 6.4 Trajectory of oscillator in two-dimensional phase space (u, v). The *vertical lines* are drawn with equal values of ds. One notices the corresponding non-uniformity of the intervals du

$$vdv + udu = 0 \tag{6.136}$$

We express the line element ds^2 in terms of u by inserting (6.136) into (6.135):

$$ds^2 = \left(\frac{2E}{2E - u^2}\right)du^2 \tag{6.137}$$

This line element ds is uniform along the circular trajectory. The velocity $\frac{ds}{dt}$ is a constant along the circle. The circular trajectory is that of a free moving object which kinetic energy is equal to the total energy of the oscillator, namely

$$\frac{1}{2}\left(\frac{ds}{dt}\right)^2 = E \tag{6.138}$$

Combining (6.138) and (6.137) yields

$$dt^2 = \left(\frac{1}{2E - u^2}\right)du^2 \tag{6.139}$$

One recovers the inverse of the metric, $g = 2(E - V(u))$. When plotting $t = \int_0^u \left(\frac{1}{2E-u^2}\right)du$ with $u = \sin t$ one converts the plot of Fig. 6.3 into that of a straight line (Fig. 6.5).

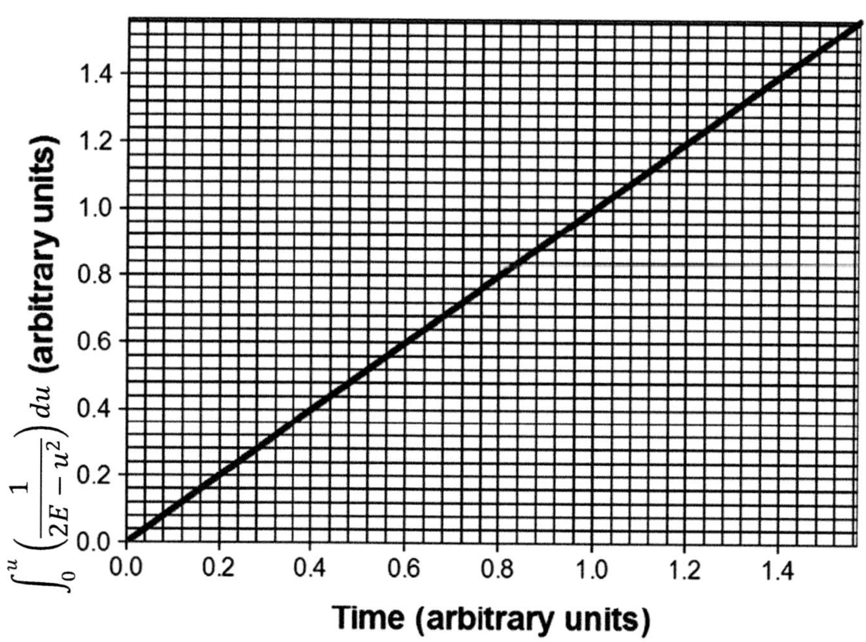

Fig. 6.5 $\int_0^u \left(\frac{1}{2E-u^2}\right)du$ versus time t. The *straight line* is the signature of a geodesic

References

1. I. El-Kady, in Phonon-based scalable quantum computing and sensing, *Proceedings of SPIE 9436, Smart Sensor Phenomena, Technology, Networks, and Systems Integration 2015*, 94360O (13 May 2015)
2. R. Ruskov, C. Tahan, Coherent phonons as a new element of quantum computing and devices. J. Phys. **398**, 012011 (2012)
3. R. Ruskov, C. Tahan, On-chip cavity quantum phonodynamics with an acceptor qubit in silicon. Phys. Rev. B **88**, 064308 (2013)
4. M.V. Gustafsson, T. Aref, A.F. Kockum, M.K. Ekström, G. Johansson, P. Delsing, Propagating phonons coupled to an artificial atom. Science **346**, 207 (2014)
5. D. Deutsch, R. Jozsa, Rapid solutions of problems by quantum computation. Proc. R. Soc. Lond. A **439**, 553 (1992)
6. K.D. Arvind, N. Mukunda, A two-qubit algorithm involving quantum entanglement, arXiv: quant-ph/0006069
7. P. Ramond, *Field Theory: A Modern Primer*, 2nd edn. (Westview Press, Boulder, 2001)
8. K. Fang, Z. Yu, S. Fan, Photonic Aharonov-Bohm effect based on dynamic modulation. Phys. Rev. Lett. **108**, 103901 (2012)
9. K. Fang, Z. Yu, S. Fan, Realizing effective magnetic field for photons by controlling the phase of dynamic modulation. Nat. Photonics **6**, 782 (2012)
10. L. Tzuang, K. Fang, P. Nussenzveig, S. Fan, M. Lipson, Non-reciprocal phase shift induced by an effective magnetic flux for light. Nat. Photonics **8**, 701 (2014)
11. E. Li, B.J. Eggleton, K. Fang, S. Fan, Photonic Aharonov-Bohm effect in photon-phonon interactions. Nat. Commun. **5**, 3225 (2014)
12. Y. Couder, E. Fort, Single-particle diffraction and interference at a macroscopic scale. Phys. Rev. Lett. **97**, 154101 (2006)
13. Y. Couder, S. Protiere, E. Fort, A. Boudaoud, Dynamical phenomena – walking and orbiting droplets. Nature **437**, 208 (2005)
14. S. Protiere, A. Boudaoud, Y. Couder, Particle-wave association on a fluid interface. J. Fluid. Mech. **554**, 85 (2006)
15. J.W.M. Bush, Pilot-wave hydrodynamics. Ann. Rev. Fluid Mech. **47**, 269 (2014)
16. A.U. Oza, D.M. Harris, R.R. Rosales, J.W.M. Bush, Pilot-wave dynamics in a rotating frame: on the emergence of orbital quantization. J. Fluid Mech. **744**, 404 (2014)
17. J. Molacek, J.W.M. Bush, Drops walking on a vibrating bath: towards a hydrodynamic pilot-wave theory. J. Fluid. Mech. **727**, 612 (2013)
18. Leighton, *The Acoustic Bubble* (Academic, San Diego, 1994)
19. B. Van der Pol, M.J.O. Strutt II, On the stability of the solutions of Mathieu's equation. Philos. Mag. **5**, 18–38 (1928)
20. P.M. Morse, H. Feshbach, *Methods of Theoretical Physics Part I* (McGraw-Hill, New York, 1953)
21. A.A. Doinikov, Translational motion of a spherical bubble in an acoustic standing wave of high intensity. Phys. Fluids **14**, 1420 (2002)
22. A.A. Doinikov, Translational motion of a bubble undergoing shape oscillation. J. Fluid Mech. **501**, 1–24 (2004)
23. A.A. Doinikov, Equations of coupled radial and translational motions of a bubble in a weakly compressible liquid. Phys. Fluids **17**, 128101 (2005)
24. R. Mettin, A.A. Doinikov, Translational instability of a spherical bubble in a standing ultrasound waves. Appl. Acoust. **70**, 1330–1339 (2009)
25. M. Visser, C. Barcelo, S. Liberati, Analogue models of and for gravity. General Relativity and Gravitation **34**, 1719 (2002)
26. G. Unruh, Experimental black-hole evaporation? Phys. Rev. Lett. **46**, 1351 (1981)
27. C. Barcelo, S. Liberati, M. Visser, Analog gravity from field theory normal modes. Class. Quant. Grav. **18**, 3595 (2001)
28. M.P. Hobson, G. Efstathiou, A.N. Lasenby, *General Relativity: An Introduction for Physicists* (Cambridge University Press, Cambridge, 2006)

Subject Index

A

Acoustic bubbles, 319, 328–336
Acoustic metamaterials, 27, 37, 38, 82, 163
Acoustic spin pumping, 262, 281–303
Acousto-hydrodynamics, 262, 319, 328, 334–336
Anticommutation, 99–101
Anti-Hermitian, 99

B

Band structure, 5, 8, 12, 15, 28, 48, 50, 52, 57, 66, 72, 85, 87, 89, 91, 95, 103, 104, 107, 110, 112, 114, 115, 117, 125, 138, 141, 142, 145, 146, 149, 164, 181, 182, 184, 185, 189, 192, 193, 196, 197, 203, 206, 207, 216, 217, 219, 220, 241, 254, 262, 277, 280, 303, 349, 353
Berry connection, 37, 39, 41, 43, 51, 53–56, 66, 72, 73, 91, 92, 139, 140
Berry phase, 39, 63, 73, 91, 97, 139–140
Bi-stable media, 225–240
Bjerknes force, 329, 331, 333
Bloch wave, 38, 67, 70, 112, 113, 116, 121, 241, 272, 278
Bone, 164, 207, 208, 212, 214, 220, 261–263
Boson-like (phonon), 85, 127
Bragg scattering, 12, 16, 26, 27
Brillouin scattering, 114, 125
Brillouin zone, 2, 5, 12, 15, 39, 49, 50, 52, 57, 63, 66, 72, 73, 94, 97, 110, 112, 115–117, 121, 124, 125, 140, 142, 151, 170, 181, 184, 185, 197, 203, 206, 216, 241, 254, 272, 278, 280, 281, 297, 298

C

Calcium wave, 164, 225–240, 261–276, 303, 310–314
Cell-to-cell conductance, 261, 266, 268, 270, 272, 314
Chiral symmetry, 81, 87
Collagen, 164, 207–217, 220, 276
Contravariant metric tensor, 340, 344, 345, 349, 354
Cosserat element, 143, 145
Cosserat model, 141
Covariant metric tensor, 340, 345, 349, 354
Creation and annihilation operators, 100
Curie temperature, 299

D

d'Alembertian, 345, 354
Density of state, 28, 29, 34, 60, 62, 66, 76–78
Deutsch-Jozsa algorithm, 321–323
Differential geometry, 320
Dirac equation, 88, 92, 93, 97, 99, 101, 120–123, 125, 127, 319, 325, 327
Dispersion relation, 2–4, 9–11, 15, 38, 48, 49, 52, 65, 69, 70, 85, 86, 90, 91, 94, 103, 113, 115, 124, 135, 141, 143, 147, 167, 184, 192, 206, 254, 273, 290, 296, 297, 299–302, 311, 314, 330, 333
Dynamical matrix, 17, 21
Dynamical phase, 40–42, 353
Dyson equation, 74–76, 78

E

Eikonal approximation, 230

P. Deymier, K. Runge, *Sound Topology, Duality, Coherence and Wave-Mixing*, Springer Series in Solid-State Sciences 188, DOI 10.1007/978-3-319-62380-1

Einstein notation, 45, 340, 342
Electric current density, 281
Endoplasmic reticulum (ER), 226, 227, 235, 264, 266
Endothelial cells, 164, 225–240, 261, 263, 266, 271
Entanglement, 82, 322, 323, 325
Euclidian space, 350, 356
Euler-Lagrange equation, 343, 350, 355–357

F

Failure of propagation, 226, 238, 240
Fast Fourier transform (FFT), 149, 182
Fermion-like (phonon), 121
Ferromagnet, 262, 281, 282, 298, 299, 302, 303
Feynman diagrams, 228
Finite Difference Time Domain (FDTD) method, 164, 214, 238, 240
Floppy modes, 66
Floquet's theorem, 330
Fourier transform, 63, 148, 149, 215, 327
Friedel phase, 37, 39, 60, 66

G

Gap junction, 232, 261, 263–272
Gauge field, 319, 325–328
General relativity, 319, 336–353
Geodesic, 351, 353, 357, 358
Geometric phase, 37–42, 44–52, 81, 116, 274
Geometrical nonlinearity, 163
Granular materials, 163
Green's function, 1, 8–10, 16–23, 25, 28–37, 39, 42, 43, 53–64, 66, 74–78, 164, 225, 227, 228, 230, 231, 233, 234, 238, 240
Group velocity, 1, 3–4, 6, 28, 58, 104, 117, 181
Gyromagnetic ratio, 284

H

Hadamard gate, 106, 321–323
Hamiltonian, 227, 283–285, 301, 325, 348, 349, 355
Hamiltonian density, 99
Helicoid, 44, 46, 47, 52
Hermitian conjugate, 99–102
Hermitian operator, 37, 39, 43, 53, 75
Hilbert space, 128–132, 137, 138, 207, 322, 323
Homotopy Perturbation Method, 164, 227, 229
Hybridization gap, 110, 115–117, 142, 254, 280, 281
Hydroxy-apatite (HAP), 164, 207, 208, 214, 216, 217
Hysteretic nonlinearity, 163

I

Inositol 1,4,5-triphosphate (IP3), 227, 263–269, 271–273, 275, 303, 304, 311, 314
Interface Response Theory (IRT), 1, 16–28, 62, 64, 104
Interfacial modes, 39, 65, 66
Intrinsic nonlinearity, 163
Inversion symmetry, 7, 39, 52, 63, 71, 81
Irrotational fluid flow, 334
Ising model, 299

K

Klein–Gordon equation, 84, 88, 92, 93, 108–109, 120, 125

L

Lagrange equations, 336
Lagrangian, 84, 97–99, 101, 325, 335, 341–345
Lagrangian density, 354–356
Landau-Lifshitz equation, 284
Laplace equation, 335
Laplace transform, 228, 231–234
Laplacian, 93, 232, 238, 354
Linear reaction diffusion model, 264, 271–276
Lippman-Schwinger propagator theory, 164
Lorentzian function, 148, 182, 183, 185
Lorentzian geometry, 341, 346

M

Magneto-elastic coupling, 262, 282–284, 286, 287, 301–303
Magnon, 262, 276, 281–283, 290, 296, 297, 300–303
Mathieu's equation, 329, 330
Maxwell lattices, 66
Minkowski, 353
Molecular dynamics (MD) method, 110, 125, 147–155, 164, 169, 181–188, 203, 212
Multi-phonon scattering, 164, 207–220
Multiple time scale perturbation theory (MTSPT), 112–115, 121, 156, 158, 166, 190–191, 242, 262, 272, 278, 283, 287, 310–314, 345, 346

N

Neumann boundary condition, 64
Nonlinear reaction diffusion model, 261, 264–271, 275

Non-reciprocal wave, 81, 117, 164
Non-separability, 81, 127, 133–138, 207, 319, 323
Number operators, 101–103

O
One-phonon-two-magnon resonant process, 300, 302
Open system, 163, 208, 214
Open system nonlinearity, 163
Oracle operator, 321, 322, 325

P
Parallel transport, 37–39, 46–48, 52, 57, 92, 97, 116, 274, 280
Parity symmetry, 81, 84, 127
Particle-hole symmetry, 81, 109, 110
Particle-wave duality, 81, 319, 328–336
Pauli matrices, 88, 93, 98, 130
Periodic boundary conditions (PBC), 147, 181, 203, 215, 266
Phase bit, 319–325
Phase velocity, 6, 181
Phonon drag, 262, 276–282
Phonon-electron scattering, 276
Phononic crystals (PCs), 27, 37, 38, 82, 120, 141, 303
Pilot wave, 319, 328
Propagator, 228, 229, 231, 232, 234, 238
Pseudospin, 320–324

Q
Quantum bit, 319–325
Quantum field theory, 88, 100, 120, 121, 327
Quasi standing waves, 127
Qubit, 319–325

R
Rayleigh-Plesset equation, 332
Reaction dynamics, 164, 225–240, 264, 268
Reaction-diffusion problem, 225, 232–234
Rotational modes, 106–108, 141–146
Rotational wave, 106–108, 141, 146
Runge-Kutta algorithm, 266

S
Schroedinger equation, 40, 262, 277, 278, 348
Seebeck effect, 276, 282
Separability, 81, 127–129, 133–138
Sokhotski's formula, 77, 296
Spatio-temporal modulation, 81, 164, 240–254
Spin exchange coupling constant, 283, 302
Spin Hall effect (SHE), 282
Spinor, 83, 87–90, 92–97, 101–106, 120, 122, 124, 127, 136, 138–140, 146, 319, 322, 324, 325
Spin Seebeck effect (SSE), 282
Spin waves, 262, 282, 283, 285, 290
Superlattice, 37–39, 48–52, 66–71, 109–115, 117–119, 164, 207–220, 240, 345–353
Surface acoustic wave (SAW), 300
Surface avoiding modes, 66

T
Targeted energy transfer, 188
Tensor product, 81, 127, 129–133, 135–138, 207, 320, 322–325
Thouless pump, 276
Time-reversal symmetry, 7, 38, 81, 82, 84, 128, 163, 262, 274
Transmission coefficient, 11, 12, 24, 26, 29, 33, 60, 62, 103–105, 117, 320
Transmission phase, 28, 29, 32, 33, 60, 66
Two-qubit states, 323

U
Umklapp processes, 170

V
Velocity potential, 334, 335, 337, 338
Verlet algorithm, 181
Vibro-acoustic disease (VAD), 275, 276

W
Webster equation, 108

Z
Zak phase, 39, 48–52, 63–66, 74

Printed in the United States
By Bookmasters